Security and Resilience in Cyber-Physical Systems

Masoud Abbaszadeh · Ali Zemouche
Editors

Security and Resilience in Cyber-Physical Systems

Detection, Estimation and Control

Springer

Editors
Masoud Abbaszadeh
GE Research
New York, NY, USA

Ali Zemouche
CRAN, IUT Henri Poincaré de Longwy
Université de Lorraine
Cosnes-et-Romain, France

ISBN 978-3-030-97168-7 ISBN 978-3-030-97166-3 (eBook)
https://doi.org/10.1007/978-3-030-97166-3

Mathematics Subject Classification: 93B52, 93C15, 93E10, 93E35

This Springer imprint is published by the registered company Springer Nature Switzerland AG
The registered company address is: Gewerbestrasse 11, 6330 Cham, Switzerland

Preface

Cyber-physical systems enable interoperability of cyber and physical worlds through control, computation, and communications. Control, optimization, monitoring, and diagnostic schemes that operate cyber-physical systems (e.g., power generation, transportation, oil & gas, computing and communication systems, healthcare systems, etc.) are increasingly connected via local networks or the Internet. As a result, these control systems have been increasingly vulnerable to threats and jamming, such as cyber-attacks (e.g., associated with a computer virus, malicious software, etc.), that could disrupt their operation, damage equipment, inflict malfunctions, etc. Many of current cybersecurity methods primarily consider attack detection and mitigation in Information Technology ("IT", such as, computers that store, retrieve, transmit, manipulate data) and Operational Technology ("OT", such as direct monitoring devices and communication bus interfaces) at the network and communication layers. Cyber-attacks can still penetrate through these protection layers and reach the physical "domain" as seen in 2010 with the Stuxnet attack. Such attacks can negatively affect the performance of a control system and may cause total shut down or catastrophic damage to the system. Currently, fewer methods are available to automatically detect, during a cyber-incident, attacks at the physical domain layer (i.e. the process level) where sensors, controllers, and actuators are located. In some cases, multiple attacks may occur simultaneously (e.g., more than one actuator, sensor, or parameter inside control system devices might be altered maliciously by an unauthorized party at the same time). Furthermore, some subtle consequences of cyber-attacks, such as stealthy attacks occurring at the physical layer, might not be readily detectable (e.g., when only one monitoring node, such as a sensor node, is used in a detection algorithm). In addition, to maintain system availability and integrity and to protect assets, attack resilience much be achieved through resilient estimation and control methodologies, which are beyond existing fault-tolerant control approaches. It may also be important to determine and distinguish when a monitoring node is experiencing a malicious attack (as opposed to a natural fault/failure) and, in some cases, exactly what type of attack is occurring. Existing approaches to protect an industrial control system, such as fault detection and diagnostics technologies, may not adequately address these problems, especially when multiple, simultaneous attacks occur, since multiple faults/failure diagnostic technologies are not designed for detecting stealthy attacks in an automatic manner. It would therefore be desirable to research and develop new theories and technologies to protect cyber-physical systems, including those in critical infrastructure, from cyber-attacks in an automatic and accurate manner even when attacks percolate through the IT and OT layers and directly harm the control systems.

This book is intended to cover some of the latest research on cyber-physical security and resilience and highlight active research directions and solutions that are currently pursued in academia and industry. A collection of book chapters are gathered from well-known experts in the field with diverse technical backgrounds of controls, estimation, machine learning, signal processing, and information theory, as well as diverse geographical representation from North America, Europe, and Asia. The book addresses a very important topic with a growing attention from the research community and critical applications and implications in industries and governments. The book chapters comprise of a blend of new theoretical results on detection, resilient estimation, and control combined with machine learning techniques, as well as important application areas such as power generation, electric power grid, autonomous systems, communication networks, and chemical plants. In the recent years, there have been multiple books published on cyber-security from various perspectives, including vulnerability and impact analysis, safety, security, privacy, networks intrusion detection and mitigation, etc. While there are synergies between the current book and the previously published books, the book complements previous publications by addressing cyber-physical security at the physical sensing and control layer of cyber-physical systems, and systems resilience under attack via resilient estimation and control. The book is not solely based on control/estimation theory but a combination of control theory and machine learning approaches to cyber-physical security and resilience, by which several chapters are taking a "Controls+AI" approach. The book is aimed for both researchers and technology developers in the academia and industry. This area is vast and rapidly growing, with crucial needs for additional research and development. The editors hope that this book will be a useful advancement in that front.

Finally, we deeply thank authors and reviewers for their contributions. We also gratefully acknowledge support from Springer's editorial and production staff. A. Zemouche would like to thank the ANR agency for the partial support of this work via the project ArtISMo ANR-20-CE48-0015.

New York, USA
Grand Est, France
October 2021

Masoud Abbaszadeh
Ali Zemouche

Contents

Contributors

Masoud Abbaszadeh GE Research, Niskayuna, NY, USA

Ania Adil King Abdullah University of Science and Technology (KAUST), Thuwal, Saudi Arabia

H. M. Sabbir Ahmad Qatar University, Doha, Qatar

Hatem Ahriz Robert Gordon University, Aberdeen, UK

M. Omar Al-Kadri Birmingham City University, Birmingham, UK

Olugbenga Moses Anubi FAMU-FSU College of Engineering, Tallahassee, FL, USA

Jean-Pierre Barbot QUARTZ Laboratory, ENSEA, Cergy-Pontoise, France

Souad Bezzaoucha Rebai EIGSI-La Rochelle, La Rochelle, France

Mohamed Darouach Research Center for Automatic Control of Nancy (CRAN), Université de Lorraine, IUT de Longwy, Cosnes et Romain, France

Christopher Edwards The University of Exeter, Exeter, UK

Hope Nkiruka Eke Robert Gordon University, Aberdeen, UK

Song Fang New York University, New York, USA

Junjie Fu School of Mathematics, Southeast University, Nanjing, People's Republic of China

Mohammad Khajenejad Arizona State University, Tempe, AZ, USA

Taous-Meriem Laleg-Kirati King Abdullah University of Science and Technology (KAUST), Thuwal, Saudi Arabia

Nader Meskin Qatar University, Doha, Qatar

Lalit K. Mestha Genetic Innovations Inc. (work performed while at GE Research), Honolulu, HI, USA

Shamila Nateghi University of Alabama in Huntsville, Huntsville, AL, USA

Mohammad Noorizadeh Qatar University, Doha, Qatar

Ibrahima N'Doye King Abdullah University of Science and Technology (KAUST), Thuwal, Saudi Arabia

Andrei Petrovski Robert Gordon University, Aberdeen, UK

Rajesh Rajamani Department of Mechanical Engineering, University of Minnesota, Minneapolis, MN, USA

Yuri Shtessel University of Alabama in Huntsville, Huntsville, AL, USA

Saurabh Sihag University of Pennsylvania, Philadelphia, PA, USA

Ali Tajer Rensselaer Polytechnic Institute, Troy, NY, USA

Holger Voos Interdisciplinary Centre for Security, Reliability and Trust (SnT), Automatic Control Research Group, University of Luxembourg, Luxembourg, Luxembourg

Bohui Wang School of Electrical and Electronic Engineering, Nanyang Technological University, Singapore, Singapore

Guanghui Wen School of Mathematics, Southeast University, Nanjing, People's Republic of China

Miaomiao Xie College of Automation Science and Technology, South China University of Technology, Guangzhou, China

Wei Xie College of Automation Science and Technology, South China University of Technology, Guangzhou, China

Yongjun Xu Institute of Computing Technology, Chinese Academy of Sciences, Beijing, China

Weizhong Yan GE Research, Niskayuna, NY, USA

Sze Zheng Yong Arizona State University, Tempe, AZ, USA

Ali Zemouche CRAN CNR-UMR 7039, IUT Henri Poincaré de Longwy, Université de Lorraine, Cosnes-et-Romain, France

Ding Zhang Department of Electronic and Computer Engineering, The Hong Kong University of Science and Technology, Clear Water Bay, Kowloon, Hong Kong, China

Fan Zhang School of Aeronautics and Astronautics (Shenzhen), Sun Yat-sen University, Guangzhou, People's Republic of China

Langwen Zhang College of Automation Science and Technology, South China University of Technology, Guangzhou, China

Yu Zheng FAMU-FSU College of Engineering, Tallahassee, FL, USA

Quanyan Zhu New York University, New York, USA

Chapter 1
Overview

Masoud Abbaszadeh and Ali Zemouche

The increasing sophistication and severity of intelligently designed cyber-attacks warrants new theoretical and technological developments beyond current detection, estimation, and control methodologies. On the other hand, cyber-physical systems are dramatically changing, incorporating new elements such as the Internet of Things (IoT) connectivity and distributed intelligence into the perspective. This transformation further expands the digital footprint of these systems, hence making them susceptible to cyber-attacks and other safety and security issues. Advanced targeted attacks against control systems have increased in the past years with evidence of high risks related to zero-day and replay attacks. In order to tackle this threat, we need advances in detection, feedback control, and estimation with built-in resilience to cyber-attacks, to maintain system integrity and reliability at all times, by providing uninterrupted, equipment-safe, and controlled operation.

This book is intended to cover some of the latest theory and technology advancements for detection and protection against cyber-attacks in cyber-physical systems. This is a very important emerging field and a very active multidisciplinary research and technology development area. The book covers some of the latest problems and research on cyber-physical security and resilience, and highlights active research directions and solutions that are currently pursued in academia and industry. The topics comprise of a blend of new theoretical results on resilient estimation and control combined with machine learning techniques, as well as important application areas such as industrial control systems, power generation and distribution,

M. Abbaszadeh (✉)
GE Research, Niskayuna, NY, USA
e-mail: masoud@ualberta.net

A. Zemouche
CRAN CNR-UMR 7039, IUT Henri Poincaré de Longwy, Université de Lorraine, Cosnes-et-Romain, France
e-mail: ali.zemouche@univ-lorraine.fr

M. Abbaszadeh and A. Zemouche (eds.), *Security and Resilience in Cyber-Physical Systems*, https://doi.org/10.1007/978-3-030-97166-3_1

autonomous systems, wireless communication networks, and chemical plants. The book comprises of a collection of chapters from well-known researchers in academia, and industrial research labs providing a comprehensive perspective of some of the latest advancements and prospects of cyber-physical security and resilience.

The book is structured as follows. It starts with an introductory chapter on cyber-physical security and resilience (Chap. 2), and continues with chapters containing theoretical results on attack detection and situational awareness, resilient estimation, and control, with case studies on power generation, transmission and distribution, sensor networks, cooperative tracking, and autonomous vehicles (Chaps. 3–11). Then, it moves to application-oriented chapters on wastewater treatment plants, oil refinery, and wireless communication networks (Chaps. 12–14). A fundamental trade-off study of stealthiness–distortion is offered in Chap. 3. This is an important topic and sets foundations for future work in this emerging space. Chapter 4 is dedicated to predictive situational awareness in which an anomaly detection and forecasting framework is proposed, combining elements from estimation theory and machine learning. Chapter 5 provides a resilient observer design solution using a concurrent learning approach, while a framework for detection of advanced persistent threats is presented in Chap. 6. Chapters 7–10 are focused on secure and resilient estimation from different perspectives. Chapter 7 addresses the resilient state estimation and attack mitigation problems for switched linear systems with stochastic and set-membership uncertainties. Chapter 8 is on state and attack estimation for nonlinear fuzzy systems with delayed measurements. Chapter 9 establishes the notion of secure estimation under imperfect attack detection and isolation decisions and studies the fundamental couplings between those decisions and the estimation problem, characterizing closed-form decision rules. Chapter 10, addresses cyber-attack reconstruction using higher-order sliding mode observers and sparse recovery methods. Chapter 11 is on resilient cooperative control over networks to achieve consensus tracking under input constraints and communication restraints. Chapter 12 is on resilient distributed estimation, addressing the resilience of wastewater treatment plants against natural disasters. A distributed attack detection algorithm is proposed in Chap. 13 for crude oil distillation columns. Chapter 14 is on resilient estimation in optical wireless communication networks for cooperative robot autonomy under actuator faults and noise jamming. The titles and abstracts of the chapters are in the following.

Chapter 2. Introduction to Cyber-Physical Security and Resilience: This chapter describes the fundamentals of the cyber-physical security and resilience approaches as well as some of the current research directions, and provides a survey of latest results in attack detection, isolation, resilient estimation and resilient control. It also makes distinctions between cyber-physical security versus adjoining and seemingly related applications such as fault detection, and data communications and network security (a.k.a, cyber-security).

Chapter 3. Fundamental Stealthiness–Distortion Tradeoffs in Cyber-Physical Systems: In this chapter, we analyze the fundamental stealthiness–distortion tradeoffs of linear Gaussian open-loop dynamical systems and (closed-loop) feedback control systems under data injection attacks using a power spectral analysis,, whereas the Kullback–Leibler (KL) divergence is employed as the stealthiness measure. Particularly, we obtain explicit formulas in terms of power spectra that characterize analytically the stealthiness–distortion trade-offs as well as the properties of the worst-case attacks. Furthermore, it is seen in general that the attacker only needs to know the input–output behaviors of the systems in order to carry out the worst-case attacks.

Chapter 4. **Predictive Situation Awareness and Anomaly Forecasting in Cyber-Physical Systems:** A new feature-based situation awareness and forecasting framework is presented for rapid detection and early warning of abnormalities in cyber-physical systems. The abnormalities may refer to intelligent cyber-attacks or naturally occurring faults and failures. Techniques presented here are aimed at protecting against unauthorized intrusions as well as fault prevention. Time series signals from system monitoring nodes are converted to features using feature discovery techniques. The feature behavior for each monitoring node is characterized in the form of decision boundaries, separating normal and abnormal space with operating data collected from the plant or by running virtual models of the plant. A set of ensemble state-space models are constructed for representing feature evolution in time domain, where the ensembles are selected using Gaussian Mixture Model (GMM) clustering. The forecasted outputs are anticipated time evolution of features, computed by applying an adaptive Kalman predictor to each ensemble model. The overall features forecast is then obtained through dynamic ensemble averaging. This is done by projecting evolution of feature vector to future times. This projection can be performed either in a receding horizon or a committed horizon fashion. The feature forecasts are compared to the decision boundary to estimate if/when the feature vectors will cross the boundary. The decision boundary is a high-dimensional manifold in the feature space learned by a neural network. The training of the neural network is based on labeled data provided either through simulation of the system digital twin or by capturing historical field data. In this chapter, we also establish a framework for situation awareness, discussing the different components to achieve full situation awareness and showing the interactions between the attack detection, isolation, and prediction modules at the system level. Simulation results in a high-fidelity GE gas turbine platform show the effectiveness of our approach for forecasting abnormalities, which can be used for protecting physical assets from abnormalities due to cyber-intrusion or natural faults.

Chapter 5. **Resilient Observer Design for Cyber-Physical Systems with Data-Driven Measurement Pruning:** Resilient observer design for Cyber-Physical Systems (CPS) in the presence of adversarial false data injection attacks (FDIA) is an active area of research. Existing state-of-the-art algorithms tend to break down as more and more knowledge of the system is built into the attack model; also as the percentage of attacked nodes increases. From the view of optimization theory, the problem is often cast as a classical error correction problem for which a theoretical limit of 50% has been established as the maximum percentage attacked nodes for which state recovery is guaranteed. Beyond this limit, the performance of ℓ_1-minimization-based schemes, for instance, deteriorates rapidly. Similar performance degradation occurs for other types of resilient observers beyond certain percentages of attacked nodes. In order to increase the corresponding percentage of attacked nodes for which state recoveries can be guaranteed, researchers have begun to incorporate prior information into the underlying resilient observer design framework. For the most pragmatic cases, this prior information is often obtained through a data-driven machine learning process. Existing results have shown a strong positive correlation between the maximum attacked percentages that can be tolerated and the accuracy of the data-driven model. Motivated by these results, this chapter examines the case for *pruning algorithms* designed to improve the *Positive Prediction Value* (PPV) of the resulting prior information, given stochastic uncertainty characteristics of the underlying machine learning model. Theoretical quantification of the achievable improvement is given. Simulation results show that the pruning algorithm significantly increases the maximum correctable percentage of attacked nodes, even for machine learning model whose prediction power is comparable to the random flip of a coin.

Chapter 6. Framework for Detecting APTs Based on Steps Analysis and Correlation: An advanced persistent threat, (APT), is an attack that uses multiple attack behavior to penetrate a system, to achieve specifically targeted and highly valuable goals within a system. This type of attack has presented an increasing concern for cyber-security and business continuity. The resource availability, integrity, and confidentiality of the operational cyber-physical systems' (CPS) state and control are highly impacted by the safety and security measures adopted. In this study, we propose a framework based on deep APT steps analysis and correlation, abbreviated as "APT-DASAC", for securing industrial control systems (ICSs) against APTs. This approach takes into consideration the distributed and multi-level nature of ICS architecture, and reflects on multi-step APT attack lifecycle. We validated the framework with three case studies: (i) network transactions between a remote terminal unit (RTU) and a master control unit (MTU) within a supervisory control and data acquisition (SCADA) gas pipeline control system, (ii) a case study of command and response injection attacks, and (iii) a scenario based on network traffic containing hybrid of the real modern normal and the contemporary synthesized attack activities of the network traffic. Based on the achieved result, we show that the proposed approach achieves a significant attacks detection capability and demonstrates that attack detection techniques that performed very well in one application domain may not yield the same result in another. Hence, robustness and resilience of operational CPS state or any system and performance are determined by the security measures in place, which is specific to the application system and domain.

Chapter 7. Resilient State Estimation and Attack Mitigation in Cyber-Physical Systems: Smart and Cyber-Physical Systems (CPS), e.g., power and traffic networks and smart homes, are becoming increasingly ubiquitous as they offer new opportunities for improved performance and functionalities. However, these often safety-critical systems have also recently become the target of cyber- or physical attacks. This chapter contributes to the area of CPS security from the perspective of leveraging the knowledge of physical system dynamics as an additional "sensor" to mitigate the effects of false data injection attacks on actuator and sensor signals as well as attacks on the switching mechanisms, e.g., circuit breakers, on the state estimation and control algorithms in these systems, in order to ensure continued safety, reliability and integrity of systems despite attacks. Specifically, we consider physical models with switched linear dynamics subject to two classes of uncertainties: (a) stochastic uncertainty (aleatoric), i.e., with (unbounded) stochastic process and measurement noise signals and (b) set-membership uncertainty (epistemic), i.e., with distribution-free bounded-norm process and measurement disturbances. In both settings, we model the system under attack as a hidden-mode switched linear system with unknown inputs (attacks) and propose multiple-model inference algorithms to perform attack-resilient state estimation with stability and optimality guarantees. Moreover, we characterize fundamental limitations to resilient state estimation (e.g., upper bound on the number of tolerable signal attacks) and discuss the topics of attack detection, identification, and mitigation under this framework. Simulation examples of switching and false data injection attacks on a benchmark system and an IEEE 68-bus test system show the efficacy of our approach to recover resilient state estimates as well as to identify and mitigate the attacks in the presence of stochastic and set-membership uncertainties.

Chapter 8. **State and Attacks Estimation for Nonlinear Takagi–Sugeno Multiple Models Systems with Delayed Measurements:** In the following contribution, a state and attacks estimation for nonlinear Takagi–Sugeno Systems with variable time-delay measurements is proposed. Based on the sector nonlinearity approach, sufficient conditions in term of Linear Matrix Inequalities ($LMIs$) formulation are given for the observer design. It is demonstrated that, despite the presence of cyber-attack (i.e., data deception attacks on both actuators and sensors), and the delayed measurements, the proposed observer is quite efficient and ensures the asymptotic convergence of the estimation errors with an $\mathscr{L}_2$ attenuation constraint.

Chapter 9. **Secure Estimation under Model Uncertainty:** The increasing scale and widespread deployment of cyber-physical systems for novel applications leave them vulnerable to malicious intrusions and potential failures. Therefore, the performance of a cyber-physical system hinges on both the successful detection and elimination of malicious behavior. More importantly, the robustness of inference algorithms in making high-quality inference decisions even under active malicious behaviors is instrumental to making reliable decisions. This chapter focuses on the robustness of state estimates in complex networks . In such systems, state estimation is the key inference task at the core of monitoring and decision-making. One key challenge when facing malicious attacks is uncertainty in the true underlying statistical model of the data collected. Such uncertainty can be an outcome of a variety of adversarial behaviors, such as false data injection attacks, denial of service (DoS) attacks, and causative attacks . In all such scenarios, the estimation algorithms operate under a distorted statistical model with respect to what they expect. Therefore, forming estimates under malicious attacks involves an additional decision pertinent to the presence of an attack and isolating the true statistical model. This chapter introduces new notions of secure estimation under the knowledge that imperfect detection and isolation decisions induce a coupling between the desired estimation performance and the auxiliary necessary detection and isolation decisions. The fundamental interplay among the different decisions involved is established and closed-form decision rules are provided.

Chapter 10. **Resilient Control of Nonlinear Cyber-Physical Systems: Higher-Order Sliding Mode Differentiation and Sparse Recovery-based Approaches:** In this chapter, we focus on a cyber-attack reconstruction and secure state estimation to facilitate the resilient control of nonlinear cyber-physical systems under sensor and/or actuator attacks. The Sliding Mode Observation/Differentiation (SMO/D) techniques, which can handle systems of arbitrary relative degree perturbed by bounded attacks of arbitrary shape, are used for online reconstruction of the attacks and secure state estimation in CPSs under attacks. The Sparse Recovery (SR) algorithm is also employed to reconstruct the stealth sensor attacks to the unprotected sensors. Next, the corrupted measurements and states are to be cleaned up online in order to prevent the attack propagation to the CPS via the feedback control signal. The case study based on the US Western Electricity Coordinating Council (WECC) power network under attack is considered. The power network performance degradation as a result of cyber-attacks to actuators and/or sensors is observed. The proposed SMO/D and SR algorithms and methodologies are applied to recover the performance of the attacked WECC power network. Simulation results illustrate the efficacy of the proposed approaches.

Chapter 11. **Resilient Cooperative Control of Input Constrained Networked Cyber-Physical Systems:** This chapter mainly studies the resilient cooperative control methods for Networked Cyber-Physical Systems (NCPS) subject to input saturation constraints. First, input constrained asymptotic consensus tracking problems for high-order triangular form NCPS are investigated. Sliding mode control methods are employed to achieve robust consensus tracking under input saturation and bounded input disturbances. Both the cases of static leader and dynamic leader are considered. Observer-based distributed controllers are further designed to reduce the relative state measurement requirement between the systems. Then, input constrained robust finite-time consensus tracking problems for high-order triangular form NCPS are studied. A switching control strategy is proposed which is shown to achieve consensus tracking in finite time under the input saturation constraint. Both the cases with relative state measurement and only relative output measurement are handled. The proposed control strategies are novel in that they are resilient to both the control input constraints, the unknown external disturbances and the possible digital communication restraints. Numerical simulation is performed and an application to the vehicle platoon control problem is given to illustrate the effectiveness of the proposed control strategies.

Chapter 12. **Optimal Subsystem Decomposition and Resilient Distributed State Estimation for Wastewater Treatment Plants:** In this work, an optimal subsystem decomposition algorithm is proposed based on the community discovery algorithm with weighted network graph and is applied to a benchmark wastewater treatment plants (WWTP) system. With the obtained subsystems, a resilient distributed state estimation method is further investigated to deal with the natural disasters (storm and rain) and the unreliable communication networks. The nodes of information graph theory are introduced to represent the state, input and output variables of the WWTP system. By defining a sensitivity of an edge, a weighted directed graph of WWTP system is constructed. The nodes are connected by weighted edges with the weight reflecting the strength of the connection between nodes. The weighted network graph can reflect both the connectivity and connection strength of the system. The community structure discovery algorithm is used to divide all variables into subsystem groups, such that the interaction between groups is strong. Then, the subsystem decomposition of complex process system is obtained. The optimal subsystem decomposition method is validated by designing a resilient distributed state estimation for WWTP system with unreliable communication networks. An information compensation strategy is proposed to coordinate the sub-estimators. Comparative study is carried out for the subsystem decomposition by physical structure and unweighted network-based method. The results show that the subsystem decomposition and distributed state estimation scheme improves the resiliency of the system, compared to a centralized scheme applied to the whole system.

Chapter 13. **Cyber-Attack Detection for a Crude Oil Distillation Column:** Industrial control systems are recently being interfaced to the cyber-domain as computing, communication, and electronics technologies continue to evolve giving rise to what is known as Cyber-Physical Systems (CPSs). Integration of cyber-domain makes these plants vulnerable to cyber-threats and hence it is indispensable to address the cyber-security of these systems. The huge worldwide demand for crude oil can make them a lucrative target for cyber-intrusions. In this chapter, a continuous binary Distillation Column (DC) plant is considered as a CPS and a distributed attack detection algorithm is proposed to enhance its security. In order to demonstrate the real-time performance of attack detection algorithm, a hybrid Hardware-In-the-Loop (HIL) testbed is developed where the DC plant is simulated in real time inside PC and the controllers as well as the detection algorithms are implemented inside Siemens PLC. Finally, the real-time performance of the developed attack detection algorithm is validated through several attack scenarios.

Chapter 14. **A Resilient Nonlinear Observer for Light-Emitting Diode Optical Wireless Communication under Actuator Fault and Noise Jamming:** Optical wireless communication is emerging as a low-power, low-cost, and high data rate alternative to acoustic and radio-frequency communications in several short to medium-range applications. However, it requires a close-to-line-of-sight link between the transmitter and the receiver. Indeed, a severe misalignment can lead to intolerable signal fades and can significantly degrade system performance. Despite recent efforts to establish a line-of-sight (LOS) between transmitter and receiver by improving system designs and active alignment, maintaining a perfect LOS between the two sides despite the robot's mobility remain a challenging task for cooperative autonomy. On the other hand, the optical wireless communication system is often hampered by noise jamming on the optical communication channel that reduces the system capacity of the wireless optical mobile networks. Additionally, a situation of an occurrence of actuator failures can occur due to malfunctions or high instantaneous torques of the actuator mechanism flexible on the receiver orientation. To address this problem, we propose a novel extended state switched-gain discrete-time nonlinear observer to simultaneously estimate the actuator fault and the optical communication system's state variables subject to noise jamming attack. Furthermore, Lyapunov function-based analysis is used to design the proposed unknown switched-gain input observer in each piecewise monotonic region of the optical communication model output functions and ensures global stability of the extended error system. Numerical simulation results are then provided to demonstrate the validity and effectiveness of the proposed extended switched-gain state observer subject to noise jamming attack on the optical communication link.

Chapter 2
Introduction to Cyber-Physical Security and Resilience

Masoud Abbaszadeh and Ali Zemouche

2.1 Introduction

Motivated by increasing demand for performance, availability, efficiency, and resilience, several sectors including energy, manufacturing, healthcare, and transportation have adopted latest advances in controls, automation, communications, and monitoring in the past decades, moving towards semi-autonomous or fully autonomous systems in some cases. The resulting integration of information, control, communication, and computation with physical systems, demands new methodologies for detailed systematic and modular analysis and synthesis of Cyber-Physical Systems (CPSs) as a means to realize the desired performance metrics of efficiency, sustainability, and safety (Dibaji et al. 2019). However, CPSs suffer from extendable vulnerabilities that are beyond classical networked systems due to the tight integration of cyber- and physical components. Sophisticated and malicious cyber-attacks continue to emerge to adversely impact CPS operation, resulting in performance degradation, service interruption, and system failure. Cyber-physical security provides a new line of defense at the physical domain layer (i.e., the process level) in addition to the network Information Technology (IT) and higher level Operational Technology (OT) solutions.

In the past few years, there has been tremendous research and development efforts in cyber-physical security and resilience. The forefront of these efforts is to develop theory and technology to detect and localize cyber-attacks, identify attack types, estimate, and reconstruct attacks, and to perform secure estimation and control under

M. Abbaszadeh (✉)
GE Research, Niskayuna, NY, USA
e-mail: masoud@ualberta.net

A. Zemouche
CRAN - Nancy Automatic Research Center IUT Henri Poincaré de Longwy, University of Lorraine, Nancy, France
e-mail: ali.zemouche@univ-lorraine.fr

M. Abbaszadeh and A. Zemouche (eds.), *Security and Resilience in Cyber-Physical Systems*, https://doi.org/10.1007/978-3-030-97166-3_2

attack. To this end, a variety of results have been proposed based on both model-based and data-driven methodologies (Abbaszadeh et al. 2018; Akowuah and Kong 2021; Alguttar et al. 2020; AlZubi et al. 2021; Ameli et al. 2018; An and Yang 2020; Ao et al. 2016; Azzam et al. 2021; Baniamerian et al. 2019; Brentan et al. 2017; Buason et al. 2019; Cao et al. 2020; Chen et al. 2021, 2016; Cómbita et al. 2020; Dibaji et al. 2018; Ding et al. 2020a, b, 2021, 2018; Dutta et al. 2021; Fang et al. 2020; Farivar et al. 2019; Ferrari and Teixeira 2017; Fillatre et al. 2017; Giraldo et al. 2018; Gu et al. 2020; Guan and Ge 2017; Han et al. 2021; Hendrickx et al. 2014; Housh and Ohar 2018; Humayed and Luo 2015; Humayed et al. 2017; Iwendi et al. 2021; Jahromi et al. 2021, 2019; Junejo and Goh 2016; Khan et al. 2020; Kim et al. 2021; Kozik et al. 2018; Krishnamurthy et al. 2014; Kumar et al. 2022; Lee et al. 2014; Li et al. 2021a, b, 2020; Loukas et al. 2019; Mestha et al. 2017; Narayanan et al. 2021; Noorizadeh et al. 2021; Olowononi et al. 2020; Orumwense and Abo-Al-Ez 2019; Paredes et al. 2021; Park et al. 2015, 2019; Pasqualetti et al. 2013; Pirani et al. 2021; Roy and Dey 2021; Sahoo et al. 2018; Semwal 2021; Shin et al. 2017; Su et al. 2020; Taheri et al. 2020; Tan et al. 2020; Teixeira et al. 2015; Tian et al. 2020; Tiwari et al. 2021; Tsiami and Makropoulos 2021; Valencia et al. 2019; Wang et al. 2021a, b, 2020; Wu et al. 2021; Xiong and Wu 2020; Yan et al. 2018, 2019; Ye et al. 2020; Zhang et al. 2021a, b, c, d, 2017; Zhang and Zhu 2020; Zhu et al. 2018).

Cyber-physical security technologies leverage dynamic models of the closed-loop control systems through utilization of first-principle or data-driven (e.g., system identification-based) modeling paradigms. This, in addition to utilizing historical operational data, enables realistic simulations of attack and fault scenarios, which, compared to normal operation data, are usually rare in the field. This in turn, enables utilization of both model-based and data-driven detectors, and in terms of data-driven detectors, enables exploiting both supervised and unsupervised machine learning approaches.

2.2 Cyber-Physical Security and Resilience Functionality Overview

Cyber-physical security and resilience generally consists of the following functionality modules:

- **Detection:** Determines if an attack has happened.
- **Isolation:** Determines what is under attack, in terms of sensor, actuator, or control nodes. It may also provide foundations for early warning generation at the system/subsystem/component level.
- **Identification:** Determines severity and impact of the attack (including attack type and magnitude), and backtracks the attack to find its source through attack forensics. It may also separate the source of the abnormality and distinguishes malicious attacks from naturally occurring faults/failures.

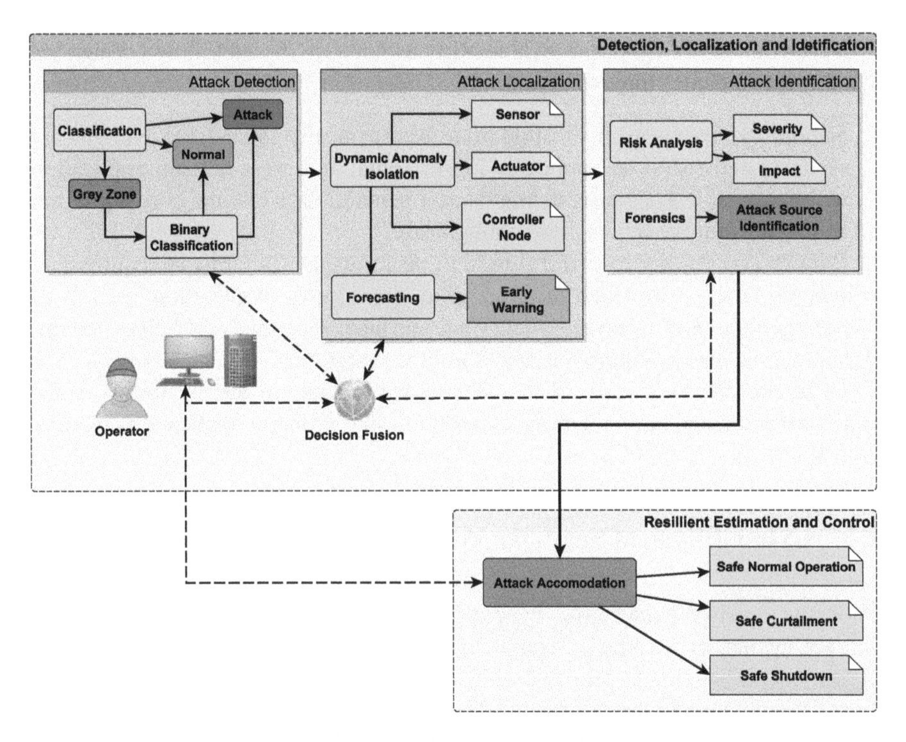

Fig. 2.1 Cyber-physical security & resilience functionality diagram

- **Resilience:** Maintains the integrity, operability, and availability of the system by accommodating (mitigating) the attack through resilient estimation and control, with/without a degraded performance (i.e., curtailment); or commands a controlled safe shutdown.

Figure 2.1 shows an example of a cyber-physical security system functionality diagram with modules as described above. The detection (and isolation) decisions may be made in one shot or in a two-step process, in which a second decision algorithm resolves the gray zones in the first decision. The system may be completely autonomous or with a human in the loop, in which case the operator may be in the loop for the whole process with the ability to override machine-made decisions. The system may also provide visual and/or textual status reports to the operator in real time through security user interfaces such as a Security Information and Event Management (SIEM) dashboard. Furthermore, to increase the decisions accuracy and speed, the detection and isolation decisions may be taken in parallel and fused with potential input from the operator.

Cyber-physical security goes beyond cyber-security, as it can provide an additional layer of defence. Attack neutralization through resilient estimation and control, helps providing the system with capabilities to overcome damage and continue operation when sensors or control signals are disrupted by adversarial threats.

Development of a cyber-physical security technology should follow a design philosophy that includes three main aspects:

1. ***Scalability:*** This is itself two-fold (a) to be organically expandable to large-scale systems, and (b) to be applicable to horizontal and cross-domain applications with reasonable system modeling/dataset generation, while the core algorithms and architecture remaining domain-agnostic.
2. ***Robustness:*** Ability to perform in high performance (in terms of requirements such as false positive and false negative rates, speed of detection, etc.) in the presence of model uncertainty, data value and label uncertainty, as well as system's operational and configuration/manufacturing variations.
3. ***Coherence:*** Having a unified architecture with modularity and flexibility to identify essential and optional modules and to fit into different application domains.

2.3 Cyber-Physical Security Versus Adjacent Fields

From the security perspective, cyber-physical security provides a new layer of defence against cyber-attacks, complementing the existing defence in the IT and higher level OT network security, and increasing the overall security posture of systems via a defence-in-depth strategy (Mosteiro-Sanchez et al. 2020). The focus of cyber-physical security is on the impact of the attack on the physical behavior of the system as opposed to monitoring data communications and network traffic. Furthermore, the attack resilience capability maintains safe operation and/or prevents system damage even at the presence of attacks which may go stealthy and undetected by the IT/OT network-layer security solutions. This increases the availability and integrity of the systems under protection.

2.3.1 Cyber-Physical Security Versus Cyber-Security

Although sounding similar, there are important distinctions between cyber-security and cyber-physical security. The IT layer cyber-security is concerned with data authenticity and integrity. Cyber-physical security, on the other hand, addresses the availability and reliability, in addition to the IT layer, and maintains system operability in an operational technology (OT) environment, at the *physical layer*. Therefore, mere access control, for example, does not help in the OT layer, e.g., the industrial communication bus in Supervisory Control, Data Acquisition (SCADA) systems or Distributed Control Systems (DCS), and physical layers. For example, in a data-only IT layer, it is possible to log out users or prevent their access to the network, but in the OT layer, operators should never be log out of the system during an emergency. Cyber-physical security complements IT and higher level OT cyber-security. While cyber-security tries to prevent a cyber-attack from happening at the first place, cyber-physical security comes into play when an attacker has already bypassed the IT and higher level OT layers, and thus, an attack has already happened. Furthermore, a cyber-security solution detects an attack through anomalous activities in a communication data network, while cyber-physical security detects an attack by analyzing its impact on the physical behavior of the system. Additionally, cyber-security detects network attacks only, while cyber-physical security, due to its interaction with the physical world, can also detect physical attacks. Finally, cyber-security is often based on static analyses (in terms of system dynamics), while cyber-physical security is essentially based on physical dynamics of the system.

2.3.2 Cyber-Physical Security Versus FDII

A fault is a natural cause, while a cyber-attack is a malicious cause, often intelligently designed and targeted towards specific aspect(s) of a system. A fault is due to a component/system natural malfunction. Therefore, it is highly unlikely that multiple independent and unrelated faults happen simultaneously. A multi-fault scenario is most often a cascaded event started by a single fault. Fault Detection, Isolation, and Identification (FDII) methods cannot detect and isolate multiple simultaneous uncorrelated faults. A cyber-attack on the other hand, is artificially designed and can target multiple places of a system or even multiple systems as the same time without any system relations. Faults usually happen in the sensors, actuators, or some other hardware nodes, while a cyber-attack may happen in any hardware (e.g., sensor or actuator) or software (e.g., inside controller) node. Software faults are rare, especially in a certified code. For example, the probability of a software fault in an airworthy code certified by DO-178 aviation standard is less than 10^{-6} (RTCA 2011). There is yet no certification against a cyber-attack. FDI often works against a pre-determined set of system faults, identified through tools such as fault tree analysis (FTA) or Failure Mode and Effect Analysis (FMEA). A cyber-attack can very much go beyond specified or even known system faults. A cyber-attack can target or randomly activate a vulnerability even unknown to the system designers. Furthermore, FDI cannot detect stealthy attacks that keep the monitored signals within normal operational ranges.

2.3.3 Cyber-Physical Security Versus Prognostics

Prognostics concerns aspects like system ageing, estimation of the remaining useful life (RUL), life optimization, condition monitoring, and condition-based maintenance. These are all categorized under industrial asset performance management (APM). Prognostics provides a solution to the APM problem, which is quite different from what cyber-physical security is all about. Due to its mission, prognostics happens at time scales much slower than what is needed for cyber-physical security. Fast response at the sampling rate of a real-time controller, as needed in cyber-physical security and resilience, is simply out of scope for prognostics. As a result, prognostics often uses steady-state or quasi-steady-state models. Cyber-physical security, on the other hand, often requires dynamic models of higher fidelity. In summary, prognostics is often a tool for gaining more financial benefit from an existing asset that would operate otherwise, anyway. However, cyber-physical security and resilience is about maintaining system operability at the first place, and therefore must enable the system to withstand and respond to existential threats.

2.4 Attack Detection, Isolation, and Identification

In this section, we provide a survey of some of the main and latest results on cyber-attack detection, isolation, and identification for cyber-physical systems.

A generic CPS architecture by considering the applications related to secure industrial control system (ICS) to explain the cyber resilience concepts is illustrated in Fig. 2.2, which is from the US DHS ICS-CERT recommended practice for defense-in-depth strategies (Dakhnovich et al. 2019; Homeland Security 2014), and based on the Purdue five-level model (Dakhnovich et al. 2019). An ICS is a set of electronic devices to monitor, control, and operate the behavior of interconnected systems. ICSs receive data from remote sensors measuring process variables, compare those values with desired values, and take necessary actions to drive (through actuators) or control the system to function at the required level of services (Galloway and Hancke 2013). Industrial networks are composed of specialized components and applications, such as programmable logic controllers (PLCs), SCADA systems, and DCS. There are other components of ICS such as remote terminal unit (RTU), intelligent electronic devices (IED), and phasor measurement units (PMU). Those devices communicate with the human–machine interface (HMI) located in the control network. With the rise of 5G and industrial IoT, the ICS architecture is becoming even more connected with lower level edge devices increasingly connected to each other and to the cloud, hence, expanding the attack surface and demanding for better cybersecurity solutions (Abosata et al. 2021). This increased connectivity and reduced latency have also enabled design of distributed architectures and distributed edge computing, creating both cybersecurity opportunities and challenges.

Cyber-attack detection is in general concerned with detecting a malicious cyber-incident in a system, while cyber-attack isolation is concerned with pinpointing specifics part(s) of the system that are under attack, and trying to trace back the entry point(s), and the root cause of the cyber-attack. Localizing the initial point(s) of cyber-incident is both critical and hard, in the sense that the attack may cause a series of cascaded events or propagate through the system, especially in feedback control systems. For cyber-physical systems, attack detection and isolation at the physical process level is based on monitoring the process variables such as sensor measurements and actuator commands in a control system. Several recent surveys on attack detection and isolation are available, covering the space from different

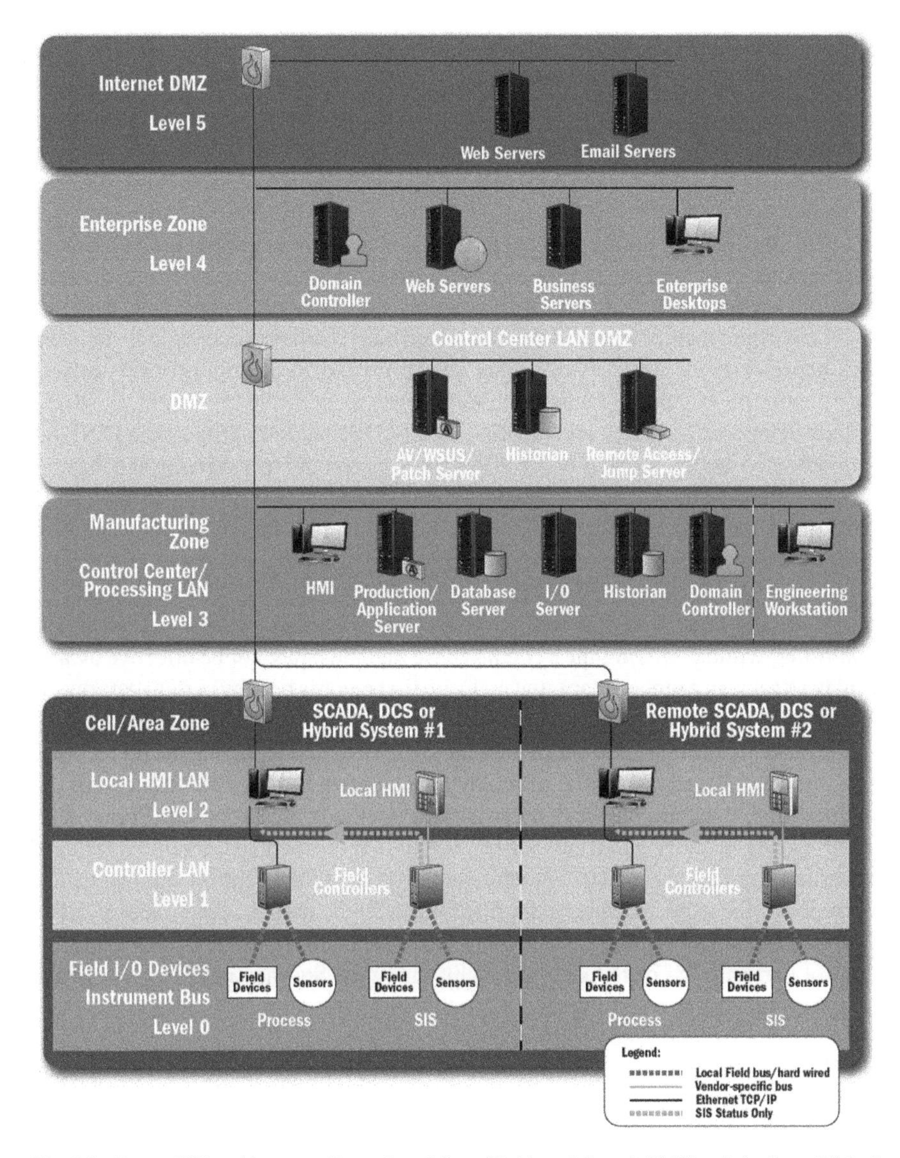

Fig. 2.2 Secure ICS architecture. Reproduced from (Dakhnovich et al. 2019), originally published under a CC BY 3.0 license, doi:10.1088/1757-899X/497/1/012006

perspectives and for different application domains including for general CPS (Ding et al. 2018; Giraldo et al. 2018; Humayed et al. 2017; Li et al. 2020; Tan et al. 2020), ICS (Zhang et al. 2021d), smart grid (Musleh et al. 2019; Peng et al. 2019), autonomous vehicles (Chowdhury et al. 2020; Grigorescu et al. 2020; Loukas et al. 2019) and energy systems (Orumwense and Abo-Al-Ez 2019).

Attack identification is concerned with providing additional insights about the nature of the attack, identifying the type of the attack, impact analysis and forensics (Long et al. 2005; Pasqualetti et al. 2013; Xuan and Naghnaeian 2021). Another important aspect of attack identification is to separate anomalies from novelties (e.g., environmental or operational changes) which can have process-level impacts, and hence, may be detected by attack detectors, and to distinguish cyber-attacks from naturally occurring faults or failures (Anwar et al. 2015; Pan et al. 2015). Attack detection, isolation, and identification (ADII) has similarities with FDII, but as mentioned before, also has major differences, especially for detecting and locating stealthy and coordinated attacks. Similar to other anomaly detection paradigms, ADII algorithms face fundamental design trade-offs among performance and robustness requirements such as false positive rate, false negative rate, and speed of detection (Ding et al. 2018; Li et al. 2020; Zhang et al. 2021d). Many of ADII algorithms are passive in the sense that they receive time-series data from sensors, actuators, and controller, without altering the system. These methods may not be effective against replay attacks. In a replay attack, the malware first records healthy system data during the normal operation, then injects malicious signals into sensors and/or actuators, while masking the real-time data to be sent to the HMI and replaying the prerecorded healthy data instead. Detection of replay attacks often requires active methods. To address this, dynamic physical watermarking methods are proposed (Porter et al. 2020; Satchidanandan and Kumar 2016, 2019). In these methods, carefully designed watermark signals are injected into the system on top of the control commands. The presence of the expected watermark fingerprints in the outputs, determines whether the system is uncompromised. These additional injections, however, may affect the control performance or reduce the stability margins. So, they need to be designed and implemented in a safe manner, through a trade-off optimization between attack detectability and control performance (Khazraei et al. 2017b, a).

The ADII algorithms may work stand-alone for monitoring and alarm generation, or may work in conjunction with an automatic attack mitigation and neutralization algorithm (Li et al. 2020; Mestha et al. 2017), or as part of a cyber-situational awareness system (Abbaszadeh et al. 2018; Chang et al. 2017; Pöyhönen et al. 2021. The ADII techniques can be categorized into two main categories: (i) model-based approaches and (ii) data-driven approaches. Next, we will provide an overview of some of the latest results in each category.

2.4.1 Model-Based ADII

Model-Based ADII utilizes a system model in the detection, isolation, and identification procedures. The model can be a simple encapsulation of domain knowledge of the system operation such as in traditional rule-based or expert systems, or can be a more formal dynamic system model, such as a state space model, developed using first-principles or system identification. Once such a model is available, an observer-based method is often used for attack detection and isolation. The most popular of such observers is the Kalman filter, providing an innovation signal between the measured outputs and the predicted outputs by the model. Detection and isolation procedures are mainly based on two threshold mechanisms over the innovation signal: (i) the chi-square distribution and (ii) the Cumulative Sum (CUSUM) (Ahmed et al. 2017; Housh and Ohar 2018; Sridhar and Govindarasu 2014). The CUSUM approach has the advantage to make a more robust decision based on a weighted sequential sum of the innovation signal as opposed to its instantaneous value, potentially reducing the false positives. However, it may induce a time delay in detecting cyber-events. The attack isolation is done mainly using two techniques, (i) a bank of observers (such as Kalman filters) running in parallel, each designed to be sensitive to a specific element of the innovation vector (Taheri et al. 2020; Ye et al. 2020; Zhang and Zhu 2020) and (ii) a hierarchical approach in which a hierarchy of detectors is designed to zoom in from the top system level into specific subsystems, components, or sensors/actuaros in a top-down manner (Karimipour and Leung 2019; Li et al. 2021a). Model-based attack identification mainly relies upon modeling different attack types and scenarios, and exploiting those attack models along with the system model (Azzam et al. 2021; Li et al. 2020; Park et al. 2019; Teixeira et al. 2015).

2.4.2 Data-Driven ADII

Many attacks detection algorithms available in the literature root back to fault detection techniques. Indeed, from the physical process perspective, cyber-attacks can be viewed as intelligent disturbances, which can affect the system in a malicious manner. To solve complex architectures of cyber-physical attacks, it is necessary to go beyond the traditional methods resulting from fault diagnosis. Novel and intelligent techniques are needed to deal with malicious attacks that appear nonlinearly in mathematical models. To this end, to avoid the need of conservative mathematical conditions, merging learning-based algorithms with standard control theory based techniques is gaining a lot of interest as a promising hybrid approach and a compelling solution.

In recent years, machine learning and deep learning methods have become popular in ADII Zhang et al. (2021c), Narayanan et al. (2021). Recent results for ICS and CPS include classification using statistical machine learning (Ameli et al. 2018; Lee et al. 2014), deep neural networks (Jahromi et al. 2021, 2019; Lee et al. 2014; Yan et al.

2018; Zhu et al. 2018), and pattern recognition (Brentan et al. 2017). Distributed machine learning methods are also proposed for large-scale systems including in IoT and edge computing (Guan and Ge 2017; Kozik et al. 2018). A challenge for adopting AI/ML techniques for CPS ADII is how to obtain the right training data sets, specially for supervised learning methods, and in particular for two-class learning, in which both normal and abnormal samples are required. To overcome this challenge, some researchers have proposed unsupervised learning methods, where no labeled data are required (Jahromi et al. 2019; Tiwari et al. 2021). Unsupervised machine learning methods have also been used in the past in anomaly and intrusion detection in communication and computer networks. However, these approaches need to go through an initial learning phase, often in-field, during which they tend to have a large false alarm rate. Their final accuracy is also often lower than those achieved by supervised learning methods. The alternative approach is to generate synthetic training data using a simulation platform of the system. To this end, digital twins have become a powerful tools to conduct controlled simulations, and to generate labeled data samples of both normal and abnormal classes, both for training and validation of the machine learning models (Abbaszadeh et al. 2018; Mestha et al. 2017; Yan et al. 2018). Digital twin simulations can be used together with available historical field data to address class imbalance (caused due to scarcity of abnormal data in the field), and also to generate data for complete coverage of normal operational and environmental conditions. Furthermore, intelligently designed experiments for digital twin simulations can reduce the need for large training datasets (Abbaszadeh et al. 2018; Yan et al. 2019).

Machine learning algorithms used for ADII are themselves susceptible to cyber-attacks, and hence, need to be secured via hardware and software protections. Robust and adversarial machine learning are active fields of research addressing the security and resilience of machine learning algorithms. A survey on secure and resilient machine learning for CPS security is given in (Olowononi et al. 2020). Besides, in order to be adopted in safety-critical and mission-critical systems, machine learning algorithms must exhibit trustworthiness, which includes certain level of explainability in a human-readable fashion. The explainability can, for example, include providing physical insights, outputting decision factors and their contributions to the overall decision, and giving decision confidence scores based on conformal prediction methods.

2.5 Attack Resilience

In this section, we provide an introduction to the notion of resilience, and a survey of some of the main results. Then in Sects. 2.6 and 2.7, we will cover some of the latest results on two major approaches towards achieving resilience for cyber-physical systems, namely, resilient estimation and resilient control.

Real-world attacks on control systems have in fact occurred in the past decade and have in some cases caused significant damage to the targeted physical processes.

One of the most popular examples is the attack on Maroochy Shire Council's sewage control system in Queensland, Australia, that happened in January 2000 Cardenas et al. (2008), Slay and Miller (2007). In this incident, an attacker managed to hack into some controllers that activate and deactivate valves causing flooding of the grounds of a hotel, a park, and a river with a million liters of sewage (Cardenas et al. 2008). Another well-known example of an attack launched on physical systems is the Stuxnet virus that targeted Siemens' supervisory control on an Iranian uranium enrichment plant targeting a commercially available PLC. Operating under a narrow set of conditions, the attackers were able to ensure the attack reached its intended recipient with limited fallout. They inserted a malware which would lie dormant in the system and go undetected (Falliere et al. 2018). This shows that even air-gapped systems are susceptible to cyber-espionage and -attack.

Given that the end-goal of CPS is a reliable and safe functioning at all times, cyber-physical resilience of CPS is a necessary requirement. It corresponds to the ability to withstand high-impact disturbances, which may occur due to either physical outages or cyber-causes, and to continue to deliver acceptable performance even under attack.

The term resilience is being discussed increasingly in the context of CPS lately, ranging from transportation (Ip and Wang 2011), power (Albasrawi et al. 2014; Zhu and Basar 2011), control systems (Rieger et al. 2009, 2013; Zhu and Basar 2011) as well as other types of systems such as ecological (Holling 1996, 1973) and biological (Kitano 2004). Resilience is often discussed concomitantly with other system-oriented notions such as robustness, reliability, and stability (Levin and Lubchenco 2008) and quite often used interchangeably with the term robustness. We argue however that these two terms are distinct. The reason is that resilience and robustness characterize fundamentally different system properties. The term robustness applies in the context of small bounded disturbances while resilience, in the context of extreme high-impact disturbances. Resilience of a CPS with respect to a class of extreme and high-impact disturbances, is the property that characterizes its ability to withstand and recover from this particular class of disturbances by being allowed to temporarily transit to a state where its performance is significantly degraded and returning within acceptable time to a state where certain minimal but critical performance criteria are met (Baros et al. 2017).

The National Academy of Sciences (NAS) (Cutter et al. 2013) defined resilience as the ability to prepare and plan for, absorb, recover from, or more successfully adapt to actual or potential adverse events. The authors in Linkov et al. (2013) used the resilience definition provided by NAS to define a set of resilience metrics spread over four operational domains: physical, information, cognitive, and social. In another work (Linkov et al. 2013), the authors applied the previous resilience framework by Linkov et al. (2013) to develop and organize useful resilience metrics for cyber-systems. In Bruneau et al. (2003), the authors have proposed a conceptual framework initially to define seismic resilience, and later in Tierney and Bruneau (2007) the R4 framework for disaster resilience is introduced. It comprises robustness (ability of systems to function under degraded performance), redundancy (identification of substitute elements that satisfy functional requirements in event of significant performance degradation), resourcefulness (initiate solutions by identifying resources based on prioritization of problems), and rapidity (ability to restore functionality in timely fashion).

The design of control and estimation algorithms that are resilient against faults and failures is certainly not a new problem. In fault-detection and identification Massoumnia et al. (1989), Blanke et al. (2006), the objective is to detect if one or more of the components of a system has failed. Traditionally, this is done by comparing the measurements of the sensors with an analytical model of the system and by forming the so-called residual signal. This residual signal is then analyzed (e.g., using signal processing techniques) in order to determine if a fault has occurred, however, in such algorithms there is in general one residual signal per failure mode and in some problems formulations, the number of failure modes can be very large and one cannot afford to generate and analyze a residual signal for each possible failure mode (Fawzi et al., 2014).

In another area, namely robust control (Zhou and Doyle 1998), one seeks to design control methods that are robust against disturbances in the model. However, these disturbances are mainly treated as natural disturbances to the system and are assumed to be bounded. This does not apply in the context of security since the disturbances will typically be adversarial and therefore cannot be assumed bounded which is also the case in stochastic control and estimation, where the disturbances are assumed to follow a certain probabilistic model, which we cannot adopt for CPSs.

Resilient or secure state estimation and control constitute effective and promising means for addressing various security-related issues of CPSs. The main objective is to keep an acceptable performance level of the CPS by resorting to different security countermeasures, including attack attenuation and mitigation, isolation, detection, and compensation. When an attack occurs, the developed secure estimation/control mechanisms possess certain capabilities to mitigate or counteract attack effects, or prevent CPSs from severe performance degradation and loss, or allow the system designers to make corrections and recover the system from any unsafe operation (Ding et al. 2020a).

Recently, there are several survey papers of security-oriented CPSs. For example, the recent progress of secure communication and control of smart grids under malicious cyber-attacks is reviewed in Peng et al. (2019), where different attack models and effects as well as security strategies are reviewed from IT protection and secure control-theoretic perspectives. A summary of detection methods of false data injection (FDI) attacks on smart grids is made in Musleh et al. (2019). The existing FDI attack detection algorithms in smart grids are classified into model-based types and data-driven types. From a systems and control perspective, the CPS security issue is evaluated in Dibaji et al. (2019), where some latest systems and control methods are reviewed and classified into prevention, resilience, detection, and isolation. An overview of security control and attack detection for industrial CPSs is conducted in Ding et al. (2018). An intensive discussion of adversarial attacks and their defenses is provided in Li et al. (2020) for sensor-based CPSs in the field of computer vision. Emerging techniques improving the safety and security of CPSs and IoT systems are surveyed in Wolf and Serpanos (2017) from two aspects: (1) design time techniques verifying properties of subsystems and (2) runtime mechanisms helpful against both failures and attacks.

2.6 Resilient Estimation

This section is devoted to a general state of the art on available resilient and secure estimation algorithms in cyber-physical systems. Before recalling existing estimation methods, we give a general introduction to emphasize the importance of resilient and secure estimation, and explain what the software sensors have to face to ensure resilience and security of the estimation. State estimation plays an important role in better understanding the real-time dynamics of CPSs and executing some specific control tasks. These states can be reconstructed based on only measured yet possibly corrupted information from sensors. Different from traditional control systems, the tight integration of physical and cyber-components, and the occurrence of various malicious attacks pose nontrivial challenges to the performance analysis and the design of state estimators or filters. Vulnerability of cyber-physical systems may come from two kind of malicious attacks, namely cyber-attacks and physical attacks:

- *Cyber-attacks:* Cyber-attacks occur on the cyber-variables of the system. They may be due to a software virus or to a corruption in communication channels. The well-known Stuxnet malware is one of the relevant examples of cyber-attacks Mishra et al. (2016), Ferrari and Teixeira (2021, Chap. 7). The attackers exploited vulnerabilities of the system such as those running over SCADA devices (Fig. 2.2) to for example, inject false data in the sensor measurements gathered by the SCADA system.
- *Physical attacks:* Physical attacks (also called kinetic attacks) are intentional offensive actions which aim to get unauthorised access to physical assets such as infrastructure, hardware, or interconnection. Sensors are among the devices most exposed to this type of attack. This will have a direct and significant impact on any estimation algorithm using measurements issued from such sensors because, in addition to susceptible manipulations on the cyber-layer, sensor readings rely on physical layer properties that can be manipulated (Taormina et al. 2016). Examples of physical attacks include manipulating gyroscopes used to stabilize drones during formation flights, spoofing LiDAR sensors used in autonomous driving, manually deactivating the pump to disconnect the network from the reservoir in modern water distribution systems, and spoofing magnetic sensors used in several applications, like anti-lock braking systems in automotive.

In the following, we classify some existing secure state estimation approaches according to performance indicators and defense strategies against cyber-attacks.

2.6.1 State of the Art on Resilient and Secure Estimation: A Glimpse on Existing Methods

This section is dedicated to a short but complete overview of existing secure and resilient estimation methods. The overview is shared into two categories (Ding et al. 2020a), namely statistical methods and Lyapunov stability-based techniques.

2.6.1.1 Resilient Estimation Based on Statistical Methods

The statistical-based state estimation aims to select appropriate gain parameters to minimize estimation error variance, hence, the structured information of cyber-attacks, such as statistical information or boundedness information, is assumed to be known. Following this idea, the main focus is then placed on disclosing or offsetting the undesirable impact from compromised data generated by malicious attacks (Ding et al. 2020a. In Ma et al. (2017), an algorithm of variance-constrained filtering over sensor networks is proposed for discrete time-varying stochastic systems and by resorting to the recursive linear matrix inequality approach, a sufficient condition is established for the existence of the desired filter satisfying the pre-specified requirements on the estimation error variance. In the framework of Kalman filtering, a distributed filter with double gains is designed in Ding et al. (2017) which can be regarded as two weight matrices reflecting the different confidence levels of the information from itself and from neighboring nodes.

Estimators or filters can be integrated in some detection mechanisms to remove the compromised data generated by malicious attacks as much as possible. Benefiting from their favorable statistical characteristics, χ^2 detector and its variants are widely adopted. In light of such a detection rule, a critical attack probability is analyzed in Yang et al. (2019) where it is shown that when the considered probability is bigger than some critical value, the steady-state solution of estimation error covariance could exceed a preset value.

It is worth noting that the estimation performance can be properly warranted if the corrupted sensor is accurately detected and effectively isolated. For example, in Mishra et al. (2016) they have estimated the state of a noisy linear dynamical system when an unknown subset of sensors is arbitrarily corrupted by an adversary. They have proposed a secure state estimation algorithm, and derived optimal bounds on the achievable state estimation error given an upper bound on the number of attacked sensors. The proposed state estimator involves Kalman filters operating over subsets of sensors to search for a sensor subset which is reliable for state estimation. When the attack subset is properly identified, the performance of the developed algorithm does not exceed the one by the worst-case Kalman estimation. The optimal secure estimation is pursued in Shoukry et al. (2017) for attacks without restrictions on their statistical properties, boundedness, and time evolution in comparison with the sparse attacks. They have presented a novel algorithm that uses a satisfiability modulo theory approach to harness the complexity of secure state estimation.

2.6.1.2 Lyapunov Theory-Based Methods

Inspired by its mature approaches, an analysis of vulnerabilities of cyber-physical systems in the face of unforeseen failures and external attacks has received increasing attention in the recent years and some preliminary results have been published in literature, see, for instance, Ao et al. (2016), Pasqualetti et al. (2013). In Pasqualetti et al. (2013), the authors have characterized fundamental monitoring limitations from system-theoretic and graph-theoretic perspectives and a Luenberger-type detection filter is designed. Similarly, detectability of attacks is explored in Ao et al. (2016) in which detectability of attacks based on linear system theory is explored and some sufficient conditions of detecting state attacks and sensor attacks are established. Then, two adaptive sliding mode observers with online parameter estimation are designed to estimate state attacks and sensor attacks with uniformly bounded errors. A co-estimation of system states and attacks inspiration from fault-tolerant state reconstruction, as an alternative scheme, is investigated in Amin et al. (2012), Shoukry and Tabuada (2015). For instance, a scheme based on an unknown input observer is developed in Amin et al. (2012) to estimate the states of SCADA systems subject to stealthy deception attacks. In Fawzi et al. (2014), the secure state estimation problem is transformed into the solvability of an l_0 optimization issue and an ℓ_1/ℓ_r optimization issue in Liu et al. (2016), or the performance analysis problem of ℓ_2, $\mathscr{H}_2$, and $\mathscr{H}_\infty$ systems in Nakahira and Mo (2018) by virtue of the classical robust control, and fault detection and isolation methods.

Employing some artificial saturation constraint on state estimators is regarded as a promising security measure for constraining attacker capability and mitigating the impulsive outlier-like effects of cyber-attacks by attenuating the effects of these attack-incurred abnormal measurements using estimators with some saturated output rejection. For example, a saturated innovation update scheme is adopted in Chen et al. (2018) for distributed state estimators with an adaptive threshold of the saturation level, and in Sun et al. (2021) for stochastic nonlinear systems with a sector bounded condition on the saturation constraint. In Xie and Yang (2018), a saturated innovation scheme with an adaptive gain coefficient and a mode switch mechanism is developed, where the mismatched unknown inputs are suppressed by resorting to the well-known $\mathscr{L}_2$-gain attenuation property. Dynamic saturations with an adaptive rule are further developed in Alessandri and Zaccarian (2018); Casadei et al. (2019). It is noted that dynamic saturations with adaptive saturation levels enjoy more flexible attack attenuation capability and less estimation performance degradation.

2.7 Resilient Control

Besides the resilient state estimation above, CPSs also need to mitigate the threat from secret attackers via various control strategies. Compared with other control applications, security control techniques for CPSs are yet in their infancy, and few results can be found in literature Ding et al. (2018). There are two main lines of research on secure control for CPSs under cyber-attacks, which are categorized as attack-tolerant control and attack-compensated control. The first category focuses on the design of a suitable control policy/law to tolerate unpredictable anomalies caused by attacks (Zhao et al. 2019). In Zhao et al. (2019), a novel observer-based PID controller is proposed and sufficient conditions are derived under which the exponentially mean square input-to-state stability is guaranteed and the desired security level is then achieved. An emphasis is then placed on examining the prescribed tolerance capability or pursuing the maximal tolerance capability for the controlled system, allowing further intervention actions to be made from the system designers. The second category deals with the design of preferable compensation schemes to prevent the system performance and stability from severe deterioration or even becoming unstable. For this purpose, it is essential to implement appropriate attack detection mechanisms to identify and locate the occurrence of cyber-attacks. With respect to networked control systems subject to various cyber-attacks, some preliminary and interesting results can be found in Dolk et al. (2016); Long et al. (2005); Zhang et al. (2016) for DoS attacks, in Amin et al. (2012), Ding et al. (2016a), Dolk et al. (2016), Ding et al. (2016b), Pang and Liu (2011), Pang et al. (2016) for deception attacks, and in Lee et al. (2014); Zhu and Martinez (2013) for replay attacks. The latest development of secure control is evaluated from three aspects: (1) centralized secure control; (2) distributed secure control; and (3) resource-aware secure control.

2.7.1 *Centralized Secure Control*

When CPSs are subjected to DoS attacks, they operate in an open-loop manner as the desired controller is not capable to receive any sensor data for feedback. To ensure the secure control for CPSs under such DoS attacks, switched system theory is deployed, allowing the system to operate in closed-loop mode during attack-free case and in open-loop mode otherwise. It is noteworthy, however, that the resulting system performance depends on the running duty cycle, which is commonly known as dwell time, between the two cases. Hence, the primary goal of secure control is to find the tolerant duration and/or attack frequency such that the desired system performance remains achievable. For example, a robustness measure against DoS attacks, which describes the tolerable maximum attack frequency and duration is investigated in De Persis and Tesi (2015), where an explicit characterization of the frequency and duration of DoS attacks under which closed-loop stability can be preserved is given. The obtained characterization is flexible enough so as to allow the designer to choose from several implementation options that can be used for trading-off performance versus communication resources. Such a robustness measure is further extended in Feng and Tesi (2017) by resorting to an impulsive controller based on a dynamic observer. A cyclic dwell-time switching strategy is proposed in Zhu and Zheng (2019) where an observer-based output feedback control problem for a class of cyber-physical systems with periodic (DoS) attacks is investigated; the attacks coexist both in the measurement and control channels in the network scenario.

By means of a cyclic piecewise linear Lyapunov function approach, the exponential stability and ℓ_2-gain analysis, and observer-based controller design are carried out for the augmented discrete-time cyclic switched system. Then, the desired observer and controller gains in piecewise linear form are determined simultaneously so as to ensure that the resulting closed-loop system is exponentially stable with a prescribed $\mathscr{H}_\infty$ performance index. Furthermore, a switching signal taking values in a finite set is employed to model the number of consecutive DoS attacks in Pessim and Lacerda (2020), where the corresponding stability criterion is derived by making use of a switching parameter-dependent Lyapunov function.

Adaptive detection of cyber-attacks offers an effective means to enhance the system's adaptation to malicious attacks. In An and Yang (2018), an adaptive switching logic is exploited to provide an online location of the real system mode via observing the variation of the traditional quadratic cost in the framework of linear quadratic control. A Kalman-based attack detector with an observation window of a given length is designed in Du et al. (2018) to remove the occurred deception attacks. When the noise level is below a threshold derived, the maximum allowable duration of deception attacks is obtained to maintain the exponential stability of the system. A common feature of the above detectors is that the duration of deception attacks is captured to describe their negative effects. Then, the maximum allowable duration threshold is examined to maintain the desired system stability.

Complete security of CPSs is generally difficult to be maintained from a control-oriented perspective. As a result, an alternative indicator, known as security in probability, is exploited (Ding et al. 2016c). A definition of security in probability is adopted to account for the transient dynamics of controlled systems. Then, a dynamic output feedback controller is designed such that the prescribed security in probability is guaranteed while obtaining an upper bound of the quadratic cost criterion and an original easy-solution scheme of desired controller gain is derived via the matrix inverse lemma.

2.7.2 Distributed Secure Control

In distributed CPSs, the subsystems are connected through communication links, which constitute a communication topology modeled by the Laplacian matrix (Chen and Shi 2017; Liu 2019). According to attack locations, the cyber-attacks in distributed CPSs are classified into two types: (1) intrasystem attacks and (2) intersystem attacks. As such, a critical concern is to design a suitable distributed secure controller to render the resulting closed-loop system survivable or recoverable from cyber-attacks by embedding attack model information (i.e., statistical or structured information). For example, in He et al. (2020) a distributed impulsive controller using a pinning strategy is redesigned, which ensures that mean square bounded synchronization is achieved in the presence of randomly occurring deception attacks, and in the presence of distributed DoS attacks, a control protocol guaranteeing scalability and robustness is proposed in Xu et al. (2019) for multi-agent systems under event-triggered communication. On the other hand, the classical fault detection and estimation approaches provide a foundation to deal with the secure control issue of CPSs with an understanding of similarities of both mathematical descriptions and practical influences between faults and certain cyber-attacks. As in Modares et al. (2019); Moghadam and Modares (2018), a distributed state predictor is employed to estimate the existing attacks, and then a resilient controller is designed to guarantee robust performance and to adaptively compensate for the influence of attacks.

2.7.3 Resource-Aware Secure Control

In, the context of communication scheduling, it is apparent that cyber-attacks can result in a tremendous data sparsity issue because less sensor/control data is adopted for achieving feedback control. This further leads to some inherent and nontrivial challenges for performance analysis and secure control design of CPSs that are beyond the capacity of the existing results on stability analysis and controller design of event-based control systems without cyber-attacks.

The time series of data transmissions or updates under communication schedules become more complex due to the interference of malicious attacks, which poses a significant challenge for continuous-time physical systems. Under the assumption that the execution period and a uniform lower bound of sleeping periods are a priori known, a sufficient condition of exponential stability is derived in Hu et al. (2018) by using a piecewise Lyapunov functional along with a reconstructed state error-dependent switched system. An event-triggered scheduling and control co-design algorithm is developed in Peng et al. (2016) to obtain both the triggering parameter and the control gain. This event-triggered scheme is improved by integrating measurement variations with a minimal trigger sleeping interval in order to avoid the well-known Zeno behavior (Hu et al. 2019; Lu and Yang 2019). Then, under a sparse observability condition, an observer in a delta domain is designed in Gao et al. (2020) to estimate the system state under sensor and actuator attacks, and a self-triggered controller is designed via iterative analysis.

In the context of distributed secure control, there are considerable results reported for CPSs under event-triggered communication scheduling. In Ding et al. (2018), an observer-based event-triggering consensus control problem is investigated for a class of discrete-time multi-agent systems with lossy sensors and cyber-attacks. A novel distributed observer is proposed to estimate the relative full states and the estimated states are then used in the feedback protocol in order to achieve the overall consensus. An event-triggered mechanism with state-independent threshold is adopted to update the control input signals so as to reduce unnecessary data communications. In Feng and Hu (2019), two elaborate interval classifications are constructed by introducing the upper bound of adjacent event intervals under DoS attacks, their duration and their launching time, and then the switched system theory is employed to derive the consensus condition. It should be noted that the presence of cyber-attacks makes the exclusion of Zeno behavior from the designed distributed event-triggered secure controllers generally difficult. This is because the interval of two consecutive data transmissions may not be that of two adjacent events invoked.

To mention a few, an event-triggered controller is designed in Dolk et al. (2016) to tolerate DoS attacks characterized by given frequency and duration properties. An optimal schedule of jamming attacks is proposed in Zhang et al. (2016) to maximize the linear quadratic Gaussian cost under energy constraints. An event-triggering consensus resilient-control with a state-independent threshold is discussed in Ding et al. (2016a) for discrete-time multi-agent systems with both lossy sensors and cyber-attacks.

Acknowledgements M. Abbaszadeh would like to thank GE Research for the partial support of this work. A. Zemouche would like to thank the ANR agency for the partial support of this work via the project ArtISMo ANR-20-CE48-0015.

References

M. Abbaszadeh, L.K. Mestha, W. Yan, Forecasting and early warning for adversarial targeting in industrial control systems, in *2018 IEEE Conference on Decision and Control (CDC)* (IEEE, 2018), pp. 7200–7205

N. Abosata, S. Al-Rubaye, G. Inalhan, C. Emmanouilidis, Internet of things for system integrity: a comprehensive survey on security, attacks and countermeasures for industrial applications. Sensors **21**(11), 3654 (2021)

C.M. Ahmed, C. Murguia, J. Ruths, Model-based attack detection scheme for smart water distribution networks, in *Proceedings of the 2017 ACM on Asia Conference on Computer and Communications Security*, (2017), pp. 101–113

F. Akowuah, F. Kong, Real-time adaptive sensor attack detection in autonomous cyber-physical systems. in *IEEE 27th Real-Time and Embedded Technology and Applications Symposium (RTAS)* (IEEE, 2021), pp. 237–250

M.N. Albasrawi, N. Jarus, K.A. Joshi, S.S. Sarvestani, Analysis of reliability and resilience for smart grids, in *2014 IEEE 38th Annual Computer Software and Applications Conference (COMPSAC)* (IEEE, 2014)

A. Alessandri, L. Zaccarian, Stubborn state observers for linear time-invariant systems. Automatica **88**, 1–9 (2018)

A. Alguttar, S. Hussin, K. Alashik, R. Yildirim, An observation of intrusion detection techniques in cyber physical systems, *Avrupa Bilim ve Teknoloji Dergisi*, (2020), pp. 277–284

A.A. AlZubi, M. Al-Maitah, A. Alarifi, Cyber-attack detection in healthcare using cyber-physical system and machine learning techniques. Soft Comput. 1–14 (2021)

A. Ameli, A. Hooshyar, E.F. El-Saadany, A.M. Youssef, Attack detection and identification for automatic generation control systems. IEEE Trans. Power Syst. **33**(5), 4760–4774 (2018)

S. Amin, X. Litrico, S. Sastry, A.M. Bayen, Cyber security of water scada systems-part i: analysis and experimentation of stealthy deception attacks. IEEE Trans. Control Syst. Technol. **21**(5), 1963–1970 (2012)

L. An, G.-H. Yang, Secure distributed adaptive optimal coordination of nonlinear cyber-physical systems with attack diagnosis (2020). arXiv:2009.12739

L. An, G.-H. Yang, LQ secure control for cyber-physical systems against sparse sensor and actuator attacks. IEEE Trans. Control Netw. Syst. **6**(2), 833–841 (2018)

A. Anwar, A. N. Mahmood, Z. Shah, A data-driven approach to distinguish cyber-attacks from physical faults in a smart grid, in *Proceedings of the 24th ACM International on Conference on Information and Knowledge Management*, (2015), pp. 1811–1814
W. Ao, Y. Song, C. Wen, Adaptive cyber-physical system attack detection and reconstruction with application to power systems. IET Control Theory Appl. **10**(12), 1458–1468 (2016)
M. Azzam, L. Pasquale, G. Provan, B. Nuseibeh, Grounds for suspicion: Physics-based early warnings for stealthy attacks on industrial control systems, in *IEEE Transactions on Dependable and Secure Computing*, vol. 09, (2021), pp. 1–1
A. Baniamerian, K. Khorasani, N. Meskin, Determination of security index for linear cyber-physical systems subject to malicious cyber attacks, in *2019 IEEE 58th Conference on Decision and Control (CDC)* (IEEE, 2019), pp. 4507–4513
S. Baros, D. Shiltz, P. Jaipuria, A. Hussain, A. Annaswamy, Towards resilient cyber-physical energy systems (2017). https://core.ac.uk/download/pdf/83232958.pdf
M. Blanke, M. Kinnaert, J. Lunze, M. Staroswiecki, *Diagnosis and Fault-Tolerant Control* (Springer, 2006)
B.M. Brentan, E. Campbell, G. Lima, D. Manzi, D. Ayala-Cabrera, M. Herrera, I. Montalvo, J. Izquierdo, E. Luvizotto Jr., On-line cyber attack detection in water networks through state forecasting and control by pattern recognition. World Environ. Water Res. Cong. **2017**, 583–592 (2017)
M. Bruneau, S. Chang, R. Eguchi, G. Lee, T. O'Rourke, A. Reinhorn, D.V. Winterfeldt, A framework to quantitatively assess and enhance the seismic resilience of communities. Earthq. Spectra **19**(4), 733–752 (2003)
P. Buason, H. Choi, A. Valdes, H.J. Liu, Cyber-physical systems of microgrids for electrical grid resiliency, in *2019 IEEE International Conference on Industrial Cyber Physical Systems (ICPS)* (IEEE, 2019), pp. 492–497
J. Cao, D. Wang, Z. Qu, M. Cui, P. Xu, K. Xue, K. Hu, A novel false data injection attack detection model of the cyber-physical power system. IEEE Access **8**, 95 109–95 125 (2020)
A. Cardenas, S. Amin, S. Sastry, Research challenges for the security of control systems, in *3rd conference on Hot topics in security* (ACM, 2008), pp. 1–6
G. Casadei, D. Astolfi, A. Alessandri, L. Zaccarian, Synchronization in networks of identical nonlinear systems via dynamic dead zones. IEEE Control Syst. Lett. **3**(3), 667–672 (2019)
E. Chang, F. Gottwalt, Y. Zhang, Cyber situational awareness for CPS, 5g and IOT, in *Frontiers in Electronic Technologies* (Springer, 2017), pp. 147–161
S. Chen, M. Wu, P. Wen, F. Xu, S. Wang, S. Zhao, A multimode anomaly detection method based on oc-elm for aircraft engine system. IEEE Access **9**, 28 842–28 855 (2021)
Y. Chen, Y. Shi, Distributed consensus of linear multiagent systems: Laplacian spectra-based method. IEEE Trans. Syst. Man Cybern.: Syst. **50**(2), 700–706 (2017)
Y. Chen, S. Kar, J.M. Moura, Dynamic attack detection in cyber-physical systems with side initial state information. IEEE Trans. Autom. Control **62**(9), 4618–4624 (2016)
Y. Chen, S. Kar, J.M. Moura, Resilient distributed estimation: sensor attacks. IEEE Trans. Autom. Control **64**(9), 3772–3779 (2018)
A. Chowdhury, G. Karmakar, J. Kamruzzaman, A. Jolfaei, R. Das, Attacks on self-driving cars and their countermeasures: a survey. IEEE Access **8**, 207 308–207 342 (2020)
L.F. Cómbita Alfonso et al., Intrusion response on cyber-physical control systems, Ph.D. dissertation, Uniandes (2020)
S.L. Cutter, J. Ahearn, B. Amadei, P. Crawford, E. Eide, G. Galloway, Disaster resilience: a national imperative. *Environment: Science and Policy for Sustainable Development*, vol. 55, no. 2, (2013), pp. 25–29
A. Dakhnovich, D. Moskvin, D. Zeghzda, An approach for providing industrial control system sustainability in the age of digital transformation, in *IOP Conference Series: Materials Science and Engineering*, vol. 497, no. 1. IOP Publishing, (2019), p. 012006
C. De Persis, P. Tesi, Input-to-state stabilizing control under denial-of-service. IEEE Trans. Autom. Control **60**(11), 2930–2944 (2015)

S.M. Dibaji, M. Pirani, A.M. Annaswamy, K.H. Johansson, A. Chakrabortty, Secure control of wide-area power systems: confidentiality and integrity threats, in *2018 IEEE Conference on Decision and Control (CDC)* (IEEE, 2018), pp. 7269–7274

S.M. Dibaji, M. Pirani, D.B. Flamholz, A.M. Annaswamy, K.H. Johansson, A. Chakrabortty, A systems and control perspective of CPS security. Ann. Rev. Control **47**, 394–411 (2019)

D. Ding, Q.-L. Han, X. Ge, J. Wang, Secure state estimation and control of cyber-physical systems: a survey. IEEE Trans. Syst. Man Cybern.: Syst. **51**(1), 176–190 (2020a)

D. Ding, Q.-L. Han, Z. Wang, X. Ge, Recursive filtering of distributed cyber-physical systems with attack detection. IEEE Trans. Syst. Man Cybern. (2020b)

S.X. Ding, L. Li, D. Zhao, C. Louen, T. Liu, Application of the unified control and detection framework to detecting stealthy integrity cyber-attacks on feedback control systems (2021), arXiv:2103.00210

D. Ding, Y. Shen, Y. Song, Y. Wang, Recursive state estimation for discrete time-varying stochastic nonlinear systems with randomly occurring deception attacks. Int. J. Gen. Syst. **45**(5), 548–560 (2016)

D. Ding, Z. Wang, G. Wei, F.E. Alsaadi, Event-based security control for discrete-time stochastic systems. IET Control TheoryAppl. **10**(15), 1808–1815 (2016)

D. Ding, Z. Wang, Q.-L. Han, G. Wei, Security control for discrete-time stochastic nonlinear systems subject to deception attacks. IEEE Trans. Syst. Man Cybern.: Syst. **48**(5), 779–789 (2016)

D. Ding, Z. Wang, D.W. Ho, G. Wei, Distributed recursive filtering for stochastic systems under uniform quantizations and deception attacks through sensor networks. Automatica **78**, 231–240 (2017)

D. Ding, Q.-L. Han, Y. Xiang, X. Ge, X.-M. Zhang, A survey on security control and attack detection for industrial cyber-physical systems. Neurocomputing **275**, 1674–1683 (2018)

V. Dolk, P. Tesi, C. De Persis, W. Heemels, Event-triggered control systems under denial-of-service attacks. IEEE Trans. Control Netw. Syst. **4**(1), 93–105 (2016)

D. Du, C. Zhang, H. Wang, X. Li, H. Hu, T. Yang, Stability analysis of token-based wireless networked control systems under deception attacks. Inf. Sci. **459**, 168–182 (2018)

A.K. Dutta, R. Negi, S.K. Shukla, Robust multivariate anomaly-based intrusion detection system for cyber-physical systems, in *International Symposium on Cyber Security Cryptography and Machine Learning* (Springer, 2021), pp. 86–93

N. Falliere, L. Murchu, E. Chien, W32. stuxnet dossier: syman-tec security response, 2018, technical Report, Symantec, https://www.symantec.com/content/en/us/enterprise/media/security_response/whitepapers/w32_stuxnet_dossier.pdf

C. Fang, Y. Qi, P. Cheng, W.X. Zheng, Optimal periodic watermarking schedule for replay attack detection in cyber-physical systems. Automatica **112**, 108698 (2020)

F. Farivar, M.S. Haghighi, A. Jolfaei, M. Alazab, Artificial intelligence for detection, estimation, and compensation of malicious attacks in nonlinear cyber-physical systems and industrial iot. IEEE Trans. Ind. Inf. **16**(4), 2716–2725 (2019)

H. Fawzi, P. Tabuada, S. Diggavi, Secure estimation and control for cyber-physical systems under adversarial attacks. IEEE Trans. Autom. control **59**(6), 1454–1467 (2014)

Z. Feng, G. Hu, Secure cooperative event-triggered control of linear multiagent systems under dos attacks. IEEE Trans. Control Syst. Technol. **28**(3), 741–752 (2019)

S. Feng, P. Tesi, Resilient control under denial-of-service: robust design. Automatica **79**, 42–51 (2017)

R.M. Ferrari, A.M. Teixeira, Detection and isolation of routing attacks through sensor watermarking, in *American Control Conference (ACC)*, vol. 2017 (IEEE, 2017), pp. 5436–5442

R.M. Ferrari, A.M. Teixeira, Safety, security, and privacy for cyber-physical systems (2021)

L. Fillatre, I. Nikiforov, P. Willett et al., Security of scada systems against cyber-physical attacks. IEEE Aerosp. Electron. Syst. Mag. **32**(5), 28–45 (2017)

B. Galloway, G. Hancke, Introduction to industrial control networks. Commun. Surv. Tutor. **15**(2), 860–880 (2013)

Y. Gao, G. Sun, J. Liu, Y. Shi, L. Wu, State estimation and self-triggered control of CPSS against joint sensor and actuator attacks. Automatica **113**(2020)

J. Giraldo, D. Urbina, A. Cardenas, J. Valente, M. Faisal, J. Ruths, N.O. Tippenhauer, H. Sandberg, R. Candell, A survey of physics-based attack detection in cyber-physical systems. ACM Comput. Surv. (CSUR) **51**(4), 1–36 (2018)

S. Grigorescu, B. Trasnea, T. Cocias, G. Macesanu, A survey of deep learning techniques for autonomous driving. J. Field Robot. **37**(3), 362–386 (2020)

C.-Y. Gu, J.-W. Zhu, W.-A. Zhang, L. Yu, Sensor attack detection for cyber-physical systems based on frequency domain partition. IET Control Theory Appl. **14**(11), 1452–1466 (2020)

Y. Guan, X. Ge, Distributed attack detection and secure estimation of networked cyber-physical systems against false data injection attacks and jamming attacks. IEEE Trans. Signal Inf. Process. Over Netw. **4**(1), 48–59 (2017)

K. Han, Y. Duan, R. Jin, Z. Ma, H. Wang, W. Wu, B. Wang, X. Cai, Attack detection method based on bayesian hypothesis testing principle in CPS. Procedia Comput. Sci. **187**, 474–480 (2021)

W. He, Z. Mo, Q.-L. Han, F. Qian, Secure impulsive synchronization in lipschitz-type multi-agent systems subject to deception attacks. IEEE/CAA J. Automatica Sinica **7**(5), 1326–1334 (2020)

J.M. Hendrickx, K.H. Johansson, R.M. Jungers, H. Sandberg, K.C. Sou, Efficient computations of a security index for false data attacks in power networks. IEEE Trans. Autom. Control **59**(12), 3194–3208 (2014)

C.S. Holling, *Engineering Resilience Versus Ecological Ressilience* (National Academy Press, 1996), ch. 3, pp. 31–43

C.S. Holling, Resilience and stability of ecological systems. Ann. Rev. Ecol. Syst. **4**, 1–23 (1973)

M. Housh, Z. Ohar, Model-based approach for cyber-physical attack detection in water distribution systems. Water Res. **139**, 132–143 (2018)

S. Hu, D. Yue, X. Xie, X. Chen, X. Yin, Resilient event-triggered controller synthesis of networked control systems under periodic dos jamming attacks. IEEE Trans. Cybern. **49**(12), 4271–4281 (2018)

S. Hu, D. Yue, Q.-L. Han, X. Xie, X. Chen, C. Dou, Observer-based event-triggered control for networked linear systems subject to denial-of-service attacks. IEEE Trans. Cybern. **50**(5), 1952–1964 (2019)

A. Humayed, B. Luo, Cyber-physical security for smart cars: taxonomy of vulnerabilities, threats, and attacks, in *Proceedings of the ACM/IEEE Sixth International Conference on Cyber-Physical Systems*, (2015), pp. 252–253

A. Humayed, J. Lin, F. Li, B. Luo, Cyber-physical systems security-a survey. IEEE Int. Things J. **4**(6), 1802–1831 (2017)

W.H. Ip, D. Wang, Resilience and friability of transportation networks: evaluation, analysis and optimization. IEEE Syst. J. **5**(2), 189–198 (2011)

C. Iwendi, S.U. Rehman, A.R. Javed, S. Khan, G. Srivastava, Sustainable security for the internet of things using artificial intelligence architectures. ACM Trans. Int. Technol. (TOIT) **21**(3), 1–22 (2021)

A.N. Jahromi, H. Karimipour, A. Dehghantanha, R.M. Parizi, Deep representation learning for cyber-attack detection in industrial iot, in *AI-Enabled Threat Detection and Security Analysis for Industrial IoT* (Springer, 2021), pp. 139–162

A.N. Jahromi, J. Sakhnini, H. Karimpour, A. Dehghantanha, A deep unsupervised representation learning approach for effective cyber-physical attack detection and identification on highly imbalanced data, in *Proceedings of the 29th Annual International Conference on Computer Science and Software Engineering*, (2019), pp. 14–23

K.N. Junejo, J. Goh, Behaviour-based attack detection and classification in cyber physical systems using machine learning, in *Proceedings of the 2nd ACM International Workshop on Cyber-Physical System Security*, (2016), pp. 34–43

H. Karimipour, H. Leung, Relaxation-based anomaly detection in cyber-physical systems using ensemble Kalman filter. IET Cyber-Phys. Syst.: Theory Appl. **5**(1), 49–58 (2019)

M.T. Khan, D. Serpanos, H. Shrobe, M.M. Yousuf, Rigorous machine learning for secure and autonomous cyber physical systems, in *2020 25th IEEE International Conference on Emerging Technologies and Factory Automation (ETFA)*, vol. 1 (IEEE, 2020), pp. 1815–1819

A. Khazraei, H. Kebriaei, F.R. Salmasi, A new watermarking approach for replay attack detection in LGG systems, in *2017 IEEE 56th Annual Conference on Decision and Control (CDC)* (IEEE, 2017), pp. 5143–5148

A. Khazraei, H. Kebriaei, F.R. Salmasi, Replay attack detection in a multi agent system using stability analysis and loss effective watermarking, in *American Control Conference (ACC)* (IEEE, 2017), pp. 4778–4783

S. Kim, Y. Eun, K.-J. Park, Stealthy sensor attack detection and real-time performance recovery for resilient cps. IEEE Trans. Ind. Inf. **17**(11), 7412–7422 (2021)

H. Kitano, Biological robustness. Nat. Rev. Gen. **5**, 826–837 (2004)

R. Kozik, M. Choraś, M. Ficco, F. Palmieri, A scalable distributed machine learning approach for attack detection in edge computing environments. J. Parallel Distrib. Comput. **119**, 18–26 (2018)

S. Krishnamurthy, S. Sarkar, A. Tewari, Scalable anomaly detection and isolation in cyber-physical systems using bayesian networks, in *Dynamic Systems and Control Conference*, vol. 46193 (American Society of Mechanical Engineers, 2014), p. V002T26A006

D. Kumar, H. Nayyar, D. Pandey, A. Hussian Khan, Cyber physical security of the critical information infrastructure, in *ISUW 2019* (Springer, 2022), pp. 275–285

D. Lee, D. Kundur, Cyber attack detection in pmu measurements via the expectation-maximization algorithm, in *2014 IEEE Global Conference on Signal and Information Processing (GlobalSIP)* (IEEE, 2014), pp. 223–227

P. Lee, A. Clark, L. Bushnell, R. Poovendran, A passivity framework for modeling and mitigating wormhole attacks on networked control systems. IEEE Trans. Autom. Control **59**(12), 3224–3237 (2014)

S.A. Levin, J. Lubchenco, Resilience, robustness, and marine ecosystem-based management. BioScience **58**(1), 27–32 (2008)

Q. Li, B. Bu, J. Zhao, A novel hierarchical situation awareness model for CBTC using SVD entropy and GRU with PRD algorithms. IEEE Access (2021a)

L. Li, W. Wang, Q. Ma, K. Pan, X. Liu, L. Lin, J. Li, Cyber attack estimation and detection for cyber-physical power systems. Appl. Math. Comput. **400** (2021b)

J. Li, Y. Liu, T. Chen, Z. Xiao, Z. Li, J. Wang, Adversarial attacks and defenses on cyber-physical systems: a survey. IEEE Int. Things J. **7**(6), 5103–5115 (2020)

I. Linkov, D. Eisenberg, M. E. Bates, D. Chang, M. Convertino, K. Plourde, J. Allen, T. Seager, *Measurable Resilience for Actionable Policy* (ACS Publications, 2013), pp. 25–29

I. Linkov, D. Eisenberg, K. Plourde, T. Seager, J. Allen, A. Kott, Resilience metrics for cyber systems. Environ. Syst. Decis. **33**(4), 471–476 (2013)

G.-P. Liu, Coordinated control of networked multiagent systems with communication constraints using a proportional integral predictive control strategy. IEEE Trans. Cybern. **50**(11), 4735–4743 (2019)

C. Liu, J. Wu, C. Long, Y. Wang, Dynamic state recovery for cyber-physical systems under switching location attacks. IEEE Trans. Control Netw. Syst. **4**(1), 14–22 (2016)

M. Long, C.-H. Wu, J.Y. Hung, Denial of service attacks on network-based control systems: impact and mitigation. IEEE Trans. Ind. Inf. **1**(2), 85–96 (2005)

G. Loukas, E. Karapistoli, E. Panaousis, P. Sarigiannidis, A. Bezemskij, T. Vuong, A taxonomy and survey of cyber-physical intrusion detection approaches for vehicles. Ad Hoc Netw. **84**, 124–147 (2019)

A.-Y. Lu, G.-H. Yang, Observer-based control for cyber-physical systems under denial-of-service with a decentralized event-triggered scheme. IEEE Trans. Cybern. **50**(12), 4886–4895 (2019)

L. Ma, Z. Wang, Q.-L. Han, H.-K. Lam, Variance-constrained distributed filtering for time-varying systems with multiplicative noises and deception attacks over sensor networks. IEEE Sens. J. **17**(7), 2279–2288 (2017)

M. Massoumnia, G. Verghese, A. Willsky, Failure detection and identification. IEEE Trans. Autom. Control **34**(3), 316–321 (1989)

L.K. Mestha, O.M. Anubi, M. Abbaszadeh, Cyber-attack detection and accommodation algorithm for energy delivery systems, in *2017 IEEE Conference on Control Technology and Applications (CCTA)* (IEEE, 2017), pp. 1326–1331

S. Mishra, Y. Shoukry, N. Karamchandani, S.N. Diggavi, P. Tabuada, Secure state estimation against sensor attacks in the presence of noise. IEEE Trans. Control Netw. Syst. **4**(1), 49–59 (2016)

H. Modares, B. Kiumarsi, F.L. Lewis, F. Ferrese, A. Davoudi, Resilient and robust synchronization of multiagent systems under attacks on sensors and actuators. IEEE Trans. Cybern. **50**(3), 1240–1250 (2019)

R. Moghadam, H. Modares, Resilient autonomous control of distributed multiagent systems in contested environments. IEEE Trans. Cybern. **49**(11), 3957–3967 (2018)

A. Mosteiro-Sanchez, M. Barcelo, J. Astorga, A. Urbieta, Securing IIOT using defence-in-depth: towards an end-to-end secure industry 4.0. J. Manuf. Syst. **57**, 367–378 (2020)

A. Musleh, G. Chen, A. Dong, A survey on the detection algorithms for false data injection attacks in smart grids. IEEE Trans. Smart Grid **11**(3), 2218–2234 (2019)

Y. Nakahira, Y. Mo, Attack-resilient $\mathscr{H}_2/\mathscr{H}_\infty$ and ℓ_1 state estimator. IEEE Trans. Autom. Control **63**(12), 4353–4360 (2018)

S.K. Narayanan, S. Dhanasekaran, V. Vasudevan, Intelligent abnormality detection method in cyber physical systems using machine learning, in *Proceedings of International Conference on Machine Intelligence and Data Science Applications* (Springer, 2021), pp. 595–606

M. Noorizadeh, M. Shakerpour, N. Meskin, D. Unal, K. Khorasani, A cyber-security methodology for a cyber-physical industrial control system testbed. IEEE Access **9**, 16 239–16 253 (2021)

U.S.D. of Homeland Security, *Recommended Practice: Improving Industrial Control Systems Cybersecurity with Defense-in-Depth Strategies* (Createspace Independent Pub, 2014), https://books.google.com/books?id=1OO8oQEACAAJ

F.O. Olowononi, D.B. Rawat, C. Liu, Resilient machine learning for networked cyber physical systems: a survey for machine learning security to securing machine learning for CPS. IEEE Commun. Surv. Tutor. **23**(1), 524–552 (2020)

E.F. Orumwense, K. Abo-Al-Ez, A systematic review to aligning research paths: energy cyber-physical systems. Cogen. Eng. **6**(1), 1700738 (2019)

S. Pan, T. Morris, U. Adhikari, Classification of disturbances and cyber-attacks in power systems using heterogeneous time-synchronized data. IEEE Trans. Ind. Inf. **11**(3), 650–662 (2015)

Z.-H. Pang, G.-P. Liu, Design and implementation of secure networked predictive control systems under deception attacks. IEEE Trans. Control Syst. Technol. **20**(5), 1334–1342 (2011)

Z.-H. Pang, G.-P. Liu, D. Zhou, F. Hou, D. Sun, Two-channel false data injection attacks against output tracking control of networked systems. IEEE Trans. Ind. Electron. **63**(5), 3242–3251 (2016)

C.M. Paredes, D. Martínez-Castro, V. Ibarra-Junquera, A. González-Potes, Detection and isolation of dos and integrity cyber attacks in cyber-physical systems with a neural network-based architecture. Electronics **10**(18), 2238 (2021)

J. Park, R. Ivanov, J. Weimer, M. Pajic, I. Lee, Sensor attack detection in the presence of transient faults, in *Proceedings of the ACM/IEEE Sixth International Conference on Cyber-Physical Systems*, (2015), pp. 1–10

G. Park, C. Lee, H. Shim, Y. Eun, K.H. Johansson, Stealthy adversaries against uncertain cyber-physical systems: threat of robust zero-dynamics attack. IEEE Trans. Autom. Control **64**(12), 4907–4919 (2019)

F. Pasqualetti, F. Dörfler, F. Bullo, Attack detection and identification in cyber-physical systems. IEEE Trans. Autom. control **58**(11), 2715–2729 (2013)

C. Peng, J. Li, M. Fei, Resilient event-triggering $\mathscr{H}_\infty$ load frequency control for multi-area power systems with energy-limited dos attacks. IEEE Trans. Power Syst. **32**(5), 4110–4118 (2016)

C. Peng, H. Sun, M. Yang, Y. Wang, A survey on security communication and control for smart grids under malicious cyber attacks. IEEE Trans. Syst. Man Cybern.: Syst. **49**(8), 1554–1569 (2019)

P.S. Pessim, M.J. Lacerda, State-feedback control for cyber-physical LPV systems under dos attacks. IEEE Control Syst. Lett. **5**(3), 1043–1048 (2020)

M. Pirani, E. Nekouei, H. Sandberg, K.H. Johansson, A game-theoretic framework for the security-aware sensor placement problem in networked control systems. IEEE Trans. Autom. Control (2021)

M. Porter, P. Hespanhol, A. Aswani, M. Johnson-Roberson, R. Vasudevan, Detecting generalized replay attacks via time-varying dynamic watermarking. IEEE Trans. Autom. Control (2020)

J. Pöyhönen, J. Rajamäki, V. Nuojua, M. Lehto, Cyber situational awareness in critical infrastructure organizations. Digit. Transform. Cyber Secur. Resil. Mod. Soc. **84**, 161 (2021)

C.G. Rieger, D.I. Gertman, M.A. McQueen, Resilient control systems: next generation design research, in *2nd Conference on Human System Interactions*, (2009), pp. 632–636

C.G. Rieger, K.L. Moore, T.L. Baldwin, Resilient control systems: a multi-agent dynamic systems perspective, in *International Conference on Electro/Information Technology (EIT)* (2013)

T. Roy, S. Dey, Security of distributed parameter cyber-physical systems: cyber-attack detection in linear parabolic PDES (2021), arXiv:2107.14159

RTCA, DO-178/EUROCAE ED-12, *Software Considerations in Airborne Systems and Equipment Certification* (2011)

S. Sahoo, S. Mishra, J.C.-H. Peng, T. Dragičević, A stealth cyber-attack detection strategy for dc microgrids. IEEE Trans. Power Electron. **34**(8), 8162–8174 (2018)

B. Satchidanandan, P.R. Kumar, Dynamic watermarking: active defense of networked cyber-physical systems. Proc. IEEE **105**(2), 219–240 (2016)

B. Satchidanandan, P. Kumar, On the design of security-guaranteeing dynamic watermarks. IEEE Control Syst. Lett. **4**(2), 307–312 (2019)

P. Semwal, A multi-stage machine learning model for security analysis in industrial control system, in *AI-Enabled Threat Detection and Security Analysis for Industrial IoT* (Springer, 2021), pp. 213–236

J. Shin, Y. Baek, Y. Eun, S.H. Son, Intelligent sensor attack detection and identification for automotive cyber-physical systems. IEEE Symp. Ser. Comput. Intell. (SSCI) **2017**, 1–8 (2017)

Y. Shoukry, P. Tabuada, Event-triggered state observers for sparse sensor noise/attacks. IEEE Trans. Autom. Control **61**(8), 2079–2091 (2015)

Y. Shoukry, P. Nuzzo, A. Puggelli, A.L. Sangiovanni-Vincentelli, S.A. Seshia, P. Tabuada, Secure state estimation for cyber-physical systems under sensor attacks: a satisfiability modulo theory approach. IEEE Trans. Autom. Control **62**(10), 4917–4932 (2017)

J. Slay, M. Miller, Lessons learned from the maroochy water breach, in *International Conference on Critical Infrastructure Protection* (Springer, 2007), pp. 73–82

S. Sridhar, M. Govindarasu, Model-based attack detection and mitigation for automatic generation control. IEEE Trans. Smart Grid **5**(2), 580–591 (2014)

Q. Su, Z. Fan, Y. Long, J. Li, Attack detection and secure state estimation for cyber-physical systems with finite-frequency observers. J. Franklin Inst. **357**(17), 12 724–12 741 (2020)

Y. Sun, D. Ding, H. Dong, H. Liu, Event-based resilient filtering for stochastic nonlinear systems via innovation constraints. Inf. Sci. **546**, 512–525 (2021)

M. Taheri, K. Khorasani, I. Shames, N. Meskin, Cyber attack and machine induced fault detection and isolation methodologies for cyber-physical systems (2020), arXiv:2009.06196

S. Tan, J.M. Guerrero, P. Xie, R. Han, J.C. Vasquez, Brief survey on attack detection methods for cyber-physical systems. IEEE Syst. J. **14**(4), 5329–5339 (2020)

R. Taormina, S. Galelli, N.O. Tippenhauer, A. Ostfeld, E. Salomons, Assessing the effect of cyber-physical attacks on water distribution systems. World Environ. Water Res. Cong. **2016**, 436–442 (2016)

A. Teixeira, F. Kupzog, H. Sandberg, K.H. Johansson, Cyber-secure and resilient architectures for industrial control systems, in *Smart Grid Security* (Elsevier, 2015), pp. 149–183

J. Tian, B. Wang, T. Li, F. Shang, K. Cao, R. Guo, Total: Optimal protection strategy against perfect and imperfect false data injection attacks on power grid cyber-physical systems. IEEE Int. Things J. **8**(2), 1001–1015 (2020)

K. Tierney, M. Bruneau, Conceptualizing and measuring resilience: a key to disaster loss reduction. TR News **250**(1), 14–17 (2007)

D.D. Tiwari, S. Naskar, A.S. Sai, V.R. Palleti, Attack detection using unsupervised learning algorithms in cyber-physical systems. Comput. Aided Chem. Eng. Elsevier **50**, 1259–1264 (2021)

L. Tsiami, C. Makropoulos, Cyber-physical attack detection in water distribution systems with temporal graph convolutional neural networks. Water **13**(9), 1247 (2021)

C.M.P. Valencia, R.E. Alzate, D.M. Castro, A.F. Bayona, D.R. García, Detection and isolation of dos and integrity attacks in cyber-physical microgrid system, in *2019 IEEE 4th Colombian Conference on Automatic Control (CCAC)* (IEEE, 2019), pp. 1–6

X. Wang, S. Li, M. Liu, Y. Wang, A.K. Roy-Chowdhury, Multi-expert adversarial attack detection in person re-identification using context inconsistency (2021a), arXiv:2108.09891

H. Wang, X. Wen, S. Huang, B. Zhou, Q. Wu, N. Liu, Generalized attack separation scheme in cyber physical smart grid based on robust interval state estimation. Int. J. Electr. Power Energy Syst. **129** (2021b)

H. Wang, X. Wen, Y. Xu, B. Zhou, J.-C. Peng, W. Liu, *Operating state reconstruction in cyber physical smart grid for automatic attack filtering*. IEEE Trans. Ind. Inf. (2020)

M. Wolf, D. Serpanos, Safety and security in cyber-physical systems and internet-of-things systems. Proc. IEEE **106**(1), 9–20 (2017)

C. Wu, W. Yao, W. Pan, G. Sun, J. Liu, L. Wu, Secure control for cyber-physical systems under malicious attacks. IEEE Trans. Control Netw. Syst. (2021)

C.-H. Xie, G.-H. Yang, Secure estimation for cyber-physical systems with adversarial attacks and unknown inputs: an l 2-gain method. Int. J. Robust Nonlinear Control **28**(6), 2131–2143 (2018)

J. Xiong, J. Wu, Construction of approximate reasoning model for dynamic CPS network and system parameter identification. Comput. Commun. **154**, 180–187 (2020)

W. Xu, G. Hu, D.W. Ho, Z. Feng, Distributed secure cooperative control under denial-of-service attacks from multiple adversaries. IEEE Trans. Cybern. **50**(8), 3458–3467 (2019)

Y. Xuan, M. Naghnaeian, Detection and identification of cps attacks with application in vehicle platooning: a generalized luenberger approach, in *American Control Conference (ACC)* (IEEE, 2021), pp. 4013–4020

W. Yan, L. Mestha, J. John, D. Holzhauer, M. Abbaszadeh, M. McKinley, Cyberattack detection for cyber physical systems security–a preliminary study, in *Proceedings of the Annual Conference of the PHM Society*, vol. 10 (2018)

W. Yan, L.K. Mestha, M. Abbaszadeh, Attack detection for securing cyber physical systems. IEEE Int. Things J. **6**(5), 8471–8481 (2019)

W. Yang, Y. Zhang, G. Chen, C. Yang, L. Shi, Distributed filtering under false data injection attacks. Automatica **102**, 34–44 (2019)

L. Ye, F. Zhu, J. Zhang, Sensor attack detection and isolation based on sliding mode observer for cyber-physical systems. Int. J. Adapt. Control Signal Process. **34**(4), 469–483 (2020)

K. Zhang, C. Keliris, T. Parisini, M.M. Polycarpou, Identification of sensor replay attacks and physical faults for cyber-physical systems. IEEE Control Syst. Lett. (2021a)

K. Zhang, C. Keliris, M.M. Polycarpou, T. Parisini, Discrimination between replay attacks and sensor faults for cyber-physical systems via event-triggered communication. Eur. J. Control (2021b)

J. Zhang, L. Pan, Q.-L. Han, C. Chen, S. Wen, Y. Xiang, Deep learning based attack detection for cyber-physical system cybersecurity: a survey. IEEE/CAA J. Automatica Sinica (2021c)

D. Zhang, Q.-G. Wang, G. Feng, Y. Shi, A. V. Vasilakos, A survey on attack detection, estimation and control of industrial cyber–physical systems. ISA Trans. (2021d)

T. Zhang, Y. Wang, X. Liang, Z. Zhuang, W. Xu, Cyber attacks in cyber-physical power systems: a case study with gprs-based scada systems, in *29th Chinese control and decision conference (CCDC)* (IEEE, 2017), pp. 6847–6852

X. Zhang, F. Zhu, Observer-based sensor attack diagnosis for cyber-physical systems via zonotope theory. Asian J. Control (2020)

H. Zhang, Y. Shu, P. Cheng, J. Chen, Privacy and performance trade-off in cyber-physical systems. IEEE Netw. **30**(2), 62–66 (2016)

D. Zhao, Z. Wang, D.W. Ho, G. Wei, Observer-based PID security control for discrete time-delay systems under cyber-attacks, in *IEEE Transactions on Systems, Man, and Cybernetics: Systems* (2019)

K. Zhou, J. Doyle, *Diagnosis and Fault-Tolerant Control* (Prentice-Hall, 1998)

Q. Zhu, T. Basar, Robust and resilient control design for cyber-physical systems with an application to power systems, in *50th IEEE Conference on Decision and Control* (IEEE, 2011)

M. Zhu, K. Ye, C.-Z. Xu, Network anomaly detection and identification based on deep learning methods, in *International Conference on Cloud Computing* (Springer, 2018), pp. 219–234

M. Zhu, S. Martinez, On the performance analysis of resilient networked control systems under replay attacks. IEEE Trans. Autom. Control **59**(3), 804–808 (2013)

Y. Zhu, W.X. Zheng, Observer-based control for cyber-physical systems with periodic dos attacks via a cyclic switching strategy. IEEE Trans. Autom. Control **65**(8), 3714–3721 (2019)

Chapter 3
Fundamental Stealthiness–Distortion Trade-Offs in Cyber-Physical Systems

Song Fang and Quanyan Zhu

3.1 Introduction

Security issues such as the presence of malicious attacks could cause severe consequences in cyber-physical systems, which are safety-critical in most cases since they are interacting with the physical world. In the trend that cyber-physical systems are becoming more and more prevalent nowadays, it is also increasingly critical to be fully aware of such systems' performance limits (Fang et al. 2017), e.g., in terms of performance degradation, after taking the security issues into consideration. Accordingly, in this chapter, we focus on analyzing the fundamental limits of resilience in cyber-physical systems, including open-loop dynamical systems and (closed-loop) feedback control systems. More specifically, we examine the fundamental trade-offs between the systems' performance degradation that can be brought about by a malicious attack and the possibility of it being detected, of which the former is oftentimes measured by the mean squared-error distortion, whereas the latter is fundamentally determined by the Kullback–Leibler (KL) divergence.

The KL divergence was proposed in Kullback and Leibler (1951) (see also Kullback (1997)), and ever since it has been employed in various research areas, including, e.g., information theory (Cover and Thomas 2006), signal processing (Kay 2020), statistics (Pardo 2006), control and estimation theory (Lindquist and Picci 2015), system identification (Stoorvogel and Van Schuppen 1996), and machine learning (Goodfellow et al. 2016). Particularly, in statistical detection theory (Poor 2013), KL divergence provides the optimal exponent in probability of error for binary hypotheses testing problems as a result of the Chernoff–Stein lemma (Cover and Thomas 2006). Accordingly, in the context of determining whether an attack signal is present

S. Fang (✉) · Q. Zhu
New York University, 370 Jay Street, Brooklyn, New York 11201, USA
e-mail: song.fang@nyu.edu

Q. Zhu
e-mail: quanyan.zhu@nyu.edu

M. Abbaszadeh and A. Zemouche (eds.), *Security and Resilience in Cyber-Physical Systems*, https://doi.org/10.1007/978-3-030-97166-3_3

or not in security problems, the KL divergence has also been employed as a measure of stealthiness for attacks (see detailed discussions in, e.g., Bai et al. (2017a, b)).

In the context of dynamical and control system security (see, e.g., Poovendran et al. (2012), Johansson et al. (2014), Sandberg et al. (2015), Cheng et al. (2017), Giraldo et al. (2018), Weerakkody et al. (2019), Dibaji et al. (2019), Chong et al. (2019) and the references therein), particularly in dynamical and control systems under injection attacks, fundamental stealthiness–distortion trade-offs (with the mean squared-error as the distortion measure and the KL divergence as the stealthiness measure) have been investigated for feedback control systems (see, e.g., Zhang and Venkitasubramaniam (2017), Bai et al. (2017b)) as well as state estimation systems (see, e.g., Bai et al. (2017a), Kung et al. (2016), Guo et al. (2018)). Generally speaking, the problem considered is: Given a constraint (upper bound) on the level of stealthiness, what is the maximum degree of distortion (for control or for estimation) that can be caused by the attacker? This is dual to the following question: Given a least requirement (lower bound) on the degree of distortion, what is the maximum level of stealthiness that can be achieved by the attacker? Answers to these questions can not only capture the fundamental trade-offs between stealthiness and distortion but also characterize what the worst-case attacks are.

In this chapter, unlike the aforementioned works in Bai et al. (2017a, b), Kung et al. (2016), Zhang and Venkitasubramaniam (2017), Guo et al. (2018), we adopt an alternative approach to this stealthiness–distortion trade-off problem using power spectral analysis. The scenarios we consider include linear Gaussian open-loop dynamical systems and (closed-loop) feedback control systems. By using the power spectral approach, we obtain explicit formulas that characterize analytically the stealthiness–distortion trade-offs as well as the properties of the worst-case attacks. It turns out that the worst-case attacks are stationary colored Gaussian attacks with power spectra that are shaped specifically according to the transfer functions of the systems and the power spectra of the system outputs, the knowledge of which is all that the attacker needs to have access to in order to carry out the worst-case attacks. In other words, the attacker only needs to know the input–output behaviors of the systems, whereas it is not necessary to know their state-space models.

The remainder of the chapter is organized as follows. Section 3.2 provides the technical preliminaries. Section 3.3 is divided into two subsections, focusing on open-loop dynamical systems and feedback control systems, respectively. Section 3.4 presents numerical examples. Concluding remarks are given in Sect. 3.5.

More specifically, Theorem 3.1, as the first main result, characterizes explicitly the stealthiness–distortion trade-off and the worst-case attack in linear Gaussian open-loop dynamical systems. Equivalently, Corollary 3.1 considers the dual problem to that of Theorem 3.1. On the other hand, Theorem 3.2, together with Corollary 3.2 (in a dual manner), provides analytical expressions for the stealthiness–distortion trade-off and the worst-case attack in linear Gaussian feedback control systems. In addition, the preliminary results on the implications in control design, as presented in the Conclusion, indicate how the explicit stealthiness–distortion trade-off formula for feedback control systems can be employed to render the controller design explicit and intuitive.

Note that this chapter is based upon (Fang and Zhu 2021), which, however, only discusses the case of open-loop dynamical systems. Meanwhile, in this chapter, we also consider (closed-loop) feedback control systems. Note also that the results presented in this book chapter are applicable to discrete-time systems.

Notation: Throughout the chapter, we consider zero-mean real-valued continuous random variables and random vectors, as well as discrete-time stochastic processes. We represent random variables and random vectors using boldface letters, e.g., $\mathbf{x}$, while the probability density function of $\mathbf{x}$ is denoted as $p_{\mathbf{x}}$. In addition, $\mathbf{x}_{0,\ldots,k}$ will be employed to denote the sequence $\mathbf{x}_0, \ldots, \mathbf{x}_k$ or the random vector $\left[\mathbf{x}_0^{\mathrm{T}}, \ldots, \mathbf{x}_k^{\mathrm{T}}\right]^{\mathrm{T}}$, depending on the context. Note in particular that, for simplicity and with abuse of notations, we utilize $\mathbf{x} \in \mathbb{R}$ and $\mathbf{x} \in \mathbb{R}^m$ to indicate that $\mathbf{x}$ is a real-valued random variable and that $\mathbf{x}$ is a real-valued m-dimensional random vector, respectively.

3.2 Preliminaries

A stochastic process $\{\mathbf{x}_k\}, \mathbf{x}_k \in \mathbb{R}$ is said to be stationary if $R_{\mathbf{x}}(i,k) := \mathbb{E}\left[\mathbf{x}_i \mathbf{x}_{i+k}\right]$ depends only on k, and can thus be denoted as $R_{\mathbf{x}}(k)$ for simplicity. The power spectrum of a stationary process $\{\mathbf{x}_k\}, \mathbf{x}_k \in \mathbb{R}$ is defined as

$$S_{\mathbf{x}}(\omega) := \sum_{k=-\infty}^{\infty} R_{\mathbf{x}}(k)\, \mathrm{e}^{-\mathrm{j}\omega k}.$$

Moreover, the variance of $\{\mathbf{x}_k\}$ is given by

$$\sigma_{\mathbf{x}}^2 = \mathbb{E}\left[\mathbf{x}_k^2\right] = \frac{1}{2\pi} \int_{-\pi}^{\pi} S_{\mathbf{x}}(\omega)\, \mathrm{d}\omega.$$

The KL divergence (see, e.g., Kullback and Leibler (1951)) is defined as follows.

Definition 3.1 Consider random vectors $\mathbf{x} \in \mathbb{R}^m$ and $\mathbf{y} \in \mathbb{R}^m$ with probability densities $p_{\mathbf{x}}(\mathbf{u})$ and $p_{\mathbf{y}}(\mathbf{u})$, respectively. The KL divergence from distribution $p_{\mathbf{x}}$ to distribution $p_{\mathbf{y}}$ is defined as

$$\mathrm{KL}\left(p_{\mathbf{y}} \| p_{\mathbf{x}}\right) := \int p_{\mathbf{y}}(\mathbf{u}) \ln \frac{p_{\mathbf{y}}(\mathbf{u})}{p_{\mathbf{x}}(\mathbf{u})} \mathrm{d}\mathbf{u}.$$

The next lemma (see, e.g., Kay (2020)) provides an explicit expression of KL divergence in terms of covariance matrices for Gaussian random vectors; note that herein and in the sequel, all random variables and random vectors are assumed to be zero mean.

Lemma 3.1 *Consider Gaussian random vectors* $\mathbf{x} \in \mathbb{R}^m$ *and* $\mathbf{y} \in \mathbb{R}^m$ *with covariance matrices* $\Sigma_{\mathbf{x}}$ *and* $\Sigma_{\mathbf{y}}$*, respectively. The KL divergence from distribution* $p_{\mathbf{x}}$ *to distribution* $p_{\mathbf{y}}$ *is given by*

$$\mathrm{KL}\left(p_{\mathbf{y}} \| p_{\mathbf{x}}\right) = \frac{1}{2}\left[tr\left(\Sigma_{\mathbf{y}}\Sigma_{\mathbf{x}}^{-1}\right) - \ln\det\left(\Sigma_{\mathbf{y}}\Sigma_{\mathbf{x}}^{-1}\right) - m\right].$$

It is clear that in the scalar case (when $m = 1$), Lemma 3.1 reduces to the following formula for Gaussian random variables:

$$\mathrm{KL}\left(p_{\mathbf{y}} \| p_{\mathbf{x}}\right) = \frac{1}{2}\left[\frac{\sigma_{\mathbf{y}}^2}{\sigma_{\mathbf{x}}^2} - \ln\left(\frac{\sigma_{\mathbf{y}}^2}{\sigma_{\mathbf{x}}^2}\right) - 1\right].$$

The KL divergence rate (see, e.g., Lindquist and Picci (2015)) is defined as follows.

Definition 3.2 Consider stochastic processes $\{\mathbf{x}_k\}$, $\mathbf{x}_k \in \mathbb{R}^m$ and $\{\mathbf{y}_k\}$, $\mathbf{y}_k \in \mathbb{R}^m$ with densities $p_{\{\mathbf{x}_k\}}$ and $p_{\{\mathbf{y}_k\}}$, respectively; note that $p_{\{\mathbf{x}_k\}}$ and $p_{\{\mathbf{y}_k\}}$ will be denoted by $p_{\mathbf{x}}$ and $p_{\mathbf{y}}$ for simplicity in the sequel. Then, the KL divergence rate from distribution $p_{\mathbf{x}}$ to distribution $p_{\mathbf{y}}$ is defined as

$$\mathrm{KL}_{\infty}\left(p_{\mathbf{y}} \| p_{\mathbf{x}}\right) := \limsup_{k\to\infty} \frac{\mathrm{KL}\left(p_{\mathbf{y}_{0,\ldots,k}} \| p_{\mathbf{x}_{0,\ldots,k}}\right)}{k+1}.$$

The next lemma (see, e.g., Lindquist and Picci (2015)) provides an explicit expression of KL divergence rate in terms of power spectra for stationary Gaussian processes.

Lemma 3.2 *Consider stationary Gaussian processes* $\{\mathbf{x}_k\}$, $\mathbf{x}_k \in \mathbb{R}$ *and* $\{\mathbf{y}_k\}$, $\mathbf{y}_k \in \mathbb{R}$ *with densities* $p_{\mathbf{x}}$ *and* $p_{\mathbf{y}}$ *as well as power spectra* $S_{\mathbf{x}}(\omega)$ *and* $S_{\mathbf{y}}(\omega)$, *respectively. Suppose that* $S_{\mathbf{y}}(\omega)/S_{\mathbf{x}}(\omega)$ *is bounded (see Lindquist and Picci (2015) for details). Then, the KL divergence rate from distribution* $p_{\mathbf{x}}$ *to distribution* $p_{\mathbf{y}}$ *is given by*

$$\mathrm{KL}_{\infty}\left(p_{\mathbf{y}} \| p_{\mathbf{x}}\right) = \frac{1}{2\pi}\int_0^{2\pi} \frac{1}{2}\left\{\frac{S_{\mathbf{y}}(\omega)}{S_{\mathbf{x}}(\omega)} - \ln\left[\frac{S_{\mathbf{y}}(\omega)}{S_{\mathbf{x}}(\omega)}\right] - 1\right\} \mathrm{d}\omega. \tag{3.1}$$

3.3 Stealthiness–Distortion Trade-Offs and Worst-Case Attacks

In this section, we analyze the fundamental stealthiness–distortion trade-offs of linear Gaussian open-loop dynamical systems and (closed-loop) feedback control systems under data injection attacks, whereas the KL divergence is employed as the stealthiness measure. Consider the scenario where attacker can modify the system input, and consequently, the system state and system output will then all be changed. From the attacker's point of view, the desired outcome is that the change in system state (as measured by state distortion) is large, while the change in system output (as measured by output stealthiness) is relatively small, so as to make the possibility of being detected low. Meanwhile fundamental trade-offs in general exist between state distortion and output stealthiness, since the system's state and output are correlated. In other words, increase in state distortion may inevitably lead to decrease in output stealthiness, i.e., increase in the possibility of being detected. How to capture such trade-offs? And what is the worst-case attack that can cause the maximum distortion given a certain stealthiness level, or vice versa? The answers are provided subsequently in terms of power spectral analysis.

3.3.1 Open-Loop Dynamical Systems

In this subsection, we focus on open-loop dynamical systems. Specifically, consider the scalar dynamical system depicted in Fig. 3.1 with state-space model given by

$$\begin{cases} \mathbf{x}_{k+1} = a\mathbf{x}_k + b\mathbf{u}_k + \mathbf{w}_k, \\ \mathbf{y}_k = c\mathbf{x}_k + \mathbf{v}_k, \end{cases}$$

where $\mathbf{x}_k \in \mathbb{R}$ is the system state, $\mathbf{u}_k \in \mathbb{R}$ is the system input, $\mathbf{y}_k \in \mathbb{R}$ is the system output, $\mathbf{w}_k \in \mathbb{R}$ is the process noise, and $\mathbf{v}_k \in \mathbb{R}$ is the measurement noise. The system parameters are $a \in \mathbb{R}$, $b \in \mathbb{R}$, and $c \in \mathbb{R}$; we further assume that $|a| < 1$ and $b, c \neq 0$, i.e., the system is stable, controllable, and observable. Accordingly, the transfer function of the system is given by

$$P(z) = \frac{bc}{z - a}. \tag{3.2}$$

(It is clear that $P(z)$ is minimum phase.) Suppose that $\{\mathbf{w}_k\}$ and $\{\mathbf{v}_k\}$ are stationary white Gaussian with variances $\sigma_{\mathbf{w}}^2$ and $\sigma_{\mathbf{v}}^2$, respectively. Furthermore, $\{\mathbf{w}_k\}$, $\{\mathbf{v}_k\}$, and $\mathbf{x}_0$ are assumed to be mutually independent. Assume also that $\{\mathbf{u}_k\}$ is stationary with power spectrum $S_{\mathbf{u}}(\omega)$. As such, $\{\mathbf{x}_k\}$ and $\{\mathbf{y}_k\}$ are both stationary, and denote their power spectra by $S_{\mathbf{x}}(\omega)$ and $S_{\mathbf{y}}(\omega)$, respectively.

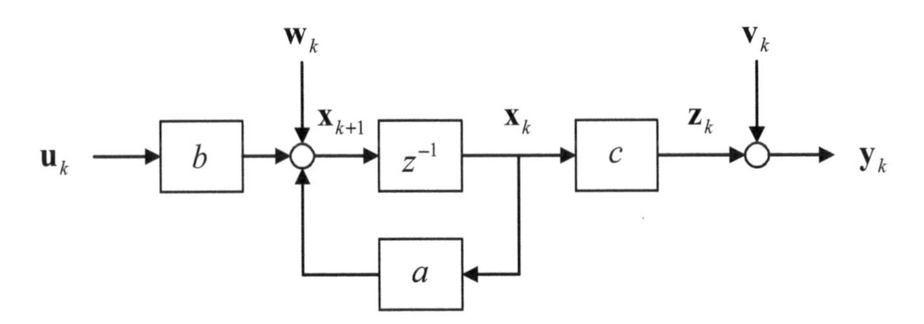

Fig. 3.1 A dynamical system

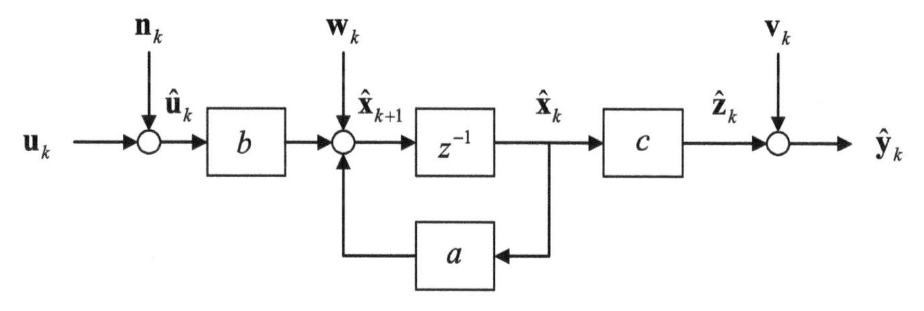

Fig. 3.2 A dynamical system under injection attack

Consider then the scenario that an attack signal $\{\mathbf{n}_k\}$, $\mathbf{n}_k \in \mathbb{R}$, is to be added to the input of the system $\{\mathbf{u}_k\}$ to deviate the system state, while aiming to be stealthy in the system output; see the depiction in Fig. 3.2. In addition, denote the true plant input under attack as $\{\widehat{\mathbf{u}}_k\}$, where

$$\widehat{\mathbf{u}}_k = \mathbf{u}_k + \mathbf{n}_k, \tag{3.3}$$

whereas the system under attack $\{\mathbf{n}_k\}$ is given by

$$\begin{cases} \widehat{\mathbf{x}}_{k+1} = a\widehat{\mathbf{x}}_k + b\widehat{\mathbf{u}}_k + \mathbf{w}_k = a\widehat{\mathbf{x}}_k + b\mathbf{u}_k + b\mathbf{n}_k + \mathbf{w}_k, \\ \widehat{\mathbf{y}}_k = c\widehat{\mathbf{x}}_k + \mathbf{v}_k. \end{cases} \tag{3.4}$$

Meanwhile, suppose that the attack signal $\{\mathbf{n}_k\}$ is independent of $\{\mathbf{u}_k\}$, $\{\mathbf{w}_k\}$, $\{\mathbf{v}_k\}$, and $\mathbf{x}_0$; consequently, $\{\mathbf{n}_k\}$ is independent of $\{\mathbf{x}_k\}$ and $\{\mathbf{y}_k\}$ as well.

The following questions then naturally arise: What is the fundamental trade-off between the degree of distortion caused in the system state (as measured by the mean squared-error distortion $\mathbb{E}\left[(\widehat{\mathbf{x}}_k - \mathbf{x}_k)^2\right]$ between the original state $\{\mathbf{x}_k\}$ and the state under attack denoted as $\{\widehat{\mathbf{x}}_k\}$) and the level of stealthiness resulted in the system output (as measured by the KL divergence rate $\mathrm{KL}_\infty\left(p_{\widehat{\mathbf{y}}} \| p_{\mathbf{y}}\right)$ between the original output $\{\mathbf{y}_k\}$ and the output under attack denoted as $\{\widehat{\mathbf{y}}_k\}$)? More specifically, to achieve a certain degree of distortion in state, what is the maximum level of stealthiness that can be maintained by the attacker? And what is the worst-case attack in this sense? The following theorem, as the first main result of this chapter, answers the questions raised above.

Theorem 3.1 *Consider the dynamical system under injection attacks depicted in Fig. 3.2. Suppose that the attacker aims to design the attack signal $\{\mathbf{n}_k\}$ to satisfy the following attack goal in terms of state distortion:*

$$\mathbb{E}\left[(\widehat{\mathbf{x}}_k - \mathbf{x}_k)^2\right] \geq D. \tag{3.5}$$

Then, the minimum KL divergence rate between the original output and the attacked output is given by

$$\inf_{\mathbb{E}[(\widehat{\mathbf{x}}_k-\mathbf{x}_k)^2]\geq D} \mathrm{KL}_\infty \left(p_{\widehat{\mathbf{y}}} \| p_{\mathbf{y}}\right) = \frac{1}{2\pi} \int_0^{2\pi} \frac{1}{2} \left\{ \frac{S_{\widehat{\mathbf{n}}}(\omega)}{S_{\mathbf{y}}(\omega)} - \ln \left[1 + \frac{S_{\widehat{\mathbf{n}}}(\omega)}{S_{\mathbf{y}}(\omega)} \right] \right\} \mathrm{d}\omega, \quad (3.6)$$

where

$$S_{\widehat{\mathbf{n}}}(\omega) = \frac{\zeta S_{\mathbf{y}}^2(\omega)}{1-\zeta S_{\mathbf{y}}(\omega)}, \quad (3.7)$$

and $S_{\mathbf{y}}(\omega)$ *is given by*

$$S_{\mathbf{y}}(\omega) = \frac{b^2c^2}{\left|\mathrm{e}^{\mathrm{j}\omega}-a\right|^2} S_{\mathbf{u}}(\omega) + \frac{c^2}{\left|\mathrm{e}^{\mathrm{j}\omega}-a\right|^2}\sigma_{\mathbf{w}}^2 + \sigma_{\mathbf{v}}^2. \quad (3.8)$$

Herein, ζ *is the unique constant that satisfies*

$$\frac{1}{2\pi} \int_{-\pi}^{\pi} \frac{\zeta S_{\mathbf{y}}^2(\omega)}{1-\zeta S_{\mathbf{y}}(\omega)} \mathrm{d}\omega = c^2 D, \quad (3.9)$$

while

$$0 < \zeta < \min_{\omega} \frac{1}{S_{\mathbf{y}}(\omega)}. \quad (3.10)$$

Moreover, the worst-case (in the sense of achieving this minimum KL divergence rate) attack $\{\mathbf{n}_k\}$ *is a stationary colored Gaussian process with power spectrum*

$$S_{\mathbf{n}}(\omega) = \frac{\left|\mathrm{e}^{\mathrm{j}\omega}-a\right|^2}{b^2c^2} \frac{\zeta S_{\mathbf{y}}^2(\omega)}{1-\zeta S_{\mathbf{y}}(\omega)}. \quad (3.11)$$

Proof To begin with, it can be verified that the power spectrum of $\{\mathbf{y}_k\}$ is given by

$$\begin{aligned} S_{\mathbf{y}}(\omega) &= \left|P\left(\mathrm{e}^{\mathrm{j}\omega}\right)\right|^2 S_{\mathbf{u}}(\omega) + \frac{1}{b^2}\left|P\left(\mathrm{e}^{\mathrm{j}\omega}\right)\right|^2 \sigma_{\mathbf{w}}^2 + \sigma_{\mathbf{v}}^2, \\ &= \frac{b^2c^2}{\left|\mathrm{e}^{\mathrm{j}\omega}-a\right|^2} S_{\mathbf{u}}(\omega) + \frac{c^2}{\left|\mathrm{e}^{\mathrm{j}\omega}-a\right|^2}\sigma_{\mathbf{w}}^2 + \sigma_{\mathbf{v}}^2. \end{aligned}$$

Note then that due to the property of additivity of linear systems, the system in Fig. 3.2 is equivalent to that of Fig. 3.3, where

$$\widehat{\mathbf{y}}_k = \mathbf{y}_k + \widehat{\mathbf{n}}_k,$$

and $\{\widehat{\mathbf{n}}_k\}$ is the output of the subsystem

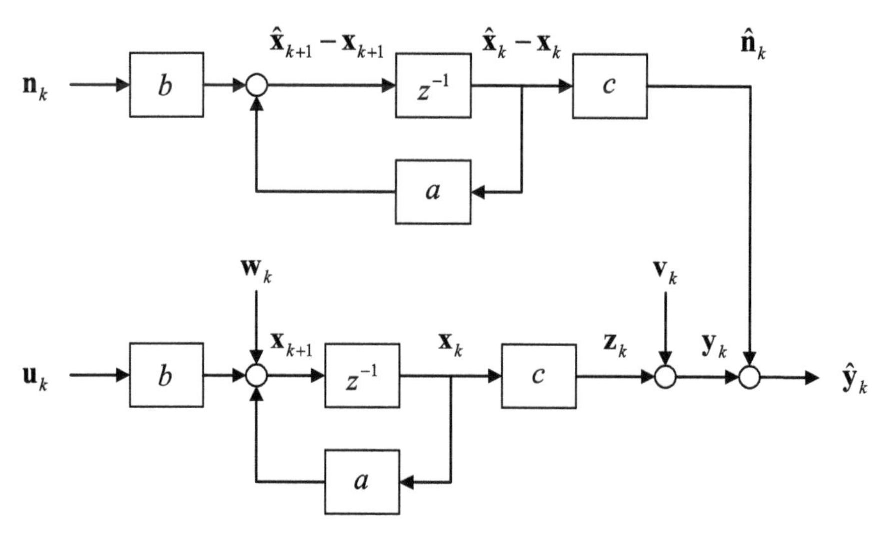

Fig. 3.3 A dynamical system under injection attack: equivalent system

$$\begin{cases} \widehat{\mathbf{x}}_{k+1} - \mathbf{x}_{k+1} = a\left(\widehat{\mathbf{x}}_k - \mathbf{x}_k\right) + b\mathbf{n}_k, \\ \widehat{\mathbf{n}}_k = c\left(\widehat{\mathbf{x}}_k - \mathbf{x}_k\right), \end{cases}$$

as depicted by the upper half of Fig. 3.3; note that in this subsystem, $(\widehat{\mathbf{x}}_k - \mathbf{x}_k) \in \mathbb{R}$ is the system state, $\mathbf{n}_k \in \mathbb{R}$ is the system input, and $\widehat{\mathbf{n}} \in \mathbb{R}$ is the system output. On the other hand, the distortion constraint

$$\mathbb{E}\left[(\widehat{\mathbf{x}}_k - \mathbf{x}_k)^2\right] \geq D$$

is then equivalent to being with a power constraint

$$\mathbb{E}\left[\widehat{\mathbf{n}}_k^2\right] \geq c^2 D,$$

since $\widehat{\mathbf{n}}_k = \widehat{\mathbf{y}}_k - \mathbf{y}_k$ and thus

$$\widehat{\mathbf{n}}_k^2 = (\mathbf{y}_k - \widehat{\mathbf{y}}_k)^2 = (c\mathbf{x}_k - c\widehat{\mathbf{x}}_k)^2 = c^2\left(\mathbf{x}_k - \widehat{\mathbf{x}}_k\right)^2.$$

Accordingly, the system in Fig. 3.3 may be viewed as a "virtual channel" modeled as

$$\widehat{\mathbf{y}}_k = \mathbf{y}_k + \widehat{\mathbf{n}}_k$$

with noise constraint

$$\mathbb{E}\left[\widehat{\mathbf{n}}_k^2\right] \geq c^2 D,$$

where $\{\mathbf{y}_k\}$ is the channel input, $\{\widehat{\mathbf{y}}_k\}$ is the channel output, and $\{\widehat{\mathbf{n}}_k\}$ is the channel noise. In addition, due to the fact that $\{\mathbf{n}_k\}$ is independent of $\{\mathbf{y}_k\}$, $\{\widehat{\mathbf{n}}_k\}$ is also independent of $\{\mathbf{y}_k\}$.

The approach we shall take herein, as developed in Cover and Thomas (2006), is to treat the multiple uses of a scalar channel (i.e., a scalar dynamic channel) equivalently as a single use of parallel channels (i.e., a set of parallel static channels). We consider first the case of a finite number of parallel static channels with

$$\widehat{\mathbf{y}} = \mathbf{y} + \widehat{\mathbf{n}},$$

where $\mathbf{y}, \widehat{\mathbf{y}}, \widehat{\mathbf{n}} \in \mathbb{R}^m$, and $\widehat{\mathbf{n}}$ is independent of $\mathbf{y}$. In addition, $\mathbf{y}$ is Gaussian with covariance $\Sigma_{\mathbf{y}}$, and the noise power constraint is given by

$$\operatorname{tr}\left(\Sigma_{\widehat{\mathbf{n}}}\right) = \mathbb{E}\left[\sum_{i=1}^{m} \widehat{\mathbf{n}}^2(i)\right] \geq c^2 D,$$

where $\widehat{\mathbf{n}}(i)$ denotes the i-th element of $\widehat{\mathbf{n}}$. In addition, according to Fang and Zhu (2020) (see Proposition 2 therein), we have

$$\mathrm{KL}\left(p_{\widehat{\mathbf{y}}} \| p_{\mathbf{y}}\right) \geq \mathrm{KL}\left(p_{\widehat{\mathbf{y}}^{\mathrm{G}}} \| p_{\mathbf{y}}\right),$$

where $\widehat{\mathbf{y}}^{\mathrm{G}}$ denotes a Gaussian random vector with the same covariance as $\widehat{\mathbf{y}}$, and equality holds if $\widehat{\mathbf{y}}$ is Gaussian. Meanwhile, it is known from Lemma 3.1 that

$$\mathrm{KL}\left(p_{\widehat{\mathbf{y}}^{\mathrm{G}}} \| p_{\mathbf{y}}\right) = \frac{1}{2}\left[\operatorname{tr}\left(\Sigma_{\widehat{\mathbf{y}}} \Sigma_{\mathbf{y}}^{-1}\right) - \ln\det\left(\Sigma_{\widehat{\mathbf{y}}} \Sigma_{\mathbf{y}}^{-1}\right) - m\right].$$

On the other hand, since $\mathbf{y}$ and $\widehat{\mathbf{n}}$ are independent, we have

$$\Sigma_{\widehat{\mathbf{y}}} = \Sigma_{\widehat{\mathbf{n}}+\mathbf{y}} = \Sigma_{\widehat{\mathbf{n}}} + \Sigma_{\mathbf{y}}.$$

Consequently,

$$\operatorname{tr}\left(\Sigma_{\widehat{\mathbf{y}}} \Sigma_{\mathbf{y}}^{-1}\right) - \ln\det\left(\Sigma_{\widehat{\mathbf{y}}} \Sigma_{\mathbf{y}}^{-1}\right) = \operatorname{tr}\left[\left(\Sigma_{\widehat{\mathbf{n}}} + \Sigma_{\mathbf{y}}\right) \Sigma_{\mathbf{y}}^{-1}\right] - \ln\det\left[\left(\Sigma_{\widehat{\mathbf{n}}} + \Sigma_{\mathbf{y}}\right) \Sigma_{\mathbf{y}}^{-1}\right].$$

Denote the eigendecomposition of $\Sigma_{\mathbf{y}}$ by $U_{\mathbf{y}} \Lambda_{\mathbf{y}} U_{\mathbf{y}}^{\mathrm{T}}$, where

$$\Lambda_{\mathbf{y}} = \operatorname{diag}\left(\lambda_1, \ldots, \lambda_m\right).$$

Then,

$$\begin{aligned}
&\operatorname{tr}\left[\left(\Sigma_{\widehat{\mathbf{n}}}+\Sigma_{\mathbf{y}}\right)\Sigma_{\mathbf{y}}^{-1}\right]-\ln\det\left[\left(\Sigma_{\widehat{\mathbf{n}}}+\Sigma_{\mathbf{y}}\right)\Sigma_{\mathbf{y}}^{-1}\right]\\
&\quad=\operatorname{tr}\left[\left(\Sigma_{\widehat{\mathbf{n}}}+U_{\mathbf{y}}\Lambda_{\mathbf{y}}U_{\mathbf{y}}^{\mathrm{T}}\right)\left(U_{\mathbf{y}}\Lambda_{\mathbf{y}}U_{\mathbf{y}}^{\mathrm{T}}\right)^{-1}\right]-\ln\det\left[\left(\Sigma_{\widehat{\mathbf{n}}}+U_{\mathbf{y}}\Lambda_{\mathbf{y}}U_{\mathbf{y}}^{\mathrm{T}}\right)\left(U_{\mathbf{y}}\Lambda_{\mathbf{y}}U_{\mathbf{y}}^{\mathrm{T}}\right)^{-1}\right],\\
&\quad=\operatorname{tr}\left[\left(\Sigma_{\widehat{\mathbf{n}}}+U_{\mathbf{y}}\Lambda_{\mathbf{y}}U_{\mathbf{y}}^{\mathrm{T}}\right)U_{\mathbf{y}}\Lambda_{\mathbf{y}}^{-1}U_{\mathbf{y}}^{\mathrm{T}}\right]-\ln\det\left[\left(\Sigma_{\widehat{\mathbf{n}}}+U_{\mathbf{y}}\Lambda_{\mathbf{y}}U_{\mathbf{y}}^{\mathrm{T}}\right)U_{\mathbf{y}}\Lambda_{\mathbf{y}}^{-1}U_{\mathbf{y}}^{\mathrm{T}}\right],\\
&\quad=\operatorname{tr}\left[U_{\mathbf{y}}U_{\mathbf{y}}^{\mathrm{T}}\left(\Sigma_{\widehat{\mathbf{n}}}+U_{\mathbf{y}}\Lambda_{\mathbf{y}}U_{\mathbf{y}}^{\mathrm{T}}\right)U_{\mathbf{y}}\Lambda_{\mathbf{y}}^{-1}U_{\mathbf{y}}^{\mathrm{T}}\right]\\
&\qquad-\ln\det\left[U_{\mathbf{y}}U_{\mathbf{y}}^{\mathrm{T}}\left(\Sigma_{\widehat{\mathbf{n}}}+U_{\mathbf{y}}\Lambda_{\mathbf{y}}U_{\mathbf{y}}^{\mathrm{T}}\right)U_{\mathbf{y}}\Lambda_{\mathbf{y}}^{-1}U_{\mathbf{y}}^{\mathrm{T}}\right],\\
&\quad=\operatorname{tr}\left\{U_{\mathbf{y}}\left[U_{\mathbf{y}}^{\mathrm{T}}\left(\Sigma_{\widehat{\mathbf{n}}}+U_{\mathbf{y}}\Lambda_{\mathbf{y}}U_{\mathbf{y}}^{\mathrm{T}}\right)U_{\mathbf{y}}\Lambda_{\mathbf{y}}^{-1}\right]U_{\mathbf{y}}^{\mathrm{T}}\right\}\\
&\qquad-\ln\det\left\{U_{\mathbf{y}}\left[U_{\mathbf{y}}^{\mathrm{T}}\left(\Sigma_{\widehat{\mathbf{n}}}+U_{\mathbf{y}}\Lambda_{\mathbf{y}}U_{\mathbf{y}}^{\mathrm{T}}\right)U_{\mathbf{y}}\Lambda_{\mathbf{y}}^{-1}\right]U_{\mathbf{y}}^{\mathrm{T}}\right\},\\
&\quad=\operatorname{tr}\left[U_{\mathbf{y}}^{\mathrm{T}}\left(\Sigma_{\widehat{\mathbf{n}}}+U_{\mathbf{y}}\Lambda_{\mathbf{y}}U_{\mathbf{y}}^{\mathrm{T}}\right)U_{\mathbf{y}}\Lambda_{\mathbf{y}}^{-1}\right]-\ln\det\left[U_{\mathbf{y}}^{\mathrm{T}}\left(\Sigma_{\widehat{\mathbf{n}}}+U_{\mathbf{y}}\Lambda_{\mathbf{y}}U_{\mathbf{y}}^{\mathrm{T}}\right)U_{\mathbf{y}}\Lambda_{\mathbf{y}}^{-1}\right],\\
&\quad=\operatorname{tr}\left[\left(U_{\mathbf{y}}^{\mathrm{T}}\Sigma_{\widehat{\mathbf{n}}}U_{\mathbf{y}}+\Lambda_{\mathbf{y}}\right)\Lambda_{\mathbf{y}}^{-1}\right]-\ln\det\left[\left(U_{\mathbf{y}}^{\mathrm{T}}\Sigma_{\widehat{\mathbf{n}}}U_{\mathbf{y}}+\Lambda_{\mathbf{y}}\right)\Lambda_{\mathbf{y}}^{-1}\right],\\
&\quad=\operatorname{tr}\left[\left(\overline{\Sigma}_{\widehat{\mathbf{n}}}+\Lambda_{\mathbf{y}}\right)\Lambda_{\mathbf{y}}^{-1}\right]-\ln\det\left[\left(\overline{\Sigma}_{\widehat{\mathbf{n}}}+\Lambda_{\mathbf{y}}\right)\Lambda_{\mathbf{y}}^{-1}\right],
\end{aligned}$$

where $\overline{\Sigma}_{\widehat{\mathbf{n}}}=U_{\mathbf{y}}^{\mathrm{T}}\Sigma_{\widehat{\mathbf{n}}}U_{\mathbf{y}}$. Denoting the diagonal terms of $\overline{\Sigma}_{\widehat{\mathbf{n}}}$ by $\overline{\sigma}_{\widehat{\mathbf{n}}(i)}^{2}$, $i=1,\ldots,m$, it is known from (Fang and Zhu 2020) (see Proposition 4 therein) that

$$\begin{aligned}
\operatorname{tr}&\left[\left(\overline{\Sigma}_{\widehat{\mathbf{n}}}+\Lambda_{\mathbf{y}}\right)\Lambda_{\mathbf{y}}^{-1}\right]-\ln\det\left[\left(\overline{\Sigma}_{\widehat{\mathbf{n}}}+\Lambda_{\mathbf{y}}\right)\Lambda_{\mathbf{y}}^{-1}\right],\\
&\geq\sum_{i=1}^{m}\left[\frac{\overline{\sigma}_{\widehat{\mathbf{n}}(i)}^{2}+\lambda_{i}}{\lambda_{i}}\right]-\sum_{i=1}^{m}\ln\left[\frac{\overline{\sigma}_{\widehat{\mathbf{n}}(i)}^{2}+\lambda_{i}}{\lambda_{i}}\right],\\
&=\sum_{i=1}^{m}\left[1+\frac{\overline{\sigma}_{\widehat{\mathbf{n}}(i)}^{2}}{\lambda_{i}}\right]-\sum_{i=1}^{m}\ln\left[1+\frac{\overline{\sigma}_{\widehat{\mathbf{n}}(i)}^{2}}{\lambda_{i}}\right],
\end{aligned}$$

where equality holds if $\overline{\Sigma}_{\widehat{\mathbf{n}}}$ is diagonal. For simplicity, we denote

$$\overline{\Sigma}_{\widehat{\mathbf{n}}}=\operatorname{diag}\left(\overline{\sigma}_{\widehat{\mathbf{n}}(1)}^{2},\ldots,\overline{\sigma}_{\widehat{\mathbf{n}}(m)}^{2}\right)=\operatorname{diag}\left(\widehat{N}_{1},\ldots,\widehat{N}_{m}\right)$$

when $\overline{\Sigma}_{\widehat{\mathbf{n}}}$ is diagonal. Then, the problem reduces to that of choosing $\widehat{N}_{1},\ldots,\widehat{N}_{m}$ to minimize

$$\sum_{i=1}^{m}\left(1+\frac{\widehat{N}_{i}}{\lambda_{i}}\right)-\sum_{i=1}^{m}\ln\left(1+\frac{\widehat{N}_{i}}{\lambda_{i}}\right)$$

subject to the constraint that

$$\sum_{i=1}^{m}\widehat{N}_{i}=\operatorname{tr}\left(\overline{\Sigma}_{\widehat{\mathbf{n}}}\right)=\operatorname{tr}\left(U_{\mathbf{y}}^{\mathrm{T}}\Sigma_{\widehat{\mathbf{n}}}U_{\mathbf{y}}\right)=\operatorname{tr}\left(\Sigma_{\widehat{\mathbf{n}}}U_{\mathbf{y}}U_{\mathbf{y}}^{\mathrm{T}}\right)=\operatorname{tr}\left(\Sigma_{\widehat{\mathbf{n}}}\right)=mc^{2}D.$$

Define the Lagrange function by

$$\sum_{i=1}^{m}\left(1+\frac{\widehat{N}_i}{\lambda_i}\right)-\sum_{i=1}^{m}\ln\left(1+\frac{\widehat{N}_i}{\lambda_i}\right)+\eta\left(\sum_{i=1}^{m}\widehat{N}_i-\widehat{N}\right),$$

and differentiate it with respect to $\widehat{N}_i$, then we have

$$\frac{1}{\lambda_i}-\frac{1}{\widehat{N}_i+\lambda_i}+\eta=0,$$

or equivalently,

$$\widehat{N}_i=\frac{1}{\frac{1}{\lambda_i}+\eta}-\lambda_i=\frac{\lambda_i}{1+\eta\lambda_i}-\lambda_i=\frac{-\eta\lambda_i^2}{1+\eta\lambda_i},$$

where η satisfies

$$\sum_{i=1}^{m}\widehat{N}_i=\sum_{i=1}^{m}\frac{-\eta\lambda_i^2}{1+\eta\lambda_i}=mc^2D,$$

while

$$-\min_{i=0,\ldots,m}\frac{1}{\lambda_i}<\eta<0.$$

For simplicity, we denote $\zeta=-\eta$, and accordingly,

$$\widehat{N}_i=\frac{\zeta\lambda_i^2}{1-\zeta\lambda_i},$$

where ζ satisfies

$$\sum_{i=1}^{m}\widehat{N}_i=\sum_{i=1}^{m}\frac{\zeta\lambda_i^2}{1-\zeta\lambda_i}=mc^2D,$$

while

$$0<\zeta<\min_{i=0,\ldots,m}\frac{1}{\lambda_i}.$$

Correspondingly,

$$\inf_{p_{\widehat{\mathbf{n}}}} \mathrm{KL}\left(p_{\widehat{\mathbf{y}}} \| p_{\mathbf{y}}\right) = \frac{1}{2}\left[\sum_{i=1}^{m}\left(1+\frac{\widehat{N}_i}{\lambda_i}\right) - \sum_{i=1}^{m} \ln\left(1+\frac{\widehat{N}_i}{\lambda_i}\right) - m\right],$$
$$= \sum_{i=1}^{m} \frac{1}{2}\left[\frac{\widehat{N}_i}{\lambda_i} - \ln\left(1+\frac{\widehat{N}_i}{\lambda_i}\right)\right].$$

Consider now a scalar dynamic channel

$$\widehat{\mathbf{y}}_k = \mathbf{y}_k + \widehat{\mathbf{n}}_k,$$

where $\mathbf{y}_k, \widehat{\mathbf{n}}_k, \widehat{\mathbf{y}}_k \in \mathbb{R}$, while $\{\mathbf{y}_k\}$ and $\{\widehat{\mathbf{n}}_k\}$ are independent. In addition, $\{\mathbf{y}_k\}$ is stationary colored Gaussian with power spectrum $S_{\mathbf{y}}(\omega)$, whereas the noise power constraint is given by $\mathbb{E}\left[\widehat{\mathbf{n}}_k^2\right] \geq c^2 D$. We may then consider a block of consecutive uses from time 0 to k of this channel as $k+1$ channels in parallel Cover and Thomas (2006). Particularly, let the eigendecomposition of $\Sigma_{\mathbf{y}_{0,\ldots,k}}$ be given by

$$\Sigma_{\mathbf{y}_{0,\ldots,k}} = U_{\mathbf{y}_{0,\ldots,k}} \Lambda_{\mathbf{y}_{0,\ldots,k}} U_{\mathbf{y}_{0,\ldots,k}}^{\mathrm{T}},$$

where

$$\Lambda_{\mathbf{y}_{0,\ldots,k}} = \mathrm{diag}\left(\lambda_0, \ldots, \lambda_k\right).$$

Then, we have

$$\min_{p_{\widehat{\mathbf{n}}_{0,\ldots,k}}:\ \sum_{i=0}^{k} \mathbb{E}\left[\widehat{\mathbf{n}}_i^2\right] \geq (k+1)c^2 D} \frac{\mathrm{KL}\left(p_{\widehat{\mathbf{y}}_{0,\ldots,k}} \| p_{\mathbf{y}_{0,\ldots,k}}\right)}{k+1} = \frac{1}{k+1}\sum_{i=0}^{k} \frac{1}{2}\left[\frac{\widehat{N}_i}{\lambda_i} - \ln\left(1+\frac{\widehat{N}_i}{\lambda_i}\right)\right],$$

where

$$\widehat{N}_i = \frac{\zeta\lambda_i^2}{1-\zeta\lambda_i},\ i = 0, \ldots, k.$$

Herein, ζ satisfies

$$\sum_{i=0}^{k} \widehat{N}_i = \sum_{i=0}^{k} \frac{\zeta\lambda_i^2}{1-\zeta\lambda_i} = (k+1)\, c^2 D,$$

or equivalently,

$$\frac{1}{k+1}\sum_{i=0}^{k} \widehat{N}_i = \frac{1}{k+1}\left(\frac{\zeta\lambda_i^2}{1-\zeta\lambda_i}\right) = c^2 D,$$

while

$$0 < \zeta < \min_{i=0,\ldots,k} \frac{1}{\lambda_i}.$$

In addition, since the processes $\{\mathbf{y}_k\}$, $\{\widehat{\mathbf{n}}_k\}$, and $\{\widehat{\mathbf{y}}_k\}$ are stationary, we have

$$\begin{aligned}
&\lim_{k\to\infty} \min_{p_{\widehat{\mathbf{n}}_{0,\ldots,k}}:\ \sum_{i=0}^{k}\mathbb{E}[\widehat{\mathbf{n}}_i^2]\geq(k+1)c^2D} \frac{\mathrm{KL}\left(p_{\widehat{\mathbf{y}}_{0,\ldots,k}}\|p_{\mathbf{y}_{0,\ldots,k}}\right)}{k+1} \\
&= \inf_{\mathbb{E}[\widehat{\mathbf{n}}_k^2]\geq c^2D} \lim_{k\to\infty} \frac{\mathrm{KL}\left(p_{\widehat{\mathbf{y}}_{0,\ldots,k}}\|p_{\mathbf{y}_{0,\ldots,k}}\right)}{k+1} = \inf_{\mathbb{E}[\widehat{\mathbf{n}}_k^2]\geq c^2D} \limsup_{k\to\infty} \frac{\mathrm{KL}\left(p_{\widehat{\mathbf{y}}_{0,\ldots,k}}\|p_{\mathbf{y}_{0,\ldots,k}}\right)}{k+1} \\
&= \inf_{\mathbb{E}[\widehat{\mathbf{n}}_k^2]\geq c^2D} \mathrm{KL}_\infty\left(p_{\widehat{\mathbf{y}}}\|p_{\mathbf{y}}\right) = \inf_{\mathbb{E}[(\widehat{\mathbf{x}}_k-\mathbf{x}_k)^2]\geq D} \mathrm{KL}_\infty\left(p_{\widehat{\mathbf{y}}}\|p_{\mathbf{y}}\right).
\end{aligned}$$

On the other hand, since the processes are stationary, the covariance matrices are Toeplitz (Grenander and Szegö 1958), and their eigenvalues approach their limits as $k \to \infty$. Moreover, the densities of eigenvalues on the real line tend to the power spectra of the processes (Gutiérrez-Gutiérrez and Crespo 2008; Lindquist and Picci 2015; Pinsker 1964). Accordingly,

$$\begin{aligned}
&\inf_{\mathbb{E}[(\widehat{\mathbf{x}}_k-\mathbf{x}_k)^2]\geq D} \mathrm{KL}_\infty\left(p_{\widehat{\mathbf{y}}}\|p_{\mathbf{y}}\right) = \lim_{k\to\infty} \frac{1}{k+1}\sum_{i=0}^{k}\frac{1}{2}\left[\frac{\widehat{N}_i}{\lambda_i} - \ln\left(1+\frac{\widehat{N}_i}{\lambda_i}\right)\right], \\
&= \frac{1}{2\pi}\int_0^{2\pi}\frac{1}{2}\left\{\frac{S_{\widehat{\mathbf{n}}}(\omega)}{S_{\mathbf{y}}(\omega)} - \ln\left[1+\frac{S_{\widehat{\mathbf{n}}}(\omega)}{S_{\mathbf{y}}(\omega)}\right]\right\}\mathrm{d}\omega,
\end{aligned}$$

where

$$S_{\widehat{\mathbf{n}}}(\omega) = \frac{\zeta S_{\mathbf{y}}^2(\omega)}{1-\zeta S_{\mathbf{y}}(\omega)},$$

and ζ satisfies

$$\lim_{k\to\infty}\frac{1}{k+1}\sum_{i=0}^{k}\widehat{N}_i = \frac{1}{2\pi}\int_{-\pi}^{\pi} S_{\widehat{\mathbf{n}}}(\omega)\,\mathrm{d}\omega = \frac{1}{2\pi}\int_{-\pi}^{\pi}\frac{\zeta S_{\mathbf{y}}^2(\omega)}{1-\zeta S_{\mathbf{y}}(\omega)}\mathrm{d}\omega = c^2D,$$

while

$$0 < \zeta < \min_{\omega}\frac{1}{S_{\mathbf{y}}(\omega)}.$$

Lastly, note that

$$S_{\widehat{\mathbf{n}}}(\omega) = \left|P\left(\mathrm{e}^{\mathrm{j}\omega}\right)\right|^2 S_{\mathbf{n}}(\omega) = \frac{b^2c^2}{\left|\mathrm{e}^{\mathrm{j}\omega}-a\right|^2}S_{\mathbf{n}}(\omega),$$

and hence

$$S_{\mathbf{n}}(\omega) = \frac{\left|e^{j\omega} - a\right|^2}{b^2 c^2} S_{\widehat{\mathbf{n}}}(\omega) = \frac{\left|e^{j\omega} - a\right|^2}{b^2 c^2} \frac{\zeta S_{\mathbf{y}}^2(\omega)}{1 - \zeta S_{\mathbf{y}}(\omega)}.$$

This concludes the proof. ■

It is clear that $S_{\mathbf{n}}(\omega)$ may be rewritten as

$$S_{\mathbf{n}}(\omega) = \frac{1}{\left|P\left(e^{j\omega}\right)\right|^2} \frac{\zeta S_{\mathbf{y}}^2(\omega)}{1 - \zeta S_{\mathbf{y}}(\omega)}. \tag{3.12}$$

This means that the attacker only needs the knowledge of the power spectrum of the original system output $\{\mathbf{y}_k\}$ and the transfer function of the system (from $\{\mathbf{n}_k\}$ to $\{\widehat{\mathbf{y}}_k\}$), i.e., $P(z)$, in order to carry out this worst-case attack. It is worth mentioning that the power spectrum of $\{\mathbf{y}_k\}$ can be estimated based on its realizations (see, e.g., Stoica and Moses (2005)), while the transfer function of the system can be approximated by system identification (see, e.g., Ljung (1999)).

Note that it can be verified (Kay 2020) that the (minimum) output KL divergence rate $\mathrm{KL}_{\infty}\left(p_{\widehat{\mathbf{y}}} \| p_{\mathbf{y}}\right)$ increases strictly with the state distortion bound D. In other words, in order for the attacker to achieve larger distortion, the stealthiness level of the attack will inevitably decrease.

On the other hand, the dual problem to that of Theorem 3.1 would be: Given a certain stealthiness level in output, what is the maximum distortion in state that can be achieved by the attacker? And what is the corresponding attack? The following corollary answers these questions.

Corollary 3.1 *Consider the dynamical system under injection attacks depicted in Fig. 3.2. Then, in order for the attacker to ensure that the KL divergence rate between the original output and the attacked output is upper bounded by a (positive) constant R as*

$$\mathrm{KL}_{\infty}\left(p_{\widehat{\mathbf{y}}} \| p_{\mathbf{y}}\right) \leq R, \tag{3.13}$$

the maximum state distortion $\mathbb{E}\left[(\widehat{\mathbf{x}}_k - \mathbf{x}_k)^2\right]$ that can be achieved is given by

$$\sup_{\mathrm{KL}_{\infty}\left(p_{\widehat{\mathbf{y}}} \| p_{\mathbf{y}}\right) \leq R} \mathbb{E}\left[(\widehat{\mathbf{x}}_k - \mathbf{x}_k)^2\right] = \frac{1}{2\pi} \int_{-\pi}^{\pi} \frac{1}{c^2} \left[\frac{\zeta S_{\mathbf{y}}^2(\omega)}{1 - \zeta S_{\mathbf{y}}(\omega)}\right] d\omega, \tag{3.14}$$

where ζ is the unique constant that satisfies

$$\frac{1}{2\pi}\int_0^{2\pi}\frac{1}{2}\left\{\frac{\frac{\zeta S_{\mathbf{y}}^2(\omega)}{1-\zeta S_{\mathbf{y}}(\omega)}}{S_{\mathbf{y}}(\omega)}-\ln\left[1+\frac{\frac{\zeta S_{\mathbf{y}}^2(\omega)}{1-\zeta S_{\mathbf{y}}(\omega)}}{S_{\mathbf{y}}(\omega)}\right]\right\}\mathrm{d}\omega$$

$$=\frac{1}{2\pi}\int_0^{2\pi}\frac{1}{2}\left\{\frac{\zeta S_{\mathbf{y}}(\omega)}{1-\zeta S_{\mathbf{y}}(\omega)}-\ln\left[\frac{1}{1-\zeta S_{\mathbf{y}}(\omega)}\right]\right\}\mathrm{d}\omega=R, \tag{3.15}$$

while

$$0<\zeta<\min_{\omega}\frac{1}{S_{\mathbf{y}}(\omega)}. \tag{3.16}$$

Note that herein $S_{\mathbf{y}}(\omega)$ is given by (3.8)*. Moreover, this maximum distortion is achieved when the attack signal $\{\mathbf{n}_k\}$ is chosen as a stationary colored Gaussian process with power spectrum*

$$S_{\mathbf{n}}(\omega)=\frac{\left|\mathrm{e}^{\mathrm{j}\omega}-a\right|^2}{b^2c^2}\frac{\zeta S_{\mathbf{y}}^2(\omega)}{1-\zeta S_{\mathbf{y}}(\omega)}. \tag{3.17}$$

3.3.2 Feedback Control Systems

We will now proceed to examine (closed-loop) feedback control systems in this subsection. Specifically, consider the feedback control system depicted in Fig. 3.4, where the state-space model of the plant is given by

$$\begin{cases}\mathbf{x}_{k+1}=a\mathbf{x}_k+b\mathbf{u}_k+\mathbf{w}_k,\\ \mathbf{y}_k=c\mathbf{x}_k+\mathbf{v}_k,\end{cases}$$

while $K(z)$ is the transfer function of the (dynamic) output controller. Herein, $\mathbf{x}_k\in\mathbb{R}$ is the plant state, $\mathbf{u}_k\in\mathbb{R}$ is the plant input, $\mathbf{y}_k\in\mathbb{R}$ is the plant output, $\mathbf{w}_k\in\mathbb{R}$ is the

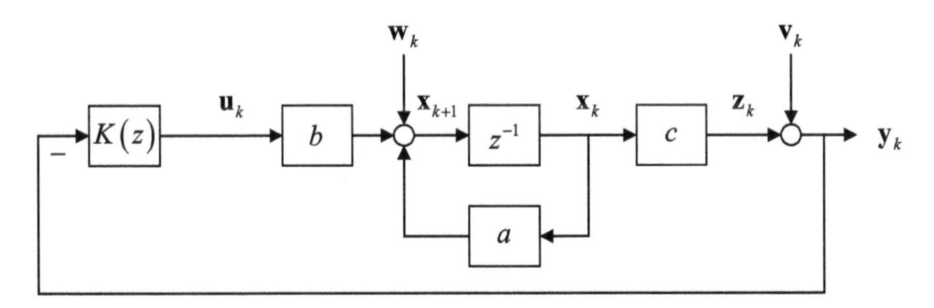

Fig. 3.4 A feedback control system

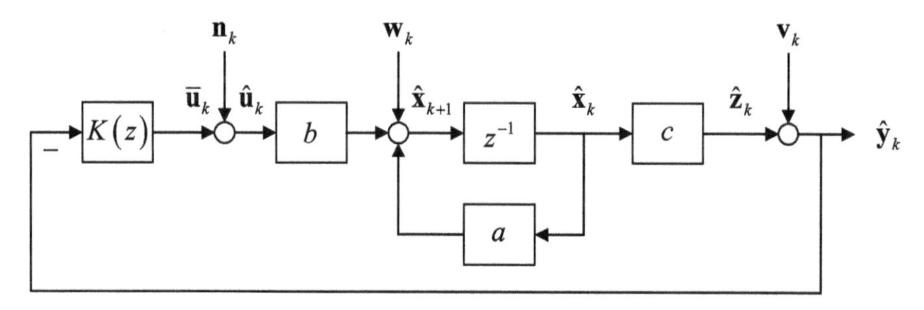

Fig. 3.5 A feedback control system under actuator attack

process noise, and $\mathbf{v}_k \in \mathbb{R}$ is the measurement noise. The system parameters are $a \in \mathbb{R}$, $b \in \mathbb{R}$, and $c \in \mathbb{R}$. Note that the plant is not necessarily stable. Meanwhile, we assume that $b, c \neq 0$, i.e., the plant is controllable and observable, and thus can be stabilized by controller $K(z)$. On the other hand, the transfer function of the plant is given by

$$P(z) = \frac{bc}{z - a}. \tag{3.18}$$

Suppose that $\{\mathbf{w}_k\}$ and $\{\mathbf{v}_k\}$ are stationary white Gaussian with variances $\sigma_{\mathbf{w}}^2$ and $\sigma_{\mathbf{v}}^2$, respectively. Furthermore, $\{\mathbf{w}_k\}$, $\{\mathbf{v}_k\}$, and $\mathbf{x}_0$ are assumed to be mutually independent. Assume also that $K(z)$ stabilizes $P(z)$, i.e., the closed-loop system is stable. Accordingly, $\{\mathbf{x}_k\}$ and $\{\mathbf{y}_k\}$ are both stationary, and denote their power spectra by $S_{\mathbf{x}}(\omega)$ and $S_{\mathbf{y}}(\omega)$, respectively.

Consider then the scenario that an attack signal $\{\mathbf{n}_k\}$, $\mathbf{n}_k \in \mathbb{R}$, is to be added to the input of the plant $\{\mathbf{u}_k\}$ to deviate the plant state, while aiming to be stealthy in the plant output; see the depiction in Fig. 3.5. In fact, this corresponds to actuator attack. Note in particular that since we are now considering a closed-loop system, the presence of $\{\mathbf{n}_k\}$ will eventually distort the original $\{\mathbf{u}_k\}$ (through feedback) as well, which is an essential difference form the open-loop system setting considered in Sect. 3.3.1, and the distorted $\{\mathbf{u}_k\}$ will be denoted as $\{\overline{\mathbf{u}}_k\}$. In addition, we denote the true plant input under attack as $\{\widehat{\mathbf{u}}_k\}$, where

$$\widehat{\mathbf{u}}_k = \overline{\mathbf{u}}_k + \mathbf{n}_k, \tag{3.19}$$

whereas the plant under attack $\{\mathbf{n}_k\}$ is given by

$$\begin{cases} \widehat{\mathbf{x}}_{k+1} = a\widehat{\mathbf{x}}_k + b\widehat{\mathbf{u}}_k + \mathbf{w}_k = a\widehat{\mathbf{x}}_k + b\overline{\mathbf{u}}_k + b\mathbf{n}_k + \mathbf{w}_k, \\ \widehat{\mathbf{y}}_k = c\widehat{\mathbf{x}}_k + \mathbf{v}_k. \end{cases} \tag{3.20}$$

Meanwhile, suppose that the attack signal $\{\mathbf{n}_k\}$ is independent of $\{\mathbf{w}_k\}$, $\{\mathbf{v}_k\}$, and $\mathbf{x}_0$; consequently, $\{\mathbf{n}_k\}$ is independent of $\{\mathbf{x}_k\}$ and $\{\mathbf{y}_k\}$ as well.

The following theorem, as the second main result of this chapter, characterizes the fundamental trade-off between the distortion in state and the stealthiness in output for feedback control systems.

Theorem 3.2 *Consider the feedback control system under injection attacks depicted in Fig. 3.5. Suppose that the attacker needs to design the attack signal* $\{\mathbf{n}_k\}$ *to satisfy the following attack goal in terms of state distortion:*

$$\mathbb{E}\left[\left(\widehat{\mathbf{x}}_k - \mathbf{x}_k\right)^2\right] \geq D. \tag{3.21}$$

Then, the minimum KL divergence rate between the original output and the attacked output is given by

$$\inf_{\mathbb{E}\left[\left(\widehat{\mathbf{x}}_k - \mathbf{x}_k\right)^2\right] \geq D} \mathrm{KL}_{\infty}\left(p_{\widehat{\mathbf{y}}} \| p_{\mathbf{y}}\right) = \frac{1}{2\pi} \int_0^{2\pi} \frac{1}{2} \left\{ \frac{S_{\widehat{\mathbf{n}}}(\omega)}{S_{\mathbf{y}}(\omega)} - \ln\left[1 + \frac{S_{\widehat{\mathbf{n}}}(\omega)}{S_{\mathbf{y}}(\omega)}\right] \right\} \mathrm{d}\omega, \tag{3.22}$$

where

$$S_{\widehat{\mathbf{n}}}(\omega) = \frac{\zeta S_{\mathbf{y}}^2(\omega)}{1 - \zeta S_{\mathbf{y}}(\omega)}, \tag{3.23}$$

and $S_{\mathbf{y}}(\omega)$ *is given by*

$$S_{\mathbf{y}}(\omega) = \left| \frac{c}{\mathrm{e}^{\mathrm{j}\omega} - a + K\left(\mathrm{e}^{\mathrm{j}\omega}\right) bc} \right|^2 \sigma_{\mathbf{w}}^2 + \left| \frac{\mathrm{e}^{\mathrm{j}\omega} - a}{\mathrm{e}^{\mathrm{j}\omega} - a + K\left(\mathrm{e}^{\mathrm{j}\omega}\right) bc} \right|^2 \sigma_{\mathbf{v}}^2. \tag{3.24}$$

Herein, ζ *is the unique constant that satisfies*

$$\frac{1}{2\pi} \int_{-\pi}^{\pi} \frac{\zeta S_{\mathbf{y}}^2(\omega)}{1 - \zeta S_{\mathbf{y}}(\omega)} \mathrm{d}\omega = c^2 D, \tag{3.25}$$

while

$$0 < \zeta < \min_{\omega} \frac{1}{S_{\mathbf{y}}(\omega)}. \tag{3.26}$$

Moreover, the worst-case attack $\{\mathbf{n}_k\}$ *is a stationary colored Gaussian process with power spectrum*

$$S_{\mathbf{n}}(\omega) = \left| \frac{\mathrm{e}^{\mathrm{j}\omega} - a + K\left(\mathrm{e}^{\mathrm{j}\omega}\right) bc}{bc} \right|^2 \frac{\zeta S_{\mathbf{y}}^2(\omega)}{1 - \zeta S_{\mathbf{y}}(\omega)}. \tag{3.27}$$

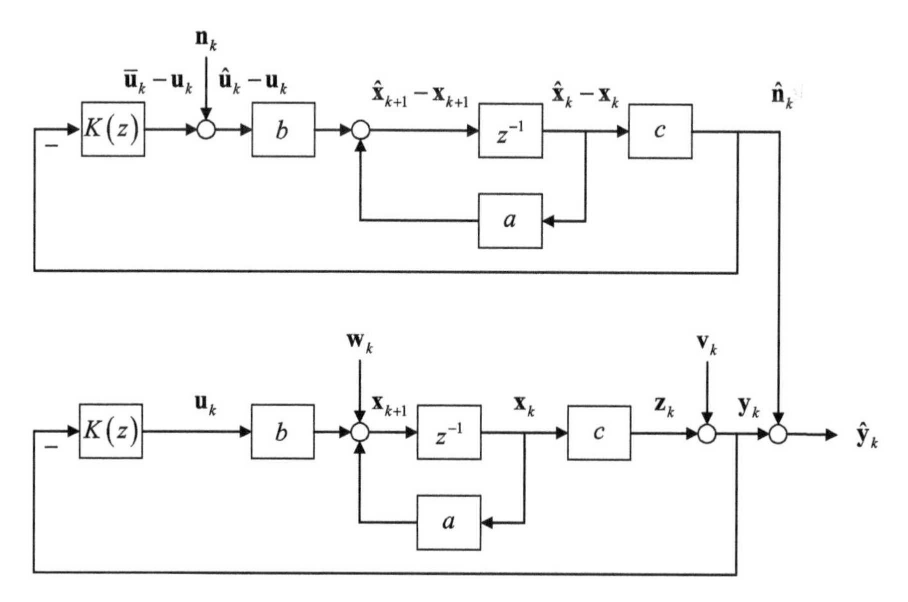

Fig. 3.6 A feedback control system under actuator attack: equivalent system

Proof Note first that when the closed-loop system is stable, the power spectrum of $\{\mathbf{y}_k\}$ is given by

$$\begin{aligned} S_{\mathbf{y}}(\omega) &= \frac{1}{b^2}\left|\frac{P\left(\mathrm{e}^{\mathrm{j}\omega}\right)}{1+K\left(\mathrm{e}^{\mathrm{j}\omega}\right)P\left(\mathrm{e}^{\mathrm{j}\omega}\right)}\right|^2\sigma_{\mathbf{w}}^2 + \left|\frac{1}{1+K\left(\mathrm{e}^{\mathrm{j}\omega}\right)P\left(\mathrm{e}^{\mathrm{j}\omega}\right)}\right|^2\sigma_{\mathbf{v}}^2, \\ &= \frac{1}{b^2}\left|\frac{\frac{bc}{\mathrm{e}^{\mathrm{j}\omega}-a}}{1+K\left(\mathrm{e}^{\mathrm{j}\omega}\right)\frac{bc}{\mathrm{e}^{\mathrm{j}\omega}-a}}\right|^2\sigma_{\mathbf{w}}^2 + \left|\frac{1}{1+K\left(\mathrm{e}^{\mathrm{j}\omega}\right)\frac{bc}{\mathrm{e}^{\mathrm{j}\omega}-a}}\right|^2\sigma_{\mathbf{v}}^2, \\ &= \left|\frac{c}{\mathrm{e}^{\mathrm{j}\omega}-a+K\left(\mathrm{e}^{\mathrm{j}\omega}\right)bc}\right|^2\sigma_{\mathbf{w}}^2 + \left|\frac{\mathrm{e}^{\mathrm{j}\omega}-a}{\mathrm{e}^{\mathrm{j}\omega}-a+K\left(\mathrm{e}^{\mathrm{j}\omega}\right)bc}\right|^2\sigma_{\mathbf{v}}^2. \end{aligned}$$

Note then that since the systems are linear, the system in Fig. 3.5 is equivalent to that of Fig. 3.6, where

$$\widehat{\mathbf{y}}_k = \mathbf{y}_k + \widehat{\mathbf{n}}_k,$$

and $\{\widehat{\mathbf{n}}_k\}$ is the output of the closed-loop system composed by the controller $K(z)$ and the plant

$$\begin{cases} \widehat{\mathbf{x}}_{k+1} - \mathbf{x}_{k+1} = a\left(\widehat{\mathbf{x}}_k - \mathbf{x}_k\right) + b\left(\overline{\mathbf{u}}_k - \mathbf{u}_k\right) + b\mathbf{n}_k, \\ \widehat{\mathbf{n}}_k = c\left(\widehat{\mathbf{x}}_k - \mathbf{x}_k\right), \end{cases}$$

as depicted by the upper half of Fig. 3.6. Meanwhile, as in the case of Fig. 3.3, the system in Fig. 3.6 may also be viewed as a "virtual channel" modeled as

$$\widehat{\mathbf{y}}_k = \mathbf{y}_k + \widehat{\mathbf{n}}_k$$

with noise constraint

$$\mathbb{E}\left[\widehat{\mathbf{n}}_k^2\right] \geq c^2 D,$$

where $\{\mathbf{y}_k\}$ is the channel input, $\{\widehat{\mathbf{y}}_k\}$ is the channel output, and $\{\widehat{\mathbf{n}}_k\}$ is the channel noise that is independent of $\{\mathbf{y}_k\}$. Then, following procedures similar to those in the proof of Theorem 3.1, it can be derived that

$$\inf_{\mathbb{E}\left[(\widehat{\mathbf{x}}_k - \mathbf{x}_k)^2\right] \geq D} \mathrm{KL}_\infty \left(p_{\widehat{\mathbf{y}}} \| p_{\mathbf{y}}\right) = \frac{1}{2\pi} \int_0^{2\pi} \frac{1}{2} \left\{ \frac{S_{\widehat{\mathbf{n}}}(\omega)}{S_{\mathbf{y}}(\omega)} - \ln\left[1 + \frac{S_{\widehat{\mathbf{n}}}(\omega)}{S_{\mathbf{y}}(\omega)}\right] \right\} \mathrm{d}\omega,$$

where

$$S_{\widehat{\mathbf{n}}}(\omega) = \frac{\zeta S_{\mathbf{y}}^2(\omega)}{1 - \zeta S_{\mathbf{y}}(\omega)},$$

and ζ is the unique constant that satisfies

$$\frac{1}{2\pi} \int_{-\pi}^{\pi} S_{\widehat{\mathbf{n}}}(\omega) \, \mathrm{d}\omega = \frac{1}{2\pi} \int_{-\pi}^{\pi} \frac{\zeta S_{\mathbf{y}}^2(\omega)}{1 - \zeta S_{\mathbf{y}}(\omega)} \mathrm{d}\omega = c^2 D,$$

while

$$0 < \zeta < \min_{\omega} \frac{1}{S_{\mathbf{y}}(\omega)}.$$

In addition, since

$$\begin{aligned} S_{\widehat{\mathbf{n}}}(\omega) &= \left| \frac{P\left(\mathrm{e}^{\mathrm{j}\omega}\right)}{1 + K\left(\mathrm{e}^{\mathrm{j}\omega}\right) P\left(\mathrm{e}^{\mathrm{j}\omega}\right)} \right|^2 S_{\mathbf{n}}(\omega) = \left| \frac{\frac{bc}{\mathrm{e}^{\mathrm{j}\omega} - a}}{1 + K\left(\mathrm{e}^{\mathrm{j}\omega}\right) \frac{bc}{\mathrm{e}^{\mathrm{j}\omega} - a}} \right|^2 S_{\mathbf{n}}(\omega), \\ &= \left| \frac{bc}{\mathrm{e}^{\mathrm{j}\omega} - a + K\left(\mathrm{e}^{\mathrm{j}\omega}\right) bc} \right|^2 S_{\mathbf{n}}(\omega), \end{aligned}$$

we have

$$S_{\mathbf{n}}(\omega) = \left| \frac{\mathrm{e}^{\mathrm{j}\omega} - a + K\left(\mathrm{e}^{\mathrm{j}\omega}\right) bc}{bc} \right|^2 S_{\widehat{\mathbf{n}}}(\omega) = \left| \frac{\mathrm{e}^{\mathrm{j}\omega} - a + K\left(\mathrm{e}^{\mathrm{j}\omega}\right) bc}{bc} \right|^2 \frac{\zeta S_{\mathbf{y}}^2(\omega)}{1 - \zeta S_{\mathbf{y}}(\omega)}.$$

This concludes the proof. ■

It is worth mentioning that the $S_{\mathbf{y}}(\omega)$ for Theorem 3.2 is given by (3.24), which differs significantly from that given by (3.8) for Theorem 3.1, although the notations are the same. Accordingly, η, $S_{\mathbf{n}}(\omega)$, and so on, will all be different between the two cases in spite of the same notations.

Note also that $S_{\mathbf{n}}(\omega)$ can be rewritten as

$$S_{\mathbf{n}}(\omega) = \left| \frac{1 + K\left(\mathrm{e}^{\mathrm{j}\omega}\right) P\left(\mathrm{e}^{\mathrm{j}\omega}\right)}{P\left(\mathrm{e}^{\mathrm{j}\omega}\right)} \right|^2 \frac{\zeta S_{\mathbf{y}}^2(\omega)}{1 - \zeta S_{\mathbf{y}}(\omega)}, \tag{3.28}$$

which indicates that the attacker only needs to know the power spectrum of the original system output $\{\mathbf{y}_k\}$ and the transfer function of the closed-loop system (from $\{\mathbf{n}_k\}$ to $\{\widehat{\mathbf{y}}_k\}$), i.e.,

$$\frac{P(z)}{1 + K(z) P(z)}, \tag{3.29}$$

in order to carry out this worst-case attack.

Again, we may examine the dual problem as follows.

Corollary 3.2 *Consider the feedback control system under injection attacks depicted in Fig. 3.5. Then, in order for the attacker to ensure that the KL divergence rate between the original output and the attacked output is upper bounded by a (positive) constant R as*

$$\mathrm{KL}_{\infty}\left(p_{\widehat{\mathbf{y}}} \| p_{\mathbf{y}}\right) \leq R, \tag{3.30}$$

the maximum state distortion $\mathbb{E}\left[(\widehat{\mathbf{x}}_k - \mathbf{x}_k)^2\right]$ *that can be achieved is given by*

$$\sup_{\mathrm{KL}_{\infty}\left(p_{\widehat{\mathbf{y}}} \| p_{\mathbf{y}}\right) \leq R} \mathbb{E}\left[(\widehat{\mathbf{x}}_k - \mathbf{x}_k)^2\right] = \frac{1}{2\pi} \int_{-\pi}^{\pi} \frac{1}{c^2} \left[\frac{\zeta S_{\mathbf{y}}^2(\omega)}{1 - \zeta S_{\mathbf{y}}(\omega)} \right] \mathrm{d}\omega, \tag{3.31}$$

where ζ *satisfies*

$$\begin{aligned} & \frac{1}{2\pi} \int_0^{2\pi} \frac{1}{2} \left\{ \frac{\frac{\zeta S_{\mathbf{y}}^2(\omega)}{1-\zeta S_{\mathbf{y}}(\omega)}}{S_{\mathbf{y}}(\omega)} - \ln \left[1 + \frac{\frac{\zeta S_{\mathbf{y}}^2(\omega)}{1-\zeta S_{\mathbf{y}}(\omega)}}{S_{\mathbf{y}}(\omega)} \right] \right\} \mathrm{d}\omega \\ & = \frac{1}{2\pi} \int_0^{2\pi} \frac{1}{2} \left\{ \frac{\zeta S_{\mathbf{y}}(\omega)}{1 - \zeta S_{\mathbf{y}}(\omega)} - \ln \left[\frac{1}{1 - \zeta S_{\mathbf{y}}(\omega)} \right] \right\} \mathrm{d}\omega = R, \end{aligned} \tag{3.32}$$

while

$$0 < \zeta < \min_{\omega} \frac{1}{S_{\mathbf{y}}(\omega)}. \tag{3.33}$$

Note that herein $S_{\mathbf{y}}(\omega)$ *is given by* (3.24). *Moreover, this maximum distortion is achieved when the attack signal* $\{\mathbf{n}_k\}$ *is chosen as a stationary colored Gaussian process with power spectrum*

$$S_{\mathbf{n}}(\omega) = \left| \frac{\mathrm{e}^{\mathrm{j}\omega} - a + K\left(\mathrm{e}^{\mathrm{j}\omega}\right) bc}{bc} \right|^2 \frac{\zeta S_{\mathbf{y}}^2(\omega)}{1 - \zeta S_{\mathbf{y}}(\omega)}. \tag{3.34}$$

3.4 Simulation

In this section, we will utilize (toy) numerical examples to illustrate the fundamental stealthiness–distortion trade-offs in linear Gaussian open-loop dynamical systems as well as (closed-loop) feedback control systems.

Consider first open-loop dynamical systems as in Sect. 3.3.1. Let $a = 0.5, b = 1, c = 1, \sigma_{\mathbf{w}}^2 = 1, \sigma_{\mathbf{v}}^2 = 1$, and $S_{\mathbf{u}}(\omega) = 1$ therein for simplicity. Accordingly, we have

$$S_{\mathbf{y}}(\omega) = \frac{2}{\left|\mathrm{e}^{\mathrm{j}\omega} - 0.5\right|^2} + 1 = \frac{2}{(\cos\omega - 0.5)^2 + \sin^2\omega} + 1.$$

In such a case, the relation between the minimum KL divergence rate $\mathrm{KL}_{\infty}\left(p_{\widehat{\mathbf{y}}} \| p_{\mathbf{y}}\right)$ (denoted as KL in the figure) and the distortion bound D is illustrated in Fig. 3.7. It is clear that KL increases (strictly) with D, i.e., in order for the attacker to achieve larger distortion, the stealthiness level of the attack will inevitably decrease.

Note that the relation between the maximum distortion $\mathbb{E}\left[(\widehat{\mathbf{x}}_k - \mathbf{x}_k)^2\right]$ and the KL divergence rate bound R in Corollary 3.1 is essentially the same as that between the distortion bound D and the minimum KL divergence rate $\mathrm{KL}_{\infty}\left(p_{\widehat{\mathbf{y}}} \| p_{\mathbf{y}}\right)$ in Theorem 3.1.

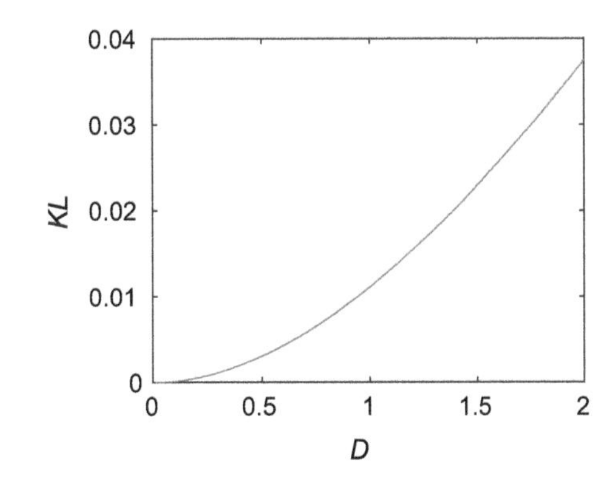

Fig. 3.7 The relation between $\mathrm{KL}_{\infty}\left(p_{\widehat{\mathbf{y}}} \| p_{\mathbf{y}}\right)$ (denoted as KL) and D in Open-Loop Dynamical Systems

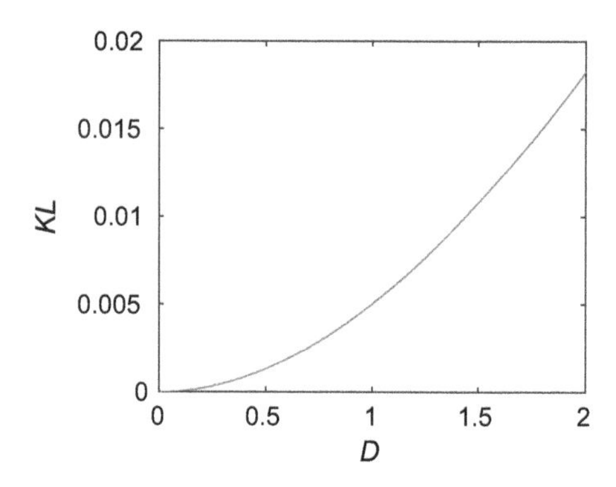

Fig. 3.8 The relation between $\mathrm{KL}_{\infty}\left(p_{\widehat{\mathbf{y}}} \| p_{\mathbf{y}}\right)$ (denoted as KL) and D in Feedback Control Systems

Consider then feedback control systems as in Sect. 3.3.2. Let $a = 2, b = 1, c = 1, \sigma_{\mathbf{w}}^2 = 1, \sigma_{\mathbf{v}}^2 = 1$, and $K(z) = 2$ therein for simplicity. Accordingly, we have

$$S_{\mathbf{y}}(\omega) = 1 + \left|\mathrm{e}^{\mathrm{j}\omega} - 2\right|^2 = 1 + (\cos\omega - 2)^2 + \sin^2\omega.$$

In such a case, the relation between the minimum KL divergence rate $\mathrm{KL}_{\infty}\left(p_{\widehat{\mathbf{y}}} \| p_{\mathbf{y}}\right)$ (denoted as KL in the figure) and the distortion bound D is illustrated in Fig. 3.8. Again, KL increases (strictly) with D, whereas the relationship between the maximum distortion $\mathbb{E}\left[(\widehat{\mathbf{x}}_k - \mathbf{x}_k)^2\right]$ and the KL divergence rate bound R in Corollary 3.2 is essentially the same as that between the distortion bound D and the minimum KL divergence rate $\mathrm{KL}_{\infty}\left(p_{\widehat{\mathbf{y}}} \| p_{\mathbf{y}}\right)$ in Theorem 3.2.

3.5 Conclusion

In this chapter, we have presented the fundamental stealthiness–distortion trade-offs of linear Gaussian open-loop dynamical systems and (closed-loop) feedback control systems under data injection attacks, and explicit formulas have been obtained in terms of power spectra that characterize analytically the stealthiness–distortion trade-offs as well as the properties of the worst-case attacks.

So why do we care about explicit formulas in the first place? One value of the explicit stealthiness–distortion trade-off formula for feedback control systems, for instance, is that they render the subsequent controller design explicit (and intuitive) as well. To be more specific, given a threshold on the output stealthiness, it is already known from Corollary 3.2 what the maximum distortion in state that can be achieved by the attacker is. Then, one natural control design criterion will be to design the controller $K(z)$ so as to minimize this maximum distortion. Mathematically, this minimax problem can be formulated as follows:

$$\inf_{K(z)} \sup_{\mathrm{KL}_{\infty}(p_{\widehat{\mathbf{y}}} \| p_{\mathbf{y}}) \leq R} \mathbb{E}\left[(\widehat{\mathbf{x}}_k - \mathbf{x}_k)^2\right] = \inf_{K(z)} \left\{ \frac{1}{2\pi} \int_{-\pi}^{\pi} \frac{1}{c^2} \left[\frac{\zeta S_{\mathbf{y}}^2(\omega)}{1 - \zeta S_{\mathbf{y}}(\omega)} \right] \mathrm{d}\omega \right\},$$

where

$$\begin{aligned} S_{\mathbf{y}}(\omega) &= \left| \frac{c}{\mathrm{e}^{\mathrm{j}\omega} - a + K(\mathrm{e}^{\mathrm{j}\omega}) bc} \right|^2 \sigma_{\mathbf{w}}^2 + \left| \frac{\mathrm{e}^{\mathrm{j}\omega} - a}{\mathrm{e}^{\mathrm{j}\omega} - a + K(\mathrm{e}^{\mathrm{j}\omega}) bc} \right|^2 \sigma_{\mathbf{v}}^2, \\ &= \left| \frac{P(\mathrm{e}^{\mathrm{j}\omega})}{1 + K(\mathrm{e}^{\mathrm{j}\omega}) P(\mathrm{e}^{\mathrm{j}\omega})} \right|^2 \frac{\sigma_{\mathbf{w}}^2}{b^2} + \left| \frac{1}{1 + K(\mathrm{e}^{\mathrm{j}\omega}) P(\mathrm{e}^{\mathrm{j}\omega})} \right|^2 \sigma_{\mathbf{v}}^2, \end{aligned}$$

whereas the infimum is taken over all $K(z)$ that stabilizes the plant $P(z)$. Herein, ζ can be treated as a tuning parameter as long as it satisfies

$$0 < \zeta < \min_{\omega} \frac{1}{S_{\mathbf{y}}(\omega)}.$$

We will, however, leave more detailed investigations of this formulation to future research.

Other potential future research directions include the investigation of such trade-offs for state estimation systems. It might also be interesting to examine the security–privacy trade-offs (see, e.g., Farokhi and Esfahani (2018), Fang and Zhu (2020, 2021)).

References

C.-Z. Bai, V. Gupta, F. Pasqualetti, On Kalman filtering with compromised sensors: attack stealthiness and performance bounds. IEEE Trans. Autom. Control **62**(12), 6641–6648 (2017)

C.-Z. Bai, F. Pasqualetti, V. Gupta, Data-injection attacks in stochastic control systems: detectability and performance tradeoffs. Automatica **82**, 251–260 (2017)

P. Cheng, L. Shi, B. Sinopoli, Guest editorial special issue on secure control of cyber-physical systems. IEEE Trans. Control Netw. Syst. **4**(1), 1–3 (2017)

M.S. Chong, H. Sandberg, A.M. Teixeira, A tutorial introduction to security and privacy for cyber-physical systems, in *Proceedings of the European Control Conference (ECC)* (2019), pp. 968–978

T.M. Cover, J.A. Thomas, *Elements of Information Theory* (Wiley, 2006)

S.M. Dibaji, M. Pirani, D.B. Flamholz, A.M. Annaswamy, K.H. Johansson, A. Chakrabortty, A systems and control perspective of CPS security. Ann. Rev. Control **47**, 394–411 (2019)

S. Fang, J. Chen, H. Ishii, *Towards Integrating Control and Information Theories: From Information-Theoretic Measures to Control Performance Limitations* (Springer, 2017)

S. Fang, Q. Zhu, Channel leakage, information-theoretic limitations of obfuscation, and optimal privacy mask design for streaming data (2020), arXiv:2008.04893

S. Fang, Q. Zhu, Fundamental limits of obfuscation for linear Gaussian dynamical systems: an information-theoretic approach, in *Proceedings of the American Control Conference* (2021)

S. Fang, Q. Zhu, Fundamental stealthiness-distortion tradeoffs in dynamical systems under injection attacks: a power spectral analysis, in *Proceedings of the European Control Conference* (2021)

S. Fang, Q. Zhu, Independent Gaussian distributions minimize the Kullback–Leibler (KL) divergence from independent Gaussian distributions (2020), arXiv: 2011.02560

F. Farokhi, P.M. Esfahani, Security versus privacy, in *Proceedings of the IEEE Conference on Decision and Control* (2018), pp. 7101–7106

J. Giraldo, D. Urbina, A. Cardenas, J. Valente, M. Faisal, J. Ruths, N.O. Tippenhauer, H. Sandberg, R. Candell, A survey of physics-based attack detection in cyber-physical systems. ACM Comput. Surv. (CSUR) **51**(4), 76 (2018)

I. Goodfellow, Y. Bengio, A. Courville, Y. Bengio, *Deep Learning* (MIT Press, 2016)

U. Grenander, G. Szegö, *Toeplitz Forms and Their Applications* (University of California Press, 1958)

Z. Guo, D. Shi, K.H. Johansson, L. Shi, Worst-case stealthy innovation-based linear attack on remote state estimation. Automatica **89**, 117–124 (2018)

J. Gutiérrez-Gutiérrez, P.M. Crespo, Asymptotically equivalent sequences of matrices and Hermitian block Toeplitz matrices with continuous symbols: applications to MIMO systems. IEEE Trans. Inf. Theory **54**(12), 5671–5680 (2008)

K.H. Johansson, G.J. Pappas, P. Tabuada, C.J. Tomlin, Guest editorial special issue on control of cyber-physical systems. IEEE Trans. Autom. Control **59**(12), 3120–3121 (2014)

S.M. Kay, *Information-Theoretic Signal Processing and its Applications* (Sachuest Point Publishers, 2020)

S. Kullback, *Information Theory and Statistics* (Courier Corporation, 1997)

S. Kullback, R.A. Leibler, On information and sufficiency. Ann. Math. Stat. **22**(1), 79–86 (1951)

E. Kung, S. Dey, L. Shi, The performance and limitations of ϵ-stealthy attacks on higher order systems. IEEE Trans. Autom. Control **62**(2), 941–947 (2016)

A. Lindquist, G. Picci, *Linear Stochastic Systems: A Geometric Approach to Modeling*. Estimation and Identification. (Springer, 2015)

L. Ljung, *System Identification: Theory For the User* (Prentice Hall, 1999)

L. Pardo, *Statistical Inference Based on Divergence Measures* (CRC Press, 2006)

M.S. Pinsker, *Information and Information Stability of Random Variables and Processes* (Holden Day, San Francisco, CA, 1964)

H.V. Poor, *An Introduction to Signal Detection and Estimation* (Springer, 2013)

R. Poovendran, K. Sampigethaya, S.K.S. Gupta, I. Lee, K.V. Prasad, D. Corman, J.L. Paunicka, Special issue on cyber-physical systems [scanning the issue]. Proc. IEEE **100**(1), 6–12 (2012)

H. Sandberg, S. Amin, K.H. Johansson, Cyberphysical security in networked control systems: an introduction to the issue. IEEE Control Syst. Mag. **35**(1), 20–23 (2015)

P. Stoica, R. Moses, *Spectral Analysis of Signals* (Prentice Hall, 2005)

A. Stoorvogel, J. Van Schuppen, System identification with information theoretic criteria, in *Identification, Adaptation, Learning: The Science of Learning Models from Data*, ed. by S. Bittanti, G. Picci (Springer, 1996)

S. Weerakkody, O. Ozel, Y. Mo, B. Sinopoli, Resilient control in cyber-physical systems: countering uncertainty, constraints, and adversarial behavior, Foundations and Trends®. Syst. Control **7**(1–2), 1–252 (2019)

R. Zhang, P. Venkitasubramaniam, Stealthy control signal attacks in linear quadratic Gaussian control systems: detectability reward tradeoff. IEEE Trans. Inf. Foren. Secur. **12**(7), 1555–1570 (2017)

Chapter 4
Predictive Situation Awareness and Anomaly Forecasting in Cyber-Physical Systems

Masoud Abbaszadeh, Weizhong Yan, and Lalit K. Mestha

4.1 Introduction

Cyber-physical systems' (CPS) security has become a critical research topic as more and more CPS applications are making increasing impacts in diverse industrial sectors. Due to the tight interaction between cyber- and physical components, CPS security requires a different strategy from the traditional Information Technology (IT) security. Cyber-Physical Systems (CPS) are an integral system featuring strong interactions between its cyber- (e.g., networks and computation) and physical components (Khaitan and McCalley 2014). CPS applications have been making great impacts on many industrial sectors, including energy, transportation, healthcare, and manufacturing. With the development of Internet of Things (IoT), more and more devices with potential security vulnerabilities are linked to CPS, which makes CPS susceptible to adversary attacks (Yan et al. 2019). While progress with machine and equipment automation has been made over the last several decades, and systems have become "smarter", the intelligence of any individual cyber-physical system to predict failures (e.g., equipment malfunction, sensor faults, etc.), outages, degradation or slow drift in performance, and cyber-threats in real time to provide early warning is difficult. Several methods have been proposed for anomaly forecast and prognostic in different industrial control systems (Abbaszadeh and Marquez 2010, 2007; Allegorico and Mantini 2014; Chandola et al. 2009; Clifton et al. 2014; Ehlers et al. 2011; Gupta et al. 2008; Lamedica et al. 1996; Pimentel et al. 2014; Rigatos et al. 2021; Sridhar and Govindarasu 2014; Xue and Yan 2007; Zaher et al. 2009; Zimek et al.

M. Abbaszadeh (✉) · W. Yan
GE Research, Niskayuna, NY, USA
e-mail: masoud@ualberta.net

W. Yan
e-mail: yan@ge.com

L. K. Mestha
Genetic Innovations Inc. (work performed while at GE Research), Honolulu, HI, USA

M. Abbaszadeh and A. Zemouche (eds.), *Security and Resilience in Cyber-Physical Systems*, https://doi.org/10.1007/978-3-030-97166-3_4

2012). Although technology exists to predict when systems fail, approaches used to predict failures from a Prognostics and Health Management (PHM) perspective are not directly applicable to situation awareness of cyber-incidents since they (1) do not model large-scale transient data incorporating fast system dynamics (i.e., have improper estimation models) and (2) do not to process multiple signals simultaneously to account for anticipated changes in future times in system behavior accurately based on current and past data (i.e., have inaccurate decision thresholds/boundaries) (Mestha et al. 2017). Especially, when it comes to forecasting cyber-attacks propagation and impact, the difficulty is further compounded by not knowing attackers' intention and their next move for exploiting weakness/vulnerabilities in the system.

There can be various types of known attacks that a system may be subjected to such as espionage attacks, eavesdropping, denial-of-service attacks, zero dynamics attack, deception attacks (e.g., covert/stealthy attack), false data injection attack, replay attack, and the like, which are just a short sampling of potential threats that exist to cyber-physical systems (Park et al. 2019). These attacks will exhibit different levels of disclosure, disruption, and knowledge to be executed successfully, corresponding to adversaries' recourses, expertise, and intent. Also, cyber-hackers always invent many new ways to create malicious code and disrupt the operation of the physical system. Present condition monitoring technology used for failure detection, prediction, and monitoring or the threat detection technologies included inside information and operational technologies (IT and OT) does not adequately provide forecasting to protect assets from such attacks. There are many examples in physical systems (e.g., electric grid, ventricular assist devices, etc.), wherein early warning of only a few seconds may be sufficient to take actions that would protect vulnerable equipment or loss of life (Kokkonen et al. 2016; Nateghi et al. 2018a, b; Skopik et al. 2015).

Proper early warning generation could thwart an attack entirely or help neutralize its effects, such as damage to equipment or sustain the operation. The goal of this chapter is to provide an innovative predictive situational awareness framework in order to maintain high levels of reliability and availability, while continuing to retain expected performance against abnormalities created by the system faults or the adversary. Building upon our previous results on anomaly detection and forecasting (Abbaszadeh et al. 2018; Mestha et al. 2017; Yan et al. 2019), the predictive situational awareness framework developed in this chapter is based on dynamic weighted averaging of multi-model ensemble forecasts both for anomaly detection and isolation. Ensemble forecasting has been proven to be very efficient in forecasting complex dynamic phenomena such as wind and other weather conditions and Internet communication traffics (Cortez et al. 2012; Gneiting and Raftery 2005). In the context of an industrial control system, we use ensembles to cover the plant variations both in operating space and ambient conditions. The ensembles are selected using GMM clustering, which provides both centroid (i.e., respective operating points) and probability membership functions. A state-space model is developed for each ensemble of each monitoring node, which is used in an adaptive multi-step Kalman predictor to provide ensemble forecast in a receding horizon fashion. Then, the ensemble forecasts are fused via dynamic averaging. Dynamic model averaging has been shown to be superior to other ensemble methods such as Markov Chain Monte Carlo (MCMC) especially for large datasets (Koop and Korobilis 2012; McCormick et al. 2012; Raftery et al. 2010). It is an effective way for estimation of fusion of ensemble models.

We carry out all key processing in a high-dimensional feature space by analyzing time-series signals received from multiple system monitoring nodes (a combination of selected control system sensors and actuators), comparing the forecasted features with anomaly decision boundaries. The decision boundaries are computed for each individual monitoring node using machine learning techniques. We use Extreme Learning Machine (ELM) as our binary classification decision boundary. ELM is a special type of flashforward neural network recently developed for fast training (Huang et al. 2012). Numerous empirical studies and recently some analytical studies as well have shown that ELM has better generalization performance than other machine learning algorithms including Support Vector Machines (SVM) and is efficient and effective for both classification and regression (Huang et al. 2012; Huang 2014; Huang et al. 2006). It is worth mentioning that the framework presented here is not limited to using Kalman predictors or ELM classifiers and can be used along with other forms of linear or nonlinear time-series models, predictors, and classifiers.

The rest of the chapter is organized as follows. In Sect. 4.2, the overall forecasting framework is described. Sections 4.3 and 4.4 provide details of the ensemble modeling and receding horizon ensemble forecasting. In Sect. 4.4, we demonstrate our algorithm in a sensor attack of a gas turbine using a high-fidelity simulation environment, followed by conclusions in Sect. 4.5.

4.2 Forecasting Framework

In this section, we discuss the framework used for anomaly forecast and early warning generation. The framework is applicable to both cyber-driven and fault-driven incidents in a unified manner.

4.2.1 Digital Twin Simulation Platform

We demonstrate our approach on a utility-scale (250 MW maximum output) power-generating gas turbine. However, the methods and techniques presented in this work are applicable to any cyber-physical system. We have created both *normal* and *abnormal* (attack and fault) datasets using GE ARTEMIS high-fidelity power plant simulation platform. The simulation environment consists of a closed loop *Digital Twin* of a utility-scale power generation gas turbine, a very complex nonlinear and time-varying physics-based model with adaptive parameters and factors such as asset performance degradation due to ageing. The closed-loop system contains multiple control loops along with their interconnections as in a real gas turbine in the field.

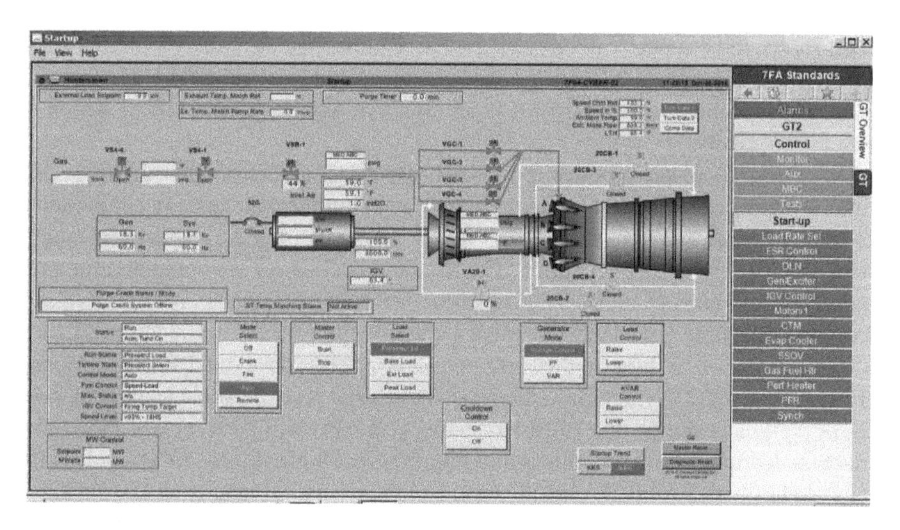

Fig. 4.1 Plant HMI used for dataset generation

The availability of such a platform enables realistic simulations of attack and fault scenarios, which, compared to normal operation data, are usually rare in the data collected from the field. This in turn enables deployment of high-performance supervised learning algorithms, as opposed to semi-supervised learning that only uses *normal* data. The *normal* dataset can be collected from the field, generated through simulations, or a combination of both. The *abnormal* dataset is synthesized utilizing the simulation platform. Our dataset consists of thousands of *normal* and *abnormal* time series of the monitoring nodes, resulting in over 2 million samples when projected into feature space. Figure 4.1 shows the HMI used for dataset generation.

4.2.2 Anomaly Forecasting Approaches

Depending on the scale of the system and outcome of the features dimensionality reduction, either the features or directly, the anomaly score may be forecasted. Each of these approaches have their pros and cons. Forecasting features make the forecasting framework independent of the decision boundary (i.e., the classifier), but it might be very difficult to do if the number of features is very large, they are highly non-linear, discontinuous, etc. On the other hand, forecasting the anomaly score directly simplifies the problem quite a bit, but makes the forecasting framework dependent on the particular anomaly classifier used (as will be described more).

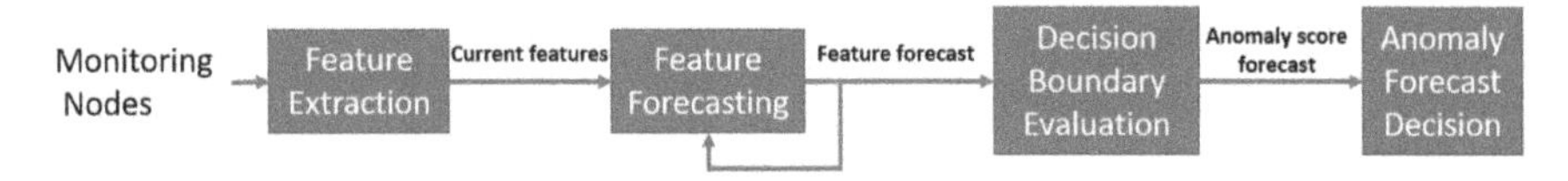

Fig. 4.2 Feature forecasting approach for anomaly prediction

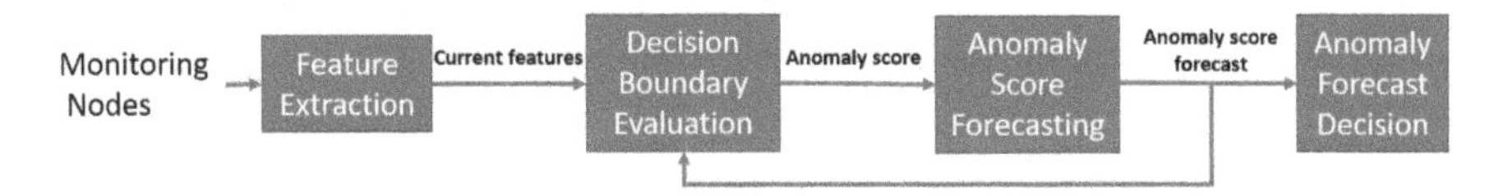

Fig. 4.3 Anomaly score forecasting approach for anomaly prediction

- **Forecasting Features:** In this approach, features are forecasted using dynamic models built for the time evolution of features, and the forecasted values are sent to classifier. A high-level depiction of the feature forecasting approach is shown in Fig. 4.2, where the feedback loop depicts the repetition of the forecasting for multi-step ahead prediction.
- **Forecasting Anomaly Score:** In this approach, the anomaly score is directly forecasted, as depicted in Fig. 4.3. Hence, instead of forecasting the features and sending the forecasted features to the classifier, the dynamic models are built for the anomaly score time series directly.

Note that assuming that there is only a single classifier for global detection and a single classifier for each local node, the anomaly score of each classification is a scalar, so such model would only have a single output. This significantly simplifies the dynamic models, reducing the number of model outputs from the number of features to only 1. Again, the anomaly score forecasting may be done both at the local and global levels. The states of the such a model may be the features or just the anomaly score. This method essentially simplifies the problem into forecasting a scalar. Note that as shown in Fig. 4.3, this brings the decision boundary into the forecasting loop. The dynamic models built in this approach will collectively represent the feature evaluation and the anomaly score evolution combined.

4.2.3 Dimensionality Reduction

Large-scale systems might have hundreds of monitoring nodes. Feature discovery techniques may lead to selection of several features for each node, resulting in a very large number of features to be forecasted. The following methods are used for dimensionality reduction in those large-scale systems.

4.2.3.1 Forecasting Features

In Feature Space

The number of features may be reduced using data dimensionality reduction methods such as PCA, ICA, and isoMap. This may be done for both the local and global levels. This enables the creation of scalable dynamic models.

In Dynamic State Space

Once the dynamic models are built, if the number of states (features and their lagged values) at each node or that of the global level is still large (normally > 50), dynamic model-order reduction techniques, such as balanced truncation or H_∞ norm-based model-order reduction, may be used to further reduce the dimensionality of the forecasting problem. The model-order reduction is performed using these two criteria:

- *Model Accuracy:* The error between the original model and the reduced-order model is less than a prescribed threshold using Hankel norm or H_∞ norm bounds. This determines the order of the reduced-order model. The error threshold may be selected by evaluating the forecasting accuracy of the reduced-order model or based on the preservation of the model observability (described below).
- *Model Observability:* The reduced-order model remains observable. In particular, in the original model, the features might be both the states and the outputs (i.e., an identity state to output mapping). Hence, the reduced-order model may have more outputs than states. The order and the model accuracy threshold then are selected in a manner to preserve the observability.

4.2.3.2 Forecasting Anomaly Score

If after dimensionality reductions in feature and/or state spaces, the order of the model is still high (normally > 50) or if the dimensionality reduction cannot be done in a way to properly satisfy the aforementioned criteria, then instead of forecasting the features, the anomaly score of the classifier is directly forecasted. In this approach, instead of forecasting the features and sending the forecasted features to the classifier, the dynamic models are built for the anomaly score time series directly. Note that the anomaly score is a scalar, so such model would only have a single output. This significantly simplifies the model reduces the number of model outputs (from the number of features to 1). Again, the anomaly score forecasting may be done both at the local and global levels. The states of such a model may be the features or just the anomaly score.

In the rest of this chapter, we will focus on forecasting using the feature forecasting approach, but the same tools and technique are applicable to the anomaly score forecasting as well.

4.2.4 *Forecasting Process*

The forecasting system is comprised of offline (training) and online (operation) modules. During the offline training, as shown in Fig. 4.4 the monitoring node datasets are used for feature engineering and decision boundary generation. To select the features, feature discovery techniques are used as described in Sect. 4.2.5. Then, state-space ensemble dynamic models are generated for the time evolution of features both at the global (for overall system status) and local (i.e., per monitoring node) levels as described in Sect. 4.3.1. At each level, dynamic forecasting models are generated for forecasting at three time scales, short term, mid-term, and long term, depending on the fundamental sampling time of the control system. Also, decision boundaries are computed both at the local and global levels as binary classifiers using machine learning techniques as described in Sect. 4.3.5.

The online module of forecasting system is shown in Fig. 4.5. First, each monitoring node signal goes through real-time feature extraction to create real-time feature time series. The features are computed using a sliding window over the monitoring node signals. In the next step, the extracted feature time series are inputted to multi-step predictors, both at the local and global levels. Using the models generated in the training phase and the multi-step predictors, future values of the feature time series are forecasted, both for local and global features, in three time scales:

1. **Short-term feature forecast:** future values of the global and local features (e.g., up to several seconds).
2. **Mid-term feature forecast:** future values of the global and local features (e.g., up to several minutes).
3. **Long-term feature forecast:** future values of the global and local features (e.g., up to several days).

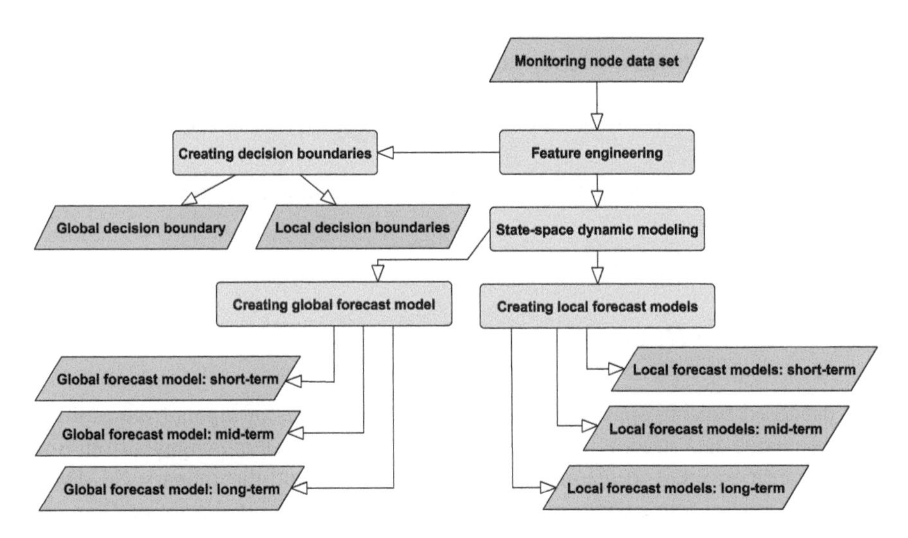

Fig. 4.4 Anomaly forecast systems: offline training

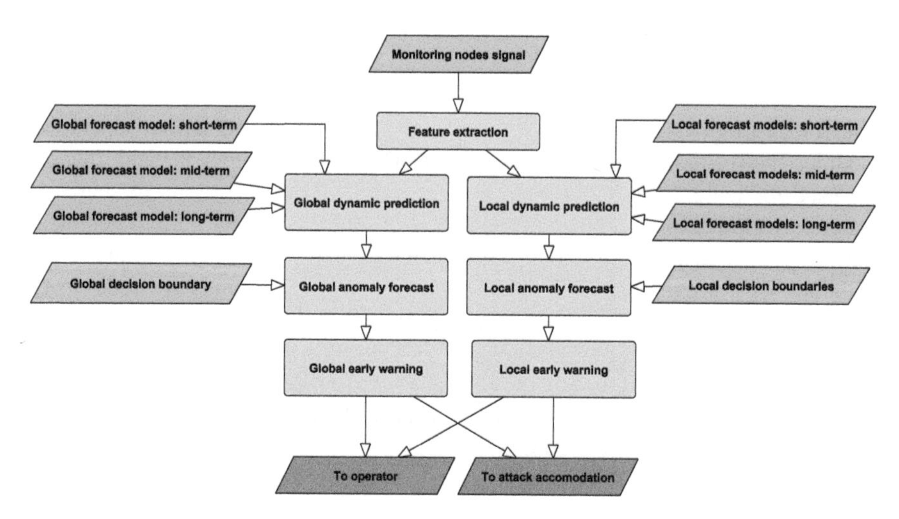

Fig. 4.5 Anomaly forecast systems: online operation

While the short-term forecast is useful for rapid detection of the incipient and transient faults and cyber-attacks, mid-term and long-term forecasts are helpful in early detection of stealthy cyber-attacks as well as component failures due to degradation. The forecasted outputs of models (*aka*, future values of the features) are compared to the corresponding decision boundaries for predictive anomaly detection. While comparing the feature vectors to the decision boundary, estimated time to cross the decision boundary will provide information for future anomaly. If a future anomaly is detected, an early warning is generated in the operator display with anticipated time to reach anomalous state and a message is sent to the automatic accommodation system (such as an attack-tolerant or fault-tolerant resilient control mechanism) for potential early engagement. The current values of the features along with the decision boundaries provide a deterministic decision of the current status of the system, while the forecasted features provide a probabilistic decision on the future system status. The global feature forecast is used for system-level anomaly detection (overall system health status) and the local feature forecasts are used for anomaly isolation (locate the abnormal nodes of the system). Using this framework a predictive situation awareness is established for the system.

4.2.5 *Feature Discovery*

The proposed sensing approach should handle many types of inputs from multiple heterogeneous data stream in complex hyper-connected systems. Signals from time domain are converted to features using multi-modal multi-disciplinary (MMMD) feature discovery framework employed as in machine learning discipline (Yan and

Yu 2015). A "feature" may refer to, for example, mathematical characterizations of data and is computed in each overlapping batch of data stream. Examples of features as applied to sensor data can be classified broadly into knowledge-based, shallow, and deep features.

Knowledge-based features use domain or engineering knowledge of physics of the system to create features. These features can be simply statistical descriptors (e.g., max, min, mean, variance), and different orders of statistical moments, calculated over a window of a time-series signal and its corresponding FFT spectrum as well. Shallow features are from unsupervised learning (e.g., k-means clustering), manifold learning, and nonlinear embedding (e.g., isoMap, locally linear embedding), low dimension projection (e.g., principal component analysis, independent component analysis), and neural networks, along with genetic programming and sparse coding. Deep learning features can be generated using deep learning algorithms which involve learning good representations of data through multiple levels of abstraction. By hierarchically learning features layer by layer, with higher level features representing more abstract aspects of the data, deep learning can discover sophisticated underlying structure and features. Still other examples include logical features (with semantic abstractions such as "yes" and "no") and interaction features.

Several methods have been proposed in the literature for feature selection and features ranking of ELM for classification and regression problems)(Wang et al. 2018; Yin et al. 2017). Machine learning-based attack and fault-detection algorithms can in general incorporate large number of features, with the number of features selected based on the Receiver Operating Characteristic (ROC) curve analysis to optimize the detection and false alarm rates. Different number of features might be selected for each individual monitoring node, however, from a systems engineering perspective, to streamline the design, it is preferred to choose the same type and number of features for all nodes, except if a particular node needs special treatment. In this work, for each monitoring node of the gas turbine, we have selected five features which are a combination of statistical and temporal features. At the system level, we have also selected multivariate features which consist of cross-correlations between critical measurements.

For the forecasting at the *global* level (i.e., the system level), the global feature vector is formed by stacking up the local feature vectors of the individual monitoring nodes. For large-scale systems with many monitoring nodes, the size of the global feature vector might be very large, and thus it can be reduced by dimensionality reduction techniques such as Principal Component Analysis (PCA).

4.3 Ensemble Forecasting

The forecasting framework described in the previous section is based on ensemble models which are used in adaptive Kalman predictors to provide ensemble feature forecasts. The ensemble feature forecasts are then averaged using dynamic weights to provide the overall feature forecast. The process described in the section is applied separately and in parallel to the local features of each individual monitoring node, as well as to the global feature vector.

4.3.1 Ensemble Modeling in Feature Space

The forecasting models at each time scale (short term, mid-term, and long term) consist of a collection of ensemble models, each providing an ensemble forecast of the features. These ensembles ensure coverage of whole operating space with operational and ambient condition variations. The operating space is partitioned through Gaussian Mixture Model clustering. A mixture model is a statistical model for representing datasets which display behavior that cannot be well described by a single standard distribution. It allows a complex probability distribution to be built from a linear superposition of simpler components. Gaussian distributions are the most common choice as mixture components because of the mathematical simplicity of parameter estimation as well as their ability to perform well in many situations (Dempster et al. 1977).

Gaussian mixture models can be used for stochastic data clustering. To select the operating point associated with each ensemble model, we use GMM clustering in the feature space. The GMM clustering partitions the operating space (projected into feature space) into multiple clusters each represented by a multivariate Gaussian process described by a mean (centroid) and a covariance matrix. The centroid of each cluster represents the operating point for each ensemble model, while its covariance matrix establishes a probabilistic membership function. The Expectation Maximization (EM) algorithm is a maximum likelihood estimation method that fits GMM clusters to the data. The EM algorithm can be sensitive to initial conditions, and therefore we repeat the GMM clustering multiple times with randomly selected initial values and choose the fit that has the largest likelihood.

Since GMM is a soft clustering method (i.e., overlapping clusters), all points in the operating space belong to all clusters with a membership probability. As an example, Fig. 4.6 shows the GMM clustering at the global level for our gas turbine dataset, where the horizontal axis is the number of clusters and the vertical axis is the Bayesian Information Criterion (BIC) computed for different covariance structures per number of clusters. BIC provides a right trade-off between model accuracy and complexity, thus avoiding over-fitting to the training dataset. The model with the lowest BIC is selected. As seen in the figure, the optimal clustering is achieved with seven clusters with Gaussian models having full (i.e., non-diagonal) unshared (i.e., individual) covariance matrices.

Remark 4.1 Note that at the local node level, GMM clustering can be done for each monitoring node separately, resulting in different number of ensembles for each monitoring node.

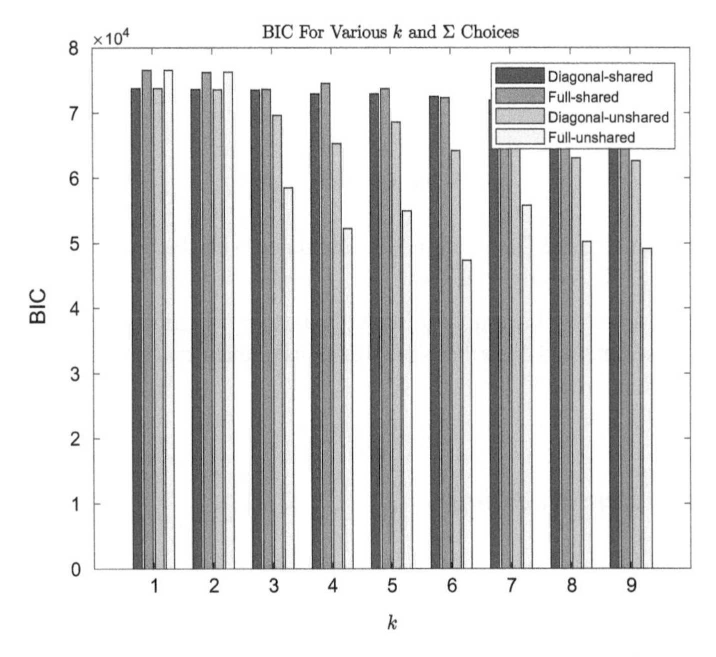

Fig. 4.6 BIC for GMM clustering. ©2018 IEEE. Reprinted, with permission, from (Abbaszadeh et al. 2018)

4.3.2 Adjusting Cluster Centroids to Physical Points

The GMM clustering may select centroid of the clusters as any arbitrary real-valued vector in the feature space. However, since centroids are deemed as operating points to create state-space models, they need to be associated with actual physical operating points of the system. This can be achieved in two ways:

- **Mixed-integer programming for EM:** GMM clustering uses Expectation Maximization (EM) algorithm for cluster optimization. Rather than running the standard EM, one can use a modified EM to enforce searching for centroids only among the points given in the training dataset (which are readily physical points of the systems). This is essentially similar to running k-medoids clustering rather than k-means clustering but in a GMM framework. This normally requires mixed-integer programming and is feasible for small- and medium-sized datasets.
- **Heuristics based:** Adjust the centroids of GMM into closest point in the dataset in post-processing. This is particularly efficient for large datasets. Moreover, since large datasets comprise of high granularity data, the distance of the initial centroid to the closest point in the data is often small and negligible. This can be further validated by putting a threshold on such point adjustments. As a result of centroid adjustment, the covariance matrices of each GMM clusters are also adjusted. Suppose that μ_i and Σ_i are the centroid and covariance of the i-th cluster, respectively,

and the closest point to μ_i is $\bar{\mu}_i$ whose Euclidean distance to μ_i in feature space is d_i, i.e., $\bar{\mu}_i - \mu_i = d_i$. Then, we have

$$\mu_i \rightarrow \bar{\mu}_i = \mu_i + d_i, \tag{4.1}$$

$$\Sigma_i \rightarrow \bar{\Sigma}_i = \Sigma_i + d_i d_i^T, \tag{4.2}$$

which means that the Gaussian model associated with the i-th cluster is adjusted from $\mathcal{N}(\mu_i, \Sigma_i)$ to $\mathcal{N}(\bar{\mu}_i, \bar{\Sigma}_i)$.

In this work, since we have a large-scale dataset with high resolution, we use the heuristics-based method described above to adjust the cluster centroids to the nearest physical point as needed.

4.3.3 Dynamic Modeling

Once the number and structure of the clusters are determined, the cluster centroids are selected as the representative operating points of the system, and a dynamic model is developed for the time series of each monitoring node of each operating point (aka ensemble models). The time-series dynamic modeling can happen using different linear or nonlinear time-series modeling techniques. The choice of linear versus nonlinear modeling can be made by assessing the feature time series through linearity tests such as those described in Harvey and Leybourne (2007). For linear time-series data, Vector Autoregressive (VAR) models are proved to be a powerful tool. For nonlinear time-series modeling, nonlinear autoregressive models, Volterra series, or recurrent neural networks (such as LSTM) could be used. In this work, due to the good fit of the feature time-series data in the linear space, the time series are modeled as VAR models. A VAR model is a multivariate autoregressive model that relates the current value of the time series to its previous values through a linear mapping plus a constant bias term. Essentially, this is not an input–output modeling but a time-series output modeling, assumed to be derived by an unknown stochastic input. VAR models are vastly used for modeling of time-series signals (Johansen 1995), similar to what we measure here from our monitoring nodes. The number of lags required for each VAR model is again determined using BIC. This determines the order of the models, which could be different among the ensembles. The parameters of the VAR models are identified, and the models are then converted into the standard state-space form for each ensemble, as follows:

$$x[k+1] = Ax[x] + Bu[k] + Qe[k], \tag{4.3}$$

$$y[k] = Cx[k] + v[k], \tag{4.4}$$

where x is the vector of monitoring node features and their lagged values, u is a fictitious Heaviside step function capturing the bias term of the VAR model, e is a zero-mean Gaussian white noise with Identity covariance, $E[ee^T] = I$, and Q is the process noise covariance. The model outputs y here are the monitoring node features with some assumed measurement noise v, whose covariance R is adaptively updated, as will be described later.

If the model is VAR(1), i.e., having one lag, then $C = I_q$, where q is the number of local features for each individual monitoring node (here, for our gas turbine application, $q = 5$). In general, for a VAR(p) model with p lags, per ensemble, per node, we have

$$x[k] = \left[x_1^f[k] \cdots x_q^f[k] \cdots x_1^f[k-p+1] \cdots x_q^f[k-p+1]\right]^T, \tag{4.5}$$

$$A = \begin{bmatrix} A_1 & A_2 & \cdots & A_{p-1} & A_p \\ I_q & 0_q & \cdots & 0_q & 0_q \\ 0_q & I_q & \cdots & 0_q & 0_q \\ \vdots & \vdots & \ddots & \vdots & \vdots \\ 0_q & 0_q & \cdots & I_q & 0_q \end{bmatrix}, \tag{4.6}$$

$$B = \Big[b \ \underbrace{0_q \cdots 0_q}_{1,\dots,p-1,\ p>1} \Big]^T, \quad C = \Big[I_q \ \underbrace{0_q \cdots 0_q}_{1,\dots,p-1,\ p>1} \Big], \tag{4.7}$$

where x_i^f, $i = 1, \dots, q$ are the local features for an individual monitoring node.

The initial value of R is set using noise characteristics of the raw measurements, linearly projected into the feature space as follows. Suppose y^r is the raw measured value of an individual monitoring node and the scalar v^r is the corresponding measurement noise, $y^r[k] = r[k] + v^r[k]$, where r is the true value of the signal and v^r is a zero-mean Gaussian white noise with variance σ. The feature vector y corresponding to this particular monitoring node is the projection of y^r in the feature space. Suppose that $\mathscr{F} : R \rightarrow R^q$ is the mapping from the raw signal measurement to its features. The raw data is projected into the feature space as

$$\left[x_1^f[k] \cdots x_q^f[k]\right]^T = Cx[k] = \mathscr{F}(r[k]). \tag{4.8}$$

Then we have

$$\begin{aligned} y[k] &= \mathscr{F}(y^r[k]) = \mathscr{F}(y[k] + v^r[k]) \\ &\simeq \mathscr{F}(r[k]) + \frac{\partial \mathscr{F}}{\partial r}_{|r=r[k]} v^r[k] \triangleq Cx[k] + J(r[k])v^r[k] \\ &\triangleq Cx[k] + v[k], \end{aligned} \tag{4.9}$$

where v is the derived measurement noise in the feature space and J is the Jacobian of $\mathscr{F}$ with respect to r. From (4.9), it is clear that the covariance of v is $\sigma J(r[k])^T J(r[k])$. Note that the scalar measurement noise of an individual monitoring node in the signal space is projected into a multivariate noise in the feature space. The linear approximation of noise maintains the noise zero-mean Gaussian white. This approximation is only used for the initial guess of the covariance, since after the initialization it is adaptively estimated.

As mentioned before, the number of such state-space models for each monitoring node equals the number of corresponding GMM clusters. The order of the state-space models remains the same within the ensembles of one particular node, but may differ from one node to another depending on the number of local features selected for each node.

4.3.4 Dynamic Ensemble Forecast Averaging

Within our proposed framework, different type of the predictors may be used to provide ensemble forecasts. This simplest predictor could be the model (4.4) itself, repeatedly executed using previous predictions through the prediction horizon. Although simple, this approach quickly leads to large prediction errors since there is no control or adjustment over the error covariance. Another simple approach is to use parametric prediction methods such as exponential smoothing. They provide certain level of parameter tuning capability but still suffer from proper error control. As such, although both approaches are applicable within our proposed framework, they are both limited to only very short prediction horizons.

In this chapter, an adaptive Kalman predictor (AKP) is applied to each ensemble model to provide ensemble forecasts. The process noise covariance of the Kalman predictor is readily available as Q as in (4.4). It is worth mentioning that for nonlinear models, an adaptive EKF or UKF can be used still in a similar fashion within the same framework. The covariance of the measurement noise of each AKP is estimated adaptively using the method proposed in Ding et al. (2007); Rutan (1991) as follows.

$$\hat{v}[k] = y[k] - C^T \hat{x}[k|k-1], \tag{4.10}$$

$$R[k] = \begin{cases} \sigma J(r[k])^T J(r[k]) & k = 1, \dots m \\ \frac{1}{m}\left[\sum_{j=1}^{m} \hat{v}[k-j]\hat{v}[k-j]\right] \dots & \\ \qquad -C^T P^e[k|k-1]C & k > m, \end{cases} \tag{4.11}$$

where $\hat{v}$ is the predictor innovation sequence, m is the width of an empirically chosen rectangular smoothing window for the innovations sequence, and P^e is the prediction error covariance matrix. The smoothing operation improves the statistical significance of the estimator for $R[k]$, as it now depends on many residuals. Figure 4.7 shows the block diagram for dynamic ensemble forecast averaging, where N is the number of ensembles corresponding to a monitoring node and P is the forecasting horizon. It is worth mentioning that the ensemble modeling (GMM clustering and state-space system identification) is performed using *normal* dataset only as the models capture the normal operational behavior of the system, while the decision

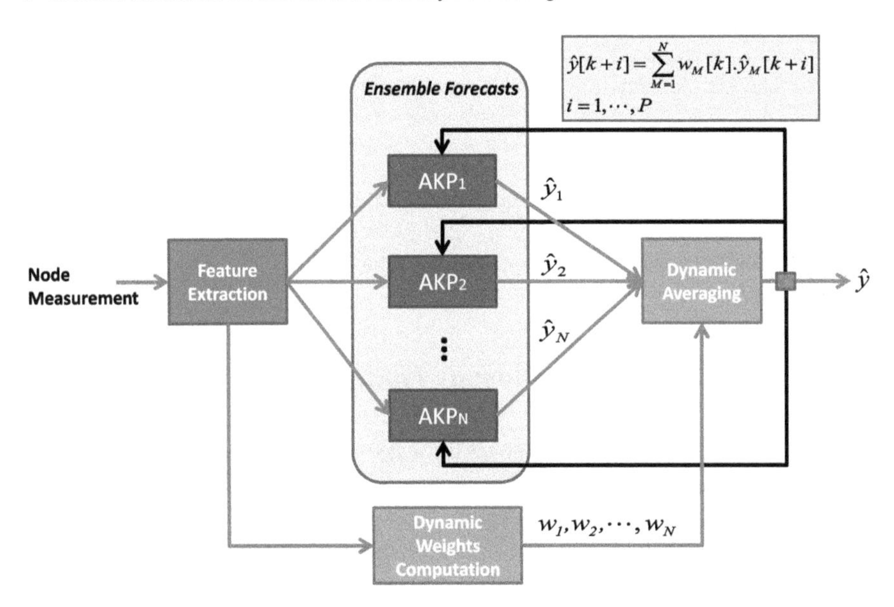

Fig. 4.7 Block diagram for dynamic ensemble forecast averaging. ©2018 IEEE. Reprinted, with permission, from (Abbaszadeh et al. 2018)

boundaries are computed using both *normal* and *abnormal* datasets. Furthermore, to emphasize the recent data, a forgetting factor is used in the covariance matrix update of each of the Kalman predictors.

The forecasting horizon of the multi-step forecasts can be determined using simulations, based on the prediction error and some threshold on the confidence interval. As the forecasting horizon extends, the confidence interval expands and eventually passes the threshold. Each AKP provides an ensemble forecast $\hat{y}_M$, $M = 1, \ldots, N$. The ensemble forecasts are dynamically averaged using weight $w_1, \ldots, w_N$. The weights are time varying and computed as normalized probabilities using the multivariate Gaussian probability density functions with mean and covariances computed during the GMM clustering. Suppose the real-time value of the feature vector is $x[k]$, and the mean and covariance of each Gaussian cluster are μ_i and Σ_i, respectively. Then we have

$$d_M[k] = \mathbf{Pr}\Big\{x[k] \mid x[k] \sim \mathcal{N}(\mu_i, \Sigma_i)\Big\}, \quad M = 1, \ldots, N,$$

$$w_M[k] = \frac{d_M[k]}{\sum_{M=1}^{N} d_M[k]}, \quad \sum_{M=1}^{N} w_M[k] = 1,$$

$$\hat{y}[k+i] = \sum_{M=1}^{N} w_M[k]\hat{y}_M[k+i], \quad i = 1, \ldots, P.$$

The ensemble averaged forecast $\hat{y}[k+i]$ is returned back to the AKPs as the next input, to provide the next-step forecast receding horizon fashion, up to the forecasting horizon.

Remark 4.2 Alternatively, the ensemble forecast of each AKP, $\hat{y}_M[k+i]$, could be fed back for multi-step forecasting, but feeding back $\hat{y}[k+i]$ to all AKPs is better, since it is a better prediction of system's true behavior.

4.3.5 Receding Horizon Anomaly Forecast

The forecasted features, $\hat{y}$, are compared to a decision boundary for anomaly forecasting in each node. At each sampling time, a P-step ahead forecast of the features is computed using the dynamic ensemble averaging method. In the next sampling time, the horizon moves forward (recedes) by one time step, and a new forecast is computed through the new forecasting horizon.

$$
\begin{aligned}
k : & \left[\hat{y}[k+1], \hat{y}[k+2], \ldots, \hat{\mathbf{y}}[\mathbf{k}+\mathbf{P}]\right], \\
k+1 : & \left[\hat{y}[k+2], \hat{y}[k+3], \ldots, \hat{\mathbf{y}}[\mathbf{k}+\mathbf{P}+1]\right], \\
k+2 : & \left[\hat{y}[k+3], \hat{y}[k+4], \ldots, \hat{\mathbf{y}}[\mathbf{k}+\mathbf{P}+2]\right], \\
& \ldots
\end{aligned}
$$

At each sampling time, the last forecast in the horizon $\hat{y}[k+P]$ is compared to the decision boundary. This is similar to the Model Predictive Control (MPC), except that in MPC, at each sampling time, the first control action in the horizon is applied to the system.

Each decision boundary is computed by training an Extreme Learning Machine (ELM) as a binary classifier in a supervised training framework. ELM is a special type of feed-forward neural networks recently introduced (Huang et al. 2012). ELM was originally developed for the single hidden layer feed-forward neural networks (SLFNs) and was later extended to the generalized SLFNs where the hidden layer need not be neuron alike (Huang et al. 2013). Unlike in traditional feed-forward neural networks where training the network involves finding all connection weights and bias, in ELM, connections between input and hidden neurons are randomly generated and fixed, that is, they do not need to be trained. Thus training an ELM becomes finding connections between hidden and output neurons only, which is simply a linear least squares problem whose solution can be directly generated by the generalized inverse of the hidden layer output matrix (Huang et al. 2012).

Because of such special design of the network, ELM training becomes very fast. The structure of a one-output ELM networks is depicted in Fig. 4.8. Assume the

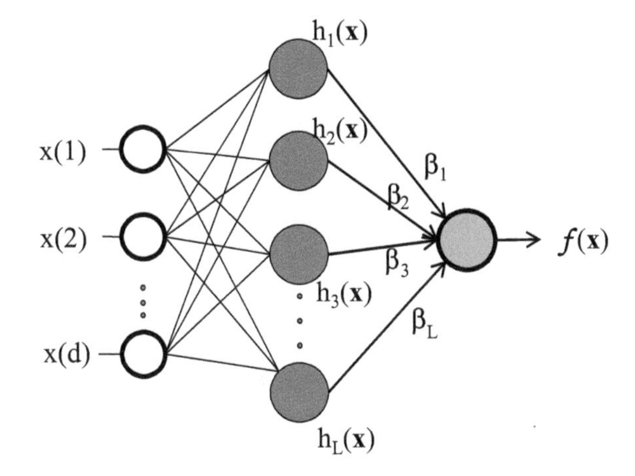

Fig. 4.8 An ELM network with one output. ©2018 IEEE. Reprinted, with permission, from (Abbaszadeh et al. 2018)

number of hidden neurons is L. Then the output function of ELM for generalized SLFNs is

$$f(x) = \sum_{j=1}^{L} \beta_j h_j(x) \triangleq \mathbf{h}(x)\boldsymbol{\beta}, \tag{4.12}$$

where $h_i(x) = G(\phi_i, b_i, x)$ is the output of jth hidden neuron with respect to the input x; $G(\phi, b, x)$ is a nonlinear piecewise continuous function satisfying ELM universal approximation capability theorems (Huang et al. 2006); β_j is the output weight vector between jth hidden neuron to the output node; and $h(x) = [h_1(x), \ldots, h_L(x)]$ is a random feature map, mapping the data from d-dimensional input space to the L-dimension random feature space (ELM feature space).

The objective function of ELM is an equality-constraint optimization problem, to minimize both the training errors and the output weights, which can be written as

$$\textit{Minimize: } \mathbf{L}_p = \frac{1}{2}\|\boldsymbol{\beta}\|^2 + \frac{1}{2}c\sum_{i=1}^{N_d} \xi_i^2 \tag{4.13}$$

$$\textit{s.t.: } \mathbf{h}(x_i)\boldsymbol{\beta} = l_i - \xi_i, \quad i = 1, \ldots, N_d, \tag{4.14}$$

where ξ_i is the training error with respect to the training sample x_i, l_i is the label of the ith sample, and N_d is the number of training samples (in the *normal* and *abnormal* datasets combined). The constant c controls the trade-off between the output weights and the training error.

Based on the Karush–Kuhn–Tucker (KKT) condition, we can have the analytic solutions for the ELM output function f for non-kernel and kernel cases, respectively (see Huang (2014) for details). Since kernel two-class ELM learns a nonlinear hyperplane, it generally works better than non-kernel two-class ELM. Therefore, we have used a kernel ELM using a Radial Basis Function (RBF) kernel.

The distance d of any point (a sample) to the hyperplane constructed by the ELM can conveniently serve as an anomaly score, that is, the larger the distance, the more likely the sample is abnormal. Here f is an anomaly score function whose sign (compared to a threshold, normally, zero) determines the binary classification decision on the system status. We have trained the ELM such that *normal* samples generate negative scores.

4.3.6 Committed Horizon Anomaly Forecast

An extension to receding horizon prediction is committed horizon prediction (Chen et al. 2019). It considers a so-called commitment level $V < P$, and instead of committing to only one estimate obtains the final predicted value at time k by combining (e.g., via a weighted average) the estimates of the V receding horizon instances from time $k + 1$ to $k + V$. Therefore, with a P-step look ahead, the effective prediction horizon is $P - V$. In other words, at each time instance, there is a delay of V sampling times to get the forecast of P steps ahead. Committed horizon prediction tends to give better estimates because it accounts for both future and past information, and also provides an additional mechanism to adjust the trade-off between delay and prediction accuracy (Chen et al. 2019). However, it reduces the effective prediction horizon, and hence its capability to generate rapid early warnings for anomaly detection applications. Nevertheless, the committed horizon prediction approach may still be effectively used for short-term forecasting, especially if the sampling rate is much faster than the system dynamics.

4.4 Predictive Situation Awareness

In general, predictive situation awareness has three main elements (Endsley 1995):

1. *Perception*: monitoring the environment.
2. *Comprehension*: understanding the current situation.
3. *Projection*: predicting the evolution of the situation.

Figure 4.9 depicts the block diagram of situation awareness modules in this work. Here, the *perception* element consists of collecting and pre-processing data from the monitoring nodes including feature extraction and any dimensionality reduction. *Comprehension* is provided by the anomaly detection module supplying the current system status, and *Projection* is performed through anomaly forecasting.

As data is processed in stream or batch modes, anomaly detection provides an instance decision on the current system status, which is either *normal* or *abnormal* (attack or fault). Before an anomaly happens, the current system status is normal

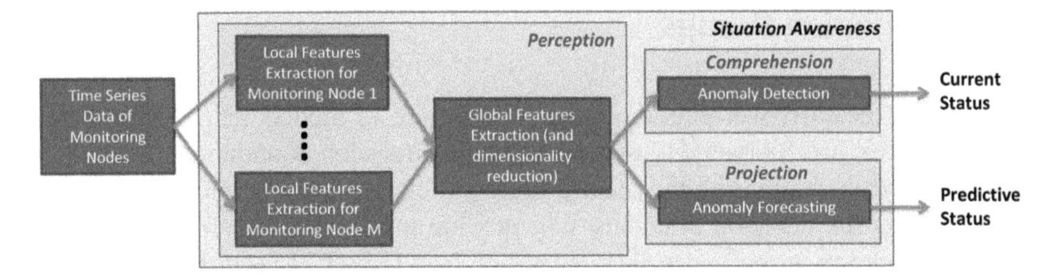

Fig. 4.9 Situation awareness block diagram

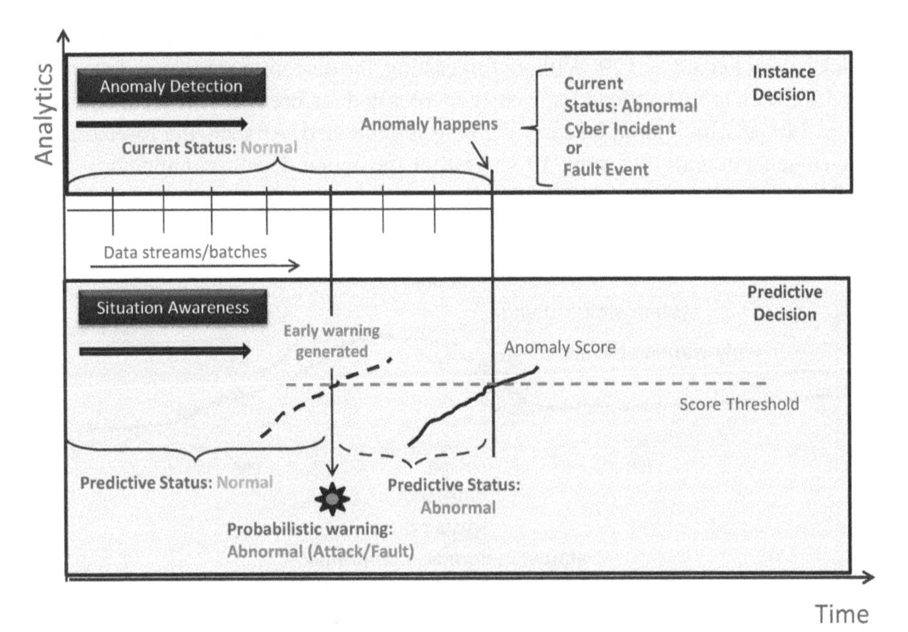

Fig. 4.10 Deterministic and probabilistic decisions for situation awareness

and it remains normal until an anomaly actually occurs. The anomaly detection algorithm detects an anomaly once it happens based on an anomaly score calculated at the current time instant passing a prescribed threshold (which could be fixed or adaptive itself). In addition, the situation awareness provides a predictive decision and generates early warnings. At each time instant, the forecasting algorithm projects the current status into future using stochastic dynamic forecasting described in the previous sections. The predictive status remains normal until the predicted anomaly score passes the threshold. Once an early warning is generated, future forecasting still continues, with a probabilistic decision on the predicted systems status based on anomaly score. The anomaly score increases between the time an early warning is generated and the time an anomaly actually happens, at which point the current status also reflects the anomaly. The concept is depicted in Fig. 4.10.

4.5 Simulation Results

To generate the early warning, the forecasted outputs of models (aka, future values of the features) are compared to the corresponding decision boundaries for anomaly detection. While comparing the feature vectors to the decision boundary, estimated time to cross the decision boundary will provide information for future anomaly. Figure 4.11 shows the early warning generation for a DWATT (gas turbine-generated power) sensor false data injection attack based on a short-term (10 s ahead) forecast. It is worth mentioning that this attack case was not included in the training dataset so this simulation represents an independent cross-validation of the algorithm. The attack is injected at $t = 129$. Without forecasting, the detection algorithm detects it at $t = 150$. With the 10-s ahead forecast, the forecasted features pass the local boundary at $t = 140$, at which point an early warning is generated. As seen, the forecasting is able to generate early warning 10 s ahead of the actual detection happening. With

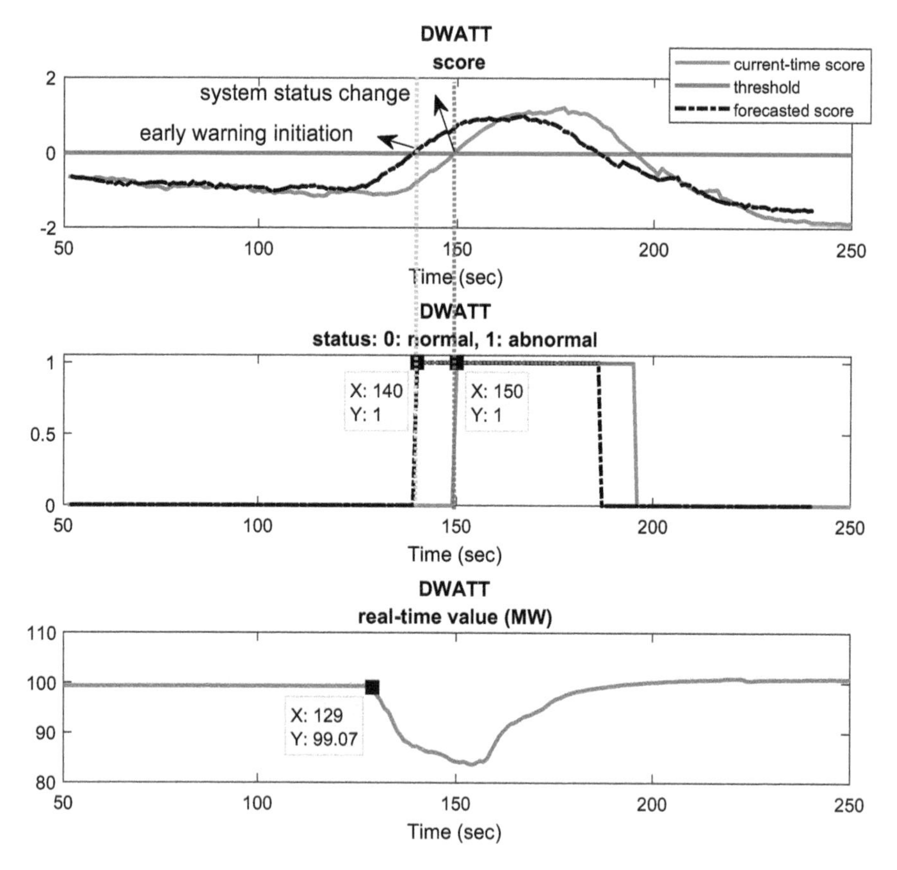

Fig. 4.11 Anomaly forecast and early warning generation for DWATT sensor. ©2018 IEEE. Reprinted, with permission, from (Abbaszadeh et al. 2018)

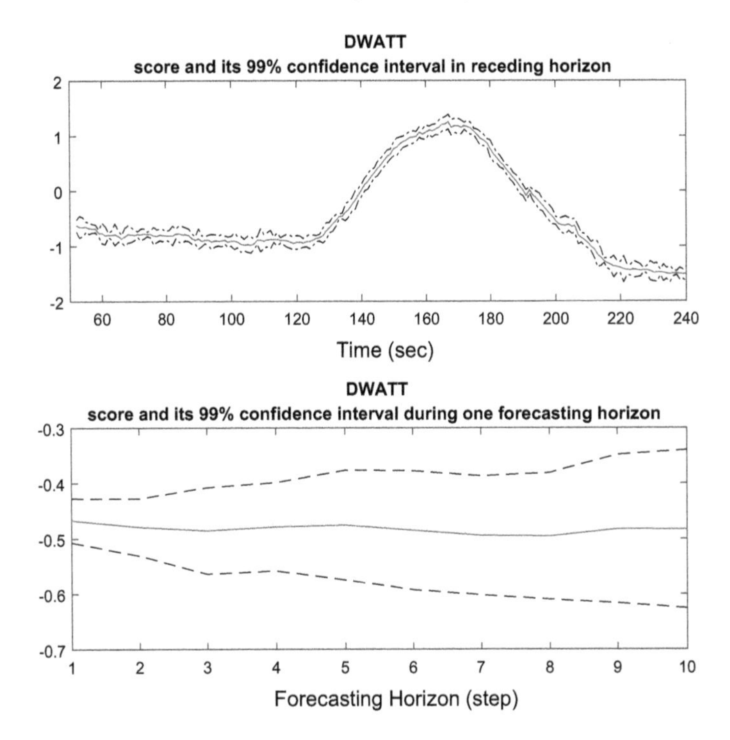

Fig. 4.12 Forecasted score for DWATT and confidence intervals for the whole simulation time and one forecasting horizon. ©2018 IEEE. Reprinted, with permission, from (Abbaszadeh et al. 2018

this technology, we are able to compensate for the delay in detection and generate early warning in the very early stage of an attack. Similarly, once the disturbance rejection control of the gas turbine brings the system back into the *normal* region, the forecasting algorithm is able to predict that before the actual system status goes back to *normal*. Note that here we are forecasting the features directly, and the anomaly score indirectly by passing the forecasted features through the decision boundary. Hence, the confidence intervals of ensemble feature forecasts are readily available from the AKPs, while those of the averaged forecasts and the anomaly score are computed using interval arithmetic (Bland and Altman 1996). The forecasted features are computed in a receding horizon with a forecasting horizon of 10 s (i.e., 10-step ahead forecasts are used for anomaly decision). In every sampling time, a 10-s forecast is computed along with its confidence interval. In the next sampling time, a new receding horizon forecast is computed, sliding the previous horizon by 1 s. Figure 4.12 shows the forecasted score for DWATT and confidence intervals for the whole simulation time and one forecasting horizon, respectively. The simulation is performed for 250 s (thus, 240 s of 10-step ahead receding horizon forecasts).

4.6 Conclusions

In this work, a framework for anomaly forecasting and early warning generation in industrial control systems is proposed based on a new feature-based dynamic ensemble forecasting method. The cyber-physical system anomalies addressed here could be either of cyber-incident or of naturally occurring faults/failures nature. The ensembles are selected via GMM clustering based on BIC criterion, each representing an operating point of the system. The cluster centroids are adjusted to the nearest physical points in the training dataset, and the associated covariance matrices are updated accordingly. Ensemble forecasts are provided by adaptive Kalman predictors applied to dynamic VAR models in the feature space, and fused through dynamic averaging, while the averaging weights are calculated using the Gaussian clusters mean and covariance matrices. The forecasts are multi-step and performed on different time scales in a receding horizon fashion. To predict the future status of the system, the forecasts are compared to decision boundaries computed using extreme learning machines. High-fidelity simulations on a GE gas turbine digital twin platform show the efficacy of our approach.

Acknowledgements This material is based on work supported by the US Department of Energy under Award Number DE-OE0000833.

Disclaimer This report was prepared as an account of work sponsored by an agency of the United States Government. Neither the United States Government nor any agency thereof, nor any of their employees, makes any warranty, express or implied, or assumes any legal liability or responsibility for the accuracy, completeness, or usefulness of any information, apparatus, product, or process disclosed, or represents that its use would not infringe privately owned rights. Reference herein to any specific commercial product, process, or service by trade name, trademark, manufacturer, or otherwise does not necessarily constitute or imply its endorsement, recommendation, or favoring by the United States Government or any agency thereof. The views and opinions of authors expressed herein do not necessarily state or reflect those of the United States Government or any agency thereof.

References

M. Abbaszadeh, H.J. Marquez, Nonlinear observer design for one-sided Lipschitz systems, in *Proceedings of the 2010 American Control Conference* (IEEE, 2010), pp. 5284–5289

M. Abbaszadeh, H.J. Marquez, Robust state observation for sampled-data nonlinear systems with exact and euler approximate models, in *American Control Conference* (IEEE, 2007), pp. 1687–1692

M. Abbaszadeh, L. K. Mestha, W. Yan, Forecasting and early warning for adversarial targeting in industrial control systems, in *2018 IEEE Conference on Decision and Control (CDC)* (IEEE, 2018), pp. 7200–7205

C. Allegorico, V. Mantini, A data-driven approach for on-line gas turbine combustion monitoring using classification models, in *Second European Conference of the Prognostics and Health Management Society* (2014), pp. 92–100

J.M. Bland, D.G. Altman, Transformations, means, and confidence intervals. BMJ: British Med. J. **312**(7038), 1079 (1996)
V. Chandola, A. Banerjee, V. Kumar, Anomaly detection: a survey. ACM Comput. Surv. (CSUR) **41**(3), 15 (2009)
H. Chen, N. Paoletti, S.A. Smolka, S. Lin, Committed moving horizon estimation for meal detection and estimation in type 1 diabetes, in *American Control Conference (ACC)* (IEEE, 2019), pp. 4765–4772
L. Clifton, D.A. Clifton, Y. Zhang, P. Watkinson, L. Tarassenko, H. Yin, Probabilistic novelty detection with support vector machines. IEEE Trans. Reliab. **63**(2), 455–467 (2014)
P. Cortez, M. Rio, M. Rocha, P. Sousa, Multi-scale internet traffic forecasting using neural networks and time series methods. Expert Syst. **29**(2), 143–155 (2012)
A.P. Dempster, N.M. Laird, D.B. Rubin, Maximum likelihood from incomplete data via the EM algorithm. J. R. Stat. Soc. Ser. B (methodological) 1–38 (1977)
W. Ding, J. Wang, C. Rizos, D. Kinlyside, Improving adaptive Kalman estimation in GPS/INS integration. J. Navig. **60**(3), 517–529 (2007)
J. Ehlers, A. van Hoorn, J. Waller, W. Hasselbring, Self-adaptive software system monitoring for performance anomaly localization, in *Proceedings of the 8th ACM International Conference on Autonomic Computing* (ACM, 2011), pp. 197–200
M.R. Endsley, Toward a theory of situation awareness in dynamic systems. Hum. Factors **37**(1), 32–64 (1995)
T. Gneiting, A.E. Raftery, Weather forecasting with ensemble methods. Science **310**(5746), 248–249 (2005)
S. Gupta, A. Ray, S. Sarkar, M. Yasar, Fault detection and isolation in aircraft gas turbine engines. Part 1: underlying concept. Proc. Inst. Mech. Eng. Part G: J. Aeros. Eng. **222**(3), 307–318 (2008)
D.I. Harvey, S.J. Leybourne, Testing for time series linearity. Econometr. J. **10**(1), 149–165 (2007)
W. Huang, N. Li, Z. Lin, G.-B. Huang, W. Zong, J. Zhou, Y. Duan, Liver tumor detection and segmentation using kernel-based extreme learning machine, in *35th annual international conference of the IEEE Engineering in medicine and biology society (EMBC)* (IEEE, 2013), pp. 3662–3665
G.-B. Huang, H. Zhou, X. Ding, R. Zhang, Extreme learning machine for regression and multiclass classification. IEEE Trans. Syst. Man Cybern. Part B (Cybernetics) **42**(2), 513–529 (2012)
G.-B. Huang, An insight into extreme learning machines: random neurons, random features and kernels. Cogn. Comput. **6**(3), 376–390 (2014)
G.-B. Huang, Q.-Y. Zhu, C.-K. Siew, Extreme learning machine: theory and applications. Neurocomputing **70**(1), 489–501 (2006)
S. Johansen, *Likelihood-Based Inference in Cointegrated Vector Autoregressive Models* (Oxford University Press, 1995)
S.K. Khaitan, J.D. McCalley, Design techniques and applications of cyberphysical systems: a survey. IEEE Syst. J. **9**(2), 350–365 (2014)
T. Kokkonen, J. Hautamäki, J. Siltanen, T. Hämäläinen, Model for sharing the information of cyber security situation awareness between organizations, in *2016 23rd International Conference on Telecommunications (ICT)* (IEEE, 2016), pp. 1–5
G. Koop, D. Korobilis, Forecasting inflation using dynamic model averaging. Int. Econ. Rev. **53**(3), 867–886 (2012)
R. Lamedica, A. Prudenzi, M. Sforna, M. Caciotta, V.O. Cencellli, A neural network based technique for short-term forecasting of anomalous load periods. IEEE Trans. Power Syst. **11**(4), 1749–1756 (1996)
T.H. McCormick, A.E. Raftery, D. Madigan, R.S. Burd, Dynamic logistic regression and dynamic model averaging for binary classification. Biometrics **68**(1), 23–30 (2012)
L.K. Mestha, O.M. Anubi, M. Abbaszadeh, Cyber-attack detection and accommodation algorithm for energy delivery systems, in *IEEE Conference on Control Technology and Applications (CCTA)* (2017), pp. 1326–1331

S. Nateghi, Y. Shtessel, J.-P. Barbot, C. Edwards, Cyber attack reconstruction of nonlinear systems via higher-order sliding-mode observer and sparse recovery algorithm, in *2018 IEEE Conference on Decision and Control (CDC)* (IEEE, 2018), pp. 5963–5968

S. Nateghi, Y. Shtessel, J.-P. Barbot, G. Zheng, L. Yu, Cyber-attack reconstruction via sliding mode differentiation and sparse recovery algorithm: Electrical power networks application, in *15th International Workshop on Variable Structure Systems (VSS)* (IEEE, 2018), pp. 285–290

G. Park, C. Lee, H. Shim, Y. Eun, K.H. Johansson, Stealthy adversaries against uncertain cyber-physical systems: threat of robust zero-dynamics attack. IEEE Trans. Autom. Control **64**(12), 4907–4919 (2019)

M.A. Pimentel, D.A. Clifton, L. Clifton, L. Tarassenko, A review of novelty detection. Signal Process. **99**, 215–249 (2014)

A.E. Raftery, M. Kárnỳ, P. Ettler, Online prediction under model uncertainty via dynamic model averaging: application to a cold rolling mill. Technometrics **52**(1), 52–66 (2010)

G. Rigatos, N. Zervos, P. Siano, P. Wira, M. Abbaszadeh, Flatness-based control for steam-turbine power generation units using a disturbance observer. IET Electr. Power Appl. (2021)

S.C. Rutan, Adaptive Kalman filtering. Anal. Chem. **63**(22), 1103A-1109A (1991)

F. Skopik, M. Wurzenberger, G. Settanni, R. Fiedler, Establishing national cyber situational awareness through incident information clustering, in *2015 International Conference on Cyber Situational Awareness, Data Analytics and Assessment (CyberSA)* (IEEE, 2015), pp. 1–8

S. Sridhar, M. Govindarasu, Model-based attack detection and mitigation for automatic generation control. IEEE Trans. Smart Grid **5**(2), 580–591 (2014)

Y.-Y. Wang, H. Zhang, C.-H. Qiu, S.-R. Xia, A novel feature selection method based on extreme learning machine and fractional-order darwinian PSO. Comput. Intell. Neurosci. **2018**, Article ID 5078268, 8 pages (2018)

F. Xue, W. Yan, Parametric model-based anomaly detection for locomotive subsystems, in *International Joint Conference on Neural Networks, IJCNN 2007* (IEEE, 2007), pp. 3074–3079

W. Yan, L. Yu, On accurate and reliable anomaly detection for gas turbine combustors: a deep learning approach, in *Proceedings of the Annual Conference of the Prognostics and Health Management Society* (2015)

W. Yan, L.K. Mestha, M. Abbaszadeh, Attack detection for securing cyber physical systems. IEEE Int. Things J. **6**(5), 8471–8481 (2019)

Y. Yin, Y. Zhao, B. Zhang, C. Li, S. Guo, Enhancing elm by markov boundary based feature selection. Neurocomputing **261**, 57–69 (2017). Advances in Extreme Learning Machines (ELM 2015)

A. Zaher, S. McArthur, D. Infield, Y. Patel, Online wind turbine fault detection through automated scada data analysis. Wind Energy **12**(6), 574–593 (2009)

A. Zimek, E. Schubert, H.-P. Kriegel, A survey on unsupervised outlier detection in high-dimensional numerical data. Stat. Anal. Data Mining: ASA Data Sci. J. **5**(5), 363–387 (2012)

Chapter 5
Resilient Observer Design for Cyber-Physical Systems with Data-Driven Measurement Pruning

Yu Zheng and Olugbenga Moses Anubi

5.1 Notation

The following notation and definitions are used throughout the whole chapter: $\mathbb{R}, \mathbb{R}^n, \mathbb{R}^{n\times m}$ denote the space of real numbers, real vectors of length n, and real matrices of n rows and m columns respectively. $\mathbb{R}_+$ denotes the space of positive real numbers. Normal-face lower-case letters (e.g., $x \in \mathbb{R}$) are used to represent real scalars, bold-face lower-case letters (e.g., $\mathbf{x} \in \mathbb{R}^n$) represent vectors, while normal-face upper-case letters (e.g., $X \in \mathbb{R}^{n\times m}$) represent matrices. $X^\top$ denotes the transpose of matrix X. $\mathbf{1}_n$ and I_n denote vector of ones and identity matrix of size n respectively. Let $\mathscr{T} \subseteq \{1, \ldots, n\}$, then for a matrix $X \in \mathbb{R}^{m\times n}$, $X_{\mathscr{T}} \in \mathbb{R}^{|\mathscr{T}|\times n}$ is the sub-matrix obtained by extracting the rows of X corresponding to the indices in $\mathscr{T}$. $\mathscr{T}^c$ denotes the complement of a set $\mathscr{T}$, and the universal set on which it is defined will be clear from the context. The support of a vector $\mathbf{x} \in \mathbb{R}^n$, a set of the indices of nonzero entries, is denoted by $\mathsf{supp}(\mathbf{x}) \triangleq \{i \subseteq \{1, \ldots, n\} \mid \mathbf{x}_i \neq 0\}$. If $|\mathsf{supp}(\mathbf{x})| = k$, we say $\mathbf{x}$ is a k-sparse vector. Moreover, $\Sigma_k \subset \mathbb{R}^n$ denotes the set of all k-sparse vectors in $\mathbb{R}^n$. The operator $\mathsf{argsort} \downarrow (\mathbf{x})$ denotes a function that returns the sorted indices of vector $\mathbf{x}$ in descending order of the magnitude of $\mathbf{x}_i$. The symbol & denotes logical "AND" operator. The symbol $*$ denotes the convolution operator for vectors. The symbol $\odot$ denotes element-wise multiplication of two vectors, $\mathbf{z} = \mathbf{x} \odot \mathbf{y} \Rightarrow \mathbf{z}_i = \mathbf{x}_i \mathbf{y}_i$. The expression $x \sim \mathscr{B}(1, p)$ means that random variable x follows the Bernoulli distribution with $\mathsf{Pr}\{x = 1\} = p$. The weighted 1-norm of a vector $\mathbf{z} \in \mathbb{R}^n$ with the weight vector $\mathbf{w} \in \mathbb{R}^n$ is given by $\|\mathbf{z}\|_{1,\mathbf{w}} \triangleq \sum_{i=1}^n \mathbf{w}_i \mathbf{z}_i$.

Y. Zheng · O. M. Anubi (✉)
FAMU-FSU College of Engineering, Tallahassee, FL 32310, USA
e-mail: oanubi@fsu.edu

Y. Zheng
e-mail: yz19b@fsu.edu

M. Abbaszadeh and A. Zemouche (eds.), *Security and Resilience in Cyber-Physical Systems*, https://doi.org/10.1007/978-3-030-97166-3_5

5.2 Introduction

As the backbone of future critical infrastructures, Cyber-Physical Systems (CPS) are complicated integration of computation, communication, and physical components. Security, within the context of CPSs, poses more challenges compared to both traditional information technology (IT) security and operational technology (OT) security due to the temporal dynamics brought by physical environment and the heterogeneous nature of operation of CPSs (Khaitan and McCalley 2014). In the context of CPS, failures induced by malicious attacks are beyond random failures studied in reliability engineering or well-defined uncertainty classes in robust control. Moreover, the coupling of computation and communication with distributed sensing and actuation components increases the vulnerability to attacks (Zetter 2015; Lee et al. 2014; Slay and Miller 2007; Chen and Abu-Nimeh 2011).

The control design for CPSs usually consists of an observer to estimate the states of the physical system and a controller to compute the control commands based on the state estimation. Thus, the control system receives diverse information from measurement substations and distributes the computed control commands to a number of actuators through a communication network (Burg et al. 2017). Thus, an elaborate attack on a CPS can be designed by considering the networked closed-loop interaction between the cyber and physical agents. Furthermore, the dispersed geographical distribution and abundance of unmanned facilities also provide malicious attackers the opportunity to construct coordinated attacks. These attacks, studied extensively in literature, either targets the system integrity (Bishop 2003), such as stealth attacks (Sui et at. 2020), replay attacks (Fang et al. 2020), covert attacks (de Sá et al. 2017), and FDIA (Zheng and Anubi 2020) or the availability (Bishop 2003), such as denial of service (DoS) (Pelechrinis et al. 2010). The locations of those attacks are shown in Fig. 5.1. It was shown in Liu et al. (2011), Guo et al. (2016), Mo and Sinopoli (2010), that if FDIA is defined properly, it can exploit certain underlying vulnerabilities of bad data detection (BDD) schemes in order to force an erroneous state estimation using sparse measurement corruption. Consequently, in this chapter, we consider the resiliency of a class of observers against FDIA. If the observer estimates, using compromised measurements, are close to the true states, then control performance can be guaranteed with any control design which is robust to estimation error.

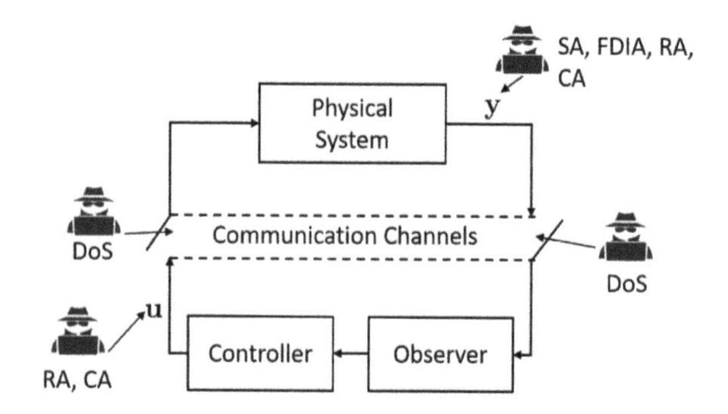

Fig. 5.1 Locations of attacks in CPS in the context of security control (SA: stealth attack, CA: covert attack, RA: replay attack, FDIA: false data injection attack, DoS: denial of service)

One of the pioneering works on resilient observer was presented in Fawzi et al. (2014), where an unconstrained ℓ_1 observer was proposed to achieve exact state recovery. A necessary and sufficient condition for exact recovery is that less than half of the system's measurements be compromised. The authors in Shoukry and Tabuada (2015) proved this condition from an interesting aspect of s-sparse observability and proposed an event-triggered Luenberger observer against FDIA. In Chong et al. (2015), the authors presented a more systematical work on the observability of the linear system under attacks and proposed a Gramian-based estimator. The authors in Pajic et al. (2015) and the authors in Lee et al. (2015) both considered resilient estimation in the presence of noise and attacks at the same time and constructed ℓ_1-ℓ_2 observers. The authors in Nakahira and Mo (2018) considered a robust estimation scheme against FDIA, in which local robust estimators and global fusion are combined to achieve resilient-robust estimation. Readers can also refer to Shoukry et al. (2017) for feasible resilient estimation methods by Satisfiability Modulo Theory (SMT) solvers. However, all the above observers would not achieve successful resilient estimation when 50%, or more, of system measurements are attacked. Equivalently, the system is not $2k$-detectable, where k is the number of attacks. This is a significant limitation since it requires that there be twice as many as needed measurement stations installed for a CPS, and the system has to be observable for every combination of 50% of the total sensors. This is a property that is currently not achieved by most critical cyber-physical critical infrastructures like the power grid.

In order to increase the corresponding percentage of attacked nodes for which state recoveries can be guaranteed, researchers have begun to incorporate prior information into the underlying resilient observer design framework. There are mainly three kinds of prior information considered in literature: state prior (Shinohara et al. 2019), measurement prior (Anubi and Konstantinou 2019; Anubi et al. 2020), and support prior (Anubi et al. 2018; Zheng and Anubi 2020). In Shinohara et al. (2019), three types of state prior were discussed: sparsity information of the estimated states, $(\alpha, \overline{n}_0)$ sparsity information, where the estimated states are assumed to have α instead of 0 in the sparsity form, and side information, which is the knowledge of the initial states from the physical attribution of the system and cannot be manipulated by malicious agents. Although the resiliency of the observer can be improved with such knowledge of the states of the system, it is very difficult to obtain such information in practice. This will require a prior determination of the state distribution for all operating conditions of an uncertain, large-scale nonlinear, and sometimes hybrid system.

Support prior is the estimated information of attack locations, which can be given by some data-driven localization algorithm or learning-based anomaly detection methods, such as watermark-based methods (Liu et al. 2020), moving-target based approach (Weerakkody and Sinopoli 2015), distributed support vector machine (Esmalifalak et al. 2014), deep learning neural network (He et al. 2017), and many more (Ozay et al. 2015; Abbaszadeh et al. 2019, Deldjoo et al. 2021, Huang et al. 2014). Although the localization algorithms can be readily defined and are very useful for monitoring purposes, using this kind of support prior to resilient estimation has two main drawbacks; imprecise classification and high training price. This limits their applicability for piratical purposes. In this chapter, we examine a class of pruning methods to generate a feasible pruned support prior with predetermined precision guarantees. Coupling the pruning algorithm with any localization algorithm can significantly improve the resulting precision, which directly improves the resiliency of the underlying resilient estimation process. This means a less precise localization algorithm can be tolerated, thus slashing the required training price. The initial pruning idea was introduced in Anubi et al. (2018), analyzed, and improved in Zheng and Anubi (2021). In this chapter, a more detailed mathematical foundation is given, in addition to improved implementation.

Measurement prior is a collection of additional auxiliary information about system measurements that is unknown to the malicious attackers. A direct use of measurement prior in resilient observer design was shown in Anubi et al. (2020), Anubi and Konstantinou (2019), Anubi et al. (2019) to improve the limit of the percentage of compromised measurement for which exact recovery is guaranteed from 50% to 80%. Also, the watermark-based detection approaches (Liu et al. 2020) and moving-average detection approaches (Weerakkody and Sinopoli 2015) both use the additional information in an authentication layer in order to detect the attacks. Thus, against measurement attacks, the measurement prior and support prior are related. An advantage of measurement prior is its expansibility to the authentication layer. The more measurement priors that can be constructed usually provide better detection precision. In this chapter, we will utilize a measurement prior constructed by using a data-driven auxiliary model between auxiliary variables and the system measurements. The attacked measurements will then be detected if they cannot be explained by both the system dynamics and the measurement model prior with high likelihood, thus reducing the resulting attack surface.

The remainder of this chapter is organized as follows: In Sect. 5.3, concurrent models of CPS, including physical model, monitoring model, thread model, prior model, and pruning algorithm are given; in Sect. 5.4, the resilient observer design with data-driven measurement pruning is given; in Sect. 5.5, numerical simulation and application examples are given to demonstrate the performance of the designed observer compared to other resilient observers in the severe adversarial environment; concluding remarks follow in Sect. 5.6.

5.3 Concurrent Models

To discuss the resilient observer design, relevant model developments are discussed in this section. Since CPS is a seamless integration of computational components, physical processes, and communication network systems, a single-layer model cannot sufficiently describe the complex characteristics of CPS. Also, as a closed-loop system, separately and independently modeling the separate layers cannot capture the tight interaction between the cyber and physical layers (Lee 2010). Concurrent modeling has been used as a good way to describe the complex operation on CPS (Derler et al. 2011), where different models in different hierarchies work concurrently. As shown in Fig. 5.2, this small version of CPS has four concurrent loops; the physical dynamical loop, prior generation loop, monitoring path, and attack injection.

The rest of the subsections are dedicated to discussing, in more detail, the modeling aspects for each layer, the underlying assumptions, and connections with the subsequent resilient observer design.

5.3.1 Physical Model and Monitor

A linear time invariant (LTI) model is considered to describe the physical behavior of the CPS in Fig. 5.2.

$$\begin{aligned} \mathbf{x}_{i+1} &= A\mathbf{x}_i \\ \mathbf{y}_i &= C\mathbf{x}_i + \mathbf{e}_i , \end{aligned} \tag{5.1}$$

where $\mathbf{x}_i \in \mathbb{R}^n$ is the internal state vector of physical model at time i which is unknown to other parts of the concurrent model, $\mathbf{y}_i \in \mathbb{R}^m$ is the measurement vector, $\mathbf{e}_i \in \mathbb{R}^m$ is the time-varying attack-noise vector. The measurement attacks and noise terms are modeled as additional error signals. Control inputs may be included in the model above. However, since control inputs are generally irrelevant to state estimation problems, we suppress it in the model considered here.

The following assumptions are used in subsequent developments:

1. The pair (A, C) is observable.
2. The measurements are redundant $(m > n)$.
3. The attack signal is possibly unbounded and sparse, $\mathbf{e}_i \in \Sigma_k$ for some $k < m$.
4. The attack-free part of $\mathbf{e}_i$ is bounded, $\sum_{i \in \mathcal{T}^c} |\mathbf{e}_i| < \varepsilon$, for some $\varepsilon > 0$.

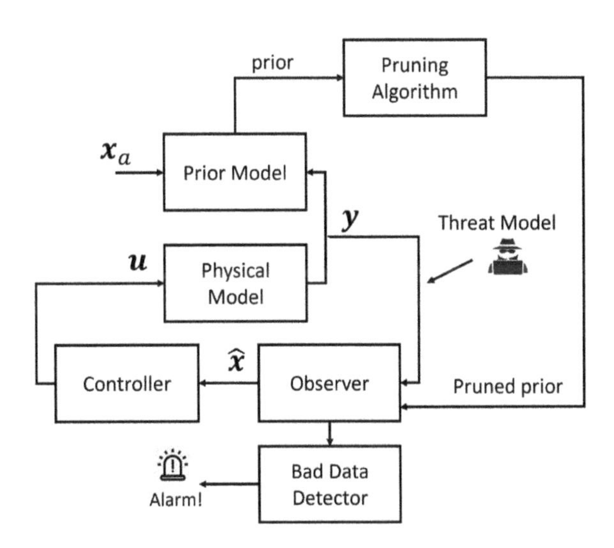

Fig. 5.2 Concurrent model on CPS ($\mathbf{x}_a$ is an auxiliary state used in the prior model)

By iterating the system model (5.1) T time steps backwards, the T-horizon observation model is given by

$$\mathbf{y}_T = H\mathbf{x}_{i-T+1} + \mathbf{e}_T, \tag{5.2}$$

where $\mathbf{y}_T = [\mathbf{y}_i^\top, \ \mathbf{y}_{i-1}^\top, \ldots, \mathbf{y}_{i-T+1}^\top]^\top \in \mathbb{R}^{Tm}$ is a sequence of observation in the moving window $[i - T + 1, \ \ i]$, $\mathbf{x}_{i-T+1} \in \mathbb{R}^n$ is the state vector at time $i - T + 1$, $\mathbf{e}_T = [\mathbf{e}_i^\top, \ \mathbf{e}_{i-1}^\top, \ldots, \mathbf{e}_{i-T+1}^\top]^\top$ is the sequence of attack-noise vectors in the same moving window, $H \in \mathbb{R}^{Tm\times n}$ is the observation matrix, $H = \begin{bmatrix} CA^{T-1} \\ \vdots \\ CA \\ C \end{bmatrix}$.

The following definitions formalize the notions of a decoder and a detector, which are used subsequently.

Definition 5.1 (*Decoder*) Given an observable pair (A, C) and a horizon parameter T, a decoder $\mathscr{D} : \mathbb{R}^{Tm} \mapsto \mathbb{R}^n$ is an operator given by

$$\hat{\mathbf{x}} = \mathscr{D}(\mathbf{y}_T \mid H) = \underset{\mathbf{x}\in\mathbb{R}^n}{\text{argmin}} \ \|\mathbf{y}_T - H\mathbf{x}\|_1, \tag{5.3}$$

where $\mathbf{y}_T = \{\mathbf{y}_i, \mathbf{y}_{i-1}, \ldots, \mathbf{y}_{i-T+1}\}$ is a moving-windowed measurement vector history and $\hat{\mathbf{x}} \in \mathbb{R}^n$ is the resulting estimated initial state vector $\mathbf{x}_{i-T+1}$. When the parameter is clear from context, they are dropped from the argument list for clarity.

Definition 5.2 (*Detector*) Given the measurements $\mathbf{y}_T \in \mathbb{R}^{Tm}$ taken in the moving window $[i - T + 1, \ \ i]$, a detector based on the ℓ_1 decoder is mapping of the form:

$$\Psi_T : \{\mathbf{y}_T\} \mapsto \{\Psi_1, \Psi_2\},$$

where $\Psi_1 \in \{0, 1\}$[1] is the first output argument indicating whether or not the measurement $\mathbf{y}_T$ is attacked, $\Psi_2 \in 2^{\{1,2,\cdots,m\}}$ is the second output argument indicating the support of attack locations.

The decoder–detector pair constitutes a monitor scheme for the system (5.1), as shown in the remark below.

Remark 5.1 (*Residual-based monitor mechanism*) Given a threshold value $\varepsilon_0 > 0$, the monitor returns $\Psi_1 = \{0\}$ in the first output argument for a given measurement vector history $\mathbf{y}_T = \{\mathbf{y}_i, \mathbf{y}_{i-1}, \ldots, \mathbf{y}_{i-T+1}\}$ if there exists a corresponding state trajectory $\hat{X}_T = \{\hat{\mathbf{x}}_i, \hat{\mathbf{x}}_{i-1}, \ldots, \hat{\mathbf{x}}_{i-T}\}$ such that

$$\begin{aligned} \|\hat{\mathbf{x}}_{j+1} - A\hat{\mathbf{x}}_j\| &\le \varepsilon_0, \quad j = i - T, \ldots, i - 1 \\ \|\mathbf{y}_j - C\hat{\mathbf{x}}_j\| &\le \varepsilon_0, \quad j = i - T + 1, \ldots, i. \end{aligned}$$

[1] 0: safe, 1: unsafe.

Otherwise, the monitor returns $\Psi_1 = \{1\}$ in the first output argument and also the support of the sparsest attack trajectory $\mathbf{e}_T = \{\mathbf{e}_i, \mathbf{e}_{i-1}, \ldots, \mathbf{e}_{i-T+1}\}$ such that

$$\begin{aligned} \|\hat{\mathbf{x}}_{j+1} - A\hat{\mathbf{x}}_j\| &\leq \varepsilon_0, \quad j = i-T, \ldots, i-1 \\ \|\mathbf{y}_j - C\hat{\mathbf{x}}_j - \mathbf{e}_j\| &\leq \varepsilon_0, \quad j = i-T+1, \ldots, i. \end{aligned}$$

5.3.2 *Threat Model*

Following the setup above, we give a formal definition of successful FDIA and prescribe conditions under which an FDIA will successfully corrupt a decoder while evading detection by the residual-based monitor. To design a successful FDIA, the following assumptions are made, which are widely used in literature Mo and Sinopoli (2010), Mo and Sinopoli (2015):

1. The attacker has perfect knowledge of the system dynamics in (5.1).
2. The attacker can inject arbitrary bias at the compromised nodes $\mathscr{T} \subset \{1, \cdots, m\}$.
3. The number of nodes the attacker can simultaneously compromise at any given time is bounded. In other words, the attackers have limited resources.

Notice, the known information of system to attacker includes the system dynamics, for example, the H matrix, and the decoder–detector scheme. All other information, such as the true state variables, are unknown to the attacker.

Definition 5.3 (*Successful FDIA* Mo and Sinopoli (2010)) Consider the CPS in (5.1) and the corresponding measurement model (5.2). Given a positive integer $k < m$, the attack sequence $\mathbf{e}_T \in \Sigma_{Tk}$ is said to be (ε, α)-successful against the decoder-detection pair described above if

$$\|\mathbf{x}^\star - \mathscr{D}(\mathbf{y}_T)\|_2 \geq \alpha, \quad \text{and } \|\mathbf{y}_T - H\mathscr{D}(\mathbf{y}_T)\|_2 \leq \varepsilon, \tag{5.4}$$

where $\mathbf{y}_T = \mathbf{y}_T^\star + \mathbf{e}_T$ with $\mathbf{y}_T^* \in \mathbb{R}^{Tm}$ being the true measurement vector, and $\mathbf{x}^\star$ is the true state vector.

In the above definition, k quantifies the attack sparsity level per time. Specifically, it is the maximum number of attacks at each time index. Given the support sequence $\mathscr{T} = \{\mathscr{T}_i \;\; \mathscr{T}_{i-1} \cdots \mathscr{T}_{i-T+1}\}$ with $|\mathscr{T}_i| \leq k$. Let $\mathbf{x}_e$ be an optimal solution of the optimization program

$$\begin{aligned} \text{Maxmize}: \quad & \|H_{\mathscr{T}}\mathbf{x}\|_1, \\ \text{Subject to}: \quad & \|H_{\mathscr{T}^c}\mathbf{x}\|_1 \leq \varepsilon. \end{aligned} \tag{5.5}$$

Then a FDIA can be defined as

$$\mathbf{e}_{\mathscr{T}} = H_{\mathscr{T}}\mathbf{x}_e, \quad \mathbf{e}_{\mathscr{T}^c} = \mathbf{0}. \tag{5.6}$$

The following theorem shows the condition under which the defined FDIA above is (ε, α)-successful for the given attack support $\mathscr{T}$.

Proposition 5.1 (Zheng and Anubi (2021)) *Suppose there exists a vector* $\mathbf{w} \in \textsf{range}(H)$ *such that*

$$\|\mathbf{w}_{\mathscr{T}}\|_1 > \|\mathbf{w}_{\mathscr{T}^c}\|_1, \tag{5.7}$$

then the FDIA in (5.6) is (ε, α)*-successful against the decoder-detector pair in Definition 5.1 and Remark 5.1 for all* $\alpha \leq \frac{\sigma_1 - 1}{\sqrt{|\mathscr{T}|}\overline{\sigma}_{\mathscr{T}} - \underline{\sigma}_{\mathscr{T}^c}}\varepsilon$*, with* $|\mathscr{T}| > \frac{\underline{\sigma}^2_{\mathscr{T}^c}}{\overline{\sigma}^2_{\mathscr{T}}}$*, where* $\overline{\sigma}_{\mathscr{T}}$ *and* $\underline{\sigma}_{\mathscr{T}^c}$ *are the largest and smallest non-zero singular values of* $H_{\mathscr{T}}$ *and* $H_{\mathscr{T}^c}$ *respectively, and*

$$\sigma_1 = \max_{\mathbf{v} \in \mathbb{R}^n \setminus \{\mathbf{0}\}} \frac{\|H_{\mathscr{T}}\mathbf{v}\|_1}{\|H_{\mathscr{T}^c}\mathbf{v}\|_1}.$$

Remark 5.2 If, in addition, $\textsf{null}(H_{\mathscr{T}^c}) \backslash \textsf{null}(H_{\mathscr{T}}) \neq \emptyset$, let $\mathbf{v}_n \in \textsf{null}(H_{\mathscr{T}^c}) \backslash \textsf{null}(H_{\mathscr{T}})$, then $\|H_{\mathscr{T}^c}\mathbf{v}_n\|_1 = 0$ but $\|H_{\mathscr{T}}\mathbf{v}_n\|_1 > 0$. Thus, $\sigma_1 \geq \frac{\|H_{\mathscr{T}}\mathbf{v}\|_1}{\|H_{\mathscr{T}^c}\mathbf{v}\|_1}$ is infinite, which implies that the FDIA in (5.6) is (ε, α)-successful for all $\varepsilon, \alpha \in \mathbb{R}_+$.

5.3.3 Data-Driven Auxiliary Measurement Prior

In this subsection, we present a data-driven auxiliary measurement prior based on a generative probabilistic regression model constructed using the Gaussian process (GP). This prior model is a mapping from the chosen auxiliary variables to the observed measurements, which plays the role of an additional authentication layer.

Given a dataset Z, Y, where $Z \in \mathbb{R}^{p \times N}$ is the matrix collecting the auxiliary states columnwise, $Y \in \mathbb{R}^{m \times N}$ is the matrix of the corresponding observed measurements, the goal is to learn the underlying function $f : \mathbb{R}^p \to \mathbb{R}^m$ such that

$$\mathbf{y}_i = f(\mathbf{z}_i) + \varepsilon, \quad i = 1, \cdots, N, \tag{5.8}$$

where $\varepsilon \sim \mathscr{N}(0, \sigma^2)$. To achieve this goal, certain restrictions have to be made on the properties of the underlying function. Otherwise, all potential functions fitting the training dataset would be equally valid. As a means of regularization, we assume that the underlying function f is restricted to a class defined by a given Gaussian process. A Gaussian process (GP) is a generalization of Gaussian probabilistic distribution Rasmussen (2003). It is a collection of random variables, every finite subset of which are jointly Gaussian (Urtasun and Darrell 2008). Gaussian process regression (GPR) uses GPs to encode prior distribution over functions f. Thus, suppose $f \in \mathbf{GP}$, then it satisfies the following distribution point-wise:

$$f(\mathbf{z}) \sim \mathscr{N}(m(\mathbf{z}), k(\mathbf{z}, \mathbf{z}')), \tag{5.9}$$

where $m(\mathbf{z}) = \mathbb{E}[f(\mathbf{z})]$ is the mean function and $k(\mathbf{z}, \mathbf{z}') = \mathbb{E}[(f(\mathbf{z}) - m(\mathbf{z}))(f(\mathbf{z}') - m(\mathbf{z}'))^\top]$ is the covariance function encoded, apriori, by the kernel function k. The model of GP contains two parts: a joint distribution model and a kernel function. Kernel functions capture the similarity between the function's (or model's) outputs, for given inputs. The design of the kernel function depends on the prior knowledge of the process that generated the data in question. For example, suppose we know that the output of the process changes slowly with respect to change in input, the smoothness of prior knowledge can be modeled in the kernel function used by the GP. One of the commonly used kernel functions is the square exponential covariance function (also called RBF), given by Liu et al. (2018)

$$k\left(\mathbf{z}, \mathbf{z}'\right) = A \exp\left\{-\frac{\|\mathbf{z} - \mathbf{z}'\|^2}{2l}\right\}, \tag{5.10}$$

where the hyperparameters A and l are amplitude coefficients and describe a single scaling factor on the influence of nearby observations, respectively. For a comprehensive summary of kernel functions, the readers are directed to Liu et al. (2018).

Given a query point $\mathbf{z}_\star \in \mathbb{R}^p$ for the auxiliary measurement, by applying Bayes's rule, the posterior distribution for j-th observed measurement $y_j = f_j(\mathbf{z})$ is given by

$$p(y_j \mid \mathbf{z}, \mathscr{D}) = \mathscr{N}(\mu_j(\mathbf{z}), \Sigma_j(\mathbf{z})), \tag{5.11}$$

where

$$\begin{aligned} \mu_j(\mathbf{z}) &= \mathbf{k}(\mathbf{z})^\top (K + \sigma_j^2 I)^{-1} Y_j^\top \\ \Sigma_j(\mathbf{z}) &= k(\mathbf{z}_\star, \mathbf{z}_\star) - \mathbf{k}(\mathbf{z})^\top (K + \sigma_j^2 I)^{-1} \mathbf{k}(\mathbf{z}), \quad j = 1, 2, \cdots, m, \end{aligned} \tag{5.12}$$

and $K = \begin{bmatrix} k(\mathbf{z}_1, \mathbf{z}_1) & \cdots & k(\mathbf{z}_1, \mathbf{z}_N) \\ \vdots & \ddots & \vdots \\ k(\mathbf{z}_N, \mathbf{z}_1) & \cdots & k(\mathbf{z}_N, \mathbf{z}_N) \end{bmatrix} \in \mathbb{R}^{N \times N}$, $\mathbf{k}(\mathbf{z}) = \begin{bmatrix} k(\mathbf{z}_1, \mathbf{z}_\star) \\ \vdots \\ k(\mathbf{z}_N, \mathbf{z}_\star) \end{bmatrix} \in \mathbb{R}^N$ are covariance matrix on training dataset, and covariance vector between training auxiliary states $\mathbf{z}_i, i = 1, 2, \cdots, N$ and the query point $\mathbf{z}_\star$ respectively.

The overall observed measurements' posterior distribution is then given by

$$p(\mathbf{y} \mid \mathbf{z}, \mathscr{D}) = \prod_{j=1}^{m} \mathscr{N}\left(\mu_j(\mathbf{z}), \Sigma_j(\mathbf{z})\right) = \mathscr{N}(\mu(\mathbf{z}), \Sigma(\mathbf{z})), \tag{5.13}$$

where $\mu(\mathbf{z}) = \begin{bmatrix} \mu_1(\mathbf{z}) \\ \vdots \\ \mu_m(\mathbf{z}) \end{bmatrix}$, $\Sigma(\mathbf{z}) = \begin{bmatrix} \Sigma_1(\mathbf{z}) & & \\ & \ddots & \\ & & \Sigma_m(\mathbf{z}) \end{bmatrix}$. Next, the localization algorithm based on the trained GPRs in (5.12), (5.13) is given in Algorithm 5.1. Based on the localization algorithm, if a measurement cannot be explained by the trained prior model, it will be recognized as being attacked. In other words, the prior model provides an additional layer of security by (1) requiring the attacker to have knowledge of the auxiliary model and the parameters and (2) limiting the magnitude of possible state corruption.

Algorithm 5.1 Localization Algorithm with Measurement Prior

I. **Inputs**: $\mathbf{y} \in \mathbb{R}^m$ (real measurement), $\mathbf{z} \in \mathbb{R}^p$ (auxiliary variables)
II. **Parameters**: m trained GPR models **GP**
III. **Posterior distribution**:

$$\mathbf{GP}_j(\mathbf{z}) \rightarrow \{\mu_j, \Sigma_j\} \;\; \forall j = 1, 2, \cdots, m$$

IV. **Calculate Z-score**:

$$\mathbf{z}_j = \frac{\mathbf{y}_j - \mu_j}{\Sigma_j}$$

V. **Calculate probability**:

$$\mathbf{p}_{c_j} = 1 - P_X(|x| \leq |\mathbf{z}_j|) = 1 - \int_{|\mathbf{z}_j|} \frac{e^{-\frac{x^2}{2}}}{\sqrt{2\pi}} \;\; \forall j = 1, 2, \cdots, m$$

VI. **Attack support prior**:

$$\mathscr{T} = \mathbf{0}_m; \;\; \mathscr{T}_j = 1 \;\text{ if }\; \mathbf{p}_{c_j} \leq 0.5 \;\; \forall j = 1, 2, \cdots, m$$

VII. **Outputs**: $\mathscr{T} \in \mathbb{R}^m$, (support prior), $\mathbf{p}_c \in \mathbb{R}^m$ (confidence)

5.3.4 Prior Pruning

As shown in the previous subsection, estimated support prior $\hat{\mathscr{T}}$ can be generated by some machine learning localization algorithms. However, there are major limitations preventing their direct usage as the prior information in resilient observer design. One is the huge amount of training often needed for high enough precision will prevent such prior from being deployed for a dynamic observer, where the real-time update is paramount. Another limitation is that the precision of data-driven results cannot be guaranteed due to their inherent uncertainties. Consequently, several fundamental questions emerge, which require significant research effort to address. For example, what is the quantitative relation between the resulting resilient estimation error bound and the auxiliary model uncertainty? In this subsection, a relationship is derived, or such connection, and a prior pruning method is considered to mend some deficiencies in order to improve the degradation due to the uncertainty of the prior model in the final estimation error bound.

Let $\mathscr{T} = \mathsf{supp}(\mathbf{e})$ be the unknown actual support of attacked channels. Let the vector $\mathbf{q} \in \{0 \;\; 1\}^{Tm}$ be an indicator of $\mathscr{T}$ defined element-wise as:

$$\mathbf{q}_i = \begin{cases} 0 & \text{if } i \in \mathcal{T} \\ 1 & \text{otherwise.} \end{cases} \tag{5.14}$$

Thus, the output of the localization algorithm $\hat{\mathcal{T}} \subseteq \{1, 2, \cdots, Tm\}$ is actually an estimate of $\mathcal{T}$, and its corresponding indicator $\hat{\mathbf{q}} \in \{0 \;\; 1\}^{Tm}$ is defined similarly to (5.14). Consequently, the precision of the support prior is evaluated using positive prediction value (Fawcett 2006) instead of true positive rate, F1 score, or other evaluation metrics. This is because the only factor affecting the resilient estimation performance is the error in the estimated prior support of safe nodes $\hat{\mathcal{T}}^c$, which is directly used in the observer.

Definition 5.4 (*Positive Prediction Value, Precision,* PPV (Fawcett (2006))) Given an estimate $\hat{\mathbf{q}} \in \{0, 1\}^{Tm}$ of an unknown attack support indicator $\mathbf{q} \in \{0, 1\}^{Tm}$, PPV is the proportion of $\mathbf{q}$ that is correctly identified in $\hat{\mathbf{q}}$. It is given by

$$\mathsf{PPV} = \frac{\|\mathbf{q} \odot \hat{\mathbf{q}}\|_{\ell_0}}{\|\hat{\mathbf{q}}\|_{\ell_0}}. \tag{5.15}$$

As will be shown in subsequent sections, the precision PPV is positively correlated to the performance of resilient estimation.

The agreement between $\hat{\mathcal{T}}$ and $\mathcal{T}$ can be described using a Bernoulli uncertainty model since $\hat{\mathcal{T}}$ can be seen as an output of binary classifier. Thus, the following uncertainty model is considered:

$$\mathbf{q}_i = \varepsilon_i \hat{\mathbf{q}}_i + (1 - \varepsilon_i)(1 - \hat{\mathbf{q}}_i), \tag{5.16}$$

where $\varepsilon_i \sim \mathcal{B}(1, \mathbf{p}_i)$, with known $\mathbf{p}_i \in \mathbb{R}_+$ based on Receiver Operating Characteristic (ROC). Here $\mathbf{p}_i = E[\varepsilon_i] = \mathsf{Pr}\{\varepsilon_i = 1\}$. Next, some initial results are given to aid in the subsequent observer development.

Lemma 5.1 *With respect to the uncertainty model in (5.16), the* PPV *defined in (5.15) can be expressed as:*

$$PPV = \frac{1}{|\hat{\mathcal{T}}^c|} \sum_{i \in \hat{\mathcal{T}}^c} \varepsilon_i. \tag{5.17}$$

Proof From (5.16), it follows that $\mathbf{q}_i \hat{\mathbf{q}}_i = \varepsilon_i \hat{\mathbf{q}}_i$. This implies that

$$\mathsf{PPV} = \frac{\|\mathbf{q} \odot \hat{\mathbf{q}}\|_{\ell_0}}{\|\hat{\mathbf{q}}\|_{\ell_0}} = \frac{\sum_{i=1}^{Tm} \mathbf{q}_i \hat{\mathbf{q}}_i}{\sum_{i=1}^{Tm} \hat{\mathbf{q}}_i} = \frac{1}{|\hat{\mathcal{T}}^c|} \sum_{i=1}^{Tm} \varepsilon_i \hat{\mathbf{q}}_i = \frac{1}{|\hat{\mathcal{T}}^c|} \sum_{i \in \hat{\mathcal{T}}^c} \varepsilon_i.$$

■

Proposition 5.2 (Zheng and Anubi (2021)) *The support estimate is better than random flip of a fair coin if and only if*

$$\sum_{i=1}^{Tm} \mathbf{p}_i > Tmp_A, \tag{5.18}$$

where $p_A \in (0, 1)$ is the expected fraction of attacked nodes. Moreover, if p_A is the maximum fraction of attacked nodes, then the conclusion is sufficient, but not necessary.

Lemma 5.2 (Fernández and Williams (2010)) *Given mutually independent Bernoulli random variables $\varepsilon_i \sim \mathscr{B}(1, \mathbf{p}_i)$, $i = 1, \cdots, N$, the following holds:*

$$Pr\left\{\sum_{i=1}^{N} \varepsilon_i = k - 1\right\} = \mathbf{r}(k), \quad k = 1, \cdots, N + 1, \tag{5.19}$$

where $\mathbf{r} = \beta \cdot \begin{bmatrix} -\mathbf{s}_1 \\ 1 \end{bmatrix} * \begin{bmatrix} -\mathbf{s}_2 \\ 1 \end{bmatrix} * \cdots * \begin{bmatrix} -\mathbf{s}_m \\ 1 \end{bmatrix}$, *with* $\beta = \prod_{i=1}^{N} \mathbf{p}_i$ *and* $\mathbf{s}_i = -\frac{1 - \mathbf{p}_i}{\mathbf{p}_i}$.

Now, we are ready to introduce the pruning method. The central idea is: if we could identify the errors in the prior information, then the precision of prior can be improved. In fact, the precision of prior will be improved by choosing an appropriate subset. However, how to achieve the best pruning performance, quantify the precision improvement, and improve resulting estimation resiliency are all essential but open questions. Here, we will give a formal definition of pruning operation, then provide some answers and give a simple algorithm to achieve sub-optimal pruning goal.

Definition 5.5 (*Pruning, Pruning Operation,* PPV_η) Given a prior support estimate $\hat{\mathscr{T}}$, Pruning Operation, with parameter η, is any operation, or sequence of operations, which returns an updated estimated support prior $\hat{\mathscr{T}}_\eta \subset \{1, \cdots, Tm\}$ such that

$$\hat{\mathscr{T}}_\eta^c \subseteq \hat{\mathscr{T}}^c.$$

Also the precision of pruned support prior $\hat{\mathscr{T}}_\eta$ is given by

$$\mathsf{PPV}_\eta = \frac{1}{|\hat{\mathscr{T}}_\eta^c|} \sum_{i \in \hat{\mathscr{T}}_\eta^c} \varepsilon_i. \tag{5.20}$$

The following proposition quantifies the resulting precision improvement through the defined pruning operation.

Proposition 5.3 *Given an estimated attack support $\hat{\mathscr{T}} \subseteq \{1, 2, \cdots, Tm\}$ with the uncertainty characteristic described in (5.16). Let $\hat{\mathscr{T}}_\eta$ be a pruned support estimate satisfying $\hat{\mathscr{T}}_\eta^c \subseteq \hat{\mathscr{T}}^c$, then, for any $\gamma \in (0, 1)$,*

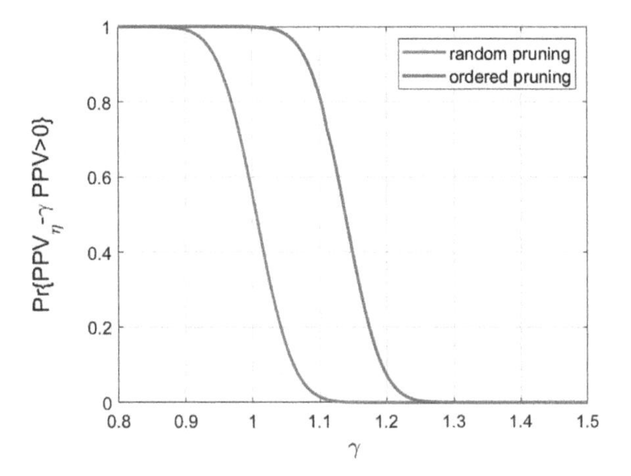

Fig. 5.3 A comparison between random pruning operation and ordered pruning operation

$$Pr\{PPV_\eta - \gamma PPV \geq 0\} \geq \sum_{j=1}^{|\hat{\mathscr{T}}_\eta^c|+1} \left(\mathbf{r}_\eta(j) \sum_{i=1}^{\Phi_{j-1}+1} \tilde{\mathbf{r}}(i) \right), \tag{5.21}$$

where

$$\mathbf{r}_\eta = \left(\prod_{i \in \hat{\mathscr{T}}_\eta^c} \mathbf{p}_i \right) \begin{bmatrix} -\mathbf{s}_{\eta,1} \\ 1 \end{bmatrix} * \begin{bmatrix} -\mathbf{s}_{\eta,2} \\ 1 \end{bmatrix} * \cdots * \begin{bmatrix} -\mathbf{s}_{\eta,|\hat{\mathscr{T}}_\eta^c|} \\ 1 \end{bmatrix}, \mathbf{s}_{\eta,i} = -\frac{1 - \mathbf{p}_{\hat{\mathscr{T}}_\eta^c,i}}{\mathbf{p}_{\hat{\mathscr{T}}_\eta^c,i}},$$

$$\tilde{\mathbf{r}} = \left(\prod_{i \in \hat{\mathscr{T}}^c \setminus \hat{\mathscr{T}}_\eta^c} \mathbf{p}_i \right) \begin{bmatrix} -\tilde{\mathbf{s}}_1 \\ 1 \end{bmatrix} * \begin{bmatrix} -\tilde{\mathbf{s}}_2 \\ 1 \end{bmatrix} * \cdots * \begin{bmatrix} -\tilde{\mathbf{s}}_{|\hat{\mathscr{T}}^c \setminus \hat{\mathscr{T}}_\eta^c|} \\ 1 \end{bmatrix}, \tilde{\mathbf{s}}_i = -\frac{1 - \mathbf{p}_{\hat{\mathscr{T}}^c \setminus \hat{\mathscr{T}}_\eta^c,i}}{\mathbf{p}_{\hat{\mathscr{T}}^c \setminus \hat{\mathscr{T}}_\eta^c,i}},$$

$$\text{and } \Phi_k = \min\left\{ \lceil [\right] \frac{|\hat{\mathscr{T}}^c|}{\gamma |\hat{\mathscr{T}}_\eta^c|} - 1 \rceil k, |\hat{\mathscr{T}}^c| - |\hat{\mathscr{T}}_\eta^c| \right\}.$$

The lower bound given by Proposition 5.3 can be expressed as $\mathbf{r}_\eta^\top \mathbf{r}_\Phi$, where $\mathbf{r}_\Phi \in [0, 1]^{|\hat{\mathscr{T}}_\eta^c|}$ is a vector whose entries are functions of $|\hat{\mathscr{T}}^c|$, $|\hat{\mathscr{T}}_\eta^c|$, γ and $\tilde{\mathbf{r}}$. Thus, given $\mathbf{p}_i$, $\hat{\mathscr{T}}$, γ and a fixed integer $l_\eta \leq |\hat{\mathscr{T}}^c|$, the pruned support $\hat{\mathscr{T}}_\eta$ can be chosen to maximize $\mathbf{r}_\eta^\top \mathbf{r}_\Phi$. However, such optimization problem is challenging and potentially NP-hard due to the index searching operation involved. But a simple heuristic of returning the indices of the channels with largest $\mathbf{p}_i$ in $\hat{\mathscr{T}}_\eta^c$ can provide a very good sub-optimal estimation. This idea is central to the pruning algorithm considered in this chapter. Figure 5.3 shows the comparison of the *ordered* pruning idea vs. randomly selecting a subset of $\hat{\mathscr{T}}$. This illustrative example clearly demonstrates that ordered operation can offer some advantages. Next, one of the ordered pruning algorithms is given in Algorithm 5.2.

Algorithm 5.2 Support Prior Pruning Algorithm

I. **Obtaining reliable trust parameter**

Given reliability level $\eta \in (0, 1)$, return the maximum size l_η such that l_η safe nodes are correctly localized with a probability of at least η:

$$\begin{aligned} l_\eta &= \max\left\{k \mid \Pr\left\{\sum_{i\in\hat{\mathcal{T}}^c} \varepsilon_i \geq k\right\} \geq \eta\right\} \\ &= \max\left\{k \mid \sum_{i=1}^{k+1} \mathbf{r}_{\hat{\mathcal{T}}^c}(i) \leq 1-\eta\right\} \end{aligned} \tag{5.22}$$

where $\mathbf{r}_{\hat{\mathcal{T}}^c}$ is given by (5.19), using the index set $\hat{\mathcal{T}}^c$. II. **Pruning**

A pruned support prior is obtained through a robust extraction:

$$\hat{\mathcal{T}}^c_\eta = \{\text{argsort} \downarrow (\mathbf{p} \odot \mathbf{p}_c)\}_1^{l_\eta}. \tag{5.23}$$

where, $\{\cdot\}_1^{l_\eta}$ is an index extraction from the first elements to l_η elements, $\mathbf{p}_c$ is the confidence vector outputted by Algorithm 5.1, $\mathbf{p}$ is the probability vector of agreement ε based on ROC.

For pragmatic reasons, it is important to ensure that $l_\eta > 0$ in (5.21). This is guaranteed if η is chosen such that at least one node is selected into the pruned set. Formally, this condition is given by:

$$\eta \leq \max_{i\in\hat{\mathcal{T}}^c} (\mathbf{p}_i). \tag{5.24}$$

Definition 5.6 (*η-successful pruning algorithm*) A η-successful pruning algorithm is any pruning operation, as defined in Definition 5.5, that achieves:

$$\Pr\{\text{PPV}_\eta = 1\} \geq \eta.$$

Proposition 5.4 (Zheng and Anubi (2021)) *Given support prior estimate $\hat{\mathcal{T}}$ generated by an underlying localization algorithm with associated uncertainty model in (5.16), the pruning algorithm in Algorithm 5.2 is η-successful.*

5.4 Pruning-Based Resilient Estimation

In this section, we will go through resilient observer designs using $\ell_0\backslash\ell_1$ minimization schemes. Firstly, the unconstrained ℓ_1 observer will be stated. Then, we will give a weighted ℓ_1 observer design and state the condition for resilient estimation with the pruned prior support. Furthermore, the quantified relationship between the precision of prior support and the resilient estimation performance will be clarified.

Researchers in compressed sensing have paid much attention to the recoverability of $\ell_0\backslash\ell_1$ minimization problem in the last decade. Most of the effort focused on finding well-defined compressed matrix satisfying Null Space Property (NSP) or Restricted Isometry Property (RIP) . Then, the complete information can be reconstructed by $\ell_0\backslash\ell_1$ minimization problem from the compressed measurements. From a mathematical aspect, the decoding process is to solve an under-determined set of equations, which does generally not have unique solutions. However, if the required solution is sparse, it can be recovered completely via ℓ_0 minimization. The condition on sparsity for exact unique recovery is also well known. However, the ℓ_0 minimization problem is an NP-hard optimization problem. However, NSP or RIP pave way for a convex relaxation via the ℓ_1 minimization problem. Interested readers are directed to Donoho (2006), Candès et al. (2006), Candes and Tao (2005), Fornasier and Rauhut (2015) for more comprehensive treatment of compressed sensing, and Candes and Tao (2007), Friedlander et al. (2011) for several extension cases.

The basic motivation for using ℓ_1 minimization for attack-resilient estimation is because the attack is possibly unbounded but is necessarily sparse. Consider the measurement model in (5.2), if a coding matrix F can be found that satisfies $FH = 0$, then a new under-determined equation $F\mathbf{y} = F\mathbf{e}$ is obtained. If the sparse attack vector is recovered, the resilient estimation goal is easily achieved. In this section, instead of finding a coding matrix F directly, we would formulate the problem within the familiar framework of linear systems theory and prove results similarly to compressed sensing literature.

5.4.1 Unconstrained ℓ_1 Observer

In this subsection, we discuss the uniqueness of resilient estimation solutions in the presence of measurement attacks and introduce the concept of Column Space Property (CSP). Furthermore, the estimation error bound is given using CSP.

Consider the system model in (5.1) and the unconstrained ℓ_1 decoder in (5.3), a formal notion of attack recovery is given as following:

Definition 5.7 (*Resilient Recovery*) k sensor attacks are correctable after T steps by $\mathscr{D} : (\mathbb{R}^m)^T \to \mathbb{R}^n$ if for any $\mathbf{x}_0 \in \mathbb{R}^n$ and any sequence of attack vectors $\mathbf{e}_0, \mathbf{e}_1, \ldots, \mathbf{e}_{T-1} \in \mathbb{R}^m$ with $\mathsf{supp}(e_t) \le k$, we have $\mathscr{D}(\mathbf{y}_0, \cdots, \mathbf{y}_{T-1}) = \mathbf{x}_0$.

The following theorem states the uniqueness of resilient estimation solution:

Theorem 5.1 *Given attack support* $\mathscr{T} = \{\mathscr{T}_i, \mathscr{T}_{i-1}, \ldots, \mathscr{T}_{i-T+1}\}$ *with* $|\mathscr{T}_i| \le k$. *Consider the noise-free version of the measurement model in (5.2). If, for any* $\mathbf{h} \in \mathsf{range}(H)$, *it is true that*

$$\|\mathbf{h}_{\mathscr{S}}\|_1 \le \|\mathbf{h}_{\mathscr{S}^c}\|_1, \quad \forall \mathscr{S} \subset \{1, 2, \cdots, Tm\}, |\mathscr{S}| \le Tk, \tag{5.25}$$

then, for each attacked measurement $\mathbf{y}_T \in \mathbb{R}^{Tm}$, *there exists an unique state vector* $\hat{\mathbf{x}} \in \mathbb{R}^n$ *and* Tk*-sparse attack vector* $\hat{\mathbf{e}}$ *which satisfy (5.2).*

Proof Let $(\mathbf{z}_1, \mathbf{e}_1), (\mathbf{z}_2, \mathbf{e}_2) \in \mathbb{R}^n \times \Sigma_{Tk}$ such that $\mathbf{y}_T = H\mathbf{z}_1 + \mathbf{e}_1 = H\mathbf{z}_2 + \mathbf{e}_2$, then $H(\mathbf{z}_1 - \mathbf{z}_2) = \mathbf{e}_2 - \mathbf{e}_1$. Thus, the uniqueness condition holds iff $\mathsf{range}(H) \cap \Sigma_{2Tk} = \{\mathbf{0}\}$. Now, given $\mathbf{h} \in \mathsf{range}(H)$ which satisfies (5.25), it suffices to show that $\|\mathbf{h}\|_0 > 2Tk$.

Suppose, for the sake of contradiction, that $\|\mathbf{h}\|_0 \leq 2Tk$. Choose $\overline{\mathscr{S}} \in \{1, 2, \cdots, Tm\}$, $|\overline{\mathscr{S}}| = Tk$ to be the indices of the largest components of $\mathbf{h}$ in absolute value.

Then, it must be that

$$\|\mathbf{h}_{\overline{\mathscr{S}}}\|_0 > \|\mathbf{h}_{\overline{\mathscr{S}}^c}\|_0 \Rightarrow \|\mathbf{h}_{\overline{\mathscr{S}}}\|_1 > \|\mathbf{h}_{\overline{\mathscr{S}}^c}\|_1,$$

which is a contradiction. Thus, (5.25) implies that $\|\mathbf{h}\|_0 > 2Tk$. ∎

Consequently, a formal definition of column space property is given as follows.

Definition 5.8 (*Column Space Property (CSP)*) A matrix $H \in \mathbb{R}^{m \times n}$ has a Column Space Property of order $s < m$ (denoted as $H \rhd \mathsf{CSP}(s)$) if there exists $\beta \in (0, 1)$ such that, for every $\mathbf{h} \in \mathsf{range}(H)$,

$$\|\mathbf{h}_{\mathscr{S}}\|_1 \leq \beta \|\mathbf{h}_{\mathscr{S}^c}\|_1, \quad \forall \mathscr{S} \subset \{1, 2, \cdots, m\}, |\mathscr{S}| \leq s. \tag{5.26}$$

The above definition is similar to the well-known *Null Space Property* but defined on the range space instead. For dynamic system (5.1), the unconstrained ℓ_1 observer is defined as a moving-horizon unconstrained ℓ_1 minimization problem:

$$\begin{aligned} \text{Minimize} \quad & \sum_{j=i-T+1}^{i} \|\mathbf{y}_j - C\mathbf{x}_j\|_1 \\ \text{Subject to} \quad & \mathbf{x}_{j+1} - A\mathbf{x}_j = 0, \quad j = i - T + 1, \ldots, i - 1. \end{aligned} \tag{5.27}$$

An equivalent optimization problem of (5.27) is given by

$$\underset{\mathbf{x} \in \mathbb{R}^n}{\text{Minimize}} \ \|\mathbf{y}_T - H\mathbf{x}\|_1. \tag{5.28}$$

The following theorem gives the conditions for resilient recovery of the state vector obtained by the above observer.

Theorem 5.2 (Resilient Recovery with CSP) *Consider the measurement model in (5.2), let $\mathscr{T} = \{\mathscr{T}_i, \mathscr{T}_{i-1}, \ldots, \mathscr{T}_{i-T+1}\}$, with $|\mathscr{T}_i| \leq k$, be the unknown sequence of the attack support. If $H \rhd \mathsf{CSP}(Tk)$, the estimation error due to the decoder in (5.28) can be upper bounded as:*

$$\|\hat{\mathbf{x}} - \mathbf{x}\|_2 \leq \frac{2(1 + \beta)}{\underline{\sigma}(1 - \beta)} \varepsilon, \tag{5.29}$$

for some $\beta \in (0, 1)$, and $\underline{\sigma}$ is the smallest singular value of H.

Proof Let $\hat{\mathbf{x}}$ be the optimal solution of (5.28), then its optimality yields

$$\begin{aligned}\|\mathbf{y}-H\hat{\mathbf{x}}\|_1 &\leq \|\mathbf{y}-H\mathbf{x}\|_1 = \|\mathbf{e}\|_1\\ \|\mathbf{y}-H\mathbf{x}+H(\mathbf{x}-\hat{\mathbf{x}})\|_1 &\leq \|\mathbf{e}\|_1.\end{aligned}$$

Let $\tilde{\mathbf{x}}=\mathbf{x}-\hat{\mathbf{x}}$, and since 1-norm is decomposable for disjoint sets, then

$$\begin{aligned}\|\mathbf{e}+H\tilde{\mathbf{x}}\|_1 &\leq \|\mathbf{e}\|_1,\\ \|\mathbf{e}_{\mathcal{T}}+H_{\mathcal{T}}\tilde{\mathbf{x}}\|_1+\|\mathbf{e}_{\mathcal{T}^c}+H_{\mathcal{T}^c}\tilde{\mathbf{x}}\|_1 &\leq \|\mathbf{e}_{\mathcal{T}}\|_1+\|\mathbf{e}_{\mathcal{T}^c}\|_1,\\ \|\mathbf{e}_{\mathcal{T}}\|_1-\|H_{\mathcal{T}}\tilde{\mathbf{x}}\|_1-\|\mathbf{e}_{\mathcal{T}^c}\|_1+\|H_{\mathcal{T}^c}\tilde{\mathbf{x}}\|_1 &\leq \|\mathbf{e}_{\mathcal{T}}\|_1+\|\mathbf{e}_{\mathcal{T}^c}\|_1.\end{aligned}$$

And let $\mathbf{h}=H\tilde{\mathbf{x}}$, it follows

$$\|\mathbf{h}_{\mathcal{T}^c}\|_1 \leq \|\mathbf{h}_{\mathcal{T}}\|_1+2\varepsilon. \tag{5.30}$$

Since $H \rhd \mathsf{CSP}(Tk)$, there exist $\beta\in(0,1)$ such that $\|\mathbf{h}_{\mathcal{T}}\|_1 \leq \beta\|\mathbf{h}_{\mathcal{T}^c}\|_1$. Thus $\|\mathbf{h}_{\mathcal{T}}\|_1 \leq \frac{2\beta}{1-\beta}\varepsilon$. Then, $\|\mathbf{h}\|_2 \leq \|\mathbf{h}_{\mathcal{T}}\|_1+\|\mathbf{h}_{\mathcal{T}^c}\|_1 \leq 2\|\mathbf{h}_{\mathcal{T}}\|_1+2\varepsilon \leq \frac{2(1+\beta)}{1-\beta}\varepsilon$. Finally, combining with $\underline{\sigma}\|\tilde{\mathbf{x}}\|_2 \leq \|\mathbf{h}\|_2$ yields the error bound in (5.29). ■

Notice that the CSP condition with $\beta\in(0,1)$ is a violation of the condition stated in (5.7), which is a guarantee of successful FDIA. The CSP condition is relevant to the sparsity of the attack vector. As shown in literature Fawzi et al. (2014), the number of attacks is one of the most important factors deciding if successful resilient estimation would be achieved. With the increasing power of FDIA, it is more likely that the CSP condition would be violated. This is one of the motivations for finding an improved resilient estimation method in the worst environment.

5.4.2 *Resilient Pruning Observer*

In this subsection, we incorporate prior information into the resilient observer design. First, support prior $\hat{\mathcal{T}}$ is generated by the localization algorithm in Algorithm 5.1. Then the pruning algorithm in Algorithm 5.2 is used to improve the precision of the support prior. Finally, a weighted ℓ_1 observer scheme is proposed to utilize the pruned support prior $\hat{\mathcal{T}}_\eta$. This process is summarized in Fig. 5.4.

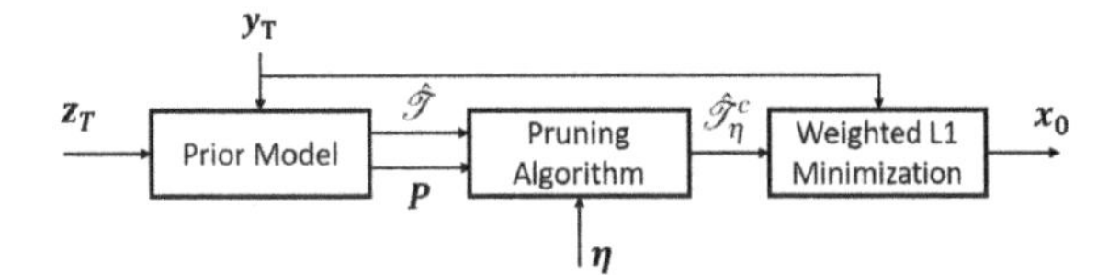

Fig. 5.4 Schematic depiction of resilient observer design with prior pruning

Consider a time horizon T and a set of attack support prior obtained by Algorithm 5.1: $\hat{\mathcal{T}} = \{\hat{\mathcal{T}}_i, \hat{\mathcal{T}}_{i-1}, \ldots, \hat{\mathcal{T}}_{i-T+1}\}$. The following weighted ℓ_1 observer is considered:

$$\begin{aligned} &\text{Minimize} \quad \sum_{j=i-T+1}^{i} \|\mathbf{y}_j - C\mathbf{x}_j\|_{1,\mathbf{w}(\hat{\mathcal{T}}_j,\omega)} \\ &\text{Subject to} \;\; \mathbf{x}_{j+1} - A\mathbf{x}_j = 0, \quad j = i - T + 1, \ldots, i - 1, \end{aligned} \tag{5.31}$$

where, for $\omega \in (0,\ 1)$, the weight vector $\mathbf{w}(\hat{\mathcal{T}}_j, \omega) \in \mathbb{R}^m$ is defined element-wise as

$$\mathbf{w}(\hat{\mathcal{T}}_j, \omega)_l = \begin{cases} \omega & \text{if } l \in \hat{\mathcal{T}}_j \\ 1 & \text{otherwise.} \end{cases} \tag{5.32}$$

The optimization problem in (5.31) is equivalent to

$$\underset{\mathbf{z}\in\mathbb{R}^n}{\text{Minimize}} \; \|\mathbf{y}_T - H\mathbf{z}\|_{1,\mathbf{w}(\hat{\mathcal{T}},\omega)}, \tag{5.33}$$

where $\mathbf{w}(\hat{\mathcal{T}}, \omega) = \begin{bmatrix} \mathbf{w}(\hat{\mathcal{T}}_i, \omega) \\ \vdots \\ \mathbf{w}(\hat{\mathcal{T}}_{i-T+1}, \omega) \end{bmatrix} \in \mathbb{R}^{Tm}$.

Theorem 5.3 (Resilient Recovery with support prior $\hat{\mathcal{T}}$) *Consider the measurement model in (5.2), let $\mathcal{T} = \{\mathcal{T}_i, \mathcal{T}_{i-1}, \ldots, \mathcal{T}_{i-T+1}\}$, with $|\mathcal{T}_i| \le k$, be the unknown support sequence of the attack vector such that $\sum_{i\in\mathcal{T}^c} |\mathbf{e}_i| < \varepsilon$. Let $\hat{\mathcal{T}} = \{\hat{\mathcal{T}}_i, \hat{\mathcal{T}}_{i-1}, \ldots, \hat{\mathcal{T}}_{i-T+1}\}$ be a support prior estimate satisfying*

$$|\hat{\mathcal{T}}| = \rho|\mathcal{T}| \text{ and } |\mathcal{T} \cap \hat{\mathcal{T}}| = \alpha|\hat{\mathcal{T}}|. \tag{5.34}$$

If $H \rhd \mathsf{CSP}(\kappa T k)$, where $\kappa = \rho + 1 - 2\alpha\rho$ with $\rho > 0, \alpha \in (0,\ 1)$, then the estimation error due to the decoder in (5.31) can be upper bounded as:

$$\|\hat{\mathbf{x}} - \mathbf{x}\|_2 \le \frac{2(1+\beta)}{\underline{\sigma}(1-\beta)}\varepsilon, \tag{5.35}$$

for some $\beta \in (0, 1)$, where $\underline{\sigma}$ is the smallest singular value of H.

Proof Let $\hat{\mathbf{x}}$ be the optimal solution of (5.33), and define $\tilde{\mathbf{x}} = \mathbf{x} - \hat{\mathbf{x}}, \mathbf{h} = H\tilde{\mathbf{x}}$. Similar to the proof of Theorem 5.2, the optimality of $\hat{\mathbf{x}}$ yields

$$\|\mathbf{e} + \mathbf{h}\|_{1,\mathbf{w}(\hat{\mathcal{T}},\omega)} \le \|\mathbf{e}\|_{1,\mathbf{w}(\hat{\mathcal{T}},\omega)}. \tag{5.36}$$

By the definition of weighted 1-norm, it follows that $\omega\|\mathbf{e}_{\hat{\mathscr{T}}} + \mathbf{h}_{\hat{\mathscr{T}}}\|_1 + \|\mathbf{e}_{\hat{\mathscr{T}}^c} + \mathbf{h}_{\hat{\mathscr{T}}^c}\|_1 \leq \omega\|\mathbf{e}_{\hat{\mathscr{T}}}\|_1 + \|\mathbf{e}_{\hat{\mathscr{T}}^c}\|_1$, then

$$\omega\|\mathbf{e}_{\hat{\mathscr{T}}\cap\mathscr{T}} + \mathbf{h}_{\hat{\mathscr{T}}\cap\mathscr{T}}\|_1 + \omega\|\mathbf{e}_{\hat{\mathscr{T}}\cap\mathscr{T}^c} + \mathbf{h}_{\hat{\mathscr{T}}\cap\mathscr{T}^c}\|_1 + \|\mathbf{e}_{\hat{\mathscr{T}}^c\cap\mathscr{T}} + \mathbf{h}_{\hat{\mathscr{T}}^c\cap\mathscr{T}}\|_1$$
$$+ \|\mathbf{e}_{\hat{\mathscr{T}}^c\cap\mathscr{T}^c} + \mathbf{h}^c_{\hat{\mathscr{T}}^c\cap\mathscr{T}}\|_1 \leq \omega\|\mathbf{e}_{\hat{\mathscr{T}}\cap\mathscr{T}}\|_1 + \|\mathbf{e}_{\hat{\mathscr{T}}\cap\mathscr{T}^c}\|_1 + \|\mathbf{e}_{\hat{\mathscr{T}}^c\cap\mathscr{T}}\|_1 + \|\mathbf{e}_{\hat{\mathscr{T}}^c\cap\mathscr{T}^c}\|_1.$$

Using the reverse triangle inequality yields

$$\omega\|\mathbf{h}_{\hat{\mathscr{T}}\cap\mathscr{T}^c}\|_1 + \|\mathbf{h}_{\hat{\mathscr{T}}^c\cap\mathscr{T}^c}\|_1 \leq \|\mathbf{h}_{\hat{\mathscr{T}}^c\cap\mathscr{T}}\|_1 + \omega\|\mathbf{h}_{\hat{\mathscr{T}}\cap\mathscr{T}}\|_1 + 2(\|\mathbf{e}_{\hat{\mathscr{T}}^c\cap\mathscr{T}^c}\|_1 + \|\mathbf{e}_{\hat{\mathscr{T}}\cap\mathscr{T}^c}\|_1).$$

Adding and subtracting $\omega\|\mathbf{h}_{\hat{\mathscr{T}}^c\cap\mathscr{T}^c}\|_1$ on the left, and $\omega\|\mathbf{h}_{\hat{\mathscr{T}}^c\cap\mathscr{T}}\|_1, \omega\|\mathbf{e}_{\hat{\mathscr{T}}^c\cap\mathscr{T}^c}\|_1$ on the right yields:

$$\omega\|\mathbf{h}_{\mathscr{T}^c}\|_1 + (1-\omega)\|\mathbf{h}_{\hat{\mathscr{T}}^c\cap\mathscr{T}^c}\|_1 \leq (1-\omega)\|\mathbf{h}_{\hat{\mathscr{T}}^c\cap\mathscr{T}}\|_1 + \omega\|\mathbf{h}_{\mathscr{T}}\|_1$$
$$+2(\omega\|\mathbf{e}_{\mathscr{T}^c}\|_1 + (1-\omega)\|\mathbf{e}_{\hat{\mathscr{T}}^c\cap\mathscr{T}^c}\|_1).$$

Again, adding and subtracting $(1-\omega)\|\mathbf{h}_{\hat{\mathscr{T}}\cap\mathscr{T}^c}\|_1$ on the left and substituting $\sum_{i\in\mathscr{T}^c} |\mathbf{e}_i| < \varepsilon$ yields:

$$\|\mathbf{h}_{\mathscr{T}^c}\|_1 \leq \omega\|\mathbf{h}_{\mathscr{T}}\|_1 + (1-\omega)(\|\mathbf{h}_{\hat{\mathscr{T}}^c\cap\mathscr{T}}\|_1 + \|\mathbf{h}_{\hat{\mathscr{T}}\cap\mathscr{T}^c}\|_1) + 2\varepsilon.$$

Let $\mathscr{T}_\alpha \triangleq (\hat{\mathscr{T}}^c \cap \mathscr{T}) \cup (\hat{\mathscr{T}} \cap \mathscr{T}^c) = \hat{\mathscr{T}} \cup \mathscr{T} \setminus \hat{\mathscr{T}} \cap \mathscr{T}$. It follows that $|\mathscr{T}_\alpha| = \kappa|\mathscr{T}| \leq \kappa Tk$. Also, since $\hat{\mathscr{T}}^c \cap \mathscr{T}$ and $\hat{\mathscr{T}} \cap \mathscr{T}^c$ are disjoint, the inequality above becomes

$$\|\mathbf{h}_{\mathscr{T}^c}\|_1 \leq \omega\|\mathbf{h}_{\mathscr{T}}\|_1 + (1-\omega)\|\mathbf{h}_{\mathscr{T}_\alpha}\|_1 + 2\varepsilon. \tag{5.37}$$

Since $H \rhd \mathsf{CSP}(\kappa Tk)$, we have

$$\|\mathbf{h}_{\mathscr{T}}\|_1 \leq \beta\|\mathbf{h}_{\mathscr{T}^c}\|_1 \tag{5.38}$$
$$\|\mathbf{h}_{\mathscr{T}_\alpha}\|_1 \leq \beta\|\mathbf{h}_{\mathscr{T}_\alpha^c}\|_1. \tag{5.39}$$

Using (5.39) and property of 1-norm yields:

$$\begin{aligned}\|\mathbf{h}_{\mathscr{T}_\alpha}\|_1 + \|\mathbf{h}_{\mathscr{T}_\alpha^c}\|_1 &= \|\mathbf{h}\|_1 \\ \|\mathbf{h}_{\mathscr{T}_\alpha}\|_1 &\leq \frac{\beta}{1+\beta}\|\mathbf{h}\|_1.\end{aligned} \tag{5.40}$$

Then, substituting (5.38) and (5.40) into (5.37) yields

$$(1-\beta\omega)\|\mathbf{h}_{\mathscr{T}^c}\|_1 \leq \frac{\beta(1-\omega)}{1+\beta}\|\mathbf{h}\|_1 + 2\varepsilon. \tag{5.41}$$

Next,

$$\|\mathbf{h}\|_1 = \|\mathbf{h}_{\mathscr{T}}\|_1 + \|\mathbf{h}_{\mathscr{T}^c}\|_1 \leq (1+\beta)\|\mathbf{h}_{\mathscr{T}^c}\|_1 \leq \frac{\beta(1-\omega)}{1-\beta\omega}\|\mathbf{h}\|_1 + \frac{2(1+\beta)}{1-\beta\omega}\varepsilon,$$

then $\|\mathbf{h}\|_2 \leq \|\mathbf{h}\|_1 \leq \frac{2(1+\beta)}{1-\beta}\varepsilon$. Finally, combining with $\underline{\sigma}\|\tilde{\mathbf{x}}\|_2 \leq \|\mathbf{h}\|_2$ yields the error bound in (5.35). ■

The estimation error bound in Theorem 5.3 is the same as the one in Theorem 5.2. The only difference is that the upper bound of the number of attacks which can be corrected by the underlying observer is governed by κ. If $\kappa < 1$, then the weighted ℓ_1 observer with prior (5.31) has better attack-resiliency compared to the unconstrained ℓ_1 observer (5.27). Furthermore, the size of κ is actually the relative size of the disagreement set $\mathscr{T}_\alpha = \hat{\mathscr{T}} \cup \mathscr{T} \setminus \hat{\mathscr{T}} \cap \mathscr{T}$ between $\mathscr{T}$ and $\hat{\mathscr{T}}$. Specifically, the quantified relationship between the precision of support prior PPV and the disagreement size κ is given by:

$$\kappa = \rho - 1 + \frac{2(1-\mathsf{PPV})(Tm - \rho|\mathscr{T}|)}{|\mathscr{T}|},$$

where ρ is given in (5.34). It is seen that the precision of support prior has a negative correlation to the disagreement size κ. Thus, it has a positive correlation to the attack-resiliency of the underlying observer. Another way to see this is to observe that the condition in Theorem 5.3 can be stated as $|\mathscr{T}_i| \leq \frac{Tk}{\kappa}$ and $H \rhd \mathsf{CSP}(Tk)$, from which it is clear that $\kappa < 1$ implies that more attacks can be accommodated by the observer with prior. This is the main motivation for the pruning algorithm. Next, the following corollary gives a better attack-resiliency of weighted ℓ_1 observer with the pruned support $\hat{\mathscr{T}}_\eta$.

Corollary 5.1 (Resilient Recovery with Pruned Prior $\hat{\mathscr{T}}_\eta$) *Given a support prior $\hat{\mathscr{T}} = \{\hat{\mathscr{T}}_i, \hat{\mathscr{T}}_{i-1}, \cdots, \hat{\mathscr{T}}_{i-T+1}\}$ generated by the localization algorithm in Algorithm 5.1. Let $\hat{\mathscr{T}}_\eta$ be the pruned support prior obtained from $\hat{\mathscr{T}}$ according to Algorithm 5.2 with a parameter $\eta \in (0, 1)$. Let the precision of $\hat{\mathscr{T}}_\eta$ be denoted by PPV_η. If $H \rhd \mathsf{CSP}(\kappa_1 Tk)$, where $\kappa_1 = \frac{|\mathscr{T}^c| + l_\eta(1-2\mathsf{PPV}_\eta)}{|\mathscr{T}|}$, then the estimation error due to (5.31) with $\hat{\mathscr{T}}_\eta$ can be upper bounded as*

$$\|\hat{\mathbf{x}} - \mathbf{x}\|_2 \leq \frac{2(1+\beta)}{\underline{\sigma}(1-\beta)}\varepsilon, \tag{5.42}$$

for some $\beta \in (0, 1)$, and $\underline{\sigma}$ is the smallest singular value of H.

Furthermore, with probability at least η, the smallest disagreement size is obtained as

$$\kappa_1 = \frac{Tm - l_\eta}{|\mathscr{T}|} - 1. \tag{5.43}$$

Proof (5.42) can be obtained by following the proof of Theorem 5.3 but using PPV_η instead. To obtain (5.43), observe that with probability at least η, $\mathsf{PPV}_\eta = 1$. ■

5.5 Simulation Results

In this section, three application examples are given in power grid, wheeled mobile robot, and water distributed system, respectively. These application examples are used to demonstrate how to implement the developed observer in previous sections. And the proposed pruning-based observer is compared to some well-known resilient observers in literature, which shows the resilience of CPS is improved using the pruning algorithm and concurrent learning prior.

5.5.1 *Resilient Power Grid*

Here, we implement the proposed pruning observer on an IEEE 14-bus system. The simulation scenario is shown in Fig. 5.5. The bus system has $n_b = 14$ buses and $n_g = 5$ generators. It is assumed that each bus in the network is equipped with IIoT sensor devices, which provide the corresponding active power injection and flow measurements.

A small signal model is constructed by linearizing the generator swing and power flow equations around the operating point. The following linearizing assumptions are made:

1. Voltage is tightly controlled at their nominal value.
2. Angular difference between each bus is small.
3. Conductance is negligible therefore the system is lossless.

By ordering the buses such that the generator nodes appear first, the admittance-weighted Laplacian matrix can be expressed as $L = \begin{bmatrix} L_{gg} & L_{lg} \\ L_{gl} & L_{ll} \end{bmatrix} \in \mathbb{R}^{N \times N}$, where $N =$

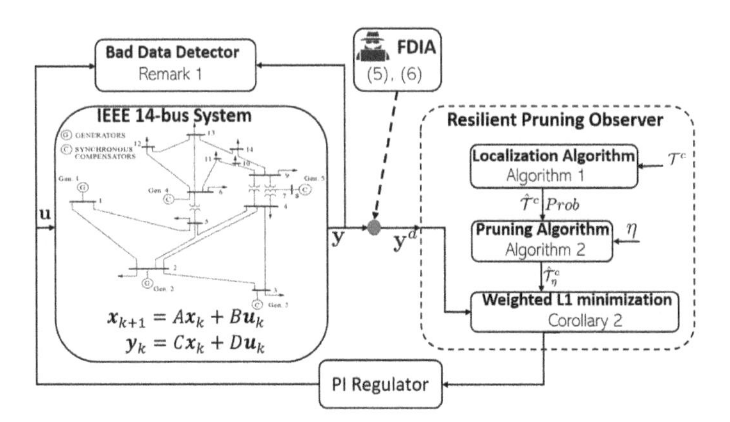

Fig. 5.5 Block diagram depiction of resilient power grid

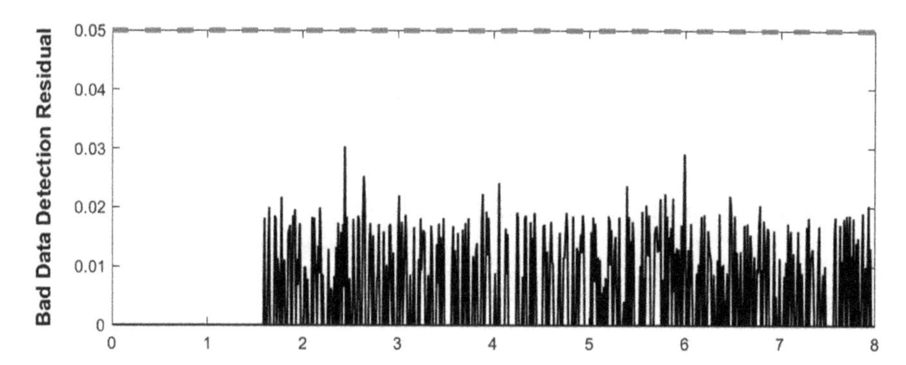

Fig. 5.6 Bad data detection result (the residual threshold is set as 0.05, 60% of measurement nodes are attacked)

$n_g + n_b$. Thus, the dynamical linearized swing equations and algebraic DC power flow equations are given by:

$$\begin{bmatrix} I & 0 & 0 \\ 0 & M & 0 \\ 0 & 0 & 0 \end{bmatrix} \dot{\mathbf{x}} = - \begin{bmatrix} 0 & -I & 0 \\ L_{gg} & D_g & L_{lg} \\ L_{gl} & 0 & L_{ll} \end{bmatrix} \mathbf{x} + \begin{bmatrix} 0 & 0 \\ I & 0 \\ 0 & I \end{bmatrix} \mathbf{u}, \tag{5.44}$$

where $\mathbf{x} = [\delta^\top \ \omega^\top \ \theta^\top]^\top \in \mathbb{R}^{2n_g+n_b}$ is the state vector containing generator rotor angle $\delta \in \mathbb{R}^{n_b}$, generator frequency $\omega \in \mathbb{R}^{n_g}$, and voltage bus angles $\theta \in \mathbb{R}^{n_b}$. $\mathbf{u} = [P_g^\top P_d^\top]^\top \in \mathbb{R}^{n_g+n_b}$ is the input vector consisting of mechanical input power from each generator $P_g \in \mathbb{R}^{n_g}$ and active power demand at each bus $P_d \in \mathbb{R}^{n_b}$, M is a diagonal matrix of inertial constants for each generator, and D_g is a diagonal matrix of damping coefficients. A PI regulator is included to regulate the generator frequency in order to control the P_g. The system in (5.44) is then simplified as follows:

$$\begin{aligned} \begin{bmatrix} \dot{\delta} \\ \dot{\omega} \end{bmatrix} &= \begin{bmatrix} 0 & I \\ -M^{-1}(L_{gg} - L_{gl}L_{ll}^{-1}L_{lg}) & -M^{-1}D_g \end{bmatrix} \begin{bmatrix} \delta \\ \omega \end{bmatrix} + \begin{bmatrix} 0 & 0 \\ M^{-1} & -M^{-1}L_{gl}L_{ll}^{-1} \end{bmatrix} \mathbf{u}, \\ \begin{bmatrix} \omega \\ P_{net} \end{bmatrix} &= \begin{bmatrix} 0 & I \\ -P_{node}L_{ll}^{-1}Llg & 0 \end{bmatrix} \begin{bmatrix} \delta \\ \omega \end{bmatrix} + \begin{bmatrix} 0 & 0 \\ P_{node}L_{ll}^{-1} & 0 \end{bmatrix} \mathbf{u}, \\ \theta &= -L_{ll}^{-1}(L_{lg}\delta - P_d), \end{aligned} \tag{5.45}$$

where P_{node} is a function of the system incidence and susceptance matrices obtained by linearizing the active power injections at the buses (Scholtz 2004), and P_{net} is the net power injected at each bus. As shown in Fig. 5.5, the FDIA designed using (5.5) and (5.6) is injected into system through the sensor channels. The bad data detection residual is then calculated after the FDIA is injected, as shown in Fig. 5.6. The figure indicates the designed FDIA can bypass the bad data detector.

The prior model is a set of trained Gaussian process regression models mapping from the real load data of New York (NY) state provided by the NY Independent System Operator (NYISO) to IEEE 14-bus model (see Power System Test Case

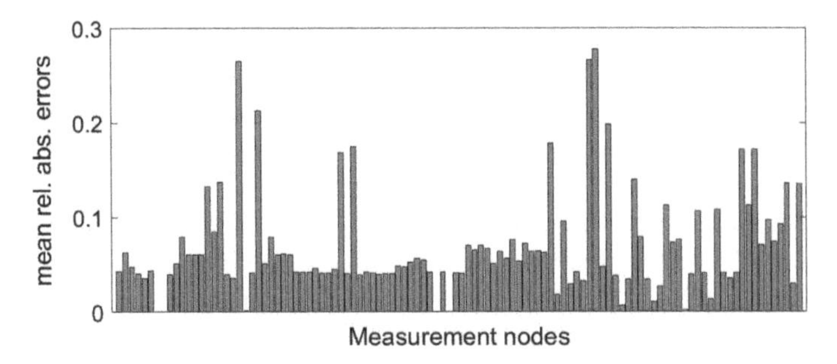

Fig. 5.7 GPRs' prediction error metrics for all measurement nodes (The mean relative absolute error is used to evaluate the prediction performance)

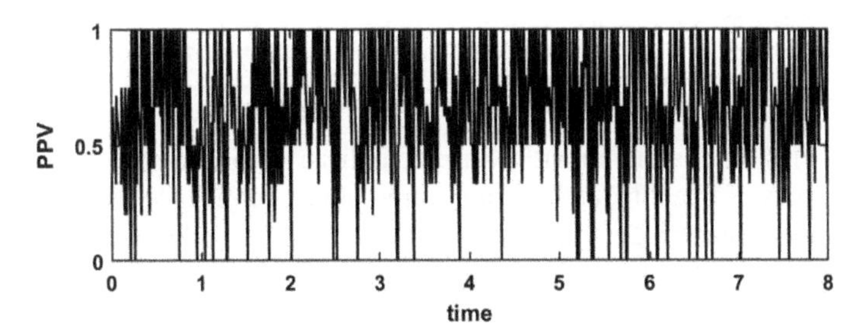

Fig. 5.8 The precision of support prior generated by the localization algorithm in Algorithm 5.1 for the power grid (The mean of precision is 0.655)

Archive 2022 for details of the model) measurements. Five-minute load data of NYISO for 3 months (between January and March) in 2017 and 2018 are used. The IEEE 14-bus model is mapped onto the NYISO transmission grid (see New York control area load zone map 2022 for details) as follows: $A \rightarrow 2$, $B \rightarrow 3$, $C \rightarrow 4$, $D \rightarrow 5$, $E \rightarrow 6$, $F \rightarrow 9$, $G \rightarrow 10$, $H \rightarrow 11$, $I \rightarrow 12$, $J \rightarrow 13$, $K \rightarrow 14$. Then, the market variables downloaded from the respective nodes of NYISO transmission grid are collected into the auxiliary vector variable $\mathbf{z} = [z_{lbmp}\ z_{mcl}\ z_{mcc}]$, where z_{lbmp} is the *locational bus marginal prices* (\$/MWh), z_{mcl} is the *marginal cost loses* (\$/MWh), and z_{mcc} is the *marginal cost congestion* (\$/MWh).

Using the load data downloaded at NY load zones for the same time period and interval as output, GPR models were trained to map the auxiliary vector $\mathbf{z}$ to each corresponding bus measurements $\mathbf{y}_j$ containing active power and reactive power of load buses. As shown in (5.12), the trained GPR models are executed to give the mean $\mu(\mathbf{z})$ and the covariance $\Sigma(\mathbf{z})$ of prior model for each of the measurements. The prediction performance of those GPR models, measured by the mean relative absolute errors (MRAE), is shown in Fig. 5.7. Finally, the localization algorithm in Algorithm 5.1 is implemented on the system model in (5.45). The precision of the generated support prior calculated at each time instance is shown in Fig. 5.8. The mean of precision is 0.655, which indicates the localization algorithm at least works better than random flip of fair coin.

Furthermore, the developed resilient observer with support prior pruning is compared with some well-known resilient observers in literature. Luenberger observer (LO) is also included to serve as a reference and to show the effectiveness of the designed FDIA. The unconstrained ℓ_1 observer (UL1O) (5.27), event-triggered Luenberger observer (ETLO) (Shoukry and Tabuada 2015), and multi-model observer (MMO) (Anubi et al. 2020) are all resilient observers included in the comparison. MMO is a ℓ_1 observer with multiple constraints including system updating law and the measurement prior in (5.12). The core optimization problem solved for the MMO is:

$$\begin{aligned} \text{Minimize} \quad & \sum_{i=k-T+1}^{k} \|\mathbf{y}_i - C\mathbf{x}_i\|_1 \\ \text{Subject to} \quad & \mathbf{x}_{i+1} - A\mathbf{x}_i - B\mathbf{u}_i = 0 \;\; j = i - T + 1, \ldots, i - 1 \\ & \|C\mathbf{x}_k - \mu(\mathbf{z}_k)\|^2_{\Sigma^{-1}(\mathbf{z})} \leq \chi^2_m(\tau), \end{aligned} \tag{5.46}$$

where $\chi^2_m(\tau)$ is the quantile function for probability τ of the chi-squared distribution with m degrees of freedom, and τ is the a pre-defined confidence threshold.

ETLO uses event-triggered projected gradient descent technique to achieve fast and reliable solution to the batch optimization problem

$$\begin{aligned} &\text{Minimize: } \|\mathbf{Y}_t - [H \;\; I]\mathbf{z}_t\|_2^2 \\ &\text{Subject to: } \mathbf{z}_t \in \mathbb{R}^n \times \Sigma_{Tk}, \end{aligned} \tag{5.47}$$

where the decision variable $\mathbf{z}_t$ is an augmented states containing desired initial states and all injected measurement error in T time horizon, $Y_t = [\mathbf{y}_1(t-T+1)^\top \;\; \mathbf{y}_1(t-T+2)^\top \;\; \cdots \;\; \mathbf{y}_1(t)^\top \;\; \cdots\cdots \;\; \mathbf{y}_m(t-T+1)^\top \;\; \mathbf{y}_m(t-T+2)^\top \cdots \;\; \mathbf{y}_m(t)^\top]^\top \in \mathbf{R}^{Tm}$ is the collection of measurements in T time horizon. A recursive solution to (5.47) is then implemented as a Luenberger-like update

$$\hat{\mathbf{z}}_t^{(m+1)} = \hat{\mathbf{z}}_t^{(m)} + 2[H \;\; I]^\top (Y_t - [H \;\; I]\hat{\mathbf{z}}_t^{(m)}), \tag{5.48}$$

alternated with a projection

$$\hat{\mathbf{z}}_\Pi = \Pi(\hat{\mathbf{z}}), \tag{5.49}$$

where $\Pi : \mathbb{R}^n \times \mathbb{R}^{Tm} \mapsto \mathbb{R}^n \times \Sigma_{Tk}$ is the associated projection operator.

Figure 5.9 shows the comparison of the bus angles estimation errors for the different observers. it is seen that the RPO has the least error of all five observers. The Luenberger observer is completely unstable as a result of the FDIA, which was designed by compromising 19 sensor measurements. For the MMO, the value of $\tau = 0.1$ was used for the confidence value. For the ETLO, the value of $\upsilon = -0.01$ was used for the decreasing level of V. According to Theorem 5.3, it is proved that the resiliency of observer can be improved by including support prior, thus, UL1O

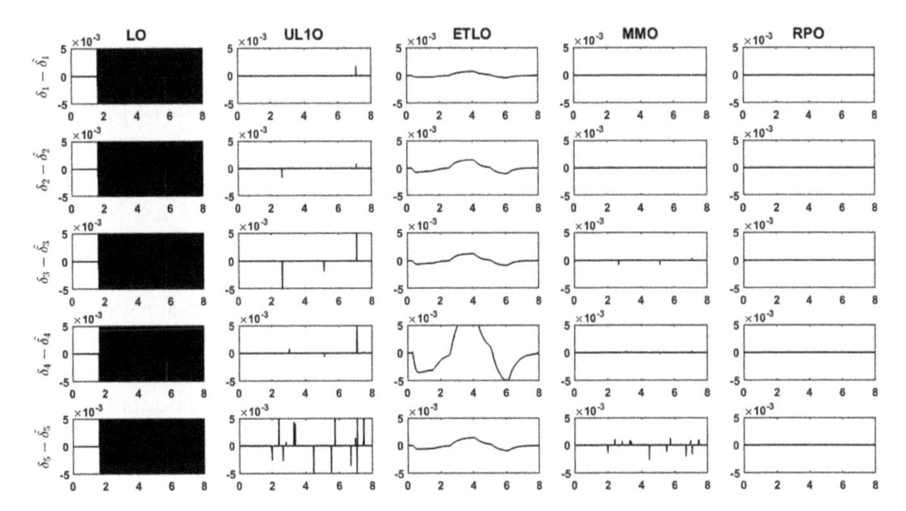

Fig. 5.9 A comparison result of estimation error of bus angles by five observers on IEEE 14-bus system (5.45) (LO: Luenberger observer, UL1O: unconstrained ℓ_1 observer, ETLO: event-triggered Luenberger observer, MMO: multi-model observer, RPO: resilient pruning observer)

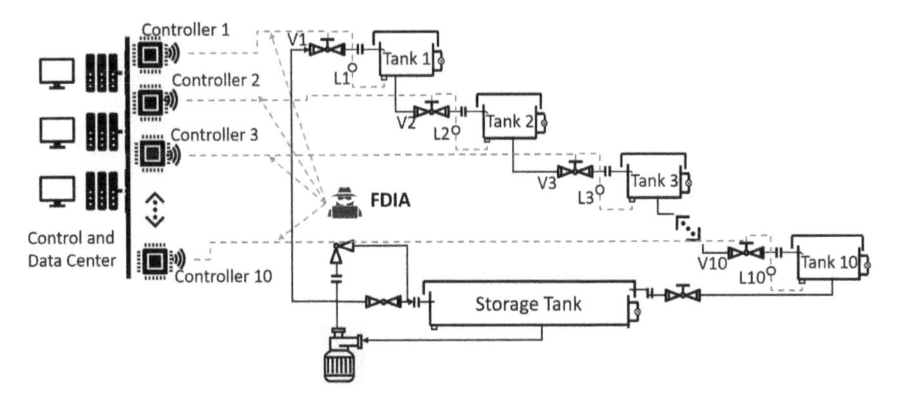

Fig. 5.10 A water distribution tank coupling control system under false data injection attacks (black solid lines are water pipelines, blue dotted lines are wireless data transmission lines for sensors data and control commands, orange dotted lines are the attack injection paths)

works worse than MMO using the measurement prior directly and RPO using the pruned support prior. Based on the Proposition 5.3, the pruning algorithm improves the precision of the prior information, thereby the localization precision of the measurement prior used in MMO is worse than the precision of the pruned support prior used in RPO. Although there is no strict theoretical proof, it can be seen in Fig. 5.9 that RPO has better resiliency than MMO. Moreover, ETLO has the most smooth estimation results since it used a projected gradient descent technique to solve the optimization program in (5.47), which scarifies partial resilient performance during recursive process.

5.5.2 Resilient Water Distribution System

In this subsection, we introduce another application example on a water tank coupling control system, shown in Fig. 5.10. The tank coupling system in Yang et al. 2020 is extended to an 11-tank system, which contains 10 operating water tanks and a storage tank. The goal is to regulate all operating tanks' water levels around desired values. The magnetic valves v at the entrance pipelines of operating tanks are controlled to adjust the tank water levels. The magnetic valve at the entrance of the storage tank is fixed at a constant opening value. It is assumed that there are water level measurement sensors and pressure sensors in the pipelines. The pressure sensors can measure the difference in water levels between adjoin tanks on each line. Thus, there are 19 measurements total. The water level adjustment process can be approximated by the LTI model:

$$\begin{aligned} \dot{\mathbf{h}} &= A\mathbf{h} + B\mathbf{v} \\ \mathbf{y} &= C\mathbf{h}, \end{aligned} \tag{5.50}$$

where $\mathbf{h}, \mathbf{v} \in \mathbb{R}^{10}$, $\mathbf{y} \in \mathbf{R}^{19}$. The system dynamics is given by

$$A = \begin{bmatrix} -0.5815 & 0 & 0 & 0 & 0 & 0 & 0 & 0 & 0 & 0 \\ 0.1870 & -0.5906 & 0 & 0 & 0 & 0 & 0 & 0 & 0 & 0 \\ 0 & 0.1870 & -0.5127 & 0 & 0 & 0 & 0 & 0 & 0 & 0 \\ 0 & 0 & 0.1870 & -0.5913 & 0 & 0 & 0 & 0 & 0 & 0 \\ 0 & 0 & 0 & 0.1870 & -0.5632 & 0 & 0 & 0 & 0 & 0 \\ 0 & 0 & 0 & 0 & 0.1870 & -0.5098 & 0 & 0 & 0 & 0 \\ 0 & 0 & 0 & 0 & 0 & 0.1870 & -0.5278 & 0 & 0 & 0 \\ 0 & 0 & 0 & 0 & 0 & 0 & 0.1870 & -0.5547 & 0 & 0 \\ 0 & 0 & 0 & 0 & 0 & 0 & 0 & 0.1870 & -0.5958 & 0 \\ 0 & 0 & 0 & 0 & 0 & 0 & 0 & 0 & 0.1870 & -0.5965 \end{bmatrix}$$

$$B = \begin{bmatrix} 0.8315 & -0.8450 & 0 & 0 & 0 & 0 & 0 & 0 & 0 & 0 \\ 0 & 0.9941 & -0.8450 & 0 & 0 & 0 & 0 & 0 & 0 & 0 \\ 0 & 0 & 0.9914 & -0.8450 & 0 & 0 & 0 & 0 & 0 & 0 \\ 0 & 0 & 0 & 0.8971 & -0.8450 & 0 & 0 & 0 & 0 & 0 \\ 0 & 0 & 0 & 0 & 0.9610 & -0.8450 & 0 & 0 & 0 & 0 \\ 0 & 0 & 0 & 0 & 0 & 0.8284 & -0.8450 & 0 & 0 & 0 \\ 0 & 0 & 0 & 0 & 0 & 0 & 0.8844 & -0.8450 & 0 & 0 \\ 0 & 0 & 0 & 0 & 0 & 0 & 0 & 0.9831 & -0.8450 & 0 \\ 0 & 0 & 0 & 0 & 0 & 0 & 0 & 0 & 0.9584 & -0.8450 \\ 0 & 0 & 0 & 0 & 0 & 0 & 0 & 0 & 0 & 0.9919 \end{bmatrix}$$

$$C = \begin{bmatrix} & & & & I_{10} & & & & & \\ 1 & -1 & 0 & 0 & 0 & 0 & 0 & 0 & 0 & 0 \\ 0 & 1 & -1 & 0 & 0 & 0 & 0 & 0 & 0 & 0 \\ 0 & 0 & 1 & -1 & 0 & 0 & 0 & 0 & 0 & 0 \\ 0 & 0 & 0 & 1 & -1 & 0 & 0 & 0 & 0 & 0 \\ 0 & 0 & 0 & 0 & 1 & -1 & 0 & 0 & 0 & 0 \\ 0 & 0 & 0 & 0 & 0 & 1 & -1 & 0 & 0 & 0 \\ 0 & 0 & 0 & 0 & 0 & 0 & 1 & -1 & 0 & 0 \\ 0 & 0 & 0 & 0 & 0 & 0 & 0 & 1 & -1 & 0 \\ 0 & 0 & 0 & 0 & 0 & 0 & 0 & 0 & 1 & -1 \end{bmatrix}$$

The model in (5.50) was discretized using Euler discretization scheme with sampling time $0.01s$. A discrete LQR controller is designed using $Q = 10^3 \times \text{diag}\{2, 1, 1, 2, 1, 1, 2, 1, 1, 2\}$ and $R = 0.2 \times I_{10}$ to obtain the feedback control gain K to regulate the water levels at $\mathbf{h}_d = 0.01 * \mathbf{1}_{10}$, The control law is given by

$$\mathbf{v} = -K(\mathbf{h} - \mathbf{h}_d) - B^{-1}A\mathbf{h}_d + B^{-1}\mathbf{h}_d.$$

The attack percentage is set as $P_A = 0.6$, and by using the designed FDIA (5.6), it can bypass the bad data detection threshold. Due to the lack of actual auxiliary data for this case, sample support prior is created by generating uniformly distributed random numbers in the interval $[0, \quad 1]$ for each measurement node. These numbers represent the localization confidence values $\mathbf{p}_i$'s used in Algorithm 5.2. The generated prior information represents a localization algorithm whose performance is comparable to the random flip of a fair coin. The reason for this is to show how the observers perform using a relatively poor localization algorithm. The precision of the generated support prior is shown in Fig. 5.11, the mean of precision is 0.5588. For a more realistic situation, possible candidate auxiliary variables include atmospheric data like temperature, humidity, atmospheric pressure, or any other values that can affect the flow of water in a long pipe. Market data and time of day are also great candidates for auxiliary variables.

Then the resilient estimation schemes described in the last subsection are also implemented for this system. The comparison of the resulting estimation errors is presented in Table 5.1, in which relative mean square error and maximum absolute error are given. Again, as seen in the table, the RPO outperforms the other observers in terms of the given error metrics.

5.5.3 *Resilient Wheeled Mobile Robot*

For this example, a nonlinear observer scheme based on prior information is given for the resilient motion control of wheeled mobile robot. Non-holonomic wheeled

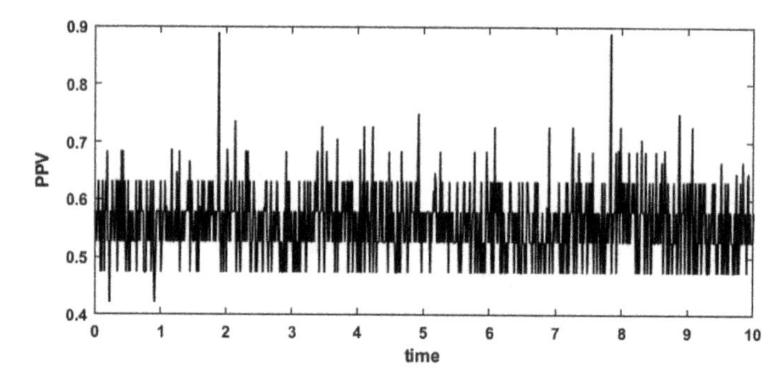

Fig. 5.11 The precision of support prior generated by the localization algorithm in Algorithm 5.1 for water tank coupling system (The mean of precision is 0.5588)

Table 5.1 Error metric values for four resilient observers on water tank coupling system (RMS Metric: relative mean square error, Max. Ans. Metric: maximum absolute error)

	RMS Metric				Max. Ans. Metric			
	LO	UL1O	MMO	RPO	LO	UL1O	MMO	RPO
e_1	1.4434	2.5657e-6	2.0794e-6	3.5704e-10	21.8421	4.6746e-5	4.6655e-5	5.7127e-9
e_2	1.5088	5.8117e-6	2.1444e-8	4.3079e-10	23.0772	1.5826e-4	5.1996e-7	5.7700e-9
e_3	0.8018	4.1172e-8	2.1381e-10	4.3901e-10	13.9374	1.1886e-6	3.4310e-9	9.6873e-9
e_4	0.7350	4.5476e-6	4.5476e-6	3.2479e-10	14.4943	1.4388e-4	1.4388e-4	4.4373e-9
e_5	0.5645	2.2444e-5	1.7216e-5	3.5302e-10	9.8116	4.7845e-4	3.7122e-4	4.8049e-9
e_6	1.0332	3.3578e-5	1.7473e-5	4.1156e-10	15.5191	5.5419e-4	3.7748e-4	8.1021e-9
e_7	1.1802	2.2776e-5	1.6834e-5	3.8149e-10	17.1720	4.1583e-4	3.7724e-4	5.5387e-9
e_8	1.2172	3.8198e-5	2.2591e-6	1.1470e-6	20.5512	0.0010	6.1343e-5	3.6289e-5
e_9	0.9802	2.2720e-5	2.2118e-5	2.0543e-6	18.1152	3.4706e-4	3.4706e-4	6.2776e-5
e_{10}	2.6151	1.0291e-4	2.0641e-6	2.3344e-7	28.5826	0.0030	6.1424e-5	7.3509e-6

mobile robot is considered with IIoT sensors, its dynamical and kinematic model can be described as Dhaouadi and Hatab (2013)

$$\begin{aligned} \dot{\mathbf{q}} &= M^{-1}(-D\mathbf{q} + B\boldsymbol{\tau}) + \mathbf{w} \triangleq g(\mathbf{x}, \mathbf{u}) + \mathbf{w} \\ \begin{bmatrix} \dot{\theta} \\ \cdots \\ \dot{\mathbf{z}} \end{bmatrix} &= \begin{bmatrix} 0 \ 1 \\ \cdots \\ C(\theta) \end{bmatrix} \mathbf{q} \triangleq \bar{C}(\theta)\mathbf{q}, \end{aligned} \tag{5.51}$$

where $\mathbf{q} = [v \ \ \omega]^\top$ is the generalized body velocities vector, $\mathbf{u} \triangleq \boldsymbol{\tau} = [\tau_R \ \ \tau_L]^\top$ is a vector of the wheels torques, and $\mathbf{z} = [x \ \ y]^\top$ is the task-space position vector, $\mathbf{x} = [\theta \ \ v \ \ \omega]^\top$ is defined as a state vector, $\mathbf{w} \sim \mathcal{N}(0, R)$ is the process noise in dynamics. The kinematic and dynamical parameters are given by:

$$M = \begin{bmatrix} m & 0 \\ 0 & md^2 + J \end{bmatrix}, D = \begin{bmatrix} 0 & -md\omega \\ md\omega & 0 \end{bmatrix}, B = \frac{1}{r}\begin{bmatrix} 1 & 1 \\ L & -L \end{bmatrix}, C(\theta) = \begin{bmatrix} \cos(\theta) & -d\sin(\theta) \\ \sin(\theta) & d\cos(\theta) \end{bmatrix}.$$

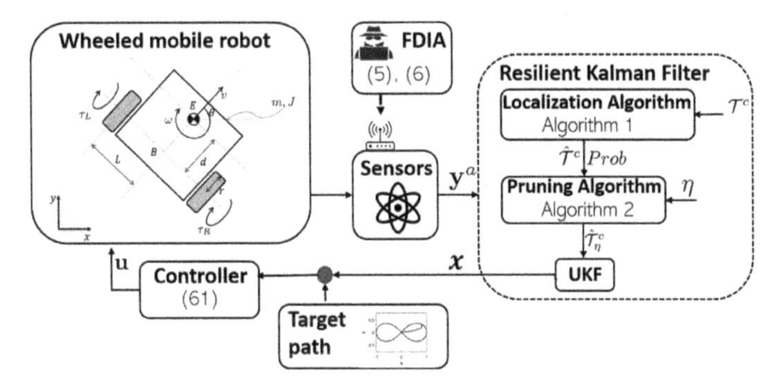

Fig. 5.12 Block diagram depiction of the resilient motion control of wheeled mobile robot

The corresponding measurement system is given by

$$\mathbf{y} = \begin{bmatrix} 1 & 0 \\ 0 & 1 \\ 1/4r & L/4r \\ 1/4r & -L/4r \\ \cos(\theta) & -d\sin(\theta) \\ \sin(\theta) & d\cos(\theta) \end{bmatrix} \cdot \mathbf{q} + \mathbf{v} \triangleq f(\mathbf{x}) + \mathbf{v} + \mathbf{e}, \tag{5.52}$$

where $\mathbf{v}$ denotes measurement noise terms, $\mathbf{e}$ denotes the attack vector.

Given a desired 2D "Fig. 5.8" path described by the continuous function:

$$\mathbf{z}_d = \begin{bmatrix} x_d(t) \\ y_d(t) \end{bmatrix} = \begin{bmatrix} \frac{a\cos(t)}{1+\sin^2(t)} \\ \frac{a\sin(t)\cos(t)}{1+\sin^2(t)} \end{bmatrix}, \quad \theta_d(t) = \arctan\left(\frac{y_d(t)}{x_d(t)}\right),$$

a stable path-tracking controller was given in Zheng and Anubi (2020) as

$$\boldsymbol{\tau} = B^{-1}(M\mathbf{u} + D\mathbf{q}), \tag{5.53}$$

where $\mathbf{u} = -k_q(\mathbf{q} - \mathbf{q}_d) + \dot{\mathbf{q}}_d - \bar{C}(\theta)^\top \widetilde{\mathbf{e}}$, with

$$\begin{aligned} \mathbf{q}_d &= C^{-1}(\theta)(\dot{\mathbf{z}}_d - k_e \mathbf{e}_\mathbf{z}), \\ \dot{\mathbf{q}}_d &= -k_e(\dot{C}^{-1}(\theta)\mathbf{e}_\mathbf{z} + \mathbf{q}) + C^{-1}(\theta)[\ddot{\mathbf{z}}_d + (k_e + C(\theta)\dot{C}^{-1}(\theta))\dot{\mathbf{z}}_d], \end{aligned}$$

and k_q, k_e are positive scalar control gains.

The next task is to design a nonlinear observer to recover the real state $\mathbf{x}$ under the compromised measurements $\mathbf{y}$, shown in Fig. 5.12. According to Theorem 5.4, the precision of $\hat{\mathcal{T}}_\eta^c$ can achieve 100% with a probability lower bound. Thus, Unscented Kalman Filter (UKF) can be used on the safe subset of measurements denoted by $\hat{\mathcal{T}}_\eta^c$. The control system with resilient Kalman filter is shown schematically in Fig. 5.12.

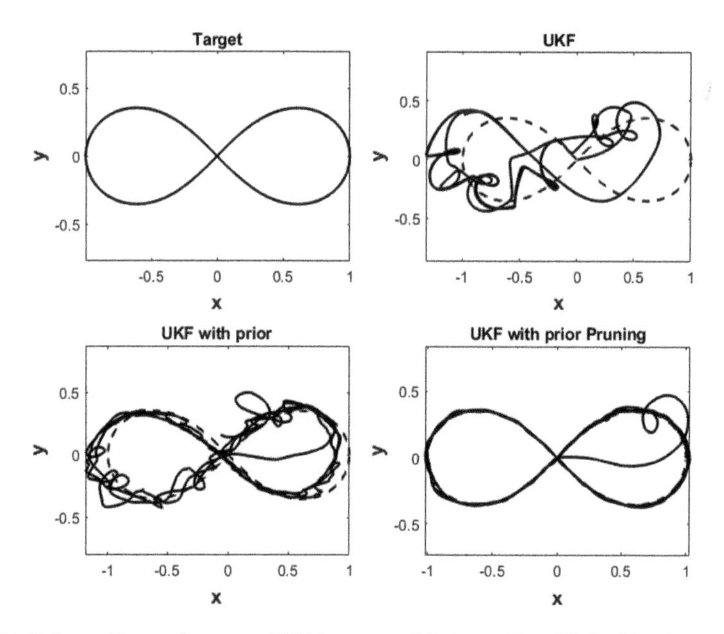

Fig. 5.13 Path tracking performance (UKF: unscented Kalman filter, UKF with prior: unscented Kalman filter with the prior generated by localization algorithm in Algorithm 5.1, UKF with prior pruning: unscented Kalman filter with pruned prior generated by Algorithm 5.2)

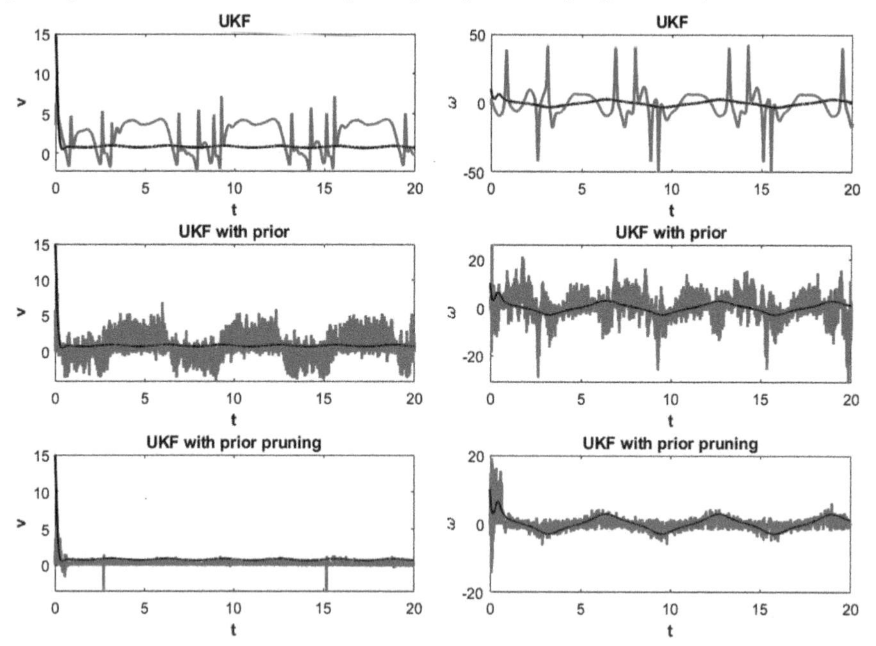

Fig. 5.14 Estimations of robot's forward velocity v and angular velocity ω by three observers (Black line is the nominal state estimation, blue line is the estimation by those three observers in presence of FDIA)

Figures 5.13 and 5.14 show the comparisons of the tracking performances between UKF, UKF with the prior, and UKF with prior and pruning. It is well known in the literature that KF cannot recover exact states in the presence of FDIA. Figures 5.13 and 5.14 confirm this fact. Specifically, it is seen that the path-tracking task and state estimation totally fail with only UKF. By adding prior information obtained by the localization algorithm whose mean of precision is around 0.6, the motion control performance is improved but has big oscillatory due to the imperfect precision. However, with the developed pruning algorithm, the robot was able to track the reference path very closely and smoothly.

5.6 Conclusion

In this chapter, a resilient observer design with prior pruning was described. First, it was shown that good support prior (better than the random flip of a fair coin) can result in significant improvement over well-known resiliency limits in literature. Next, a pruning algorithm was given to improve the resulting localization precision without additional training effort. This makes the support information more useful for estimation purposes. Finally, a pruning-based observer scheme was given and analyzed. It was shown that the resulting observer outperforms well-known resilient observers in literature. Moreover, other minor contributions of this chapter include a formal definition of successful FDIA and associated optimization-based FDIA design.

References

M. Abbaszadeh, L.K. Mestha, C. Bushey, D.F. Holzhauer, Automated attack localization and detection. U.S. Patent No. 10,417,415 (2019)

O.M. Anubi, C. Konstantinou, R. Roberts, Resilient optimal estimation using measurement prior (2019). arXiv: 1907.13102

O.M. Anubi, C. Konstantinou, C.A. Wong, S. Vedula, Multi-model resilient observer under false data injection attacks, in *2020 IEEE Conference on Control Technology and Applications (CCTA)* (IEEE, 2020), pp. 1–8

O.M. Anubi, L. Mestha, H. Achanta, Robust resilient signal reconstruction under adversarial attacks (2018). arXiv:1807.08004

O.M. Anubi, C. Konstantinou, Enhanced resilient state estimation using data-driven auxiliary models. IEEE Trans. Ind. Inf. **16**(1), 639–647 (2019)

M. Bishop, What is computer security? IEEE Sec. Priv. **1**(1), 67–69 (2003)

A. Burg, A. Chattopadhyay, K.Y. Lam, Wireless communication and security issues for cyber-physical systems and the Internet-of-Things. Proc. IEEE **106**(1), 38–60 (2017)

E.J. Candes, T. Tao, Decoding by linear programming. IEEE Trans. Inf. Theory **51**(12), 4203–4215 (2005)

E. Candes, T. Tao, The Dantzig selector: statistical estimation when p is much larger than n. Ann. Stat. **35**(6), 2313–2351 (2007)

E.J. Candès, J. Romberg, T. Tao, Robust uncertainty principles: exact signal reconstruction from highly incomplete frequency information. IEEE Trans. Inf. Theory **52**(2), 489–509 (2006)

T.M. Chen, S. Abu-Nimeh, Lessons from stuxnet. Computer **44**(4), 91–93 (2011)

M.S. Chong, M. Wakaiki, J.P. Hespanha, Observability of linear systems under adversarial attacks, in *2015 American Control Conference (ACC)*. (IEEE, 2015), pp. 2439–2444

A.O. de Sá, L.F.R. da Costa Carmo, R.C. Machado, Covert attacks in cyber-physical control systems. IEEE Trans. Ind. Inf. **13**(4), 1641–1651 (2017)

Y. Deldjoo, T.D. Noia, F.A. Merra, A survey on adversarial recommender systems: from attack/defense strategies to generative adversarial networks. ACM Comput. Surv. (CSUR) **54**(2), 1–38 (2021)

P. Derler, E.A. Lee, A.S. Vincentelli, Modeling cyber-physical systems. Proc. IEEE **100**(1), 13–28 (2011)

R. Dhaouadi, A.A. Hatab, Dynamic modelling of differential-drive mobile robots using lagrange and newton-euler methodologies: A unified framework. Advances in Robotics & Automation **2**(2), 1–7 (2013)

D.L. Donoho, Compressed sensing. IEEE Trans. Inf. Theory **52**(4), 1289–1306 (2006)

M. Esmalifalak, L. Liu, N. Nguyen, R. Zheng, Z. Han, Detecting stealthy false data injection using machine learning in smart grid. IEEE Syst. J. **11**(3), 1644–1652 (2014)

C. Fang, Y. Qi, P. Cheng, W.X. Zheng, Optimal periodic watermarking schedule for replay attack detection in cyber-physical systems. Automatica **112**, 108698 (2020)

T. Fawcett, An introduction to ROC analysis. Pattern Recogn. Lett. **27**(8), 861–874 (2006)

H. Fawzi, P. Tabuada, S. Diggavi, Secure estimation and control for cyber-physical systems under adversarial attacks. IEEE Trans. Autom. Control **59**(6), 1454–1467 (2014)

Manuel Fernández, Stuart Williams, Closed-form expression for the poisson-binomial probability density function. IEEE Trans. Aerosp. Electron. Syst. **46**(2), 803–817 (2010)

M. Fornasier, H. Rauhut, Compressive Sensing. Handbook of Math. Methods Imaging **1**, 187–229 (2015)

M.P. Friedlander, H. Mansour, R. Saab, Ö. Yilmaz, Recovering compressively sampled signals using partial support information. IEEE Trans. Inf. Theory **58**(2), 1122–1134 (2011)

Z. Guo, D. Shi, K.H. Johansson, L. Shi, Optimal linear cyber-attack on remote state estimation. IEEE Trans. Control Netw. Syst. **4**(1), 4–13 (2016)

Y. He, G.J. Mendis, J. Wei, Real-time detection of false data injection attacks in smart grid: a deep learning-based intelligent mechanism. IEEE Trans. Smart Grid **8**(5), 2505–2516 (2017)

Y. Huang, J. Tang, Y. Cheng, H. Li, K.A. Campbell, Z. Han, Real-time detection of false data injection in smart grid networks: an adaptive CUSUM method and analysis. IEEE Syst. J. **10**(2), 532–543 (2014)

S.K. Khaitan, J.D. McCalley, Design techniques and applications of cyber physical systems: a survey. IEEE Syst. J. **9**(2), 350–365 (2014)

E.A. Lee, CPS foundations, in *Design Automation Conference* (IEEE, 2010), pp. 737–742

C. Lee, H. Shim, Y. Eun, Secure and robust state estimation under sensor attacks, measurement noises, and process disturbances: Observer-based combinatorial approach, in *2015 European Control Conference (ECC)*, (IEEE, 2015), pp. 1872–1877

R.M. Lee, M.J. Assante, T. Conway, German steel mill cyber attack. Ind. Control Syst. **30**, 62 (2014)

Y. Liu, P. Ning, M.K. Reiter, False data injection attacks against state estimation in electric power grids. ACM Trans. Inf. Syst. Sec. (TISSEC) **14**(1), 1–33 (2011)

M. Liu, G. Chowdhary, B.C. Da Silva, S.Y. Liu, J.P. How, Gaussian processes for learning and control: a tutorial with examples. IEEE Control Syst. Mag. **38**(5), 53–86 (2018)

H. Liu, Y. Mo, J. Yan, L. Xie, K.H. Johansson, An online approach to physical watermark design. IEEE Trans. Autom. Control **65**(9), 3895–3902 (2020)

Y. Mo, B. Sinopoli, False data injection attacks in control systems, in *Preprints of the 1st workshop on Secure Control Systems* (2010), pp. 1–6

Y. Mo, B. Sinopoli, On the performance degradation of cyber-physical systems under stealthy integrity attacks. IEEE Trans. Autom. Control **61**(9), 2618–2624 (2015)

Y. Nakahira, Y. Mo, Attack-Resilient $\mathscr{H}_2$, $\mathscr{H}_\infty$, and ℓ_1 state estimator. IEEE Trans. Autom. Control **63**(12), 4353–4360 (2018)

New York control area load zone map. [Online]. https://www.nyiso.com/documents/20142/1397960/nyca_zonemaps.pdf

N. Y. I. S. Operator, "Load Data," [Online]. https://www.nyiso.com/load-data

M. Ozay, I. Esnaola, F.T.Y. Vural, S.R. Kulkarni, H.V. Poor, Machine learning methods for attack detection in the smart grid. IEEE Trans. Neural Netw. Learn. Syst. **27**(8), 1773–1786 (2015)

M., Pajic, P. Tabuada, I. Lee, G.J. Pappas, Attack-resilient state estimation in the presence of noise, in *2015 54th IEEE Conference on Decision and Control (CDC)* (IEEE, 2015), pp. 5827–5832

F. Pasqualetti, F. Dörfler, F. Bullo, Attack detection and identification in cyber-physical systems. IEEE Trans. Autom. Control **58**(11), 2715–2729 (2013)

K. Pelechrinis, M. Iliofotou, S.V. Krishnamurthy, Denial of service attacks in wireless networks: the case of jammers. IEEE Commun. Surv. Tutor. **13**(2), 245–257 (2010)

Power System Test Case Archive, 14 bus power flow test case. [Online]. http://labs.ece.uw.edu/pstca/pf14/pg_tca14bus.htm

C.E. Rasmussen, Gaussian processes in machine learning, in *Summer school on machine learning* (Springer, Berlin, Heidelberg, 2003), pp. 63–71

Scholtz, E. (2004). Observer-based monitors and distributed wave controllers for electromechanical disturbances in power systems (Doctoral dissertation, Massachusetts Institute of Technology)

T. Shinohara, T. Namerikawa, Z. Qu, Resilient reinforcement in secure state estimation against sensor attacks with a priori information. IEEE Trans. Autom. Control **64**(12), 5024–5038 (2019)

Y. Shoukry, P. Tabuada, Event-triggered state observers for sparse sensor noise/attacks. IEEE Trans. Autom. Control **61**(8), 2079–2091 (2015)

Y. Shoukry, P. Nuzzo, A. Puggelli, A.L. Sangiovanni-Vincentelli, S.A. Seshia, P. Tabuada, Secure state estimation for cyber-physical systems under sensor attacks: a satisfiability modulo theory approach. IEEE Trans. Autom. Control **62**(10), 4917–4932 (2017)

J. Slay, M. Miller, Lessons learned from the maroochy water breach, in *International Conference on Critical Infrastructure Protection.* (Springer, Boston, MA, 2007), pp. 73–82

T. Sui, Y. Mo, D. Marelli, X.M. Sun, M. Fu, The Vulnerability of Cyber-Physical System under Stealthy Attacks. IEEE Trans. Autom. Control (2020)

R. Urtasun, T. Darrell, Sparse probabilistic regression for activity-independent human pose inference, in *2008 IEEE Conference on Computer Vision and Pattern Recognition* (IEEE, 2008), pp. 1–8

S. Weerakkody, B. Sinopoli, Detecting integrity attacks on control systems using a moving target approach, in *2015 54th IEEE Conference on Decision and Control (CDC)* (IEEE, 2015), pp. 5820–5826

J. Yang, C. Zhou, Y.C. Tian, C. An, A Zoning-Based Secure Control Approach Against Actuator Attacks in Industrial Cyber-Physical Systems. IEEE Trans. Industr. Electron. **68**(3), 2637–2647 (2020)

K. Zetter, A cyber attack has caused confirmed physical damage for the second time ever (2015)

Y. Zheng, O.M. Anubi, Attack-resilient observer pruning for path-tracking control of wheeled mobile robot, in *2020 ASME Dynamic Systems and Control(DSC) Conference, ASME* (2020), pp. 1–9

Y. Zheng, O.M. Anubi, Attack-resilient weighted ℓ_1 observer with prior pruning, in *2021 American Control Conference (ACC)* (2021)

Chapter 6
Framework for Detecting APTs Based on Steps Analysis and Correlation

Hope Nkiruka Eke, Andrei Petrovski, Hatem Ahriz, and M. Omar Al-Kadri

6.1 Introduction

Safety and security measures in place in terms of maintaining resource availability, integrity, and confidentiality of the operational CPS state against cyber-threat such as APT remain one of the biggest challenges facing organizations and industries at various levels of operation (Eke et al. 2020).

The CPS systems are composed of computer and subsystems that are interconnected based on the context within which an exchange of vital information through computer network takes place (Monostori et al. 2016; Cardenas et al. 2009; Jazdi 2014; Petrovski et al. 2015). CPS such as distributed control system (DCS) and SCADA contain control systems that are used in critical infrastructures such as nuclear power plants (Eke et al. 2020; Kim et al. 2000), water, sewage, and irrigation systems (Humayed et al. 2017).

An APT, presented in Fig. 6.1, is an attack that navigates around defences, breach networks, and evades detection, due to APTs stealthy characteristics and sophisticated levels of expertise and significant resources of contemporary attackers (Eke et al. 2019). While APTs have been attracting an increasing attention from the industrial security community, the current APTs best practices require a wide range of security countermeasures, resulting in a multi-layered defence approach that opens new research directions (Majdani et al. 2020). This type of attacks has drawn special attention to the possibilities of APT attacks on CPS devices, such as SCADA-based system. There have been few cases of successful attacks on ICS as recorded in NJC-

H. N. Eke (✉) · A. Petrovski · H. Ahriz
Robert Gordon University, Garthdee Road, Aberdeen AB10 7GJ, UK
e-mail: h.eke@rgu.ac.uk

A. Petrovski
e-mail: a.petrovski@rgu.ac.uk

H. Ahriz
e-mail: h.ahriz@rgu.ac.uk

M. O. Al-Kadri
Birmingham City University, Birmingham B4 7XG, UK
e-mail: omar.alkadri@bcu.ac.uk

M. Abbaszadeh and A. Zemouche (eds.), *Security and Resilience in Cyber-Physical Systems*, https://doi.org/10.1007/978-3-030-97166-3_6

CIC (2017) and Slowik (2019), these led to several attempts in developing methods to detect intrusions within network and isolated devices.

Most of these approaches focus on detection of APT attack with respect to a specific domain. Work by authors in Nissim et al. (2015) detects malicious PDFs based on whitelists and their compatibility as viable PDF files while study in Chandra et al. (2016) that focus on "Tokens" and utilizes mathematical and computational analysis to filter spam emails focus on detection of only one step of APT lifecycle.

The computer systems used to control physical functions of the operating systems are not immune to the threat of today's sophisticated cyber-attacks and can be potentially vulnerable (Linda et al. 2009). Potential threats can affect ICS devices at different level. Hence, security of each component within each level is extremely important to avoid compromise on any level (Harris and Hunt 1999).

APT attacks on a control system can be considered as stealthy disturbances, carefully designed with highly sophisticated combination of different techniques to achieve a specifically targeted and highly valuable goal by attackers (Eke et al. 2020). These attackers are known to possess sophisticated levels of expertise and significant resources which allow them to create opportunities to achieve their objectives by using multiple attack vectors such as cyber, physical, and deception. However, a well-designed control system may repel against external disturbances such as Reconnaissance. The unknown and dynamic nature of designed disturbance rules poses a security threat to CPS, which can be vulnerable to various types of cyber-attacks without any sign of system component failure (Wu et al. 2016). Examples of these could be noticeable time delays and serious control system degradation as a result of control systems been vulnerable to a denial-of-service (DoS) attack.

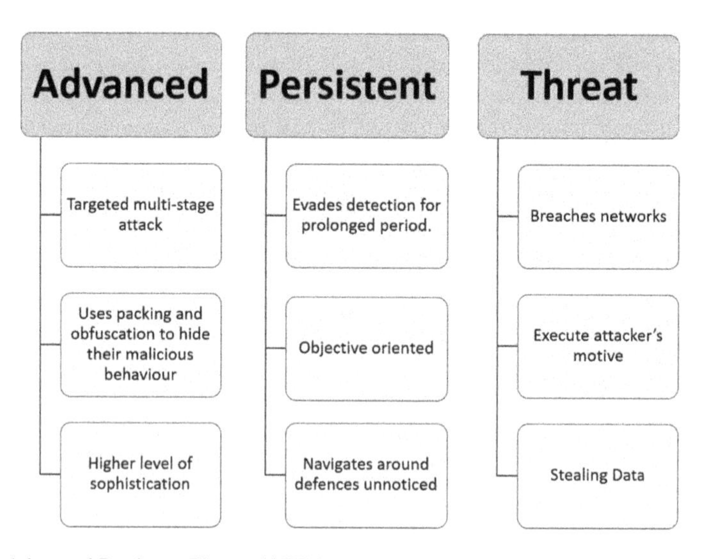

Fig. 6.1 Advanced Persistent Threats (APTs)

The successful removal or mitigating existing vulnerabilities, assessing whether a control system is experiencing any form of attack, and maintaining a secure and stable system state are the main CPS security.

6.1.1 Targeted APT Attack on CPSs

APT attacks have affected many organizations as far back as 1998, with the first public recorded targeted attack named Moonlight Maze (Thakur et al. 2016). This Moonlight Maze attack targeted Pentagon, National Aeronautics and Space Administration (NASA), the US Energy Department, research laboratories, and private universities by successfully compromised Pentagon computer networks, and accessed tens of thousands of file (Smiraus and Jasek 2011). Past years have seen an increase in the number of organizations coming forward, admitting they have been targeted. Unfortunately, in the bid to protect organization's image and to avoid providing hackers with feedback, majority of those organization are not willing to share the attack details.

However, the four main recorded targeted attacks malware tailored against ICSs are STUXNET, BLACKENERGY 2, HAVEX, and CRASHOVERRIDE (Lee et al. 2017; Domović 2017). STUXNET is the first ever recorded attack aimed at disrupting physical industrial processes resulting in violation of system availability, while CRASHOVERRIDE is the second and also the first known to specifically target the electric grid (NJCCIC 2017; Slowik 2019). CRASHOVERRIDE is not unique to any vendor or configuration but utilizes the knowledge of grid operations and network communications to cause disruptions resulting in electric outages (Lee et al. 2017; Hemsley and Fisher 2018).

6.1.2 Safety of Cyber-Physical Systems (CPSs)

CPS utilizes diverse communication platforms and protocols to increase efficiency and productivity. This is to reduce operational costs and further improve organization's support model (Odewale 2018). The complexity of the ICS architecture and the increased efforts of controlling physical functions in processing and analyzing data has led to an intensified interaction between control and business networks (Odewale 2018; Nazarenko and Safdar 2019). The possibility of deliberate targeted attacks as examined in Pasqualetti et al. (2015) on control systems and the daily operational challenges due to this increased cyber-physical interaction are on the high side (Humayed et al. 2017; Nazarenko and Safdar 2019).

Ensuring the security of these systems is critical in order to avoid any operational disruption. However, this requires a complex approach to identify and mitigate security vulnerabilities or compromise at all levels within the ICS to maintain resource availability, safety, integrity, and confidentiality, as well as becoming resilient against attacks (Cazorla et al. 2016). We have suggested and implemented a multi-layered security model based on ensemble deep neural networks approach to secure ICSs.

The contribution of this chapter can be summarized as follows:

- We discuss APT characteristics, lifecycle, and give examples of the most significant confirmed cases of attack on CPS devices.
- We propose a novel approach using ensemble deep neural networks for realizing multi-layered security detection for ICS devices. This approach takes RNNs variants to learn features from raw data in order to capture the malicious sequence patterns which reduce the cost of artificial feature engineering.
- We designed and implemented Deep APT Steps Analysis and Correlation (APT-DASAC)—a multi-layered security detection approach, that takes into consideration the distributed and multi-level nature of ICS architecture and reflects on the four main SCADA-based cyber-attacks. We further used stacked ensemble for APT-DASAC to combine networks' results for optimizing detection accuracy.
- A series of evaluation experiment, including individual APT step detection and attack-type classification, were carried out. The achieved results suggest that the proposed approach has got the attack detection capability and demonstrated that performance of attack detection techniques applied can be influenced by the nature of network transactions with respect to the domain of application.

6.1.3 Organization of Book Chapter

The remainder of this book chapter is organized as follows. Section 6.2 contains an overview of APT and APT lifecycle, brief discussion of related work directed toward the security of CPS. In Sect. 6.3, a detailed description of our proposed approach "architectural design of APT-DASAC" is discussed. The implementation of our APT-DASAC approach and the datasets used are discussed in Sect. 6.4. Experimental results are discussed in Sect. 6.5. Section 6.6 presents the conclusion of this book chapter.

6.2 Advanced Persistent Threats (APTs)

APTs and the actors behind them constitute a serious global threat. This type of attacks differs from common threats that seek to gain immediate advantage. APTs are broad in their targeting and processing. An APT is also very

- *resourceful*;
- *with well-defined objectives and purpose*;
- *uses sophisticated methods and technology*; and
- *substantially funded.*

6.2.1 Characteristics of APTs

An APT threat process follows a staged approach to target, penetrate, and exploit its target. Understanding the advanced, sophisticated, and persistent nature of APT is unavoidable in defending against such attacks.

- ***Advanced*** - The advanced nature of APT provides the attackers with the capability of maintaining prolonged existence through stealthy approach inside an organization once they successfully breach security controls. Attackers use sophisticated tools and techniques such as malware, if the malware is detected and removed, they change their tactics to secondary attack strategies as necessary (Giura and Wang 2012).
- ***Persistent*** - The meaning of "Persistent" is expanded to persistently launching spear-phishing attacks against the targets by navigating a victim's network from system to system, obtaining confidential information, monitoring network activity, and adapting to be resilient against new security measures while maintaining a stealthy approach to reach its target (Siddiqi and Ghani 2016). The mode of attack indicates the main functions of the APT-type malware, which usually placed more focus on spying instead of financial gain.
- ***Threat*** - The actors also have the capability of gaining access to electronically stored sensitive information other than the purpose of collecting national secrets or political espionage, based on the functions discovered, it is believed that this type of threats can also be applied to the cases in business or industrial espionage, spying acts, or even unethical detective investigations (Brand et al. 2010; Shashidhar and Chen 2011).

Examining the APT methods used to breach today's ICS security, it boils down to a basic understanding that attackers, especially those who have significant financial motivation, have devised an effective attack strategies centered on penetrating some of the most commonly deployed security controls. Most often it uses custom or dynamically generated malware for the initial breach and data-gathering step. The "Advanced" and "Persistent" are major features that differentiate APT from other cyber-attacks.

6.2.2 Life Cycle of APTs Attack

APT attacks are generally known to utilize a zero-day exploits of unpublished vulnerabilities in computer programs or operating systems in combination with social engineering techniques. This is to maximize the effectiveness of the exploits that target unpatched vulnerabilities. Launching an APT attack involves numerous hacking tools, a sophisticated pattern, high-level knowledge, and varieties of resources and processes. APTs proved extremely effective at infiltrating their targets and going undetected for extended periods of time, increasing their appeal to hackers who target businesses as highlighted in several large-scale security breaches (McClure et al. 2010; Alperovitch 2011; Villeneuve et al. 2013).

Although each attack is customized with respect to attacker's target and aims at various stages of the kill chain, the patterns of APT attacks are similar in most cases but differ in the techniques used at each stage. For this study, we will describe six basic APT attack phases as used in our study, based on the literature review in combination with the "Intrusion Kill Chain (IKC)" model, described in Giura and Wang (2012), Singh et al. (2019), Hutchins et al. (2011).

1. ***Reconnaissance and Weaponization*** **-** This stage involves information gathering about the target. This could be, but not limited to, about organizational environment, employees' personal details, the type of network, and defence target in use. The information gathering can be done through social engineering techniques, port scanning, and open-source intelligence (OSINT) tools.
2. ***Delivery*** **-** At this stage, attackers utilize the information gathered from reconnaissance stage to execute their exploits either directly or indirectly to the targets. In direct delivery, the attackers apply social engineering such as spear phishing by sending phishing email to target. While in indirect delivery, attacker will compromise a trusted third party, which could be a vendor or frequently visited website by the target and uses these to deliver an exploit.
3. ***Initial Intrusion and Exploitation*** **-** At this stage, attacker gains access to target's network by utilizing the credential information gathered through social engineering. The malware code delivered at this stage is downloaded, installed, and activate backdoor malware, creating a command and control (C&C) connection between the target machine and a remote attacker's machine. Once a connection to the target machine has been secured, the attacker continues to gather more relevant information such as security configuration, user names, and sniff passwords from target network while maintaining a stealthy behavior in preparation for next attack.
4. ***Lateral Movement and Operation*** **-** At this stage, once the attacker establishes communication between the target's compromised systems and servers, the attacker moves horizontally within the target network, identify the servers storing the sensitive information on users with high access privileges. This is to elevate their privileges to access sensitive data. This makes their activities undetectable or even untraceable due to the level of access they have. Attackers also create strategy to collect and export the obtained information.
5. ***Data Collection*** **-** This stage involves utilizing the privileged users credentials captured during the previous stage to gain access to the targeted sensitive data. With the attackers having a privileged access, they will now create redundant copies of C&C channels should there be any change in security configuration. Once the target information has been accessed, redundant copies are created at several staging points where the gathered information is packaged and encrypted before exfiltration.
6. ***Exfiltration*** **-** At this stage, once an attacker has gained full control of target systems, they proceed with the theft of intellectual property or other confidential data. The stolen information is transferred to attackers' external servers in the

form of encrypted package, password-protected zip files, or through clear web mail. The idea of transferring information to multiple servers is an obfuscation strategy to stop any investigation from discovering the final destination of the stolen data.

6.2.3 *Related Work*

Diverse approaches have been proposed and successfully implemented to address different types of attacks. These proposed methods have led to a significant pool of solutions geared toward addressing security and resilience of CPS devices. Most of these approaches focus on detection of attack with respect to a specific domain.

6.2.3.1 Attack Detection

One of this detection model is intrusion cyber-kill chain (IKC). This was created by Lockheed Martin analysts in 2011 to support a better detection and response to attacker's intrusions by applying the IKC model to describe different stages of intrusion (Hutchins et al. 2011; Assante and Lee 2015). Although this model is not directly applicable to the ICS-custom cyber-attacks, it serves as a great building foundation and concept to start with (Hutchins et al. 2011). Few other approaches in the literature include, but not limited to, the attack detection based on communication channels, a notion of stealthiness, false data injection attacks (FDI), and network information flow analysis.

Work in Carvalho et al. (2018) made use of the possibility of unprotected communication channels for sensor and actuator signals in plant, which may allow attackers to potentially inject false signals into the system. The authors model an approach to capture the vulnerabilities and the consequences of an attack on the ICSs, being focused on "The closed-loop control system architecture", where the plant is controlled by the supervisor through sensors and actuators in a traditional feedback loop. Their approach aims at detecting an active online attack and disables all controllable events after detecting the attack, preventing thereby the system from reaching a predefined set of unsafe states. This work is a complementary study to another work in Paoli et al. (2011), where the authors investigated an online active approach using a multiple-supervisor architecture that actively counteracts the effect of faults and introduces the idea of safe controllability in active fault-tolerant systems to characterize the conditions that must be satisfied when dealing with the issue of fault tolerance.

Other proposed approaches that mainly focus on APT detection based on network information flow analysis that is not specific for CPS as reviewed for this work include an APT attack detection method based on deep learning using information flows to analyze network traffic into IP-based network flows, reconstruct the IP information flow, and use deep learning models to extract features for detecting

APT attack IPs from other IPs (Do Xuan et al. 2022). The authors in Shang et al. (2021) propose an approach to detect the hidden C&C channel of unknown APT attacks using network flow-based C&C detection method as inspired from the belief that: (i) different APT attacks share the same intrusion techniques and services, (ii) unknown malware evolves from existing malware, and (iii) different malware groups share the same attributes resulting to hidden shared features in the network flows between the malware and the C&C server within different attacks. They applied deep learning techniques to deal with unknown malicious network flows and achieved an $f1 - score$ of 96.80%.

6.2.3.2 Attack Mitigation

Authors in Bai et al. (2017) considered a notion of stealthiness for stochastic CPS that is independent of the attack detection algorithm to quantify the difficulty of detecting an attack from the measurements. With the belief that the attacker knows the system parameters and noise statistics and can hijack and replace the nominal control input by characterizing the largest degradation of Kalman filtering induced by stealthy attacks. The study reveals that the nominal control input is the only critical piece of information to induce the largest performance degradation for right-inverting systems, while providing an achievability result that lower bounds of performance degradation that an optimal stealthy attack can achieve within non-right-inverting systems. While Miloševič et al in (2017) examined the presence of bias injection attacks for state estimation problem for stochastic linear dynamical system against the Kalman filter as an estimator equipped with the chi-squared been used as a detector of anomalies. This work suggests that the issue of finding a worst-case bias injection attack can be controlled to a certain degree.

Also, Xu et al. (2020) focus on a stealthy estimation attack that can modify the state estimation result of the CPS to evade detection. In their study, the chi-square statistic was used as a detector. A signaling game with evidence (SGE) was used to find the optimal attack and defense strategies that can mitigate the impact of the attack on the physical estimation, guaranteeing thereby CPS stability.

Furthermore, study on industrial fault diagnosis using deep Boltzmann machine and multi-grained scanning forest ensemble was done by Hu et al. (2018) and FDI (Eke et al. 2020). Also, the possibility of accurately reconstructing adversarial attacks using estimation and control of linear systems when sensors or actuators are corrupted (Fawzi et al. 2014) is studied in the quest for CPS security and more resilience against targeted attacks. The authors in Shi et al. (2021) considered the case of the FDI attack detection issue as a binary classification case and propose a statistical FDI attack detection approach based on a new dimensionality reduction method using a Gaussian mixture model and a semi-supervised learning algorithm to examine the coordinates of the data under the newly orthogonal axes obtained to establish FDI attacks if the outputs of the Gaussian mixture model exceed the pre-determined threshold.

6.3 APT Detection Framework

In this section, we present the description of our proposed APT-DASAC framework architectural design for APT intrusion detection. APT attack purposefully launched to target critical infrastructures, such as SCADA network as highlighted in Eke et al. (2019), is a multi-step attack. The detection of a single step of an APT itself does not imply detecting an APT attack (Eke et al. 2020). Hence, APT detection systems should be able to detect every single possible step applied by an APT attacker during the attack process.

6.3.1 Architectural Design of APT-DASAC

The design of our proposed model for APT intrusion detection system (IDS) is built to run through three stages. This involves implementing a multi-layered security detection approach based on Deep Leaning (DL) that takes into consideration the distributed and multi-level nature of the ICS architecture and reflect on the APT lifecycle for the four main SCADA cyber-attacks as suggested in Eke et al. (2020).

The implementation of our design model shown in Fig. 6.2 consists of three stages:

Stage 1: Data input and probing layer.
Stage 2: Data analysis layer.
Stage 3: Decision layer.

6.3.2 Three Layers of APT-DASAC

The processes taken to implement our proposed model "APT-DASAC" are discussed as follows.

For the purpose of this model explanation and illustration, the New Gas Pipeline (NGP) and University of New South Wales (UNSW-NB15) datasets were used. The specific step-by-step pseudocode for APT-DASAC and the detection process are described in the following subsection.

The first stage of this approach *"Data input and probing layer"* involves data gathering and pre-processing sample data by transforming the data into an appropriate data format ready to be used in the second stage *"Data analysis Layer"*. This second stage applies the core process of APT-DASAC, which takes stacked recurrent neural network (RNN) variant to learn the behavior of APT steps from the sequence data. These steps reflect the pattern of APT attack steps. In the final stage *"Decision Layer"*, we use ensemble RNN variants to integrate the output and make a final prediction result.

6.3.2.1 Step-by-Step Pseudocode for APT-DASAC Layers

The experimental implementation pseudocode of our proposed framework in Fig. 6.2 is represented by Algorithm 6.2–6.3 as used to build the proposed model:

- *Pseudocode for data pre-processing.*
- *Pseudocode for data analysis.*
- *Pseudocode for detection and prediction process.*

The ***pre-processing data*** stage takes raw network traffic data as an input from a specific problem domain, processes, and transforms the data into a meaningful data format that the algorithm requires by converting any symbolic attributes into usable features and deals with null values using ***Step 1 to Step 7c in Algorithm 6.2***. The output from this stage is a ***new transformed data*** containing valuable information that the analyses stage will utilize.

6.3.2.2 Data Input and Probing Layer

This layer consists of two modules: (i) Data Input and (ii) Probing Module. Algorithm 6.2 shows the steps for this module process.

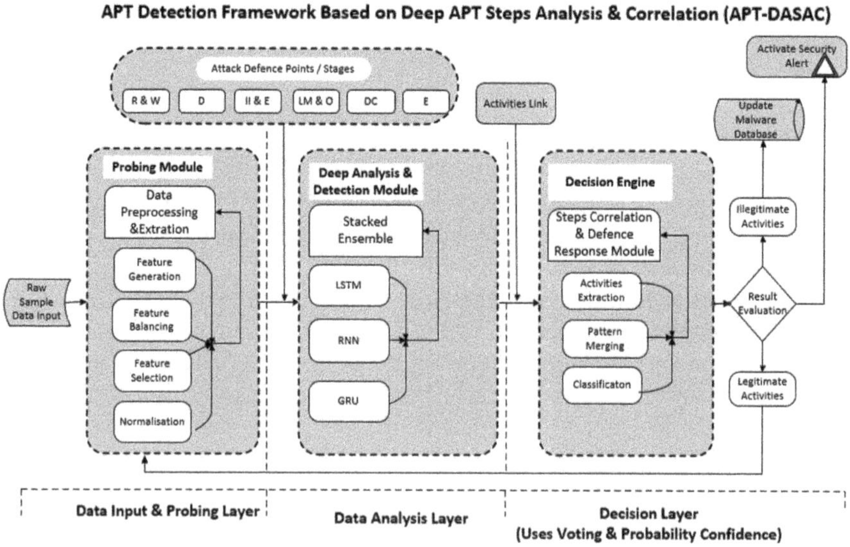

Fig. 6.2 Detection framework based on deep APT steps analysis and correlation (APT-DASAC)

1. ***Data Input*** involves data gathering, raw sample/simulated synthetic data been introduced into the system and transfer the collected data to probing module.
2. ***Probing Module*** involves data pre-processing and feature transformation which runs through four stages. Here all the data that has been collected and introduced into the module are encoded into numerical vector by the pre-processor ready to go through the neural network.

   a. ***Feature Transformation:*** UNSW-NB15 dataset consists of 42 features with three of these features been categorical (proto, service, and state) data. These three features need to be encoded into numeric feature vector as it goes to the neural network for analysis, classification, detection, and prediction. For this reason, Pandas $get_dummies()$ function was used, this function creates new dummy columns for each individual categorical feature. This leads to increase in the number of columns from 42 to 196 features available for onward analysis.
   b. ***Balancing Training and Testing Data Features:*** Both training and testing data contain different number of categorical features, this implies that $get_dummies()$ function will generate different number of columns for training and testing data. However, the number of features in both sets need to be the same. In this case, we deployed $set().union()$ function to balance the training and testing datasets.
   c. ***Normalization:*** At this stage, the $ZScore$ method of standardization is used to normalize all numerical features to preserve the data range, to introduce the dispersion of the series, and to improve model convergence speed during training.

6.3.2.3 Analysis Layer

The rate of attack detection is affected by the parameters used as these parameters have direct impact on attack detection. Based on this, several experiments with different network configuration were implemented to find the best optimal values for parameters such as learning rate and network structure.

Also, to achieve a good detection rate for rare attack steps while maintaining overall good model performance, two issues need to be considered—the rare attack class distribution and the difficulty of correctly classifying the rare class. When considering the class distribution, more emphasis should be placed on the classes with fewer examples. Secondly, more emphasis should be given to examples that are difficult to be correctly classified.

At this layer, the processed data are used to build a model that analyzes and distinguishes attack(s) from normal activities, taken note of the identified issues with class distribution and classification of rare attacks. The result of this layer is passed to Decision Engine layer.

Algorithm 6.1 Data Input and Probing Layer Pseudocode

- Pseudocode for Data Pre-processing

```
Step 1: Input the sample dataset
Step 2: Convert the symbolic attributes features
Step 3: Return new set of data
Step 4: Separate the instances of dataset into classes
        (y)
Step 5: Scale & normalize data (x_(t)) into values from
        [0 to 1]
Step 6: Split dataset into training and testing data
Step 7: Prepare and store transformed training and testing data
    Step 7a: Balance & reshape the training & testing
             data features
    Step 7b: Return balanced & reshaped training &
             testing data
    Step 7c: Pickle transformed data into a byte stream
             and store it in a file/database (.pki)
```

Algorithm 6.2 Analysis Layer Pseudocode

- Pseudocode for Sequence Data Training and Testing

```
During the training and testing stage, steps 8a-8e
are followed in each iteration.

Step 8: Train the model with this new training dataset
    Step8a: Sequentially fetch a sample data (x_(t))
            from the training set
    Step8b: Estimate the probability (p) that the
            example should be used for training
    Step8c: Generate a uniform random real number Âµ
            between 0 and 1
    Step8d: If Âµ < p, then use x_(t)to update the RNN by
            (5) for any training sample (x_(i), y_(i))
    Step8e: Repeat steps 1-4 (Algorithm 6.1) until there is no
            sample left in the training set
Step 9: Test model with testing data from Step 7b
Step10: Compute and evaluate the model performance
        accuracy output - classification, detection
        and prediction
```

6.3.2.4 Decision Layer

This layer operates using three approaches: firstly, it receives information from the analysis layer and extract the attack step present. Secondly, it processes this information and links it to the related attack steps. Lastly, it uses voting and probability confidence to establish if the attack is a potential chain of attack campaign is found, and if it is consistent with other attack campaigns.

Algorithm 6.3 Decision Layer Pseudocode

- *Pseudocode for Analysis, Detection and Prediction*

```
In analysis detection and prediction stage, steps
11-16 are followed in each iteration.
Step11:  Set ip_units, lstm_units, op_units and
         optimizer to define LST Network (DL)
Step12:  Fetch the processed data (x_(i))
         #pre-processed data through steps 1-7 (Algorithm 6.1)
Step13:  Select specified training window size (tws)
         and arrange x_(i) accordingly
Step14a: for n_epochs and batch_size do #each iteration
Step14b: Take the input vector within specified
         training window size (x_(tws)) at time (t)
         together with previous information,
         initially set to 0
Step14c: Train the Network L with x_(tws+1))
Step14d: end for
Step15:  Run Predictions using L
Step16:  Calculate the categorical_loss_function L(o,y)
Step17:  Output result
     Step17a: Percentage detection rate of individual
              attacks detected
     Step17b: Overall detection rate
     Step17c: Confirmation if there is any existence
              or complete APT steps (full APT scenario)
```

6.3.2.5 Attack Step Impacts

The attack impact is determined at this stage through the decision engine by correlating the output from the analysis layer using probability confidence to check for any presence of security risks. If an attack or security risk is present, it requests the defence response module to raise a security alert. This is checked with the previously detected step to see if this could be related to the newly discovered security risk alert. This is to reconstruct APT attack campaign steps, and hence highlights an APT campaign scenario so that an appropriate action can be taken.

The impact of an attack can be considered as low depending on the attack activity stage. However, if this stage can be linked with other attack steps to show that it is part of that attack campaign, forming a full APT step cycle, then the impact at this stage can be considered as high. With this information in mind an appropriate response can be taken.

6.4 Implementation of APT-DASAC Approach

In this section, we describe the platform and the approach taken to implement the APT-DASAC. These include the implementation setup, the hyperparameter settings used, and the datasets used.

6.4.1 Implementation Setup

The ensemble RNN-based attack detection models as explained in Eke et al. (2020) were implemented. The network topology and payload information values of the NGP dataset containing 214,580 Modbus network packets with 60,048 packets that are associated with cyber-attacks were used. These attacks are placed into 7 different categories with 35 different specific attack types as explained in Turnipseed (2020), Morris and Gao (2014). These attack categories align with APT lifecycle. Figures 6.3 and 6.4 show the number of records in each of the categories and the main four types of attacks as contained in the NGP data. During the experimental setup, the first task was focused on deriving hyperparameter values for best performance model. Secondly, the best hyperparameter values were implemented in measuring the model performance.

The standard data mining processes such as data cleaning and pre-processing, normalization, visualization, and classification were implemented in Python. The batch size of 124–300 epochs is run with a learning rate set in the range of 0.01–0.5 on a GPU-enabled TensorFlow network architecture. All the 17 features were used as input vector with 70% as training set and 30% as validation set for the multi-attack classification. The training dataset was normalized from 0 to 1. This was trained using sigmoid activation function through time with ADAM optimizer, sigmoid function was used on all the three gates and categorical cross-entropy as loss function for error rate. Also, these tasks were carried out with traditional machine learning (ML) classification algorithms—Decision Tree (DT). The ML classification result was compared to stacked Deep ensemble RNNs-LSTM result in order to further evaluate the APT steps detection capability of the experimental approach. Result evaluation is discussed in Sect. 6.5.

6.4.1.1 Hyperparameters Settings

- Batch sizes: 64 and 128.
- Learning rate: 0.0002–0.00005 with polynomial decay over all the epochs.
- Epochs: 100–300 epochs.
- Neural network: Four layers were used.
- Each of the hidden layers has a ***sigmoid/ReLU*** activation function applied to it to produce nonlinearity. This transforms the input into values usable by the output layer.

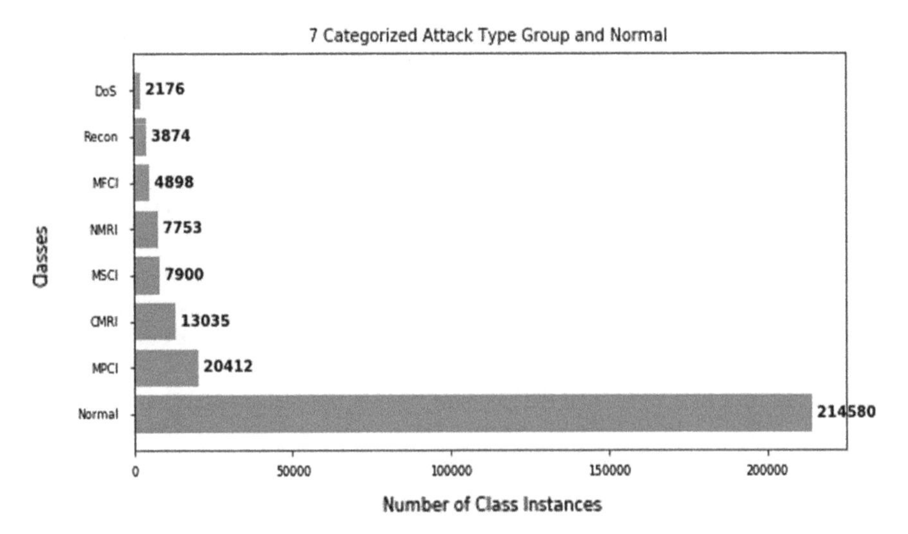

Fig. 6.3 NGP dataset records

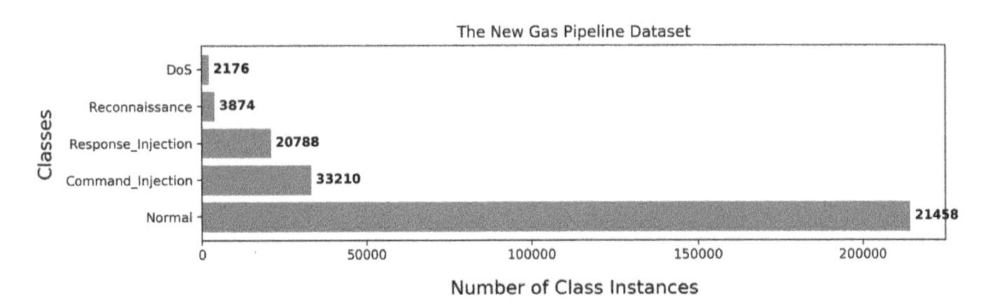

Fig. 6.4 Four main attack group and normal classes

- The ***softmax*** function is applied to the output layer to get probabilities of categories. This also helps in learning with ***cross-entropy loss*** function.
- Adaptive Moment Estimation, ***(ADAM)*** optimizer is used for the backpropagation to minimize the loss of categorical cross-entropy.
- The ***dropout*** is used to alleviate the over-fitting (used as regularization technique used to prevent over-fitting in neural networks. This randomly removes the units along with connections.

6.4.2 Implementation Dataset

Due to the specific dynamic nature of APT attack that does not follow a unique pattern, availability and accessibility of dataset containing realistic APT scenario have become a challenging issue when testing and comparing APT detection models. For the implementation of our approach, the NGP[1] and UNSW-NB15[2] datasets were used. Both datasets are available for research purposes.

6.4.2.1 New Gas Pipeline Dataset (NGP) Explained

The NGP data is generated through network transactions between a RTU and a MTU within a SCADA-based gas pipeline at Mississippi State University. This data was collected by simulating real attacks and operator activity on a gas pipeline using a novel framework for attack simulation as described in Turnipseed (2020) and Morris et al. (2015). The data contains three separate main categories of features—the network information, payload information, and labels.

The *network topologies* and the *payload information* values of SCADA systems are very important to understand the SCADA system performance and detecting if the system is in an out-of-bounds or critical state.[3]

6.4.2.2 Three Main Features of NGP dataset

- ***Network Information*** **-** This category provides a communication pattern for an IDS to train against. In SCADA systems, network topologies are fixed with repetitive and regular transactions between the nodes. This static behavior favors IDS in anomalous activities detection.

---

[1] https://sites.google.com/a/uah.edu/tommy-morris-uah/ics-data-sets.

[2] https://www.unsw.adfa.edu.au/unsw-canberra-cyber/cybersecurity/ADFA-NB15-Datasets/.

[3] http://www.simplymodbus.ca/TCP.htm.Accessedon10/03/2021.

- ***Payload Information*** **-** This provides an important information about the gas pipeline's state, settings, and parameters, which helps to understand the system performance and detecting if the system is in a critical or out-of-bounds state.
- ***Labels*** **-** It is attached to each line in data to indicate if the transaction within the system activity is normal or malicious activities.

6.4.2.3 Identified Cyber-Threats in NGP dataset

The original gas pipeline data as in Morris and Gao (2014) was improved to create a new NGP data by

- *parameterizing* and *randomizing* the order in which the attacks were executed;
- executing *all the attacks* as contained in the original data created by Gao Morris and Gao (2014);
- implementing all the attacks in a *man-in-the-middle* fashion;
- to include all the *four types of attacks* as shown below:
  - ***Interception*** **-** In this type of attack, attacks are sent to both the attacker and to the initial receiver. These types of attacks enable gaining system information such as normal system operation, each protocol node, the brand and model of the RTUs that the system is using.
  - ***Interruption*** **-** This type of attack is used to block all communication between two nodes in a system—e.g., DoS between the MTU and an RTU slave device in the gas pipeline.
  - ***Modification*** **-** This type of attacks allows an attacker to modify parameters (set point parameter exclusively and leave all other parameters untouched) or states in a system, such as the gas pipeline.
  - ***Fabrication*** **-** Attackers execute this type of attack creating a new packet to be sent between the MTU and RTU.

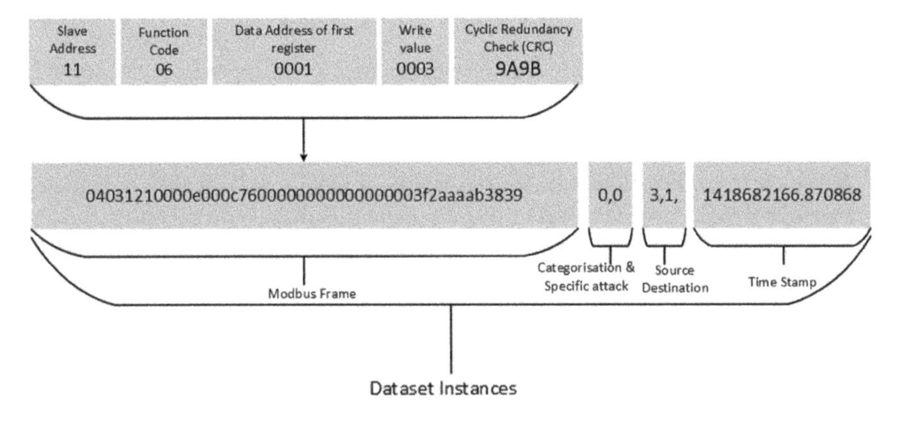

Fig. 6.5 The instances within NGP raw dataset

6.4.2.4 Raw Dataset

In this subsection, we will use Fig. 6.5 to describe and illustrate the instance features as contained within the NGP dataset.

- ***The first feature*** represents the Modbus frame as received either by the master or slave device. All valuable information from the network, state, and parameters of the gas pipeline are also contained in this Modbus frame.
- ***The second and third feature*** represent the attack category and specific attack that were executed. In case of Modbus frame normal operation, both of these features will report a zero. Both are useful to train a supervised learning algorithm, as they allow the algorithm to learn the behavior of these attack patterns.
- ***The fourth and fifth features*** represent the source and destination of the frame. There are only three possible values for the source and destination feature. The value can be a *"1"* indicates that the master device sent the packet, *"2"*, meaning the man-in-the-middle computer sent the packet, or *"3"* indicates that the slave device sent the packet.
- ***The last feature (6th)*** in the raw data contains a time stamp which can be used to calculate the time interval between change. In system normal operation, slight change may be observed between time intervals, however any modification or malicious activity such as malicious command injection may lead to noticeable time interval change.

6.4.2.5 Cyber-Attacks as Contained in the NGP Dataset Record

The NGP data contains 214,580 Modbus network packets with 60,048 packets associated with cyber-attacks. Each record contains 17 features in each network packet. These attacks are placed into 7 different attack categories with 35 different specific type of attacks. These attack categories and the individual specific attack as represented in Fig. 6.3 and Table 6.1 will be used to demonstrate an APTs steps detection with our proposed APTs detection framework in line with APTs lifecycle as described in Eke et al. (2019).

These seven attack categories are further grouped into four overall categories to align with APT lifecycle and the four identified types of cyber-attacks as described below.

- ***Response injection attacks*** contains two types of attacks, naïve malicious response injection (NMRI) (which occurs when the malicious attacker do not have sufficient information about the physical system process) and complex malicious response injection (CMRI) (these type of attack designs attacks that mimic certain normal behaviors using physical process information making it more difficult to detect).
- ***Command injection attacks*** contains three attacks, malicious state command injection (MSCI), malicious parameter command injection (MPCI), and malicious function code injection attacks (MFCI). These attacks inject control configuration commands to modify the system state and behavior, resulting to (a) loss of process

Table 6.1 Attack categories with normal records type

Attack categories	Abbreviation	Values	APTs step
Normal	Normal	0	Not applicable
Naïve malicious response injection	NMRI	1	Delivery
Complex malicious response injection	CMRI	2	Exploitation, Exfiltration
Malicious state command injection	MSCI	3	Data collection, Exploitation
Malicious parameter command injection	MPCI	4	Data collection, Exploitation
Malicious function code injection	MFCI	5	Data collection, exploitation, exfiltration
Denial of service	Dos	6	Data collection, exploitation, exfiltration
Reconnaissance	Recon	7	Reconnaissance

control, (b) device communication interruption, unauthorized modification of (c) process set points, and (d) device control.

- ***DoS attacks*** disrupt communications between the control and the process through interruption of wireless networks or network protocol exploits.
- ***Reconnaissance*** collects network and system information through passive gathering or by forcing information from a device.

6.4.2.6 UNSW-NB15 Dataset

UNSW-NB15 dataset as represented in Figs. 6.6 and 6.7 was created by Australian Centre for Cyber-Security (ACCS)[4] in their Cyber-Security Lab. A hybrid of the modern normal and abnormal network traffic features of UNSW-NB15 data was created using the IXIA PerfectStorm tools[5] to simulate nine families of attack categories as follows: Fuzzers, Analysis, Backdoors, DoS, Exploits, Generic, Reconnaissance, Shellcode, and Worms. In other to identify an attack on a network system, a comprehensive dataset that contains normal and abnormal behaviors are required to carry out a proper evaluation of network IDS effectiveness and performance (Gogoi et al. 2012). Hence, the UNSW-NB15 dataset (Moustafa and Slay 2015) was chosen for this study as the IXIA PerfectStorm tool used to generate the data contains all

[4] https://www.unsw.adfa.edu.au/unsw-canberra-cyber.

[5] https://www.ixiacom.com/products/perfectstorm.

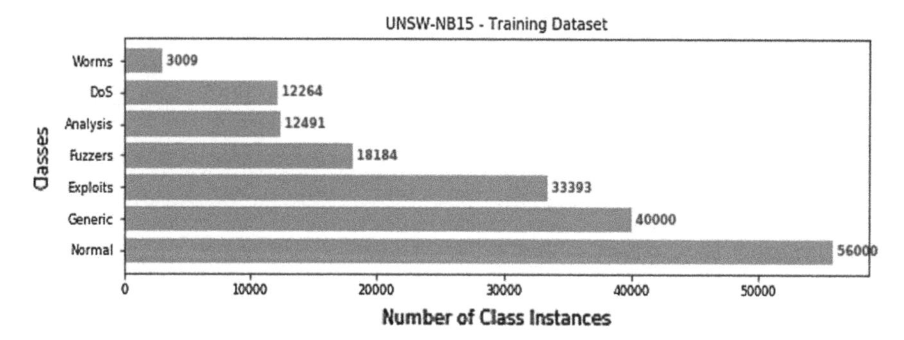

Fig. 6.6 UNSW-NB15 train dataset

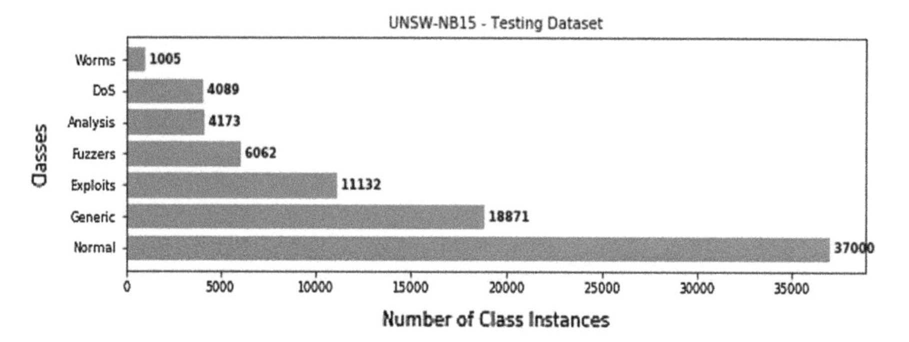

Fig. 6.7 UNSW-NB15 test dataset

information about new attacks on CVE website,[6] which is the dictionary of publicly known information security vulnerability and exposure and is updated continuously as stated in Moustafa and Slay (2015).

6.5 Experimental Evaluation of APT-DASAC Approach

Generally, accuracy is used as a traditional way of measuring classification performance. This metric measure is no longer appropriate when dealing with multi-class imbalance data since the minority class has little or no contribution when compared to majority classes toward accuracy (Sun et al. 2009). For these reasons, we applied synthetic minority oversampling technique (SMOTE) for handling data imbalance as explained in Eke et al. (2020).

[6] https://cve.mitre.org/.

Evaluation Metrics: We used *precision, recall, f1-score, overall accuracy*, area under the curve (AUC) receiver operating characteristic (ROC), and *confusion matrix* to validate the performance of implementing APT-DASAC for attack detection and clearer understanding of the output.

6.5.1 Result and Discussion

In our previous study (Eke et al. 2020), we implemented a DL multi-layered security detection approach which focused on detecting command injection (CI) and response injection (RI) attacks. We noticed a higher detection rate of CI to RI, although CI has more connection records and obtained a significant detection rate with 0% False Positive Rate (FPR) and True Positive Rate (TPR) of 96.50%. Based on the outcome of our analysis, we arrived on the conclusion that performance of attack detection techniques applied can be influenced by the nature of the network transactions with respect to the domain of application and made suggestion for further investigation in different domain.

We acknowledge the need to investigate this further in other to ascertain this claim. We implemented the application of stacked ensemble-LSTM variants for APT-DASAC. This approach combines networks' results as to optimize attack detection rate. To validate this approach for detecting APT step attacks, statistical metrics such as *precision, recall, f1-score,* AUC-ROC, *and overall accuracy* are calculated (i) to evaluate the ability of this approach to accurately detect and classify an abnormal network as an attack, (ii) to check the ability of this model to detect different type of attacks accurately, and (iii) to get a clearer understanding of the output.

Figures 6.8 and 6.9 contain the statistical classification report obtained from implementing deep ensemble-LSTM variants and ML-DT on NGP dataset, respectively. These reports show that our approach achieved an average P, R, and $f1$ of 88%, 86%, and 82%, respectively, with overall detection accuracy of 85% and macro-f1 of 62%, while the implemented ML-DT obtains 95% for P, R, and $f1$ with overall detection accuracy of 94% in detecting attacks.

Considering the fact that the proposed approach detects APT step activities in different stages, we generated ROC curves score for the stages as shown in Fig. 6.10.

Fig. 6.8 Classification—report for ensemble-LSTM variants on NGP dataset

	precision	recall	f1-score	support
Command_Injection	0.97	0.51	0.67	10959
DoS	0.99	0.44	0.61	718
Normal	0.85	1.00	0.92	70812
Reconnaissance	0.94	0.93	0.94	1279
Response_Injection	1.00	0.02	0.03	6860
avg / total	0.88	0.86	0.82	90628

The average of the five-step curves is evaluated and consolidated into a single graph representing their respective $AUCcurve$ and obtain micro-average ROC curve area of 91% and macro-average ROC curve area of 72%. It is evident from Fig. 6.10 that the classification of APT attack detection in class 3 stage has the ROC curve area of 93% , this is largely attributed to the number of connection record exhibited in this stage, while the class 4 stage has the lowest ROC curve area of 51%. Our proposed approach seems to achieve a good performance since the weighted average of the ROC curve area is closer to 1. A high area under the curve represents both high recall and high precision, an ideal model with high precision and high recall will return many results, with all results labeled correctly.

The results shown in Figs. 6.11 and 6.12 are the visual representation of each algorithm's validation accuracy and loss rate on each epochs. There are some spikes in the validation accuracy and loss, following the individual model accuracy and loss per epoch, achieving training and validation accuracy of 85.59%, 85.88% with validation loss of 33% for LSTM; 85.97%, 85.16% with validation loss of 35% for RNN; and 86.13%, 85.71% with validation loss of 34% for GRU. It is worth noting that the value of training and validation accuracy are quite close to each other, indicating that the model is not over-fitting with overall average mean detection accuracy and validation average accuracy of 85%.

We also implemented the same approach with UNSW-NB15 data, the average detection accuracy of 93.67% as recorded in Table 6.2, which is slightly higher than 85% obtained when NGP data was implemented.

6.5.1.1 The Proposed Approach and Other Works on APTs Detection

Few proposed APT detection approach recorded in Table 6.3 as reviewed for this chapter includes, work in Do Xuan et al. (2022), an APT attack detection method based on Bidirectional Long Short-Term Memory (BiLSTM) and Graph Convolu-

Fig. 6.9 Classification—report for ML-DT on NGP dataset

	precision	recall	f1-score	support
Command_Injection	0.98	0.96	0.97	10959
DoS	0.96	0.93	0.95	718
Normal	0.96	0.97	0.97	70812
Reconnaissance	0.98	0.97	0.98	1279
Response_Injection	0.72	0.67	0.69	6860
avg / total	0.95	0.95	0.95	90628

Fig. 6.10 AUC-ROC—vreport for ensemble-LSTM variants on NGP dataset

micro-average ROC curve (area = 0.91)
macro-average ROC curve (area = 0.72)
ROC curve of class 0 (area = 0.76)
ROC curve of class 1 (area = 0.73)
ROC curve of class 2 (area = 0.69)
ROC curve of class 3 (area = 0.93)
ROC curve of class 4 (area = 0.51)

tional Networks (GCN) to analyze network traffic into IP-based network flows. This approach achieved 98.24% of normal IPs and 68.89% of APT attack IPs using Malware Capture CTU-13 data warehouse dataset. The authors in Shang et al. (2021), tackled APT attack detection using network flow-based C&C detection method to detect the hidden C&C channel of unknown APT attacks and achieved an $f1 - score$ of 96.80% but did not provide the actual detection rate for their approach. Also, the author in Zimba et al. (2020) proposed a detection framework based on an enhanced SNN algorithm using semi-supervised learning approach on LANL dataset to scores suspicious APTs-related activities at three different stages of APT attack lifecycle given a high weight rank to hosts depicting characteristics of data exfiltration with the believe that main APT attack is data exfiltration. This study faced a higher computational overhead cost.

In our previous work in Eke et al. (2019), we proposed an approach using deep neural networks for APT multi-step detection which takes stacked LSTM-RNNs networks to automatically learn features from the raw data to capture the malicious patterns of APT activities using KDDCup99 dataset. This approach achieved a detection rate of 99.90%, see Table 6.3. The current chapter proposes a framework named APT-DASAC based on stacked ensemble-LSTM variants, taken into consideration the distributed and multi-level nature of ICS architecture and reflect on the four main SCADA cyber-attacks which are interception, interruption, modification and

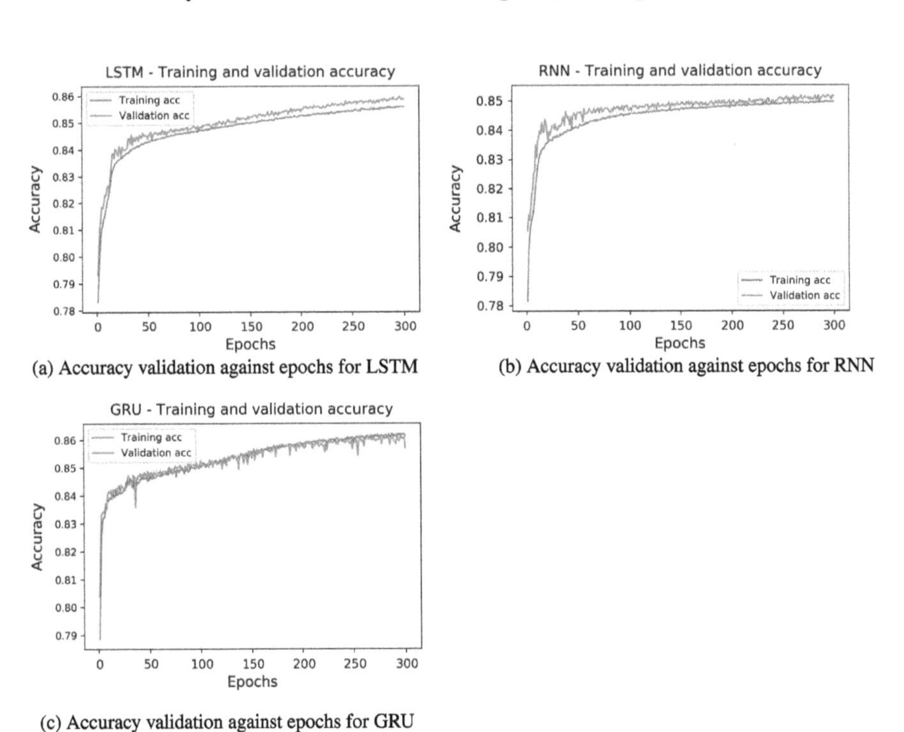

(a) Accuracy validation against epochs for LSTM

(b) Accuracy validation against epochs for RNN

(c) Accuracy validation against epochs for GRU

Fig. 6.11 Validation accuracy against epochs on NGP dataset

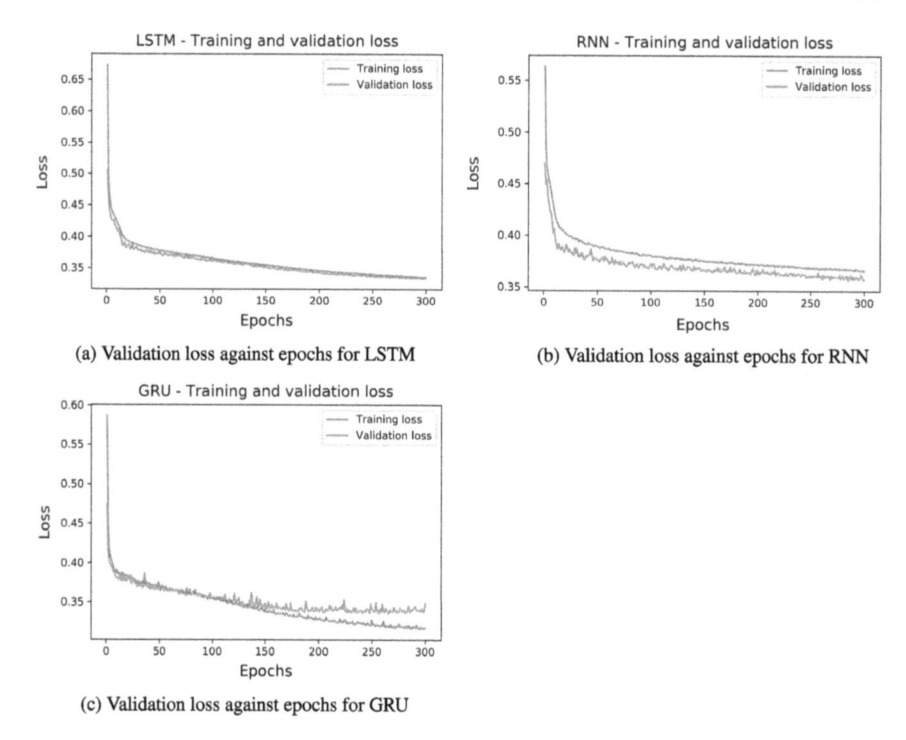

(a) Validation loss against epochs for LSTM

(b) Validation loss against epochs for RNN

(c) Validation loss against epochs for GRU

Fig. 6.12 Validation loss against epochs on NGP dataset

Table 6.2 Performance report for ensemble-LSTM variants on UNSW-NB15 dataset

Algorithm	Average accuracy (%)	Validation accuracy (%)	Validation loss (%)
LSTM	93.74	82.29	21.82
RNN	92.88	81.43	20.50
GRU	94.41	82.11	20.46
Ensemble-LSTM variants	93.67	84.94	20.47

fabrication as recorded in Turnipseed (2020) to demonstration the ability of this approach in detecting different stages of APT activities. This approach achieved an overall detection rate of 85% for NGP dataset and 93.67% for UNSW-NB15 dataset. Also, when ML-DT were implemented within our approach, we obtained 95% on both NGP and UNSW-NB15 datasets.

All the reviewed approach on this study have demonstrated a significant APT attack detection capability, however, none of these approach used the same dataset (see Table 6.3), making it difficult to rank the performance of these approaches. Also, the unavailability of a standard dataset or suitable public accessible dataset is a huge challenge in the field of cyber-security, making it unfavorable to compare an APT detection system performance so as to choose an appropriate model for any given domain.

6.6 Conclusion

In this study, to overcome the issue of detecting APT dynamics attack lifecycle, we have used supervised learning approach and a multi-layered attack detection framework that takes into consideration the distributed and multi-level nature of ICS architecture and reflects on the four main SCADA-based cyber-attacks. Therefore, a detection framework based on stacked ensemble-LSTM variants algorithm has been proposed and evaluated. This accounts as one of the contributions of this chapter. Due to the dynamic nature of APT lifecycle, APT attack cannot be detected automatically, and hence this model serves as a supplement to automated IDS. The implemented algorithms achieved a competitive overall detection rate of 85%, 93.67%, and 95% with micro-average ROC curve area of 91%. These results suggest that both stacked ensemble-LSTM variants and ML-DT approach are good candidates to be considered for developing an APT detection system.

From Fig. 6.8, the value of *recall* achieved also illustrates that when DL is used within the proposed approach, it did struggle to identify the relevant cases of command injection attack, DoS, and Response Injection attacks within the NGP dataset. The class with more connection records seems to be learnt properly without confusing their identity while those with fewer connection records during training did not show good true positive rate as it was had to identify them. This indicates a data imbalance problem. However, this was not the case when ML was used in place of DL as the system achieved good *precision* and *recall* as evidenced in Table 6.3. Also, if the output from this study is compared to our previous work in Eke et al. (2019), where we have implemented the same procedure with KDDCup99 dataset, the average detection rate achieved is 99.9% (see Table 6.3).

Table 6.3 Our proposed approach and other works on APTs detection

Proposed method	Approach	Dataset	Outcome	Reference
Enhanced $SNN_{algorithm}$	Semi-supervised learning approach	LANL	90.50%	Zimba et al. (2020)
BiLSTM&GCN	Network flow analysis	Malware capture CTU-13 data warehouse	68.89% (APT IPs attack)	Do Xuan et al. (2022)
Network flow based on C&C detection method	DL techniques	Contagio blog malware	96.80% (f-score)	Shang et al. (2021)
Stacked $RNN_{variants}$	DL techniques	KDDCup99	99.90%	Eke et al. (2019)
APT-DASAC	ML–DT	NGP & UNSW-NB15	95%	This chapter
APT-DASAC	Ensemble $LSTM_{variants}$	NGP & UNSW-NB15	85%	This chapter

We can see that this approach performed very well on KDDCup99 dataset as the feature set contained within this data is highly distinguishable in nature. The result is slightly higher when both NGP and UNSW-NB15 dataset were used. This account as an identified issue from this study when it comes to comparing performance of various proposed detection framework with regard to accessibility and availability of suitable data/network flow information in security industries with respect to domain of interest.

Considering the different results obtained with three different datasets from diverse domains, our implemented approach showed a significant attack detection capability. This has also demonstrated that performance of attack detection approach applied can be influenced by the nature of network connections with respect to the domain of application. This suggests that the ability and resilience of operational CPS state to withstand attack and maintain system performance are regulated by the safety and security measures in place, which is specific to that CPS devices or application domain. Hence, there is every need to investigation the nature of the network flow information within any system in mind to determine the security measures that will be suitable for that system.

References

D. Alperovitch, *Revealed: operation shady RAT*, vol. 3. McAfee (2011), p. 2011

M.J. Assante, R.M. Lee, The industrial control system cyber kill chain. SANS Institute InfoSec Reading Room, 1 (2015)

C.Z. Bai, F. Pasqualetti, V. Gupta, Data-injection attacks in stochastic control systems: detectability and performance tradeoffs. Automatica **82**, 251–260 (2017)

M. Brand, C. Valli, A. Woodward, Malware forensics: discovery of the intent of deception. J. Digit. Forensic Sec. Law **5**(4), 2 (2010)

A. Cardenas, S. Amin, B. Sinopoli, A. Giani, A. Perrig, S. Sastry, Challenges for securing cyber physical systems, in *Workshop on Future Directions in Cyber-Physical Systems Security*, vol. 5, No. 1 (2009)

L.K. Carvalho, Y.C. Wu, R. Kwong, S. Lafortune, Detection and mitigation of classes of attacks in supervisory control systems. Automatica **97**, 121–133 (2018)

L. Cazorla, C. Alcaraz, J. Lopez, Cyber stealth attacks in critical information infrastructures. IEEE Syst. J. **12**(2), 1778–1792 (2016)

J.V. Chandra, N. Challa, S.K. Pasupuleti, A practical approach to E-mail spam filters to protect data from advanced persistent threat, in *2016 International Conference on Circuit, Power and Computing Technologies (ICCPCT)* (IEEE, 2016), pp. 1–5

C. Do Xuan, H.D. Nguyen, M.H. Dao, APT attack detection based on flow network analysis techniques using deep learning. J. Intell. Fuzzy Syst. (Preprint) (2022), pp. 1–17

R. Domović, Cyber-attacks as a Threat to Critical Infrastructure. Integrating Ictin Society (2017), 259

H.N. Eke, A. Petrovski, H. Ahriz, The use of machine learning algorithms for detecting advanced persistent threats, in *Proceedings of the 12th International Conference on Security of Information and Networks* (2019), pp. 1–8

H. Eke, A. Petrovski, H. Ahriz, Detection of false command and response injection attacks for cyber physical systems security and resilience, in *13th International Conference on Security of*

Information and Networks (SIN 2020), November 4–7, 2020, Merkez, Turkey (ACM, NewYork, NY, USA, 2020), 8 p. https://dl.acm.org/doi/10.1145/3433174.3433615
H. Eke, A. Petrovski, H. Ahriz, Handling minority class problem in threats detection based on heterogeneous ensemble learning approach. Int. J. Syst. Softw. Sec. Protect. (IJSSSP) **11**(2), 13–37 (2020)
T. Fawcett, An introduction to ROC analysis. Pattern Recognit. Lett. **27**(8), 861–874 (2006)
H. Fawzi, P. Tabuada, S. Diggavi, Secure estimation and control for cyber-physical systems under adversarial attacks. IEEE Trans. Autom. Control **59**(6), 1454–1467 (2014)
W. Gao, T. Morris, B. Reaves, D. Richey, On SCADA control system command and response injection and intrusion detection, in *2010 eCrime Researchers Summit* (IEEE, 2010), pp. 1–9
P. Giura, W. Wang, A context-based detection framework for advanced persistent threats, in *2012 International Conference on Cyber Security* (IEEE, 2012), pp. 69–74
P. Gogoi, M.H. Bhuyan, D.K. Bhattacharyya, J.K. Kalita, Packet and flow based network intrusion dataset, in *International Conference on Contemporary Computing* (Springer, Berlin, Heidelberg, 2012), pp. 322–334
M.T. Hagan, O. De Jesús, R. Schultz, L. Medsker, L.C. Jain, Training recurrent networks for filtering and control. Chapter 12, (1999), pp. 311–340
G. Haixiang, L. Yijing, J. Shang, G. Mingyun, H. Yuanyue, G. Bing, Learning from class-imbalanced data: Review of methods and applications. Expert Syst. Appl. **73**, 220–239 (2017)
B. Harris, R. Hunt, TCP/IP security threats and attack methods. Comput. Commun. **22**(10), 885–897 (1999)
K.E. Hemsley, E. Fisher History of industrial control system cyber incidents (No. INL/CON-18-44411-Rev002). Idaho National Lab.(INL), Idaho Falls, ID (United States) (2018)
G. Hu, H. Li, Y. Xia, L. Luo, A deep Boltzmann machine and multi-grained scanning forest ensemble collaborative method and its application to industrial fault diagnosis. Comput. Ind. **100**, 287–296 (2018)
A. Humayed, J. Lin, F. Li, B. Luo, Cyber-physical systems security-a survey. IEEE Internet Things J. **4**(6), 1802–1831 (2017)
E.M. Hutchins, M.J. Cloppert, R.M. Amin, Intelligence-driven computer network defense informed by analysis of adversary campaigns and intrusion kill chains. Leading Issues Inf. Warfare Sec. Res. **1**(1), 80 (2011)
N. Jazdi, Cyber physical systems in the context of Industry 4.0, in *2014 IEEE International Conference on Automation, Quality and Testing, Robotics* (IEEE, 2014), pp. 1–4
A.D. Kent, Cybersecurity data sources for dynamic network research in *Dynamic Networks in Cybersecurity* (2015)
H.S. Kim, J.M. Lee, T. Park, W.H. Kwon, Design of networks for distributed digital control systems in nuclear power plants, in *International Topical Meeting on Nuclear Plant Instrumentation, Controls, and Human-Machine Interface Technologies (NPIC&HMIT 2000)* (2000)
R.M. Lee, M.J. Assante, T. Conway, CRASHOVERRIDE: analysis of the threat to electric grid operations. Dragos Inc. (2017). https://www.dragos.com/wp-content/uploads/CrashOverride-01.pdf
O. Linda, T. Vollmer, M. Manic, Neural network based intrusion detection system for critical infrastructures, in *2009 International Joint Conference on Neural Networks* (IEEE, 2009), pp. 1827–1834
F.A. Majdani, L. Batik, A. Petrovski, S. Petrovski, Detecting malicious signal manipulation in smart grids using intelligent analysis of contextual data, in *ACM Digital Library: Proceedings of the 13 International Conference on Security of Information and Networks* (2020), pp. 1–8
S. McClure, S. Gupta, C. Dooley, V. Zaytsev, X.B. Chen, K. Kaspersky, R. Permeh, *Protecting your critical assets-lessons learned from operation aurora* (Tech, Rep, 2010)
T. Mikolov, M. Karafiát, L. Burget, J. Černocký, S. Khudanpur, Recurrent neural network based language model, in *Eleventh Annual Conference of the International Speech Communication Association* (2010)

J. Miloševič, T. Tanaka, H. Sandberg, K.H. Johansson, Analysis and mitigation of bias injection attacks against a Kalman filter. IFAC-PapersOnLine **50**(1), 8393–8398 (2017)

L. Monostori, B. Kádár, T. Bauernhansl, S. Kondoh, S. Kumara, G. Reinhart, K. Ueda, Cyber-physical systems in manufacturing. Cirp Ann. **65**(2), 621–641 (2016)

T. Morris, W. Gao, Industrial control system traffic data sets for intrusion detection research, in *International Conference on Critical Infrastructure Protection* (Springer, Berlin, Heidelberg, 2014), pp. 65–78

T.H. Morris, Z. Thornton, I. Turnipseed, Industrial control system simulation and data logging for intrusion detection system research, in *7th Annual Southeastern Cyber Security Summit* (2015), pp. 3–4

N. Moustafa, J. Slay, UNSW-NB15: a comprehensive data set for network intrusion detection systems (UNSW-NB15 network data set), in *2015 Military Communications and Information Systems Conference (MilCIS)* (IEEE, 2015), pp. 1–6

A.A. Nazarenko, G.A. Safdar, Survey on security and privacy issues in cyber physical systems [J]. AIMS Electron. Electr. Eng. **3**(2), 111–143 (2019)

N. Nissim, A. Cohen, C. Glezer, Y. Elovici, Detection of malicious PDF files and directions for enhancements: a state-of-the art survey. Comput. Sec. **48**, 246–266 (2015)

NJCCIC, CRASHOVERRIDE NJCCIC Threat Profile, official site of the state of new jersey Original Release Date: 2017-08-10 and Accessed 3 June 21 (2017). https://www.cyber.nj.gov/threat-center/threat-profiles/ics-malware-variants/crashoverride

NJCCIC, CRASHOVERRIDE NJCCIC Threat Profile, official site of the state of new jersey Original Release Date: 2017-08-10 and Accessed 16 July 20. NJCCIC (2017)

A. Odewale, Implementing secure architecture for industrial control systems, in *Proceedings of the 27th COREN Engineering Assembly, Abuja, Nigera* (2018), pp. 6–8

A. Paoli, M. Sartini, S. Lafortune, Active fault tolerant control of discrete event systems using online diagnostics. Automatica **47**(4), 639–649 (2011)

F. Pasqualetti, F. Dorfler, F. Bullo, Control-theoretic methods for cyberphysical security: Geometric principles for optimal cross-layer resilient control systems. IEEE Control Syst. Mag. **35**(1), 110–127 (2015)

A. Petrovski, P. Rattadilok, S. Petrovski, Designing a context-aware cyber physical system for detecting security threats in motor vehicles, in *Proceedings of the 8th International Conference on Security of Information and Networks* (2015), pp. 267–270

J. Sen, (Ed.) Cryptography and Security in Computing. BoD–Books on Demand (2012)

L. Shang, D. Guo, Y. Ji, Q. Li, Discovering unknown advanced persistent threat using shared features mined by neural networks. Comput. Netw. **107937** (2021)

N. Shashidhar, L. Chen, A phishing model and its applications to evaluating phishing attacks (2011)

H. Shi, L. Xie, L. Peng, Detection of false data injection attacks in smart grid based on a new dimensionality-reduction method. Comput. Electr. Eng. **91**, 107058 (2021)

M.A. Siddiqi, N. Ghani, Critical analysis on advanced persistent threats. Int. J. Comput. Appl. **141**(13), 46–50 (2016)

S. Singh, P.K. Sharma, S.Y. Moon, D. Moon, J.H. Park, A comprehensive study on APT attacks and countermeasures for future networks and communications: challenges and solutions. J. Supercomput. **75**(8), 4543–4574 (2019)

J. Slowik, Evolution of ICS attacks and the prospects for future disruptive events. Threat Intelligence Centre Dragos Inc (2019)

M. Smiraus, R. Jasek, Risks of advanced persistent threats and defense against them, in *Annals of DAAAM & Proceedings* (2011), p. 1589

Y. Sun, A.K. Wong, M.S. Kamel, Classification of imbalanced data: a review. Int. J. Pattern Recognit. Artif. Intell. **23**(04), 687–719 (2009)

K. Thakur, M.L. Ali, N. Jiang, M. Qiu, Impact of cyber-attacks on critical infrastructure, in *2016 IEEE 2nd International Conference on Big Data Security on Cloud (BigDataSecurity), IEEE International Conference on High Performance and Smart Computing (HPSC), and IEEE International Conference on Intelligent Data and Security (IDS)* (IEEE, 2016), pp. 183–186

I.P. Turnipseed, A new SCADA dataset for intrusion detection system research (Doctoral dissertation, Mississippi State University) (2020)

N. Villeneuve, J.T. Bennett, N. Moran, T. Haq, M. Scott, K. Geers, Operation Ke3chang: Targeted Attacks Against Ministries of Foreign Affairs. FireEye, Incorporated. villeneuve2013operation (2013)

G. Wu, J. Sun, J. Chen, A survey on the security of cyber-physical systems. Control Theory Technol. **14**(1), 2–10 (2016)

Z. Xu, A. Easwaran, A game-theoretic approach to secure estimation and control for cyber-physical systems with a digital twin, in *2020 ACM/IEEE 11th International Conference on Cyber-Physical Systems (ICCPS)* (IEEE, 2020), pp. 20–29

A. Zimba, H. Chen, Z. Wang, M. Chishimba, Modeling and detection of the multi-stages of Advanced Persistent Threats attacks based on semi-supervised learning and complex networks characteristics. Futur. Gener. Comput. Syst. **106**, 501–517 (2020)

Chapter 7
Resilient State Estimation and Attack Mitigation in Cyber-Physical Systems

Mohammad Khajenejad and Sze Zheng Yong

7.1 Introduction

Cyber-Physical Systems (CPS), e.g., power grids, autonomous vehicles, medical devices, etc., are systems in which computational and communication components are deeply intertwined and interacting with each other in several ways to control physical entities. While the cyber-physical coupling introduces new functions to control systems and improves their performance, these systems also become exposed to new cyber-vulnerabilities. Such *safety-critical* systems, if jeopardized or malfunctioning, can cause serious detriment to their operators and users, as well as the controlled physical components. A need for CPS security and for new designs of resilient estimation, attack mitigation and control has been accentuated by recent incidents of attacks on CPS, e.g., the Iranian nuclear plant, the Ukrainian power grid, and the Maroochy water service (Cárdenas et al. 2008; Farwell and Rohozinski 2011; Richards 2008; Slay and Miller 2007; Zetter 2016). Specifically, mode and false data injection attacks are among the most serious types of attacks on CPS, where malicious and/or strategic attackers compromise the true mode (i.e., discrete state) of the system and/or inject counterfeit data signals into the sensor measurements and actuator signals to cause damage, steal energy, etc. Hence, reliable estimates of modes, (continuous) states, and unknown inputs (attacks) are indispensable and useful for the sake of attack identification and mitigation and resilient control. Similar state and input estimation problems can be found across a wide range of disciplines, from input estimation in physiological systems (De Nicolao et al. 1997), to fault detection and diagnosis (Patton et al. 1989), to the estimation of mean areal precipitation (Kitanidis 1987).

M. Khajenejad · S. Z. Yong (✉)
Arizona State University, Tempe, AZ 85287, USA
e-mail: szyong@asu.edu

M. Khajenejad
e-mail: mkhajene@asu.edu

M. Abbaszadeh and A. Zemouche (eds.), *Security and Resilience in Cyber-Physical Systems*, https://doi.org/10.1007/978-3-030-97166-3_7

7.1.1 Literature Review

Characterization of undetectable attacks as well as attack detection and identification techniques have been extensively studied in the literature, which range from data-driven approaches (e.g., the use of data time-stamps in Zhu and Martínez (2013), Wasserstein metric in Li and Martínez (2020) or higher-order moments in Renganathan et al. 2021) to the works seeking closed-form solutions for selecting various types of detector thresholds (e.g., Murguia and Ruths 2016; Milošević et al. 2018) to anomaly detection methods using residuals (e.g., Mo and Sinopoli 2010; Weimer et al. 2012; Kwon et al. 2013) with empirically chosen thresholds to trade-off between false alarms and probability of anomaly/attack detection. On the other hand, attack mitigation can be preventive and/or reactive (Cómbita et al. 2015). Preventive attack mitigation identifies and removes system vulnerabilities to prevent exploitation (e.g., Dan and Sandberg 2010), while reactive attack mitigation, which is mainly studied using either game theory (e.g., Ma et al. 2013; Zhu and Martínez 2011; Zhu and Basar 2015) or adaptive learning and control architectures for mitigating sensor and actuator attacks (e.g., Jin et al. 2017; Yadegar et al. 2019; Jin and Haddad 2019, 2020), initiates countermeasures after detecting an attack.

The ability to reliably estimate the true system states despite attacks (i.e., resilient estimates) is also desirable in addition to attack detection or the resulting attack mitigation, because the availability of resilient state estimates would allow for continued operation with the same controllers as in the case without attacks or for pricing/prediction based on the real unbiased/compatible state information despite attacks. This problem has been addressed for both static systems (e.g., Liu et al. 2011; Kosut et al. 2011; Liang et al. 2017 and references therein) and dynamic systems (e.g., Mishra et al. 2015; Cárdenas et al. 2008; Mo and Sinopoli 2010; Pasqualetti et al. 2013; Fawzi et al. 2014; Pajic et al. 2014, 2015; Yong et al. 2016a; Dahleh and Diaz-Bobillo 1994; Shamma and Tu 1999; Blanchini and Sznaier 2012; Yong 2018; Yong et al. 2018).

In particular, resilient state estimators for *deterministic* linear dynamic systems under actuator and sensor signal attacks (e.g., via false data injection Cárdenas et al. 2008; Mo and Sinopoli 2010; Pasqualetti et al. 2013), have been proposed as a relaxed ℓ_0 optimization problem in Fawzi et al. (2014), and extensions in Pajic et al. (2014), Pajic et al. (2015) compute the worst-case bound on the state estimate errors in the presence of additive noise errors with known bounds, while Yong et al. (2016a) propose the resilient state estimators that are robust to bounded multiplicative and additive modeling and noise errors. On the other hand, our previous work Yong et al. (2015), Yong (2018) proposed to use a simultaneous input and state estimation (see, e.g., Yong 2018; Gillijns and De Moor 2007a, b; Yong et al. 2016b, 2017) approach for resilient state estimation , where we modeled the data injection attacks as unknown inputs of dynamical systems and derived stability and optimality properties for our estimators, as well as their relationship to strong detectability (Yong et al. 2016b).

In addition, a serious CPS security concern has emerged more recently from the attacks that alter the CPS network topology or exploit the switching vulnerability of CPS, e.g., attacks on the power system network topology (Weimer et al. 2012), or on the circuit breakers of a smart grid (Liu et al. 2013), on the meter/sensor data network topology (Kim and Tong 2013) or on the logic mode (e.g., failsafe mode) of a traffic infrastructure (Ghena et al. 2014). To address this concern, our previous works (Yong et al. 2021, 2018; Khajenejad and Yong 2019) proposed inference algorithms that estimate hidden modes, unknown inputs (attacks) and states simultaneously as a means to obtain resilient state estimation despite switching (mode/topology) attacks as well as attacks on actuator and sensor signals. This framework is inspired by the *multiple-model* approach (see e.g., Bar-Shalom et al. 2004; Mazor et al. 1998 and references therein) and can be viewed as a generalization of the robust control-inspired approach in Nakahira and Mo (2018) that considers resilient state estimation against sparse data injection attacks on only the sensors.

In the context of reactive attack mitigation, the work in Ma et al. (2013) utilized a Markov game analysis for attack-defense in power systems, while a leader–follower (Stackelberg) game formulation was developed in Zhu and Martínez (2011) to model the interdependency between the operator and adversaries and solved using a receding-horizon Stackelberg control law to maintain the closed-loop system stability and some performance specifications. Further, a cross-layer coupled design was presented in a hybrid game-theoretic framework in Zhu and Basar (2015), where the occurrence of unanticipated events was modeled by stochastic switching , and deterministic uncertainties were represented by disturbances with a known range, and a robust controller was then designed at the physical layer to take into account risks of failures due to the cyber-system.

In this chapter, assuming different models for uncertainties/noise signals, we propose resilient state estimation algorithms that output reliable estimates of the true system states despite false data injection attacks and switching attacks. Our resilient estimation algorithms address switching attacks as well as actuator and sensor attacks in the presence of stochastic and/or set-valued noise signals. Our approach is built upon a general purpose inference algorithm developed and applied in our previous works (Yong et al. 2021, 2018; Khajenejad and Yong 2019) for hidden-mode stochastic/bounded error switched linear systems with unknown inputs (attacks). We model switching and false data injection attacks on Cyber-Physical Systems (CPS) in the presence of stochastic/distribution-free noise signals as an instance of this system class. By doing so, we show that unbiased and set-valued state estimates (i.e., resilient state estimates) can be (asymptotically) recovered with the algorithms in Yong et al. (2021), Khajenejad and Yong (2019). Secondly, we characterize fundamental limitations to resilient estimation that is useful for preventative mitigation, such as the upper bound on the number of correctable/tolerable attacks, and consider the subject of attack detection. In addition, we provide sufficient conditions for designing unidentifiable attacks (from the attacker's perspective) and also sufficient conditions to obtain resilient state estimates even when the attacks are not identified (from the system operator/defender's perspective). Finally, we design an attack-mitigating and stabilizing dynamic $\mathscr{H}_\infty$-controller that contributes to the literature on non-game-theoretic reactive attack mitigation.

An earlier manuscript appeared in Yong et al. (2018), where we addressed the resilient state estimation problem under switching and false data injection attacks for *stochastic* hidden-mode CPS only, while in this chapter, we also consider the uncertainties that are set-valued and further present a novel *dynamic* $\mathcal{H}_\infty$-optimal controller design for attack mitigation. Further, we provide *necessary* conditions for the attack signal to be unidentifiable to add to the previously derived sufficient conditions in Yong et al. (2018).

Notation: $\mathbb{R}^n$ denotes the n-dimensional Euclidean space and $\mathbb{N}$ nonnegative integers. For a vector $v \in \mathbb{R}^n$ and a matrix $M \in \mathbb{R}^{p\times q}$, $\|v\|_2 \triangleq \sqrt{v^\top v}$, $\|v\|_\infty \triangleq \max_{1\leq i\leq n} \mid v_i \mid$ and $\|M\|_2$, and $\sigma_{\min}(M)$ denote their induced 2-norm and non-trivial least singular value, respectively.

7.2 Problem Formulation

7.2.1 Attack Modeling

Similar to Yong et al. (2018), two different classes of possibly time-varying attacks on Cyber-Physical Systems (CPS) are considered:

Data Injection Attacks: Attacks on actuator and sensor signals via manipulation or injection with "false" signals of unknown *magnitude* and *location* (i.e., subset of attacked actuators or sensors). In other words, signal attacks consist of both *signal magnitude attacks* and *signal location attacks*. *Examples:* Denial-of-service, deceptive attacks via data injection (Cárdenas et al. 2008; Pasqualetti et al. 2013).

Switching Attacks: Attacks on the switching mechanisms that change the system's *mode* of operation, or on the sensor data or interconnection network *topology*, which we will also refer to as *mode attacks*. *Examples:* Attack on circuit breakers (Liu et al. 2013), power network topology (Weimer et al. 2012), sensor data network (Kim and Tong 2013) and logic switch of a traffic infrastructure (Ghena et al. 2014).

7.2.1.1 Data Injection Attacks

For clarity, we assume for the moment that there is only one mode of operation, and that the linear system dynamics is not perturbed by any noise signals:

$$x_{k+1} = A_k x_k + B_k(u_k + d_k^a), \quad y_k = C_k x_k + D_k(u_k + d_k^a) + d_k^s,$$

where $x_k \in \mathbb{R}^n$ is the continuous state, $y_k \in \mathbb{R}^\ell$ is the sensor output, $u_k \in \mathbb{R}^m$ is the known input, $d_k^a \in \mathbb{R}^m$ and $d_k^s \in \mathbb{R}^\ell$ are attack signals that are injected into the actuators and sensors, respectively. The attack signals are sparse, i.e., if sensor $i \in \{1, \dots, \ell\}$ is not attacked then necessarily $d_k^{s,(i)} = 0$ for all time steps k; otherwise

$d_k^{s,(i)}$ can take any value. Since we do not know which sensor is attacked, we refer to this uncertainty as the *signal location attack*, and the arbitrary values that $d_k^{s,(i)}$ can take as the *signal magnitude attack*. This holds similarly for attacks on actuators d_k^a.

If we have additional knowledge of which of the actuators and sensors are vulnerable to data injection attacks, we will use $\overline{G}_k$ and $\overline{H}_k$ to incorporate this information, resulting in the following system dynamics

$$x_{k+1} = A_k x_k + B_k u_k + \overline{G}_k d_k^a, \quad y_k = C_k x_k + D_k u_k + \overline{D}_k d_k^a + \overline{H}_k d_k^s.$$

If no such information is available, $\overline{G}_k = B_k, \overline{D}_k = D_k$, and $\overline{H}_k = I$. Further, in some cases, the actuator and sensor attack signals are coupled and cannot be separated. In order to take this into consideration, we represent the potentially coupled attack signals with d_k and introduce corresponding G_k and H_k matrices to obtain

$$x_{k+1} = A_k x_k + B_k u_k + G_k d_k, \quad y_k = C_k x_k + D_k u_k + H_k d_k.$$

The special case where the actuator and sensor attack signals are independent can be obtained with $d_k = \left[(d_k^a)^\top \ (d_k^s)^\top\right]^\top$, $G_k = \left[\overline{G}_k \ 0\right]$ and $H_k = \left[\overline{D}_k \ \overline{H}_k\right]$, which will be made more precise in Sect. 7.2.2.

7.2.1.2 Switching Attacks

A system may have multiple modes of operation, denoted by the set $\mathscr{Q}^m$ of cardinality $t_m \triangleq |\mathscr{Q}^m|$, due to the presence of switching mechanisms or different configurations/topologies of the sensor data or interconnection network, where each mode $q \in \mathscr{Q}^m$ has its corresponding set of system matrices, $\{A_k^q, B_k^q, C_k^q, D_k^q, G_k^q, H_k^q\}$. A *switching attack* or *mode attack* then refers to the change of the mode of operation q by an adversary without the knowledge of the system operator/defender.

7.2.1.3 Attacker Model Assumptions

The malicious *signal magnitude attack* may be a signal of any type (random or strategic) or model, and we assume that no 'useful' knowledge of the dynamics of d_k is available (uncorrelated with $\{d_\ell\}$ for all $k \neq \ell$, $\{w_\ell\}$ and $\{v_\ell\}$ for all ℓ).

7.2.2 System Description

Our role as a system operator/defender is to obtain resilient/reliable state estimates. Thus, we model the system in a way that facilitates this. In other words, we model the switching and false data injection attacks on a "noisy" dynamic system using a *hidden-mode switched linear discrete-time system with unknown inputs* (i.e., a dynamical system with multiple modes of operation where the system dynamics in each mode is linear and *uncertain*, and the mode and some inputs are not known/measured):

$$\begin{aligned}(x_{k+1}, q_k) &= (A_k^q x_k + B_k^q u_k^q + G_k^q d_k^q + w_k^q, q), && x_k \in \mathscr{C}_q,\\ (x_k, q)^+ &= (x_k, \delta^q(x_k)), && x_k \in \mathscr{D}_q,\\ y_k &= C_k^q x_k + D_k^q u_k^q + H_k^q d_k^q + v_k^q,\end{aligned} \tag{7.1}$$

where $x_k \in \mathbb{R}^n$ is the continuous system state and $q \in \mathscr{Q} = \{1, 2, \ldots, \mathfrak{N}\}$ is the hidden discrete state or *mode*, which a malicious attacker can influence, while $\mathscr{C}_q$ and $\mathscr{D}_q$ are flow and jump sets, and $\delta^q(x_k)$ is the mode transition function. More details on the hybrid systems formalism can be found in Goebel et al. (2009). For each mode q, $u_k^q \in U_q \subset \mathbb{R}^m$ is the known input, $d_k^q \in \mathbb{R}^p$ the unknown input or *attack signal*[1] and $y_k \in \mathbb{R}^l$ the output, whereas the corresponding process noise $w_k^q \in \mathbb{R}^n$ and measurement noise $v_k^q \in \mathbb{R}^l$ satisfy one of the following sets of assumptions for the system uncertainties:

Assumption 7.1 (*Aleatoric Uncertainty*) The system is perturbed by random (unbounded) process and measurement noise signals with process noise w_k^q and measurement noise v_k^q that are mutually uncorrelated, zero-mean Gaussian white random signals with known covariance matrices, $Q_k^q = \mathbb{E}[w_k^q {w_k^q}^\top] \succeq 0$ and $R_k^q = \mathbb{E}[v_k^q {v_k^q}^\top] \succ 0$, respectively. Moreover, x_0 is independent of v_k^q and w_k^q for all k.

Assumption 7.2 (*Epistemic Uncertainty*) The system is perturbed by uncertain, bounded process and measurement noise signals, where the corresponding process noise w_k^q and measurement noise v_k^q are distribution-free uncertain bounded signals with known bounds, i.e., $\|w_k^q\| \le \eta_w$ and $\|v_k^q\| \le \eta_v$, respectively (thus, they are ℓ_∞ sequences), where η_w^q and η_v^q are known parameters. We also assume an estimate $\hat{x}_0$ of the initial state x_0 is available, where $\|\hat{x}_0 - x_0\| \le \delta_0^{q,x}$ with known $\delta_0^{q,x}$.

Assumption 7.3 (*Aleatoric + Epistemic Uncertainty*) The system is perturbed by random and bounded process and measurement noise signals, where the corresponding process noise w_k^q and measurement noise v_k^q are mutually uncorrelated, zero-mean "truncated" Gaussian white random signals with known covariance matrices, $Q_k^q = \mathbb{E}[w_k^q {w_k^q}^\top] \succeq 0$ and $R_k^q = \mathbb{E}[v_k^q {v_k^q}^\top] \succ 0$, and bounded norms, i.e., $\|w_k^q\| \le \eta_w^q$ and $\|v_k^q\| \le \eta_v^q$, respectively, where η_w^q and η_v^q are known. Moreover, x_0 is independent of v_k^q and w_k^q for all k, and an estimate $\hat{x}_0$ of the initial state x_0 is available, where $\|\hat{x}_0 - x_0\| \le \delta_0^{q,x}$ with known $\delta_0^{q,x}$.

In the case of the stochastic/aleatoric uncertainty (i.e., if Assumption 7.1 holds and consequently, the uncertainty is characterized using probability distributions), the emphasis is on *expected/average* performance of the resilient state estimator. In this case, CPS safety/resilience is guaranteed based on probability of violation/chance constraints. On the other hand, in the case of set-valued/epistemic uncertainty (i.e., if Assumption 7.2 holds and hence the uncertainty is characterized by sets), the emphasis would be on the *best worst-case* performance and the CPS safety/resilience is

[1] Note that while the unknown inputs may also be used to represent uncertainties or noise that are unbounded or have unknown bounds, we primarily use this term to represent attack signals in this chapter and thus, we often use the terms unknown inputs and attacks interchangeably.

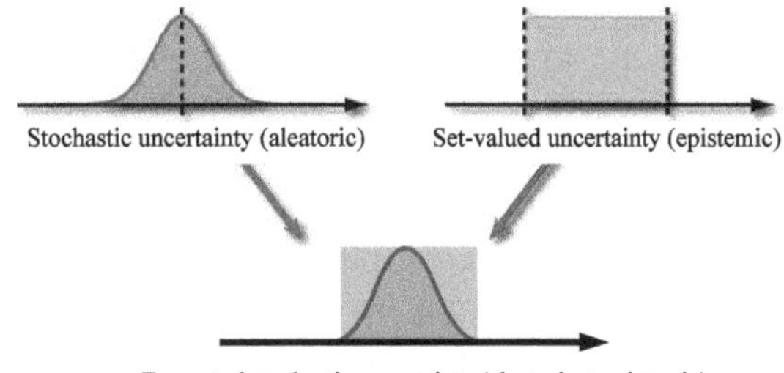

Fig. 7.1 Different assumptions on the considered uncertainty in System (7.1)

guaranteed in the worst case, including rare events/corner cases. Finally, if Assumption 7.3 holds, we can combine the information of the stochastic uncertainties and the set-membership uncertainties from Assumptions 7.1 and 7.2 to benefit from the advantages of both. Figure 7.1 illustrates the aforementioned system uncertainty models/assumptions.

Both *categorical* and *continuous* natures of the uncertainties introduced by the switching and data injection attacks to the system of interest can be captured by the Cyber-Physical System (CPS) model in (7.1). The categorical nature of the switching and data injection attacks (*mode attack* and *signal location attack*) is modeled using the *hidden mode*, whereas the unknown input captures the continuous nature of the *signal magnitude attacks*. At any particular time k, the stochastic/bounded-error CPS is in precisely one of its modes, which is not measured, hence *hidden*.

Similar to Yong et al. (2018), we consider the model set $\mathcal{Q} \triangleq \mathcal{Q}^m \times \mathcal{Q}^d$ (whose cardinality will be characterized in Theorem 7.2 in Sect. 7.3.3.2) that include

(i) the modes of operation, $\mathcal{Q}^m$ (representing attacked switching mechanisms (e.g., circuit breakers, relays) via access to the jump set $\mathcal{D}_q$ and the mode transition function $\delta^q(\cdot)$, or the possible interconnection network topologies that affect the system matrices, A_k^q and B_k^q, and the sensor data network topologies, C_k^q and D_k^q) that an attacker can choose (*mode attack*), as well as
(ii) the different hypotheses for each mode, $\mathcal{Q}^d$, about which actuators and sensors are attacked or not attacked, represented by G_k^q and H_k^q, where our approach specifies which actuators and sensors are *not attacked*, in contrast to the approach in Mishra et al. (2015), which removes *attacked* sensor measurements and is not applicable for actuator attacks. (*signal location attack*).

More precisely, for sparse false data injection attacks, we let $G_k^q \triangleq \mathcal{G}_k \mathcal{I}_G^q$ and $H_k^q \triangleq \mathcal{H}_k \mathcal{I}_H^q$ for some input matrices $\mathcal{G}_k \in \mathbb{R}^{n \times t_a}$ and $\mathcal{H}_k \in \mathbb{R}^{\ell \times t_s}$, where t_a and t_s are the number of actuator and sensor signals that are *vulnerable*, respectively and encode the sparsity using $\mathcal{I}_G^q \in \mathbb{R}^{t_a \times p}$ and $\mathcal{I}_H^q \in \mathbb{R}^{t_s \times p}$ as index matrices such that $d_k^{a,q} \triangleq \mathcal{I}_G^q d_k$ and $d_k^{s,q} \triangleq \mathcal{I}_H^q d_k$ are subvectors of $d_k \in \mathbb{R}^p$ representing *signal magnitude attacks* on the actuators and sensors, respectively. These matrices provide a means to incorporate information about how the attacks affect the system, e.g., if the same attack is injected to an actuator and a sensor, or if some signals are *not* attacked, according to a particular hypothesis/mode q about the signal attack location.

The following are some examples from Yong et al. (2018) for choosing $\mathscr{G}_k$, $\mathscr{H}_k$, $\mathscr{I}_G^q$, and $\mathscr{I}_H^q$ to encode additional information about the nature/structure of data injection attacks.

Example 7.1 For a two-state system with two vulnerable actuators and one vulnerable sensor, if the same attack signal is injected into the first actuator and the sensor under the hypothesis corresponding to mode q, then $\mathscr{G}_k = I_2$, $\mathscr{H}_k = 1$, $\mathscr{I}_G^q = I_2$ and $\mathscr{I}_H^q = \begin{bmatrix} 1 & 0 \end{bmatrix}$. In this case, we obtain $G_k^q = I_2$ and $H_k^q = \begin{bmatrix} 1 & 0 \end{bmatrix}$.

Example 7.2 For a three-state system with three actuators and two sensors, if the first actuator and the second sensor are not vulnerable and there are three attacks according to the hypothesis corresponding to mode q, then $\mathscr{G}_k = \begin{bmatrix} 0 & 0 \\ 1 & 0 \\ 0 & 1 \end{bmatrix}$, $\mathscr{H}_k = \begin{bmatrix} 1 \\ 0 \end{bmatrix}$, $\mathscr{I}_G^q = \begin{bmatrix} 1 & 0 & 0 \\ 0 & 1 & 0 \end{bmatrix}$ and $\mathscr{I}_H^q = \begin{bmatrix} 0 & 0 & 1 \end{bmatrix}$. In this case, we have $G_k^q = \begin{bmatrix} 0 & 0 & 0 \\ 1 & 0 & 0 \\ 0 & 1 & 0 \end{bmatrix}$ and $H_k^q = \begin{bmatrix} 0 & 0 & 1 \\ 0 & 0 & 0 \end{bmatrix}$.

Note that we assume that $p_a^q \leq t_a \leq m$ (i.e., the number of *attacked* actuator signals p_a^q under mode/hypothesis q cannot exceed the number of *vulnerable* actuators and in turn cannot exceed the total number of actuators m_a) and $p_s^q \leq t_s \leq \ell$ (with p_s^q *attacked* sensors from t_s *vulnerable* sensors out of ℓ measurements). Moreover, we assume that the maximum total number of attacks is $p \triangleq p_a^q + p_s^q \leq p^*$, where p^* is the maximum number of asymptotically correctable signal attacks (cf. Theorem 7.1 for its characterization).

7.2.2.1 System Assumptions

We require that the system is *strongly detectable*[2] in each mode. In fact, strong detectability is *necessary* for each mode in order to asymptotically correct the unknown attack signals, as shown in Yong et al. (2018) [Theorem 4.3] and is also necessary for deterministic systems [Sundaram and Hadjicostis (2007), Theorem 6]. Note that similar to the detectability property, strongly detectable systems need not be stable (cf. example in the proof of Theorem 7.1), but rather that the strongly undetectable modes of such systems are stable.

7.2.2.2 Knowledge of the System Operator/Defender

The matrices A_k^q, B_k^q, G_k^q, C_k^q, D_k^q, and H_k^q are known and the system (A_k^q, G_k^q, C_k^q, H_k^q) is strongly detectable in each mode. Further, the defender only knows (i) the

[2] A linear system is *strongly detectable* if $y_k = 0\ \forall k \geq 0$ implies $x_k \to 0$ as $k \to \infty$ for all initial states x_0 and input sequences $\{d_i\}_{i \in \mathbb{N}}$ (see [Yong et al. (2016b), Sect. 3.2] for necessary and sufficient conditions for this property).

upper bound on the *number* of actuators/sensors that can be attacked, p, and (ii) the switching mechanisms/topologies that may be compromised. The upper bound p allows the defender, in the worst case, to enumerate all possible combinations of G_k^q and H_k^q, while the latter assumption allows the defender to consider all possible topologies/modes of operations, representing A_k^q, B_k^q, C_k^q and D_k^q.

In addition, note that the above assumption of strong detectability can be viewed as recommendations or guidelines for system designers/operators to secure their systems as a preventative attack mitigation measure, since without strong detectability, resilient (i.e., unbiased or bounded) state estimates cannot be guaranteed. In other words, the requirement of strong detectability allows system designers to determine which actuators or sensors need to be safeguarded to guarantee resilient estimation.

7.2.3 *Security Problem Statement*

With the above modeling framework, the resilient state estimation problem can be posed as a problem of mode, state and input estimation, where the unknown inputs represent the unknown signal magnitude attacks and each mode/model represents an *attack mode* (resulting from the unknown mode attacks and unknown signal attack locations). The *objective* of this chapter is:

Problem 7.1 *Given an uncertain Cyber-Physical System (CPS) described by* (7.1),

1. *Design a* resilient estimator *that asymptotically recovers* unbiased *estimates of the system state and attack signal in the presence of aleatoric/stochastic uncertainty (i.e., if Assumption 7.1 holds), or finds the set-valued estimates of compatible states and unknown inputs in the presence of epistemic uncertainty (i.e., if Assumption 7.2 holds), irrespective of the location or magnitude of attacks on its actuators and sensors as well as switching mechanism/topology (mode) attacks.*
2. *Investigate the fundamental limitations of the estimation algorithms, specifically the maximum number of asymptotically correctable signal attacks and the maximum number of required models with our multiple-model approach.*
3. *Find the conditions under which attacks can be detected and under which the attack strategy can be identified.*
4. *Design attack mitigation tools via $\mathscr{H}_\infty$-control with attack rejection.*

7.3 Resilient State Estimation

Similar to a previous approach for stochastic systems in Yong et al. (2021), we propose the use of a *multiple-model* estimation approach to solve Problem 7.1.1. Then, we will consider Problem 7.1.2 and characterize some fundamental limitations to resilient estimation in Sect. 7.3.3.

7.3.1 *Multiple-Model State and Input Filtering/Estimation Algorithm*

Inspired by the multiple-model filtering algorithms for hidden-mode hybrid systems with *known* inputs (e.g., Bar-Shalom et al. (2004); Mazor et al. (1998) and references therein), our multiple-model (MM) framework (see Fig. 7.2) consists of three components: (i) a bank of mode-matched filters/observers, (ii) a mode estimator that finds the most likely or compatible modes, and (iii) a global fusion estimator that combines/fuses states and unknown input (attack) estimates from (i) based on the estimated modes in (ii), which are described in greater detail below.

7.3.1.1 Mode-Matched Filters/Observers

The bank of filters/observers is comprised of $\mathfrak{N}$ simultaneous state and input filters/observers, one for each mode, that differ based on the assumptions on system uncertainties and noise signals. If Assumption 7.1 (the aleatoric/stochastic uncertainty model) holds, the optimal recursive filter developed in Yong et al. (2016b) can be applied, while if Assumption 7.2 (the epistemic/set-valued uncertainty model) holds, the recursive set-valued observer developed in Yong (2018) can be utilized. Both variants are recursive and involve the same three-step structure as follows:

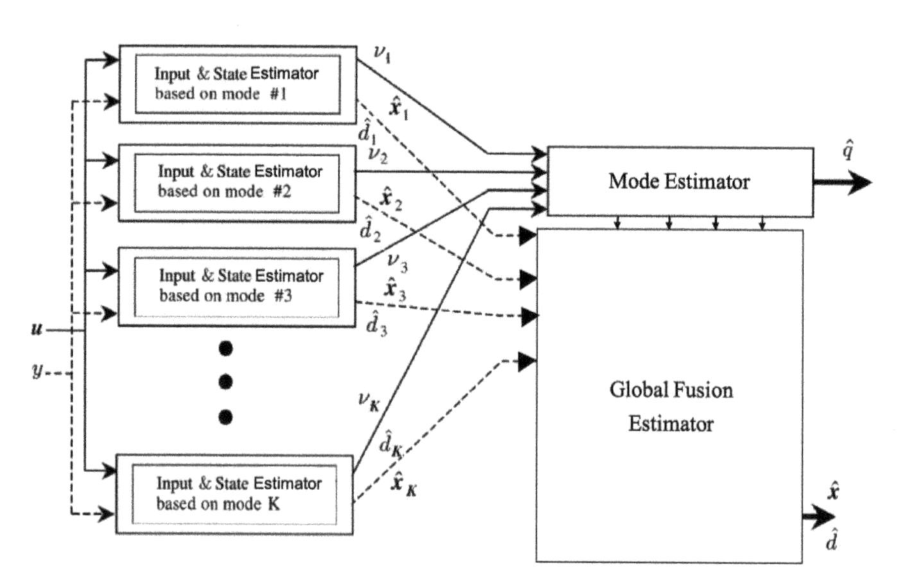

Fig. 7.2 Multiple-model framework for hidden mode, input and state estimation, which consists of a (i) bank of mode-matched filters/observers, (ii) a mode estimator and (iii) a global fusion estimator

Unknown Input Estimation:

$$\begin{aligned}\hat{d}^q_{1,k} &= M^q_{1,k}(z^q_{1,k} - C^q_{1,k}\hat{x}^q_{k|k} - D^q_{1,k}u^q_k),\\ \hat{d}^q_{2,k-1} &= M^q_{2,k}(z^q_{2,k} - C^q_{2,k}\hat{x}^q_{k|k-1} - D^q_{2,k}u^q_k),\\ \hat{d}^q_{k-1} &= V^q_{1,k-1}\hat{d}^q_{1,k-1} + V^q_{2,k-1}\hat{d}^q_{2,k-1}.\end{aligned} \tag{7.2}$$

Time Update:

$$\begin{aligned}\hat{x}^q_{k|k-1} &= A^q_{k-1}\hat{x}^q_{k-1|k-1} + B^q_{k-1}u^q_{k-1} + G^q_{1,k-1}\hat{d}^q_{1,k-1},\\ \hat{x}^{\star,q}_{k|k} &= \hat{x}^q_{k|k-1} + G^q_{2,k-1}\hat{d}^q_{2,k-1}.\end{aligned} \tag{7.3}$$

Measurement Update:

$$\hat{x}^q_{k|k} = \hat{x}^{\star,q}_{k|k} + \tilde{L}^q_k(z^q_{2,k} - C^q_{2,k}\hat{x}^{\star,q}_{k|k} - D^q_{2,k}u^q_k), \tag{7.4}$$

where $\hat{x}^q_{k-1|k-1}$, $\hat{d}^q_{1,k-1}$, $\hat{d}^q_{2,k-1}$ and $\hat{d}^q_{k-1}$ denote the optimal *point* estimates of x^q_{k-1}, $d^q_{1,k-1}$, $d^q_{2,k-1}$ and d^q_{k-1}, respectively, if Assumption 7.1 holds (cf. Algorithm 7.1 that summarizes the optimal filter for mode q in the presence of stochastic (aleatoric) uncertainty) and denote the centroids of the *hyperball-valued* estimates of x^q_{k-1}, $d^q_{1,k-1}$, $d^q_{2,k-1}$ and d^q_{k-1}, respectively, if Assumption 7.2 holds (cf. Algorithm 7.3 that finds the $\mathscr{H}_\infty$-optimal set-valued state and input estimates for mode q in the presence of distribution-free (epistemic) uncertainty).

The rest of the notations are clarified in the context of the system transformation described in Appendix 7.1.1. For details of the filter/observer derivation of both variants, as well as necessary and sufficient conditions for filter stability and optimality of the mode-matched filters/observers, the reader is referred to Yong et al. (2016b) and Yong (2018) for the aleatoric and epistemic uncertainty models, respectively.

It is worth mentioning that in the case that Assumption 7.3 holds (i.e., with a combination of aleatoric and epistemic uncertainties), we can compute (in parallel) both the point estimates corresponding to aleatoric/stochastic uncertainty and the set-valued estimates corresponding to the epistemic/bounded-error uncertainty, and utilize their combination as described in the following subsections.

7.3.1.2 Mode Estimator

The mode estimator seeks to determine the most likely or all compatible modes based on the observations. For this purpose, we consider three cases:

(a) **Aleatoric Uncertainty.** In this case, Assumption 7.1 holds and consequently, a *mode probability computation* is performed for all modes as described in Yong et al. (2018). The multiple-model approach computes the probability of each mode by exploiting the whiteness property [Yong et al. (2021), Theorem 1] of the *generalized innovation* sequence, ν^q_k, defined as

$$\nu^q_k \triangleq \tilde{\Gamma}^q_k(z^q_{a,2,k} - C^q_{a,2,k}\hat{x}^{\star,q}_{a,k|k} - D^q_{a,2,k}u^q_k), \tag{7.5}$$

i.e., $\nu_k^q \sim \mathcal{N}(0, S_k^q)$ (a multivariate normal distribution) with covariance $S_k^q \triangleq \mathbb{E}[\nu_k^q \nu_k^{q\top}] = \tilde{\Gamma}_k^q \tilde{R}_{2,k}^{\star,q} \tilde{\Gamma}_k^{q\top}$ and where $\tilde{\Gamma}_k^q$ is chosen such that S_k^q is invertible and $\tilde{R}_{2,k}^{\star,q}$ is given in Algorithm 7.1. This generalized innovation represents a residual signal with false data injection attacks removed that can be used to define the *likelihood function* for each mode q at time k conditioned on all prior measurements Z^{k-1}:

$$\mathcal{L}(q|z_{2,k}) \triangleq \mathcal{N}(\nu_k^q; 0, S_k^q) = \frac{\exp(-\frac{1}{2}\nu_k^{q\top}(S_k^q)^{-1}\nu_k^q)}{\sqrt{|2\pi S_k^q|}}. \tag{7.6}$$

Then, the posterior probability μ_k^j for each mode j is recursively computed from the prior probability μ_{k-1}^j using Bayes' rule as follows:

$$\mu_k^j = P(q = j|z_{1,k}, z_{2,k}, Z^{k-1}) = \frac{\mathcal{N}(\nu_k^j; 0, S_k^j)\mu_{k-1}^j}{\sum_{i=1}^{\mathfrak{N}} \mathcal{N}(\nu_k^i; 0, S_k^i)\mu_{k-1}^j}. \tag{7.7}$$

Furthermore, to keep the modes "alive" in case of a switch in the attacker's strategy, a heuristic lower bound on all mode probabilities is imposed.

(b) **Epistemic Uncertainty.** In the presence of distribution-free and bounded norm noise signals, i.e., when Assumption 7.2 holds, a *mode elimination* process is performed to eliminate the modes that are incompatible with observations, which results in a set of compatible modes. The mode elimination approach relies on the checking of some *residual* signals against some thresholds. We first define the residual signal r_k^q for each mode q at time step k as:

$$r_k^q \triangleq z_{e,2,k}^q - C_{e,2,k}^q \hat{x}_{a,k|k}^{\star,q} - D_{e,2,k}^q u_k^q. \tag{7.8}$$

Then, leveraging an approach in Khajenejad and Yong (2019), if the residual signal of a particular mode exceeds its upper bound conditioned on this mode being true, we can conclusively rule it out as incompatible. To do so, for each mode q, we compute a *tractable* upper bound ($\hat{\delta}_{r,k}^q$; cf. Proposition 7.2) for the 2-norm of its corresponding residual at time k, conditioned on q being the *true* mode. Then, comparing the 2-norm of residual signal in (7.8) with $\hat{\delta}_{r,k}^q$, we can eliminate mode q if the residual's 2-norm is strictly greater than the upper bound, i.e., if $\|r_k^q\|_2 > \hat{\delta}_{r,k}^q$. This can be formalized using the following proposition (cf. [Khajenejad and Yong (2019), Proposition 1 and Theorem 2] for more details and a formal proof of this result).

Proposition 7.1 *Consider mode q and its residual signal r_k^q at time step k. Assume that $\delta_{r,k}^{q,*}$ is any signal that satisfies $\|r_k^{q|*}\|_2 \le \delta_{r,k}^{q,*}$, where $r_k^{q|*}$ is the true mode's residual signal (i.e., $q = q^*$, where q^* denotes the true mode), defined as follows:*

$$r_k^{q|*} \triangleq z_{e,2,k}^{q*} - C_{e,k,2}^q \hat{x}_{e,k|k}^{\star,q} - D_{e,k,2}^q u_k^q = T_{e,k,2}^{q*} y_k - C_{e,k,2}^q \hat{x}_{e,k|k}^{\star,q} - D_{e,k,2}^q u_k^q. \tag{7.9}$$

Then, mode q is not the true mode, i.e., can be eliminated at time k, if

$$\|r_k^q\|_2 > \delta_{r,k}^{q,*}. \tag{7.10}$$

Note that by [Khajenejad and Yong (2019), Lemmas 1 and 2], the sequence $\{\delta_{r,k}^{q,*}\}_{k=0}^{\infty}$ is uniformly bounded and admits a finite valued upper sequence. Although computing the tightest possible residual norm's upper sequence potentially can eliminate the most possible number of modes, it requires to the solution a *norm maximization* problem over the intersection of level sets of lower dimensional norm functions that is NP-hard [Bodlaender et al. (1990)]. Thus, by applying [Khajenejad and Yong (2019), Theorem 3], we instead compute a tractable over-approximation of the residual norm's upper bound sequence, denoted by $\{\hat{\delta}_{r,k}^q\}_{k=0}^{\infty}$, i.e., $\forall k \in \{0, \ldots, \infty\}$, $\delta_{r,k}^{q,*} \leq \hat{\delta}_{r,k}^q$, and use this upper bound sequence as a tractable mode elimination criterion as follows (cf. [Khajenejad and Yong (2019), Theorem 3] for more details):

Proposition 7.2 *Mode q is not the true mode, i.e., can be eliminated at time k, if*

$$\|r_k^q\|_2 > \hat{\delta}_{r,k}^q \triangleq \min\{\delta_{r,k}^{q,inf}, \delta_{r,k}^{q,tri}\}, \tag{7.11}$$

where $\delta_{r,k}^{q,inf}$ and $\delta_{r,k}^{q,tri}$ are two tractable computed upper bounds for the residual norm and are given in Appendix 7.1.2.

(c) **Combined Uncertainty.** In the presence of truncated Gaussian noise signals, i.e., if Assumption 7.3 holds, both mode probability computation procedure (described in (7.3.1.2)) and mode elimination approach (described in (7.3.1.2)) are applicable and can be combined. Specifically, we first apply the mode elimination algorithm from Khajenejad and Yong (2019) to obtain a set of compatible modes, and then compute mode probabilities for only the "non-eliminated" modes using (7.7).

7.3.1.3 Global Fusion Estimator

Finally, the global fusion estimator combines the estimates from the bank of mode-matched state and input estimators and mode observer, under the three different system uncertainty models, as follows:

(a) **Aleatoric Uncertainty.** Based on the posterior mode probabilities in (7.7), the most likely mode at each time k, $\hat{q}_k$, and the associated state and input estimates and covariances, $\hat{x}_{a,k|k}$, $\hat{d}_{a,k}$, $P_{k|k}^x$ and P_k^d, can be determined:

$$\begin{aligned} \hat{q}_k &= j^* = \arg\max_{j\in\{1,\ldots,\mathfrak{N}\}} \mu_k^j, \\ \hat{x}_{a,k|k} &= \hat{x}_{a,k|k}^{j^*}, \quad \hat{d}_{a,k} = \hat{d}_{a,k}^{j^*}, \\ P_{k|k}^x &= P_{k|k}^{x,j^*}, \quad P_k^d = P_k^{d,j^*}. \end{aligned} \tag{7.12}$$

(b) **Epistemic Uncertainty.** Using the computed residuals (7.9) and their upper bound sequences (7.11), our proposed global fusion observer finds all modes that are not eliminated and computes the input and state set-valued estimates, $\hat{D}_{k-1}$ and $\hat{X}_k$, by taking the union of the mode-matched state and unknown input (attack) set estimates over the compatible modes:

$$\begin{aligned} \hat{\mathscr{Q}}_k &= \{q \in \mathscr{Q} \mid \|r^q\|_2 \leq \hat{\delta}^q_{r,k}\}, \\ \hat{D}_{k-1} &= \cup_{q \in \hat{\mathscr{Q}}_k} D^q_{k-1}, \\ \hat{X}_k &= \cup_{q \in \hat{\mathscr{Q}}_k} X^q_k. \end{aligned} \tag{7.13}$$

(c) **Combined Uncertainty.** In this case, after eliminating all modes that satisfy (7.11), the most likely mode and its associated state and input estimates and covariances at each time can be determined using only the set of non-eliminated modes (instead of all modes as in the case of aleatoric uncertainty), i.e.,

$$\begin{aligned} \hat{\hat{q}}_k &= j^{**} = \arg\max_{j \in \hat{\mathscr{Q}}} \mu^j_k, \\ \hat{x}_{c,k|k} &= \hat{x}^{j^{**}}_{c,k|k}, \quad \hat{d}_{c,k} = \hat{d}^{j^{**}}_{c,k}, \\ P^x_{c,k|k} &= P^{x,j^{**}}_{k|k}, \quad P^d_{c,k} = P^{d,j^{**}}_k. \end{aligned} \tag{7.14}$$

The multiple-model approach is summarized in Algorithms 7.1–7.4 for the aleatoric/stochastic and epistemic/set-valued uncertainties, respectively.

7.3.2 Properties of the Resilient State Estimator

Our previous results in Yong et al. (2021); Khajenejad and Yong (2019); Yong (2018) show that the resilient state estimator has nice properties, which can be summarized as follows.

7.3.2.1 Optimality

Given the attacked switched linear system with hidden modes in (7.1), if Assumption 7.1 holds (aleatoric uncertainty), the resilient state estimator (i.e., Algorithms 7.1 and 7.2) is *asymptotically optimal*, i.e., the state and input estimates in (7.12) converge on average to optimal state and input estimates in the minimum variance unbiased sense [Yong et al. (2021), Corollary 13]. On the other hand, if Assumption 7.2 holds (epistemic uncertainty), the resulting set-valued estimates in (7.13) are uniformly bounded [Yong (2018), Lemma 1] and the resilient state and input observer is stable and optimal in the $\mathscr{H}_\infty$-norm sense [Yong (2018) [Theorem 2]]. Further, in the presence of truncated Gaussian noise signals, i.e., if Assumption 7.3 is satisfied, it can be shown that the set-valued estimates are uniformly bounded, but the resilient state estimates obtained from Algorithms 7.3 and 7.4, may not be asymptotically optimal.

7.3.2.2 Mode Detectability

Given the attacked switched linear system with hidden modes in (7.1), in the presence of aleatoric/stochastic uncertainty, i.e., if Assumption 7.1 holds, the resilient state estimator is *mean consistent*, i.e., the geometric mean of the mode probability for the true model $q^* \in \mathcal{Q}$ asymptotically converges to one for all initial mode prob-

---

Algorithm 7.1 OPT- FILTER finds the optimal state and input estimates for mode q in the presence of stochastic (aleatoric) uncertainty

---

Input: $q, \hat{x}^q_{k-1|k-1}, \hat{d}^q_{1,k-1}, P^{x,q}_{k-1|k-1}, P^{xd,q}_{1,k-1}, P^{d,q}_{1,k-1}$
[superscript "q" and subscript "a" (referring to aleatoric uncertainty) omitted in the following]
$\triangleright$ Estimation of $d_{2,k-1}$ and d_{k-1}
$\hat{A}_{k-1} = A_{k-1} - G_{1,k-1}M_{1,k-1}C_{1,k-1}$;
$\hat{Q}_{k-1} = G_{1,k-1}M_{1,k-1}R_{1,k-1}M^\top_{1,k-1}G^\top_{1,k-1} + Q_{k-1}$;
$\tilde{P}_k = \hat{A}_{k-1}P^x_{k-1|k-1}\hat{A}^\top_{k-1} + \hat{Q}_{k-1}$;
$\tilde{R}_{2,k} = C_{2,k}\tilde{P}_k C^\top_{2,k} + R_{2,k}$;
$P^d_{2,k-1} = (G^\top_{2,k-1}C^\top_{2,k}\tilde{R}^{-1}_{2,k}C_{2,k}G_{2,k-1})^{-1}$;
$M_{2,k} = P^d_{2,k-1}G^\top_{2,k-1}C^\top_{2,k}\tilde{R}^{-1}_{2,k}$;
$\hat{x}_{k|k-1} = A_{k-1}\hat{x}_{k-1|k-1} + B_{k-1}u_{k-1} + G_{1,k-1}\hat{d}_{1,k-1}$;
$\hat{d}_{2,k-1} = M_{2,k}(z_{2,k} - C_{2,k}\hat{x}_{k|k-1} - D_{2,k}u_k)$;
$\hat{d}_{k-1} = V_{1,k-1}\hat{d}_{1,k-1} + V_{2,k-1}\hat{d}_{2,k-1}$;
$P^d_{12,k-1} = M_{1,k-1}C_{1,k-1}P^x_{k-1|k-1}A^\top_{k-1}C^\top_{2,k}M^\top_{2,k} - P^d_{1,k-1}G^\top_{1,k-1}C^\top_{2,k}M^\top_{2,k}$;
$P^d_{k-1} = V_{k-1}\begin{bmatrix} P^d_{1,k-1} & P^d_{12,k-1} \\ P^{d\top}_{12,k-1} & P^d_{2,k-1} \end{bmatrix} V^\top_{k-1}$;
$\triangleright$ Time update
$\hat{x}^\star_{k|k} = \hat{x}_{k|k-1} + G_{2,k-1}\hat{d}_{2,k-1}$;
$P^{\star x}_{k|k} = G_{2,k-1}M_{2,k}R_{2,k}M^\top_{2,k}G^\top_{2,k} + (I - G_{2,k-1}M_{2,k}C_{2,k})\tilde{P}_k(I - G_{2,k-1}M_{2,k}C_{2,k})^\top$;
$\tilde{R}^\star_{2,k} = C_{2,k}P^{\star x}_{k|k}C^\top_{2,k} + R_{2,k} - C_{2,k}G_{2,k-1}M_{2,k}R_{2,k} - R_{2,k}M^\top_{2,k}G^\top_{2,k-1}C_{2,k}$;
$\triangleright$ Measurement update
$\breve{P}_k = P^{\star x}_{k|k}C^\top_{2,k} - G_{2,k-1}M_{2,k}R_{2,k}$;
$\tilde{L}_k = \breve{P}_k\tilde{R}^{\star\dagger}_{2,k}$;
$\hat{x}_{k|k} = \hat{x}^\star_{k|k} + \tilde{L}_k(z_{2,k} - C_{2,k}\hat{x}^\star_{k|k} - D_{2,k}u_k)$;
$P^x_{k|k} = \tilde{L}_k R^\star_{2,k}\tilde{L}^\top_k - \tilde{L}_k\breve{P}^\top_k - \breve{P}_k\tilde{L}^\top_k$;
$\triangleright$ Estimation of $d_{1,k}$
$\tilde{R}_{1,k} = C_{1,k}P^x_{k|k}C^\top_{1,k} + R_{1,k}$;
$M_{1,k} = \Sigma^{-1}_k$;
$P^d_{1,k} = M_{1,k}\tilde{R}_{1,k}M_{1,k}$;
$\hat{d}_{1,k} = M_{1,k}(z_{1,k} - C_{1,k}\hat{x}_{k|k} - D_{1,k}u_k)$;
return $\tilde{R}^{\star,q}_{2,k}, \hat{x}^{\star,q}_{k|k}$

Algorithm 7.2 RESILIENT STATE ESTIMATOR (STATIC- MM- ESTIMATOR) finds resilient state estimates corresponding to most likely mode in the presence of stochastic (aleatoric) uncertainty

---

Input: $\forall j \in \{1, 2, \ldots, \mathfrak{N}\}$: $\hat{x}^j_{0|0}$; μ^j_0;
[subscript "a" (referring to aleatoric uncertainty) omitted in the following]
$\hat{d}^j_{1,0} = (\Sigma^j_0)^{-1}(z^j_{1,0} - C^j_{1,0}\hat{x}^j_{0|0} - D^j_{1,0}u_0)$;
$P^{d,j}_{1,0} = (\Sigma^j_0)^{-1}(C^j_{1,0}P^{x,j}_{0|0}C^{j\top}_{1,0} + R^j_{1,0})(\Sigma^j_0)^{-1}$;
for $k = 1$ *to* N **do**
  for $j = 1$ *to* $\mathfrak{N}$ **do**
    ▷ Mode-Matched Filtering Run OPT- FILTER(j,$\hat{x}^j_{k-1|k-1}$, $\hat{d}^j_{1,k-1}$, $P^{x,j}_{k-1|k-1}$, $P^{d,j}_{1,k-1}$);
    $\overline{\nu}^j_k \triangleq z^j_{2,k} - C^j_{2,k}\hat{x}^{\star,j}_{k|k} - D^j_{2,k}u_k$;
    $\mathscr{L}(j|z^j_{2,k}) = \frac{1}{(2\pi)^{p^j_{\tilde{R}}/2}|\tilde{R}^{j,\star}_{2,k}|^{1/2}_+} \exp\left(-\frac{\overline{\nu}^{j\top}_k \tilde{R}^{j,\star\dagger}_{2,k}\overline{\nu}^j_k}{2}\right)$;
  for $j = 1$ *to* $\mathfrak{N}$ **do**
    ▷ Mode Probability Update (small $\epsilon > 0$)
    $\overline{\mu}^j_k = \max\{\mathscr{L}(j|z^j_{2,k})\mu^j_{k-1}, \epsilon\}$;
  for $j = 1$ *to* $\mathfrak{N}$ **do**
    ▷ Mode Probability Update (normalization)
    $\mu^j_k = \frac{\overline{\mu}^j_k}{\sum_{\ell=1}^{\mathfrak{N}} \overline{\mu}^\ell_k}$;
    ▷ Output
    Compute (7.12);
return $\hat{x}_{k|k}$, $P^x_{k|k}$

---

abilities [Yong et al. (2021), Theorem 8]. Furthermore, in the case of epistemic/set-valued uncertainty, i.e., if Assumption 7.2 holds, the resilient state estimator is *mode detectable* by [Khajenejad and Yong (2019), Theorem 4], i.e., there exists a natural number $K > 0$, such that for all time steps $k \geq K$, all false modes are eliminated, if either the whole observation/measurement and state spaces are bounded or the unknown input/attack signal has an *unlimited energy*, as well as some additional mild conditions hold (cf. [Khajenejad and Yong (2019), Assumptions 1&2, Lemmas 3–5 and Theorem 4] for more details). Similarly, if Assumption 7.3 holds, all false modes (except for the true mode) will be eliminated after some large enough finite time under the same assumption of bounded state spaces or unlimited energy, and the unique true mode will have probability one.

7.3.3 Fundamental Limitations of Attack-Resilient Estimation

Next, to address Problem 1.2, we characterize fundamental limitations of the attack-resilient estimation problem and of our multiple mode filtering/estimation approach.

Algorithm 7.3 OPT- OBSERVER finds the $\mathscr{H}_\infty$-optimal set-valued state and input estimates for mode q in the presence of distribution-free (epistemic) uncertainty

Input: $q, \hat{x}^q_{k-1|k-1}, \hat{d}^q_{k-1}$
[superscript "q" and subscript "e" (referring to the epistemic (set-valued) uncertainty) omitted in the following]
▷ Estimation of $d_{2,k-1}$ and d_{k-1}
$M_{1,k} = \Sigma_k^{-1}$,
$M_{2,k} = (C_{2,k}G_{2,k})^\dagger$,
$\hat{A}_{k-1} = A_{k-1} - G_{1,k-1}M_{1,k-1}C_{1,k-1}$;
$\Phi_k = I - G_{2,k}M_{2,k}C_{2,k}$;
$\overline{A}_k = \Phi_k \hat{A}_k$;
$V_{e,k} = V_{1,k}M_{1,k}C_{1,k} + V_{2,k}M_{2,k}C_{2,k}\hat{A}_k$;
$A_{e,k} = (I - \tilde{L}_k C_{2,k})\overline{A}_k$;
$B_{e,w,k} = (I - \tilde{L}_k C_{2,k})\Phi_k$;
$B_{e,v_1,k} = -(I - \tilde{L}_k C_{2,k})\Phi_k G_{1,k}M_{1,k}T_{1,k}$;
$B_{e,v_2,k} = -((I - \tilde{L}_k C_{2,k})G_{2,k}M_{2,k} + \tilde{L}_k)T_{2,k}$;
$\hat{x}_{k|k-1} = A_{k-1}\hat{x}_{k-1|k-1} + B_{k-1}u_{k-1} + G_{1,k-1}\hat{d}_{1,k-1}$;
$\hat{d}_{2,k-1} = M_{2,k}(z_{2,k} - C_{2,k}\hat{x}_{k|k-1} - D_{2,k}u_k)$;
$\hat{d}_{k-1} = V_{1,k-1}\hat{d}_{1,k-1} + V_{2,k-1}\hat{d}_{2,k-1}$;
$\delta^d_{k-1} = \delta^x_0\|V_{e,k}A^{k-1}_{e,k}\| + \eta_w(\sum_{i=0}^{k-2}\|V_{e,k}A^{k-2-i}_{e,k}B_{e,w,k}\| + \|V_{2,k}M_{2,k}C_{2,k}\|) +$
$\eta_v(\|V_{2,k}M_{2,k}T_{2,k}\| + \|V_{e,k}A^{k-2}_{e,k}B_{e,v_1,k}\| + \|V_{e,k}B_{e,v_2,k} + (V_{1,k} - V_{2,k}M_{2,k}C_{2,k}G_{1,k})M_{1,k}T_{1,k}\| +$
$\sum_{i=1}^{k-2}\|V_{e,k}A^{k-2-i}_{e,k}(B_{e,v_1,k} + A_{e,k}B_{e,v_2,k})\|)$;
$\hat{D}_{k-1} = \{d \in \mathbb{R}^l : \|d - \hat{d}_{k-1}\| \le \delta^d_{k-1}\}$;
▷ Time update
$\hat{x}^\star_{k|k} = \hat{x}_{k|k-1} + G_{2,k-1}\hat{d}_{2,k-1}$;
▷ Measurement update
$\hat{x}_{k|k} = \hat{x}^\star_{k|k} + \tilde{L}_k(z_{2,k} - C_{2,k}\hat{x}^\star_{k|k} - D_{2,k}u_k)$;
$\delta^x_k = \delta^x_0\|A^k_{e,k}\| + \eta_w\sum_{i=0}^{k-1}\|A^i_{e,i}B_{e,w,i}\| + \eta_v(\|B_{e,v_2,k}\| + \|A^{k-1}_{e,k}B_{e,v_1,k}\| +$
$\sum_{i=0}^{k-2}\|A^i_{e,i}(B_{e,v_1,i} + A_{e,k}B_{e,v_2,i})\|)$;
$\hat{X}_k = \{x \in \mathbb{R}^n : \|x - \hat{x}_{k|k}\| \le \delta^x_k\}$;
▷ Estimation of $d_{1,k}$
$\hat{d}_{1,k} = M_{1,k}(z_{1,k} - C_{1,k}\hat{x}_{k|k} - D_{1,k}u_k)$;
return $\hat{X}^q_k, \hat{D}^q_{k-1}$

Note that these fundamental limitations apply to all hidden-mode switched linear systems with unknown inputs (attacks) (7.1), regardless of the assumptions about the system uncertainties. First, under the assumption that there is only false data injection attacks (no switching attacks), we find an upper bound on the number of correctable signal attacks/errors (i.e., signal attacks whose effects can be negated or cancelled). Then, we characterize the maximum number of models that is required by our multiple-model approach to obtain resilient estimates despite attacks.

Algorithm 7.4 RESILIENT MODE, STATE AND INPUT ESTIMATOR simultaneously finds compatible sets of modes, unknown inputs (attacks) and states in the presence of distribution-free (epistemic) uncertainties

Input: $\mathcal{Q} \triangleq \{1, 2, \ldots, \mathfrak{N}\}$, $\forall j \in \{1, 2, \ldots, \mathfrak{N}\}$: $\hat{x}^j_{0|0}$;
[subscript "e" (referring to the epistemic (set-valued) uncertainty) omitted in the following]
$\hat{\mathcal{Q}}_0 = \mathcal{Q}$;
for $k = 1$ *to* N **do**
  for $q \in \hat{\mathcal{Q}}_{k-1}$ **do**
    ▷Mode-Matched State and Input Set-Valued Estimates
    Run OPT- OBSERVER(q,$\hat{x}^q_{k-1|k-1}$, $\hat{d}^q_{k-1}$);
    $z^q_{2,k} = T^q_2 y_k$;
    ▷Mode Observer via Elimination
    $\hat{\mathcal{Q}}_k = \hat{\mathcal{Q}}_{k-1}$;
    Compute r^q via (7.8)
    and $\hat{\delta}^q_{r,k}$ via Proposition 7.2;
    if $\|r^q\|_2 > \hat{\delta}^q_{r,k}$ **then**
      $\hat{\mathcal{Q}}_k = \hat{\mathcal{Q}}_k \backslash \{q\}$;
  ▷State and Input Estimates
  $\hat{X}_k = \cup_{q \in \hat{\mathcal{Q}}_k} \hat{X}^q_k$;
  $\hat{D}_{k-1} = \cup_{q \in \hat{\mathcal{Q}}_k} \hat{D}^q_{k-1}$;
return $\hat{\mathcal{Q}}_k$, $\hat{D}_{k-1}$, $\hat{X}_k$

7.3.3.1 Number of Asymptotically Correctable Signal Attacks

We begin by defining the notion of correctable signal attacks in the setting with only data injection attacks, which is itself an interesting CPS security research problem.

Definition 7.1 (Correctable Signal Attacks) We say that p actuators and sensors signal attacks are correctable, if for any initial state $x_0 \in \mathbb{R}^n$ and signal attack sequence $\{d_j\}_{j \in \mathbb{N}}$ in $\mathbb{R}^p$, we have an estimator/observer such that the estimate bias asymptotically/exponentially tends to zero (under aleatoric uncerainty, cf. Assumption 7.1), i.e., $\mathbb{E}[\hat{x}_{a,k|k} - x_k] \to 0$ (and $\mathbb{E}[\hat{d}_{a,k-1} - d_{k-1}] \to 0$) as $k \to \infty$ or if the set estimation errors are ultimately uniformly bounded sequences (under epistemic uncertainty, cf. Assumption 7.2).

To derive an estimation-theoretic upper bound on the maximum number of signal attacks that can be asymptotically corrected, we assume that the true model or mode ($q = q^*$) is known. Thus, depending on the type of uncertainty, the resilient state estimation problem is identical to the state and input estimation problem in Yong et al. (2016b) or Yong (2018), where the unknown inputs represent the attacks on the actuator and sensor signals. It has been shown in Yong et al. (2016b) and Yong (2018) that the system property of strong detectability is a necessary condition for obtaining uniformly bounded estimates (cf. Yong et al. (2016b); Yong (2018) for more details, e.g., regarding filter/observer stability and existence). Thus, we will use this necessary system property to find an upper bound on the maximum number of signal attacks that can be corrected, similar to Yong et al. (2018), as follows:

Theorem 7.1 (Maximum Correctable Data Injection Attacks) *The maximum number of correctable actuators and sensors signal attacks, p^*, for system* (7.1) *is equal to the number of sensors, l, i.e., $p^* \leq l$ and the upper bound is achievable.*

Proof A necessary and sufficient condition for strong detectability (with the true model $q = q^*$) is given in Yong et al. (2016b); Yong (2018) as

$$\text{rk} \begin{bmatrix} zI - A^* & -G^* \\ C^* & H^* \end{bmatrix} = n + p^*, \ \forall z \in \mathbb{C}, |z| \geq 1. \tag{7.15}$$

Since the above system matrix has only $n + l$ rows, it follows that its rank is at most $n + l$. Thus, from the necessary condition for (7.15), we obtain $n + p^* \leq n + l \Rightarrow p^* \leq l$. The upper bound is achievable using the example of the discrete-time equivalent model (with time step $\Delta t = 0.1s$) of the smart grid case study in Liu et al. (2013), as shown in Yong et al. (2018) [Theorem 4.3]. ■

The above result means that for each mode, the total number of vulnerable actuators and sensors must not exceed the number of measurements, which can serve as a guide for *preventative attack mitigation*, where the actuators or sensors that need to be safeguarded to guarantee resilient estimation can be determined. Note that the result in Theorem 7.1 is stronger than the standard and well-known result in the literature (e.g., in Fawzi et al. 2014, Proposition 3), where the maximum number of correctable attacks is at most equal to half of the number of sensors, presumably since we only require strong detectability instead of strong observability.

7.3.3.2 Number of Required Models for Estimation Resilience

Next, returning to the more general case with false date injection as well as switching attacks, i.e., the hidden-mode switched linear system in (7.1), we characterize the maximum number of models $\mathfrak{N}^*$ that are needed with the multiple-model approach in Sect. 7.3.1, which is independent of the size of the system, e.g., the number of buses in a power system, as well as the type/model of system uncertainty:

Theorem 7.2 (Maximum Number of Models/Modes) *Suppose there are t_a actuators and t_s sensors, and at most $p \leq l$ of these signals are attacked. Suppose also that there are t_m possible attack modes (*mode attack*). Then, the combinatorial number of all possible models, and hence the maximum number of models that need to be considered with the multiple-model approach, is*

$$\mathfrak{N}^* = t_m \binom{t_a + t_s}{p} = t_m \binom{t_a + t_s}{t_a + t_s - p}.$$

Proof It is sufficient to consider only models corresponding to the maximum number of attacks p. All models with strictly less than p attacks are contained in this set of models with the attack vectors having some identically zero elements for which our estimation algorithm is still applicable. Thus, we only need to consider combinations of p attacks among $t_a + t_s$ sensors and actuators for each of the t_m attack modes of operation/topologies. Note that this number is the maximum because resilience may be achievable with less models: For instance, when $t_m = 1$, $t_a = 0$ and $t_s = 2 = l$, $p = 1$, $A = \begin{bmatrix} 0.1 & 1 \\ 0 & 0.2 \end{bmatrix}$ and $C = I_2$, we have $\mathfrak{N}^* = 2$, but it can be verified that with $G = 0_{2\times 2}$ and $H = I_2$ (only one model, i.e., $1 = \mathfrak{N} < \mathfrak{N}^*$), the system is strongly detectable. ■

Note that the number of required models may change if additional knowledge about the data injection attack strategies is available. For instance, if we know that there are at most $n_a \le t_a$ and $n_s \le t_s$ attacks on the actuators and sensors, respectively, with a total of p attacks (where $p \le l$ and $p \le n_a + n_s$), then the maximum number of models that are required,

$$\mathfrak{N}^* = t_m \sum_{i=0}^{\min\{n_a, p\}} \binom{t_a}{i} \binom{t_s}{\min\{p-i, n_s\}}$$

is less than the number required in combinatorial case in Theorem 7.2.

On the other hand, the number of models may actually increase with less vulnerable actuators or sensors, as shown in the following example with $t_m = 1$ (one mode of operation), $n_a = 0$ (no attacks on actuators), $A = \begin{bmatrix} 0.1 & 1 \\ 0 & 1.2 \end{bmatrix}$ and $C = I$. If only one of the two sensors is vulnerable ($n_s = p = 1 < l = 2$), we have two models with $G = \begin{bmatrix} 0 \\ 0 \end{bmatrix}$, $H_1 = \begin{bmatrix} 1 \\ 0 \end{bmatrix}$ and $H_2 = \begin{bmatrix} 0 \\ 1 \end{bmatrix}$, but if both sensors are vulnerable ($n_s = p = 2$), only one model is required with $G = 0$ and $H = I$. Note that the latter case is not strongly detectable with zeros at $\{0.1, 1.2\}$, thus this system violates the necessary condition in Yong et al. (2016b); Yong (2018) for obtaining resilient estimates. However, both systems in the former case can be verified to be strongly detectable, thus, resilient estimates can be obtained in this case with less vulnerable sensors, as one may expect.

7.4 Attack Detection and Identification

Next, we address Problem 1.3 by investigating how the properties of the resilient state estimation algorithm in Sect. 7.3.2 affect attack detection and identification.

To begin, it is worth recalling that the resilient state estimation algorithms in the previous section are indifferent about whether the switching and false data injection attacks on the system are strategic. Nonetheless, it is critical to understand how our algorithms can detect or identify strategic attacks. In particular, we consider strategic attackers who aim to deceive the system operator/defender into believing that the mode of operation is $q \in \mathcal{Q}, q \neq q^*$, by means of selecting data injection signals d_k and the true mode $q^* \in \mathcal{Q}$. We call an attack *unidentifiable*, if the system operathor is not able to reconstruct/identify it. Moreover, the attack is *undetectable*, if it is *unidentifiable* and is unnoticeable. Below, we formally define the concepts of attack detection and attack identification, which are extensions of their counterparts in Yong et al. (2018) [Definitions 5.1 & 5.2].

Definition 7.2 (Switching and Data Injection Attack Detection) A switching and data injection attack is detected if the true mode $q^* \in \mathcal{Q}$ (chosen by attacker) has the maximum mean probability when using the resilient state estimation algorithm in Algorithm 7.2 or is not distinguishable from another mode $q \in \mathcal{Q}, q \neq q^*$ (chosen by defender) on average, in the presence of the stochastic/aleatoric uncertainty (i.e., if Assumption 7.1 holds), or if it is not eliminated by applying Algorithm 7.4 in the presence of the set-valued/epistemic uncertainty (i.e., if Assumption 7.2 holds).

Definition 7.3 (Switching and Data Injection Attack Identification) A switching and data injection attack strategy is identified if the attack is detected and in addition, the true mode $q^* \in \mathcal{Q}$ is uniquely determined on average (under aleatoric/stochastic uncertainty) or all false modes are eliminated (under epistemic/set-valued uncertainty), which reveals that the *mode attack* and *signal attack location*, and asymptotically unbiased estimates and/or uniformly bounded set-valued estimates of attack signals d_k can be obtained, i.e., the *signal magnitude attack* is reliably estimated.

It is obvious from the definitions above that if an attack is undetectable, it is also unidentifiable. Equivalently, if an attack is identifiable, then it is detectable. It is worth noting however that attack detection or identification is not required for calculating resilient state estimates. For example, in the simple case where there are no attacks, i.e., $d_k = 0$ for all k, the performance of state estimates of all models will be equally good, meaning that the attacks need not be detected or identified in order to obtain resilient state estimates.

7.4.1 Attack Detection

Our resilient state estimation approach (i.e., Algorithms 7.2 and 7.4) guarantees that an attack will always be detected by Definition 7.2 for all three uncertainty models. This is formally stated through the following theorem, which is a generalization of [Yong et al. (2018), Theorem 5.3].

Theorem 7.3 (Attack Detection) *The resilient state estimation algorithms in Algorithms 7.2 (with ratios of prior being identically 1) and 4 guarantee that switching and data injection attacks are always detectable, for all three uncertainty models.*

Proof First, note that if Assumption 7.2 holds, i.e., in the presence of distribution-free and norm-bounded noise signals, by (7.9), (7.10) and Proposition 7.1, $\|r_k^{q*}\|_2 \leq \delta_{r,k}^{q,*} \leq \hat{\delta}_{r,k}^{q*}$, i.e., (7.11) never holds for $q = q^*$ and hence, q^* is never eliminated. On the other hand, if Assumption 7.1 holds, i.e., in the presence of Gaussian noise signals, since the Kullback Leibler divergence $D(f_\ell^* \| f_\ell^q)$ is greater than or equal to zero with equality if and only if $f_\ell^* = f_\ell^q$ ([Kullback and Leibler (1951), Lemma 3.1]), with $j = q^* \in \mathcal{Q}$ as the true model and $i \in \mathcal{Q}, i \neq q^*$, the summand in the exponent of the ratio of geometric means whose expression is given in Yong et al. (2021)[Lemma 14] is always non-negative, i.e., $D(f_\ell^* \| f_\ell^i) - D(f_\ell^* \| f_\ell^*) = D(f_\ell^* \| f_\ell^i) \geq 0$. In other words, the ratio of the true model mean probability to the model mean probabilities of any other mode ($i \in \mathcal{Q}, i \neq q^*$) cannot decrease and can at best remain the same as the ratio of their priors being one by assumption. Thus, either the true model is identified or both modes are indistinguishable and a flag can be raised for attack detection. ■

7.4.2 *Attack Identification*

A combination of switching and false data injection attacks may not be identifiable, even if it is detectable. On the other hand, it directly follows from Definition 7.3 that the mode detectability/mean consistency is sufficient to identify an attack strategy/action. This is formalized via the following theorem.

Theorem 7.4 (Attack Identification) *Suppose mode detectability and/or mean consistency, i.e., Yong et al. (2021), Theorem 8 and/or Khajenejad and Yong (2019), Theorem 4 hold (and hence Yong et al. 2021, Corollary 13 also holds). Then, the switching and data injection attack strategy can be identified using the resilient state estimation algorithms in Algorithms 7.1–7.4.*

7.4.2.1 Sufficient or Necessary Condition for Unidentifiable Attacks

Under the stochastic uncertainty model (cf. Assumption 7.1), if the true mode is in the set of models and even if the estimator is not mean consistent, a sufficient condition for an attack signal to be unidentifiable was derived in our previous work (Yong et al. (2018)), which we recap here for the sake of completeness (for more details, see [Yong et al. (2018), Sect. 5.2]).

Theorem 7.5 (Unidentifiable Attack) *[Yong et al. (2018), Theorem 5.5] If Assumption 7.1 or 7.3 hold, $\tilde{\Gamma}_k^q T_{a,2,k}^q H_k^*$ has linearly independent rows and there exists $q \neq q^* \in \mathcal{Q}$ such that*

$$\mathcal{D}_k^s \triangleq (\tilde{\Gamma}_k^q T_{a,2,k}^q H_k^*)^\dagger (S_k^* - \tilde{\Gamma}_k^q T_{a,2,k}^q (\mathbb{E}[\mu_k^{q|*} \mu_k^{q|*\top}] + R_k)(\tilde{\Gamma}_k^q T_{a,2,k}^q)^\top))(\tilde{\Gamma}_k^q T_{a,2,k}^q H_k^*)^{\dagger\top} \tag{7.16}$$

is positive definite ($\succeq 0$) for all k. Moreover, we assume that $\mu_0^* = \mu_0^q$. Then, the attack is unidentifiable if the attacker chooses this mode $q^* \neq q$ as well as the attack signal d_k as a Gaussian sequence

$$d_k \sim \mathscr{N}(d_k^d, \mathscr{D}_k^s), \quad \forall k \tag{7.17}$$

with $\mathscr{D}_k^s$ *defined in* (7.16) *and* d_k^d *is given by*

$$\begin{aligned} d_k^d \triangleq \mathbb{E}[d_k] &= -(\tilde{\Gamma}_k^q T_{a,2,k}^q H_k^*)^\dagger \tilde{\Gamma}_k^q T_{a,2,k}^q (C_k^{q^*} \mathbb{E}[x_k] - C_k^q \hat{x}_{a,k|k}^{\star,q} + (D_k^{q^*} - D_k^q)\mathbb{E}[u_k]) \\ &= -(\tilde{\Gamma}_k^q T_{a,2,k}^q H_k^*)^\dagger \tilde{\Gamma}_k^q T_{a,2,k}^q (C_k^{q^*} \hat{x}_{a,k|k}^{q^*} - C_k^q \hat{x}_{a,k|k}^{\star,q} + (D_k^{q^*} - D_k^q)\mathbb{E}[u_k]), \ \forall k. \end{aligned} \tag{7.18}$$

The above theorem highlights that an unidentifiable attack strategy often must rely on the existence of system "vulnerabilities" as well as the computational capability and system knowledge that are comparable to that of the system operator/defender. For the former factor, a system designer can consider these conditions as preventative mitigation guides for securing the system.

On the other hand, if Assumption 7.2 or 7.3 hold (i.e., epistemic/set-valued uncertainty is present), we provide a necessary condition for the attack signals to be unidentifiable, i.e., a condition that the attacker must ensure in order to guarantee that the attack signals are not identifiable.

Theorem 7.6 (A Necessary Condition for Unidentifiable Attacks) *Suppose Assumption 7.2 or 7.3 holds and* $T_{e,2,k}^q \neq T_{a,2,k}^{q'}, \forall k \geq 0, \forall q, q' \in \mathscr{Q}, q \neq q'$. *Then, a necessary condition for the attack signal to be unidentifiable is that it has limited energy when* $q = q^*$, *i.e.,* $\lim_{k\to\infty} \|d_{0:k}^{q*}\|_2 < \infty$, *where* $d_{0:k}^{q*} \triangleq \left[d_k^{q*\top} \ d_{k-1}^{q*\top} \ \ldots d_0^{q*\top}\right]^\top$.

Proof Using contraposition, suppose the attack signal has unlimited energy. Then, by [Khajenejad and Yong (2019), Theorem 4], all false modes will be eliminated after some large enough time step K and hence, the system is mode detectable (cf. Sect. 7.3.2.2). Thus, by Theorem 7.4, the attack strategy can be identified using the resilient state estimation algorithm and consequently, the attack signal cannot be unidentifiable. ■

This result has the important implication that attack signals must have limited energy to remain unidentifiable, and in this case, the harm that an attacker can inflict on a Cyber-Physical Systems (CPS) may also be limited. Note that the attack impact could still be catastrophic in this case, which incentives us to design attack mitigation approach in Sect. 7.5.

7.4.2.2 A Sufficient Condition for Resilient State Estimation

Finally, under the assumption of stochastic/aleatoric uncertainty (cf. Assumption 7.1), a sufficient condition can be found in Yong et al. (2018) to ensure that the state estimates are unbiased, even when the true mode is not uniquely determined and the attack signal cannot be estimated/identified, which is restated below.

Theorem 7.7 (Resilience Guarantee) *[Yong et al. (2018), Theorem 5.7] Suppose $H_k^q = H_k$ and $D_k^q = D_k$ for all $q \in \mathscr{Q}$. Moreover, for all $q, q' \in \mathscr{Q}$, if there exists T such that for all $k \geq T$ and the following hold*

(i) $\text{rank}\left[\tilde{\Gamma}_k^q T_{a,2,k}^q C_k^{q'} \; \tilde{\Gamma}_k^q T_{a,2,k}^q C_k^q\right] = 2n$, *if* $C_k^q \neq C_k^{q'}$,

(ii) $\text{rank}(\tilde{\Gamma}_k^q T_{a,2,k}^q C_k^{q'}) = \text{rank}(\tilde{\Gamma}_k^q T_{a,2,k}^q C_k^q) = n$, *if* $C_k^q = C_k^{q'}$,

then the state estimates obtained using Algorithm 7.2 are guaranteed to be resilient (i.e., asymptotically unbiased).

7.5 Attack Mitigation

We now move on to the challenge of minimizing the impact of attacks, i.e., attack mitigation (Problem 1.4), which is a step beyond attack detection and identification. In particular, we investigate the problem of rejecting/canceling data injection attacks assuming that the attack mode can be detected (thus, the superscript q is omitted throughout this section), while using the resilient state estimates for $\mathscr{H}_\infty$ controller synthesis, in the sense of guaranteeing the boundedness of the expected/worst case states and minimizing the effect of the attack signals. To this end, we consider a linear *dynamic controller* with attack/disturbance rejection terms in the following form, where $\hat{x}_{k|k}, \hat{d}_{1,k}, \hat{d}_{2,k-1}$ are obtained from Algorithms 7.1 or 7.3:

$$\begin{aligned} x_{k+1}^c &= A_k^c x_k^c + B_k^c \tilde{y}_k, \\ u_k &= C_k^c x_k^c + D_k^c \tilde{y}_k, \end{aligned} \tag{7.19}$$

with $K_k^c \triangleq \begin{bmatrix} A_k^c & B_k^c \\ C_k^c & D_k^c \end{bmatrix}$ being the dynamic controller gain that will be designed, $\tilde{y}_k \triangleq \left[\hat{x}_{k|k}^\top \; \hat{d}_{1,k}^\top \; \hat{d}_{2,k-1}^\top\right]^\top$, $B_k^c \triangleq \left[B_{x,k}^c \; B_{d_1,k}^c \; B_{d_2,k}^c\right]$ and $D_k^c \triangleq \left[D_{x,k}^c \; D_{d_1,k}^c \; D_{d_2,k}^c\right]$. Note that we have used a delayed estimate of $d_{2,k-1}$ given in (7.2), which is the only estimate we can obtain in light of [Yong et al. (2016b), Eq. (6)]. Before designing K_k^c for the purpose of attack mitigation and stabilization, we first show that there exists a separation principle for linear discrete-time systems with unknown inputs (attacks), i.e., when the true mode is known, which allows us to design the controller gain K_k^c independently of the observer gain $\tilde{L}_k$.

Lemma 7.1 (Separation Principle) *The state feedback controller gain K_k^c in* (7.19) *can be designed independently of the state and input estimator gains $\tilde{L}_k$, $M_{1,k}$ and $M_{2,k}$ in Algorithms 7.1 and 7.3.*

Proof Using the dynamic controller (7.19) and the filter/observer equations in (7.2), (7.3) and (7.4), it can be verified that the system and controller states and the estimator error dynamics are given by

$$
\begin{bmatrix} x^c_{k+1} \\ x_{k+1} \\ \tilde{x}_{k+1|k+1} \end{bmatrix} = \begin{bmatrix} A^c_k & B^c_{x,k} & -B^c_{x,k} \\ B_k C^c_k & A_k + B_k D^c_{k,x} & -B D^c_{k,x} \\ 0 & 0 & (I - \tilde{L}_{k+1} C_{2,k})\overline{A}_k \end{bmatrix} \begin{bmatrix} x^c_k \\ x_k \\ \tilde{x}_{k|k} \end{bmatrix}
+ \begin{bmatrix} B^c_{d_1,k} & B^c_{d_2,k} \\ G_{1,k} + B_k D^c_{d_1,k} & G_{2,k} + B_k D^c_{d_2,k} \\ 0 & 0 \end{bmatrix} \begin{bmatrix} d_{1,k} \\ d_{2,k} \end{bmatrix} + \begin{bmatrix} -B^c_{d_1,k} & -B^c_{d_1,k} \\ -B_k D^c_{d_1,k} & -B_k D^c_{d_2,k} \\ 0 & 0 \end{bmatrix} \begin{bmatrix} d_{1,k} - \hat{d}_{1,k} \\ d_{2,k} - \hat{d}_{2,k-1} \end{bmatrix} \tag{7.20}
$$

$$
+ \begin{bmatrix} 0 & 0 & 0 \\ I & 0 & 0 \\ (I - \tilde{L}_{k+1} C_{2,k+1}) & -(I - \tilde{L}_{k+1} C_{2,k+1}) & -(I - \tilde{L}_{k+1} C_{2,k+1}) \\ (I - G_{2,k} M_{2,k+1} C_{2,k+1}) & (I - G_{2,k} M_{2,k+1} C_{2,k+1}) G_{1,k} M_{1,k} & G_{2,k} M_{2,k+1} - \tilde{L}_{k+1} \end{bmatrix} \mathbf{w}_k,
$$

where $\mathbf{w}_k \triangleq \begin{bmatrix} w_k^\top & v_{1,k}^\top & v_{2,k+1}^\top \end{bmatrix}^\top$ and $\overline{A}_k \triangleq (I - G_{2,k-1} M_{2,k} C_{2,k})(A_k - G_{1,k} M_{1,k} C_{1,k})$. Since the state matrix has a block upper triangular structure, the eigenvalues of the controller and estimator are independent of each other, thus K^c_k and $\tilde{L}_k$ can be designed separately. ■

Armed with the above lemma, we present an $\mathscr{H}_\infty$ controller design for determining the controller gain matrix K^c_k that stabilizes the closed-loop system and mitigates the effects of attack signals.

Theorem 7.8 (Attack-Mitigating and Stabilizing $\mathscr{H}_\infty$Controller) *Suppose the system* (7.1) *is controllable in the true mode $q \in \mathscr{Q}$ (known or detected). Then, the dynamic controller in* (7.19) *mitigates the effects of data injection attacks and minimizes the $\mathscr{H}_\infty$-gain from the augmented noise signal $\tilde{w}_k$ to the state as the desired output, i.e., $\tilde{z}_k = x_k$, using feedback based on estimates $\tilde{y}_k \triangleq \begin{bmatrix} \hat{x}_k^\top & \hat{d}_{1,k}^\top & \hat{d}_{2,k-1}^\top \end{bmatrix}^\top$, where the gain matrix K^c_k is the $\mathscr{H}_\infty$-controller gain matrix that can be synthesized (e.g., using* `hinfsyn` *in MATLAB) for the following augmented system:*

$$
\begin{aligned}
\xi_{k+1} &= \tilde{A}_k \xi_k + \tilde{B}_{1,k} \tilde{w}_k + \tilde{B}_{2,k} u_k, \\
\tilde{z}_k &= \tilde{C}_{1,k} \xi_k + \tilde{D}_{11,k} \tilde{w}_k + \tilde{D}_{12,k} u_k, \\
\tilde{y}_k &= \tilde{C}_{2,k} \xi_k + \tilde{D}_{21,k} \tilde{w}_k + \tilde{D}_{22,k} u_k,
\end{aligned} \tag{7.21}
$$

where $\tilde{A}_k \triangleq \begin{bmatrix} A_k & G_{1,k} & G_{2,k} \\ 0 & 0 & 0 \\ 0 & 0 & 0 \end{bmatrix}$, $\tilde{B}_{1,k} \triangleq \begin{bmatrix} I\,0\,0\,0\,0\,0 \\ 0\,I\,0\,0\,0\,0 \\ 0\,0\,I\,0\,0\,0 \end{bmatrix}$, $\tilde{B}_{2,k} \triangleq \begin{bmatrix} B_k \\ 0 \\ 0 \end{bmatrix}$, $\tilde{C}_{1,k} \triangleq \begin{bmatrix} I\,0\,0 \end{bmatrix}$, $\tilde{C}_{2,k} \triangleq \begin{bmatrix} I\,0\,0 \\ 0\,I\,0 \\ 0\,0\,I \end{bmatrix}$, $\tilde{D}_{11,k} \triangleq \begin{bmatrix} 0\,0\,0\,0\,0\,0\,0 \end{bmatrix}$, $\tilde{D}_{12,k} \triangleq 0$, $\tilde{D}_{21,k} \triangleq \begin{bmatrix} 0\,0\,0\,I\,0\,0 \\ 0\,0\,0\,0\,I\,0 \\ 0\,0\,0\,0\,0\,I \end{bmatrix}$ *and* $\tilde{D}_{22,k} \triangleq \begin{bmatrix} 0\,0\,0 \end{bmatrix}^\top$.

Proof By Lemma 7.1, the state feedback gain, K^c_k, can be independently designed with no effect on the stability of the resilient state estimator/observer. In other words, K^c_k can be chosen optimally, in the sense of an $\mathscr{H}_\infty$-controller such that the augmented closed-loop system is stable, and that the effects of the augmented noise $\tilde{w}_k$ on the desired controlled output $\tilde{z}_k \triangleq x_k$ are minimized. To achieve this, we consider the following augmented system:

$$\begin{aligned} x_{k+1} &= A_k x_k + B_k u_k + G_{1,k} d_{1,k} + G_{2,k} d_{2,k} + w_k, \\ d_{1,k+1} &= \tilde{w}_{1,k}, \\ d_{2,k+1} &= \tilde{w}_{2,k}, \end{aligned} \tag{7.22}$$

with the augmented state $\xi_k \triangleq \left[x_k^\top \; d_{1,k}^\top \; d_{2,k}^\top\right]^\top$, where the goal is to use the dynamic controller (7.19) with estimates/"observations" $\tilde{y}_k \triangleq \left[\hat{x}_k^\top \; \hat{d}_{1,k}^\top \; \hat{d}_{2,k-1}^\top\right]^\top$ to stabilize the desired output/state $\tilde{z}_k \triangleq x_k$, while minimizing the effect of the augmented noise signal $\tilde{w}_k \triangleq \left[w_k^\top \; \tilde{w}_{1,k}^\top \; \tilde{w}_{2,k}^\top \; \tilde{x}_{k|k} \; \tilde{d}_{1,k}^\top \; \tilde{d}_{2,k}^\top\right]^\top$. Then, by plugging the control input u_k from (7.19) into (7.22), we obtain (7.21), where an $\mathscr{H}_\infty$-controller can be synthesized to achieve the minimum $\mathscr{H}_\infty$ performance. It is worth re-emphasizing that the control synthesis process is completely independent of the observer gains $\tilde{L}_k$, $M_{1,k}$, $M_{2,k}$. ■

Remark 7.1 The dynamic feedback gain K_k^c can be synthesized using the command

$$[K_k^c, CL_k, \gamma_k] = \texttt{hinfsyn}(P, \text{size}(D_{22,k}, 1), \text{size}(D_{22,k}, 2))$$

in MATLAB, where $P \triangleq \begin{bmatrix} \tilde{A}_k & \tilde{B}_{1,k} & \tilde{B}_{2,k} \\ \tilde{C}_{1,k} & \tilde{D}_{11,k} & \tilde{D}_{12,k} \\ \tilde{C}_{2,k} & \tilde{D}_{21,k} & \tilde{D}_{22,k} \end{bmatrix}$.

7.6 Simulation Examples

7.6.1 Benchmark System (Signal Magnitude Location Attacks)

The resilient state estimation problem for a system (modified from Yong et al. (2016b) and has been used as a benchmark for several state and input filters/observers) is considered in this example, where there exists only one mode of operation ($t_m = 1$) as well as possible attacks on the actuator and four of the five sensors ($t_a = 1, t_s = 4$):

$$A=\begin{bmatrix}0.5 & 2 & 0 & 0 & 0\\ 0 & 0.2 & 1 & 0 & 1\\ 0 & 0 & 0.3 & 0 & 1\\ 0 & 0 & 0 & 0.7 & 1\\ 0 & 0 & 0 & 0 & 0.1\end{bmatrix};\quad B=G=\begin{bmatrix}1\\ 0.1\\ 0.1\\ 1\\ 0\end{bmatrix};\quad C=\begin{bmatrix}1 & 0 & 0 & 0 & 0\\ 0 & 1 & -0.1 & 0 & 0\\ 0 & 0 & 1 & -0.5 & 0.2\\ 0 & 0 & 0 & 1 & 0\\ 0 & 0.25 & 0 & 0 & 1\end{bmatrix};$$

$$H=\begin{bmatrix}1 & 0 & 0 & 0\\ 0 & 1 & 0 & 0\\ 0 & 0 & 1 & 0\\ 0 & 0 & 0 & 1\\ 0 & 0 & 0 & 0\end{bmatrix};\quad Q=10^{-4}\begin{bmatrix}1 & 0 & 0 & 0 & 0\\ 0 & 1 & 0.5 & 0 & 0\\ 0 & 0.5 & 1 & 0 & 0\\ 0 & 0 & 0 & 1 & 0\\ 0 & 0 & 0 & 0 & 1\end{bmatrix};\quad R=10^{-4}\begin{bmatrix}1 & 0 & 0 & 0.5 & 0\\ 0 & 1 & 0 & 0 & 0.3\\ 0 & 0 & 1 & 0 & 0\\ 0.5 & 0 & 0 & 1 & 0\\ 0 & 0.3 & 0 & 0 & 1\end{bmatrix}.$$

We consider the known input $u_k = \begin{cases} 2, & 100 \le k \le 300 \\ -2, & 500 \le k \le 700 \\ 0, & \text{otherwise} \end{cases}$, whereas the unknown inputs (attacks) are as depicted in Fig. 7.4. Moreover, we assume that there are at most $p = 4$ attacks with no constraints on n_a and n_s, and consequently, there are $\mathfrak{N} = 1 \cdot \binom{5}{4} = 5$ models. The signal attack locations alternate between $q = 3$ (attack on actuator and sensors 1, 3, 4) and $q = 2$ (attack on actuator and sensors 1, 2, 4) every 350s, i.e., the dwell time is 350s.

From the top plot in Fig. 7.3 that depicts the computed mode probabilities (under aleatoric Gaussian uncertainties), we observe that except during the short transients after $t = 350$s and $t = 700$s due to switching, the mode probabilities converge to their true values ($q^* = 3 \to q^* = 2 \to q^* = 3$). On the other hand, Fig. 7.3 (bottom) depicts the values of mode indicator index, $q \times i_q$ for each mode, over time, assuming epistemic bounded-norm distribution-free uncertainties, with i_q defined as

$$i_q \triangleq \begin{cases} 0, & \text{if } q \text{ is eliminated,} \\ 1, & \text{otherwise,} \end{cases} \qquad \forall q \in \mathcal{Q}.$$

Hence, $q \times i_q$ equals q if the mode q is not eliminated and is zero otherwise. As expected, it can be observed from Fig. 7.3 (bottom) that except for $q = 3$ and $q = 2$, the other modes are eliminated after some time steps.

Figure 7.4 shows computed state and unknown attack *point* estimates for the case of aleatoric (stochastic) uncertainty model, as well *set-valued* sate and unknown input (attack) estimates, when epistemic (distribution-free and norm-bounded) uncertainty model is assumed. The point estimates are seen to be close to the true values, even before the mode probabilities converge, and both the point estimates and the actual values of the states and unknown inputs (attacks) are within the set estimates, which are uniformly bounded and convergent set sequences, as expected. Similar results (not shown for brevity) are obtained for all other attack modes, $q = 1$ (attack on actuator and sensors 1, 2, 3), $q = 4$ (attack on actuator and sensors 2, 3, 4) and $q = 5$ (attack on sensors 1, 2, 3, 4). Thus, this example illustrates that when switching

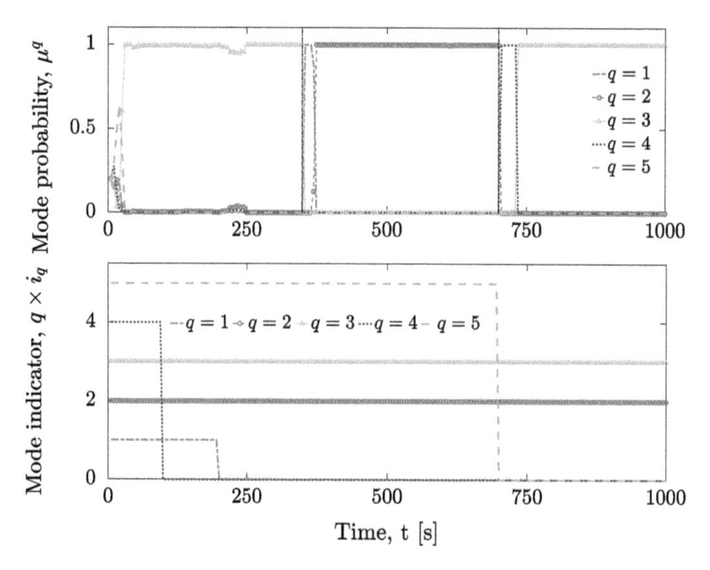

Fig. 7.3 Mode probabilities (top) assuming aleatoric/stochastic uncertainty model, as well as mode indicators (bottom) assuming epistemic/set-valued uncertainty model for the system in Sect. 7.6.1 with alternating switchings between $q = 3$ and $q = 2$ every 350s

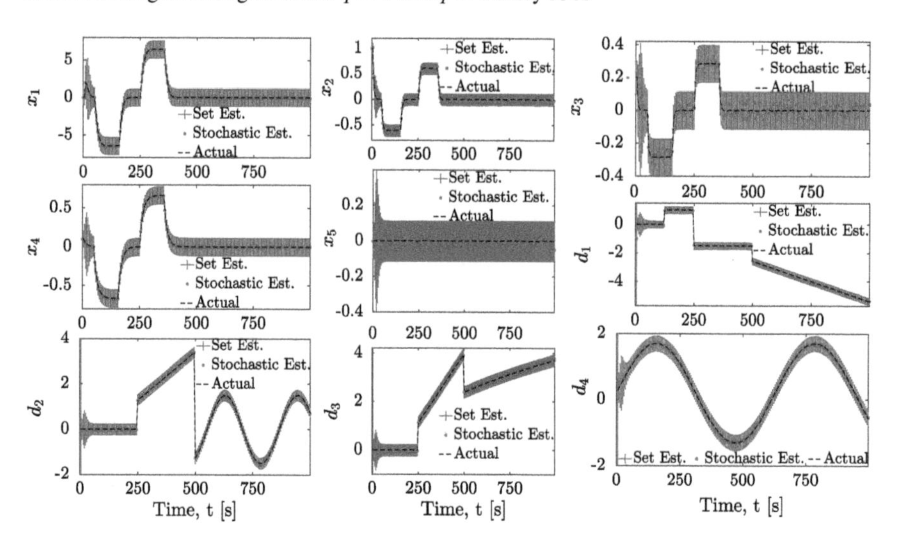

Fig. 7.4 State and attack magnitude estimates in Sect. 7.6.1 with switching between $q = 3$ and $q = 2$ with the dwell time 350 s

attacks and signal location attacks do not change quickly/frequently, i.e., the dwell time is large enough, our proposed methods work well.

7.6.2 IEEE 68-Bus Test System (Mode and Signal Magnitude Attacks)

The proposed algorithms are also applied to the IEEE 68-bus test system shown in [Yong et al. (2018), Fig. 7] to demonstrate their scalability to large systems, as well as to apply our attack mitigation approach.

An undirected graph $(\mathscr{V}, \mathscr{E})$ with the set of nodes (buses), $\mathscr{V} \triangleq \{1, \ldots, N\}$ and the set of edges (transmission/tie lines) $\mathscr{E} \subseteq \mathscr{V} \times \mathscr{V}$ is often used to describe a power network, where the busses may represent generator buses $i \in \mathscr{G}$, or load buses $i \in \mathscr{L}$. $\mathscr{S}_i \triangleq \{j \in \mathscr{V} \setminus \{i\} | (i, j) \in \mathscr{E}\}$ denotes the set of neighboring buses of $i \in \mathscr{V}$. In particular, there are 16 generator buses and 52 load buses for the IEEE 68-bus test system, i.e., $|\mathscr{G}| = 16$, $|\mathscr{L}| = 52$ and $|\mathscr{V}| = 68$. Similar to [Wood et al. (2013), Chap. 10], the dynamics of each bus, $i \in \mathscr{V}$, can be described by the following dynamical system:

$$\begin{aligned} &\dot{\theta}_i(t) = \omega_i(t), \\ &\dot{\omega}_i(t) = -\tfrac{1}{m_i}[D_i\omega_i(t) + \textstyle\sum_{j\in\mathscr{S}_i} P_{tie}^{ij}(t) - (P_{M_i}(t) + d_{a,i}(t)) + P_{L_i}(t) + w_i(t)], \end{aligned} \tag{7.23}$$

with the system states being the phase angle $\theta_i(t)$ and angular frequency $\omega_i(t)$ (hence, the state space dimension is $n = 136$) and an actuator attack signal $d_{a,i}(t)$. The power flow between neighboring buses i, j, such that $(i, j) \in \mathscr{E}$, is given by $P_{tie}^{ij}(t) = -P_{tie}^{ji}(t) = t_{ij}(\theta_i(t) - \theta_j(t))$, while $P_{M_i}(t)$ and $P_{L_i}(t)$ denote the mechanical power and power demand, respectively. The mechanical power $P_{M_i}(t)$ is the control input for the generator bus $i \in \mathscr{G}$ and is zero at load bus $i \in \mathscr{L}$. On the other hand, power demand $P_{L_i}(t)$ is taken as a known input since it can be calculated using load forecasting methods (e.g., Alfares and Nazeeruddin 2002). We assume that the noise $w_i(t)$ is a zero-mean truncated Gaussian signal (satisfying Assumption 7.3) with covariance matrix $Q_i(t) = 0.01$, $\eta_w = 0.03$ and the system parameters being adopted from Kundur et al. (1994) [p. 598]: $D_i = 1$, $t_{ij} = 1.5$ for all $i \in \mathscr{V}$, $j \in \mathscr{S}_i$ and $t_{ij} = 0$ otherwise. Angular momentums are $m_i = 10$ for $i \in \mathscr{G}$ and a larger value $m_i = 100$ for load buses $i \in \mathscr{L}$.

The measurements are sampled at discrete times (with sampling time $\Delta t = 0.01$s), satisfying the following output equation:

$$y_{i,k} = \left[P_{elec,i,k}\ \theta_{i,k}\ \omega_{i,k}\right]^\top + v_{i,k}, \tag{7.24}$$

where $P_{elec,i,k} = D_i\omega_{i,k} + P_{L_i,k}$ is the electrical power output and $v_{i,k}$ is a truncated zero-mean Gaussian noise signal with covariance matrix $R_i(t) = 0.01^4 I_3$ and $\eta_v =$

0.03. The continuous system dynamics (7.23) is also discretized with a sampling time of $\Delta t = 0.01$s. Furthermore, in this example, we choose the control inputs $P_{M_i,k}$ and $P_{L_i,k}$ through synthesizing an $\mathscr{H}_\infty$-optimal dynamic controller in the form of (7.19), as described in Theorem 7.8, to regulate the phase angles to $\theta_i = 10$ rad and mitigate the effect of the unknown attack signal.

As shown in Yong et al. (2018) [Fig. 7], the attacker could inject false data into the actuators and attack the transmission lines. Eight potential attack modes ($|\mathscr{Q}| = 8$) are considered:

Mode $q = 1$: Lines {27,53},{53,54},{60,61} & actuator $G1$.
Mode $q = 2$: Lines {18,49},{18,50} & actuator $G2$.
Mode $q = 3$: Line {40,41} & actuator $G3$.
Mode $q = 4$: Lines {18,49},{18,50},{27,53},{53,54},{60,61} & actuator $G4$.
Mode $q = 5$: Lines {27,53},{40,41},{53,54},{60,61} & actuator $G5$.
Mode $q = 6$: Lines {18,49},{18,50},{40,41} & actuator $G6$.
Mode $q = 7$: Lines {18,49},{18,50},{27,53},{40,41},{53,54},{60,61} & actuator $G7$.
Mode $q = 8$: Actuator $G8$.

We study a time-varying attack scenario where the attack mode is $q = 2$ for $t = [0, 2.5)$s followed by $q = 5$ for $t = [2.5, 5)$s, while the actuator attack signal is given in Fig. 7.6. Our goal is to demonstrate that attack signals can be detected, identified, and mitigated by our proposed approach. To synthesize the attack-mitigating dynamic controller in the form of (7.19), we consider three cases, depending on the three different assumptions on possible uncertainty models: (i) aleatoric/stochastic uncertainty (cf. Assumption 7.1), where we use $\hat{x}_{a,k|k}$ and $\hat{d}_{a,k-1}$ (i.e., the most likely estimates among all mode-matched estimates) returned by Algorithm 7.2 in (7.19), (ii) epistemic/bounded norm uncertainty (cf. Assumption 7.2), where we plug $\hat{x}_{e,k|k}$ and $\hat{d}_{e,k-1}$ (i.e., the centroids of the union of all the set-estimates that correspond to non-eliminated modes) returned by Algorithm 7.4 in (7.19), and (iii) combined uncertainty (cf. Assumption 7.3), where we use the most likely point (stochastic) estimates among all the ones that correspond to the non-eliminated modes, as described in Sect. 7.3.1.

Figure 7.5 demonstrates that attacks are detected almost instantaneously, and the attack modes are quickly identified. Further, Fig. 7.6 depicts the successful identification of the actuator attack signal and estimation of all system states (not depicted for brevity). Finally, the proposed attack mitigation is shown to be effective in regulating the phase angles at 10 rad/s despite attacks, while without attack mitigation, attackers can drastically influence the phase angles as shown in Fig. 7.6.

7.7 Conclusion

We addressed the problem of resilient state estimation for switching (mode/topology) attacks and attacks on actuator and sensor signals of Cyber-Physical Systems

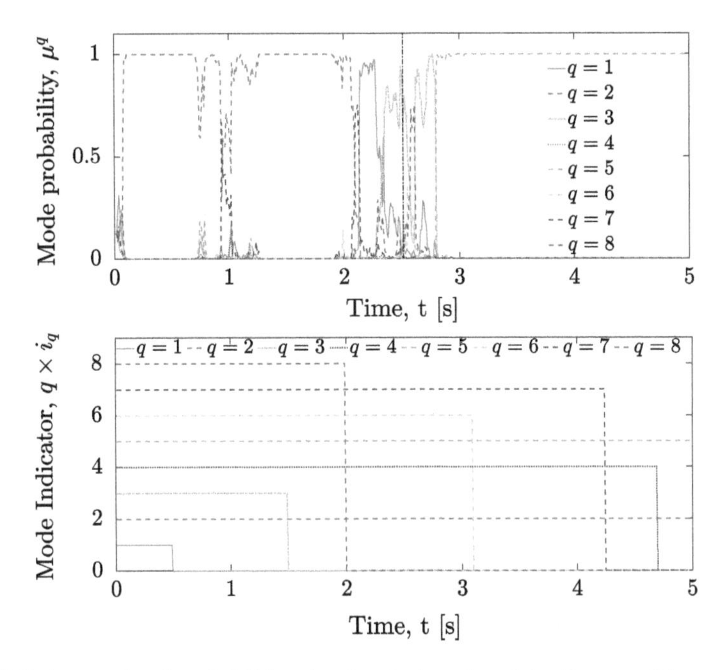

Fig. 7.5 Estimates of mode probabilities when the attack mode switches from $q = 2$ to $q = 5$ at 2.5s assuming stochastic uncertainties, as well as mode indicators assuming bounded norm uncertainties in Sect. 7.6.2

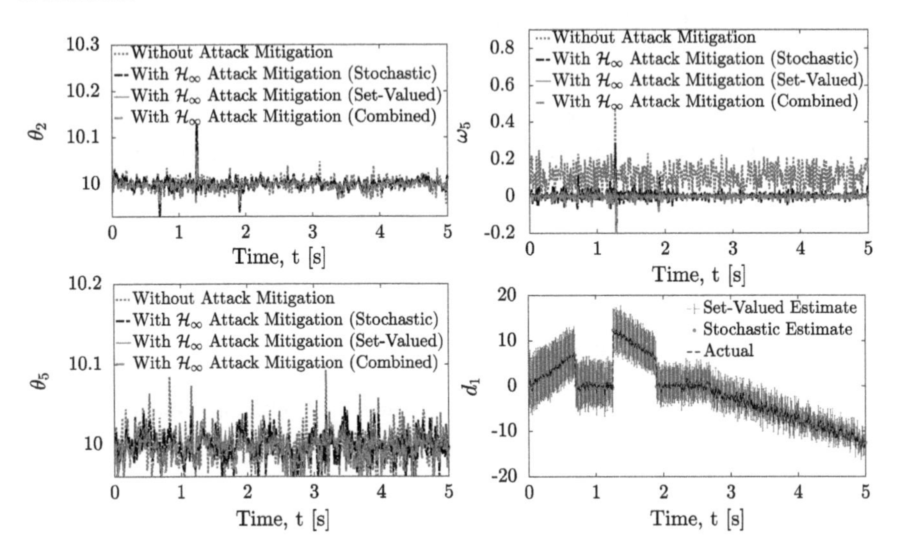

Fig. 7.6 A comparison of system states with and without the proposed attack mitigation, as well as the attack signal and its point-valued (stochastic) and set-valued (bounded-error) estimates

(CPS). We modeled the problem as a hidden-mode switched linear system with unknown inputs, where we considered three uncertainty models for the noise signals: (a) aleatoric/stochastic, (b) epistemic/set-valued and distribution-free, (c) truncated Gaussian uncertainties. We showed that the multiple-model inference algorithm in Yong et al. (2021); Khajenejad and Yong (2019) is a good solution to these problems. Furthermore, for the multiple-model approach, we presented an achievable upper bound on the maximum number of correctable signal attacks, as well as the maximum number of required models. We also derived sufficient conditions for attack (un-)detectability and identification and necessary conditions for the attack signal to be unidentifiable. Moreover, we designed an attack-mitigating $\mathscr{H}_\infty$-controller to minimize the effects of the attack signals. The effectiveness of our methods for resilient estimation, attack detection, and mitigation was demonstrated in simulations, including using an IEEE 68-bus test system.

Appendix

System Transformation

To obtain the mode-matched input and state estimator (7.2)–(7.4), we will consider a system transformation for the continuous system dynamics and output equation in (7.1) for each mode q (Yong et al. 2016b). First, we rewrite the direct feedthrough matrix H_k using singular value decomposition as $H_k = \begin{bmatrix} U_{1,k} & U_{2,k} \end{bmatrix} \begin{bmatrix} \Sigma_k & 0 \\ 0 & 0 \end{bmatrix} \begin{bmatrix} V_{1,k}^\top \\ V_{2,k}^\top \end{bmatrix}$, where $\Sigma_k \in \mathbb{R}^{p_{H_k} \times p_{H_k}}$ is a diagonal matrix of full rank, $U_{1,k} \in \mathbb{R}^{l \times p_{H_k}}$, $U_{2,k} \in \mathbb{R}^{l \times (l - p_{H_k})}$, $V_{1,k} \in \mathbb{R}^{p \times p_{H_k}}$ and $V_{2,k} \in \mathbb{R}^{p \times (p - p_{H_k})}$ with $p_{H_k} := \text{rk}(H_k)$, while $U_k := \begin{bmatrix} U_{1,k} & U_{2,k} \end{bmatrix}$ and $V_k := \begin{bmatrix} V_{1,k} & V_{2,k} \end{bmatrix}$ are unitary matrices. When there is no direct feedthrough, Σ_k, $U_{1,k}$ and $V_{1,k}$ are empty matrices,[3] and $U_{2,k}$ and $V_{2,k}$ are arbitrary unitary matrices.

Further, we define two orthogonal components of the unknown input d_k given by

$$d_{1,k} \triangleq V_{1,k}^\top d_k, \quad d_{2,k} \triangleq V_{2,k}^\top d_k. \tag{7.25}$$

Since V_k is unitary, $d_k = V_{1,k} d_{1,k} + V_{2,k} d_{2,k}$. Thus, the continuous system dynamics and output equation in (7.1) for each mode q can be rewritten as

$$x_{k+1} = A_k x_k + B_k u_k + G_{1,k} d_{1,k} + G_{2,k} d_{2,k} + w_k, \tag{7.26}$$
$$y_k = C_k x_k + D_k u_k + H_{1,k} d_{1,k} + v_k, \tag{7.27}$$

[3] We adopt the convention that the inverse of an empty matrix is also an empty matrix and assume that operations with empty matrices are possible.

where $G_{1,k} := G_k V_{1,k}$, $G_{2,k} := G_k V_{2,k}$, and $H_{1,k} := H_k V_{1,k} = U_{1,k}\Sigma_k$. Next, we decouple the output y_k using a nonsingular transformation $T_{a,k} = \begin{bmatrix} T_{a,1,k}^\top & T_{a,2,k}^\top \end{bmatrix}^\top = \begin{bmatrix} I_{p_{H_k}} & -U_{1,k}^\top R_k U_{2,k}(U_{2,k}^\top R_k U_{2,k})^{-1} \\ 0 & I_{(l-p_{H_k})} \end{bmatrix} \begin{bmatrix} U_{1,k}^\top \\ U_{2,k}^\top \end{bmatrix}$ in the presence of aleatoric uncertainty, i.e., if Assumption 7.1 holds, $T_{e,k} = \begin{bmatrix} T_{e,1,k}^\top & T_{e,2,k}^\top \end{bmatrix}^\top = \begin{bmatrix} U_{1,k} & U_{2,k} \end{bmatrix}^\top$ in the presence of epistemic uncertainty, i.e., if Assumption 7.2 holds, and both in the presence of truncated Gaussian uncertainty, i.e., if Assumption 7.3 holds. Consequently, we obtain $z_{t,1,k} \in \mathbb{R}^{p_{H_k}}$ and $z_{t,2,k} \in \mathbb{R}^{l-p_{H_k}}$, $\forall t \in \{a, e\}$, as

$$\begin{aligned} z_{t,1,k} &\triangleq T_{t,1,k} y_k = C_{t,1,k} x_k + D_{t,1,k} u_k + \Sigma_k d_{1,k} + v_{t,1,k}, \\ z_{t,2,k} &\triangleq T_{t,2,k} y_k = C_{t,2,k} x_k + D_{t,2,k} u_k + v_{t,2,k}, \end{aligned} \tag{7.28}$$

where $C_{t,1,k} \triangleq T_{t,1,k} C_k$, $C_{t,2,k} \triangleq T_{t,2,k} C_k = U_{2,k}^\top C_k$, $D_{t,1,k} \triangleq T_{t,1,k} D_k$, $D_{t,2,k} \triangleq T_{t,2,k} D_k = U_{2,k}^\top D_k$, $v_{t,1,k} \triangleq T_{t,1,k} v_k$, and $v_{t,2,k} \triangleq T_{t,2,k} v_k = U_{2,k}^\top v_k$. This system transformation essentially decouples the output equation involving y_k into two components, one with a full rank direct feedthrough matrix and the other without direct feedthrough. The transformation is also chosen such that in the case of aleatoric uncertainty, the measurement noise terms for the decoupled outputs are uncorrelated. The covariances of $v_{1,k}$ and $v_{2,k}$ are

$$\begin{aligned} R_{1,k} &\triangleq \mathbb{E}[v_{1,k} v_{1,k}^\top] = T_{a,1,k} R_k T_{a,1,k}^\top \succ 0, \\ R_{2,k} &\triangleq \mathbb{E}[v_{2,k} v_{2,k}^\top] = T_{a,2,k} R_k T_{a,2,k}^\top = U_{2,k}^\top R_k U_{2,k} \succ 0, \\ R_{12,k} &\triangleq \mathbb{E}[v_{1,k} v_{2,k}^\top] = T_{a,1,k} R_k T_{a,2,k}^\top = 0, \\ R_{12,(k,i)} &\triangleq \mathbb{E}[v_{1,k} v_{2,i}^\top] = T_{a,1,k} \mathbb{E}[v_k v_i^\top] T_{a,2,i}^\top = 0, \ \forall k \neq i. \end{aligned} \tag{7.29}$$

Moreover, $v_{1,k}$ and $v_{2,k}$ are uncorrelated with the initial state x_0 and process noise w_k. Further, in the case of bounded-norm uncertainty, the transform is also chosen such that $\| \begin{bmatrix} v_{1,k}^\top & v_{2,k}^\top \end{bmatrix}^\top \| = \| \begin{bmatrix} U_{1,k} & U_{2,k} \end{bmatrix}^\top v_k \| = \|v_k\|$.

Residual Upper Bounds

The upper bounds on the residual signal in Proposition 7.2 can be found as in Khajenejad and Yong (2019) [Theorem 3]:

$$\begin{aligned} \delta_{r,k}^{q,inf} &\triangleq \|\mathbb{A}_k^q t_k^\star\|_2, \\ \delta_{r,k}^{q,tri} &\triangleq \delta_0^{x,q} \|C_{e,2,k}^q \overline{A}_k^q {A_{e,k}^q}^{k-1}\|_2 + \eta_w(\|C_{e,2,k}^q \overline{A}_k^q {A_{e,k}^q}^{k-2}\|_2 + \|C_{e,2,k}^q B_{e,w,k}^{\star,q}\|_2) \\ &+ \textstyle\sum_{i=1}^{k-2} [\eta_w \|C_{e,2,i}^q \overline{A}_i^q {A_{e,i}^q}^i B_{e,w,i}^q\|_2 + \eta_v \|C_{e,2,i}^q \overline{A}_i^q {A_{e,i}^q}^i (B_{e,v_1,i}^q + A_{e,i}^q B_{e,v_2,i}^q)\|_2] \\ &+ \eta_v(\|C_{e,2,k}^q \overline{A}_k^q {A_{e,k}^q}^{k-2} B_{e,v_1,k}^q\|_2 + \|C_{e,2,k}^q (B_{e,v_1,k}^{q,\star} + \overline{A}_k^q B_{e,v_2,k}^q)\|_2 \\ &+ \|C_{e,2,k}^q B_{e,v_2,k}^{q,\star} + T_{e,2,k}^q\|_2), \end{aligned} \tag{7.30}$$

where $t_k^\star$ is a vertex of the following hypercube:

$$\mathscr{X}_k^q \triangleq \left\{x \in \mathbb{R}^{(n+l)(k+1)} \;:\; |x(i)| \leq \begin{cases} \delta_0^x, 1 \leq i \leq n \\ \eta_w, n+1 \leq i \leq n(k+1) \\ \eta_v, n(k+1)+1 \leq i \leq (n+l)(k+1) \end{cases} \right\},$$

i.e.,

$$t_k^\star(i) \in \begin{cases} \{-\delta_0^x, \delta_0^x\}, 1 \leq i \leq n, \\ \{-\eta_w, \eta_w\}, n+1 \leq i \leq n(k+1), \\ \{-\eta_v, \eta_v\}, n(k+1)+1 \leq i \leq (n+l)(k+1) \end{cases} \quad \text{and}$$

$$\overline{A}_k \triangleq \Phi_k \hat{A}_k,\; V_{e,k} \triangleq V_{1,k} M_{1,k} C_{1,k} + V_{2,k} M_{2,k} C_{2,k} \hat{A}_k,\; A_{e,k} \triangleq (I - \tilde{L}_k C_{2,k}) \overline{A}_k,$$
$$B_{e,w,k} \triangleq (I - \tilde{L}_k C_{2,k}) \Phi_k,\; B_{e,v_1,k} \triangleq -(I - \tilde{L}_k C_{2,k}) \Phi_k G_{1,k} M_{1,k} T_{1,k},$$
$$B_{e,v_2,k} \triangleq -((I - \tilde{L}_k C_{2,k}) G_{2,k} M_{2,k} + \tilde{L}_k) T_{2,k}.$$

References

H. Alfares, M. Nazeeruddin, Electric load forecasting: literature survey and classification of methods. Int. J. Syst. Sci. **33**(1), 23–34 (2002)

Y. Bar-Shalom, X. Li, T. Kirubarajan, *Estimation with Applications to Tracking and Navigation: Theory Algorithms and Software* (Wiley, 2004)

Y. Bar-Shalom, X. Li, T. Kirubarajan, *Estimation with Applications to Tracking and Navigation: Theory Algorithms and Software* (Wiley, 2004)

F. Blanchini, M. Sznaier, A convex optimization approach to synthesizing bounded complexity ℓ^∞ filters. IEEE Trans. Autom. Control **57**(1), 216–221 (2012)

H. Bodlaender, P. Gritzmann, V. Klee, J. Van Leeuwen, Computational complexity of norm-maximization. Combinatorica **10**(2), 203–225 (1990)

A.Cárdenas, S. Amin, S. Sastry, Research challenges for the security of control systems, in *Proceedings of the 3rd Conference on Hot Topics in Security*, ser. HOTSEC'08 (2008), pp. 6:1–6:6

A.Cárdenas, S. Amin, S. Sastry, Secure control: towards survivable cyber-physical systems, in *International Conference on Distributed Computing Systems Workshops* (2008), pp. 495–500

L. Cómbita, J. Giraldo, A. Cárdenas, N. Quijano, Response and reconfiguration of cyber-physical control systems: a survey, in *IEEE Colombian Conference on Automatic Control (CCAC)* (2015), pp. 1–6

M. Dahleh, I. Diaz-Bobillo, *Control of Uncertain Systems: a Linear Programming Approach* (Prentice-Hall, Inc., 1994)

G. Dan, H. Sandberg, Stealth attacks and protection schemes for state estimators in power systems, in *IEEE International Conference on Smart Grid Communications (SmartGridComm)* (2010), pp. 214–219

G. De Nicolao, G. Sparacino, C. Cobelli, Nonparametric input estimation in physiological systems: problems, methods, and case studies. Automatica **33**(5), 851–870 (1997)

J. Farwell, R. Rohozinski, Stuxnet and the future of cyber war. Survival **53**(1), 23–40 (2011)
H. Fawzi, P. Tabuada, S. Diggavi, Secure estimation and control for cyber-physical systems under adversarial attacks. IEEE Trans. Autom. Control **59**(6), 1454–1467 (2014)
B. Ghena, W. Beyer, A. Hillaker, J. Pevarnek, J. Halderman, Green lights forever: Analyzing the security of traffic infrastructure, in *8th USENIX Workshop on Offensive Technologies*, vol. 14, pp. 7–7 (2014)
S. Gillijns, B. De Moor, Unbiased minimum-variance input and state estimation for linear discrete-time systems. Automatica **43**(1), 111–116 (2007)
S. Gillijns, B. De Moor, Unbiased minimum-variance input and state estimation for linear discrete-time systems with direct feedthrough. Automatica **43**(5), 934–937 (2007)
R. Goebel, R. Sanfelice, A. Teel, Hybrid dynamical systems. IEEE Control Syst. Mag. **29**(2), 28–93 (2009)
X. Jin, W. Haddad, An adaptive control architecture for leader-follower multi-agent systems with stochastic disturbances and sensor and actuator attacks. Int. J. Control **92**(11), 2561–2570 (2019)
X. Jin, W. Haddad, Adaptive control for multi-agent systems with sensor-actuator attacks and stochastic disturbances. J. Guid. Control Dyn. **43**(1), 15–29 (2020)
X. Jin, W. Haddad, T. Yucelen, An adaptive control architecture for mitigating sensor and actuator attacks in cyber-physical systems. IEEE Trans. Autom. Control **62**(11), 6058–6064 (2017)
M. Khajenejad, S.Z. Yong, Simultaneous mode, input and state set-valued observers with applications to resilient estimation against sparse attacks, in *2019 IEEE 58th Conference on Decision and Control (CDC)* (IEEE, 2019), pp. 1544–1550
J. Kim, L. Tong, On topology attack of a smart grid: undetectable attacks and countermeasures. IEEE J. Sel. Areas Commun. **31**(7), 1294–1305 (2013)
P. Kitanidis, Unbiased minimum-variance linear state estimation. Automatica **23**(6), 775–778 (1987)
O. Kosut, L. Jia, R. Thomas, L. Tong, Malicious data attacks on the smart grid. IEEE Trans. Smart Grid **2**(4), 645–658 (2011)
S. Kullback, R. Leibler, On information and sufficiency. Ann. Math. Stat. **22**, 49–86 (1951)
P. Kundur, N.J. Balu, M.G. Lauby, *Power System Stability and Control* (McGraw-Hill New York, 1994)
C. Kwon, W. Liu, I. Hwang, Security analysis for cyber-physical systems against stealthy deception attacks, in *IEEE American Control Conference (ACC)* (2013), pp. 3344–3349
D. Li, S. Martínez, High-confidence attack detection via Wasserstein-metric computations. IEEE Control Syst. Lett. **5**(2), 379–384 (2020)
G. Liang, J. Zhao, F. Luo, S. Weller, Z.Y. Dong, A review of false data injection attacks against modern power systems. IEEE Trans. Smart Grid **8**(4), 1630–1638 (2017)
Y. Liu, P. Ning, M. Reiter, False data injection attacks against state estimation in electric power grids. ACM Trans. Inf. Syst. Secur. (TISSEC) **14**(1), 13 (2011)
S. Liu, S. Mashayekh, D. Kundur, T. Zourntos, K. Butler-Purry, A framework for modeling cyber-physical switching attacks in smart grid. IEEE Trans. Emerging Top. Comput. **1**(2), 273–285 (2013)
C. Ma, D. Yau, X. Lou, N. Rao, Markov game analysis for attack-defense of power networks under possible misinformation. IEEE Trans. Power Syst. **28**(2), 1676–1686 (2013)
E. Mazor, A. Averbuch, Y. Bar-Shalom, J. Dayan, Interacting multiple model methods in target tracking: a survey. IEEE Trans. Aerosp. Electron. Syst. **34**(1), 103–123 (1998)
J. Milošević, D. Umsonst, H. Sandberg, K. Johansson, Quantifying the impact of cyber-attack strategies for control systems equipped with an anomaly detector, in *European Control Conference (ECC)* (IEEE, 2018), pp. 331–337

S. Mishra, Y. Shoukry, N. Karamchandani, S. Diggavi, P. Tabuada, Secure state estimation: optimal guarantees against sensor attacks in the presence of noise, in *IEEE International Symposium on Information Theory (ISIT)* (2015), pp. 2929–2933

Y. Mo, B. Sinopoli, False data injection attacks in control systems, in *First Workshop on Secure Control Systems, CPS Week* (2010)

C. Murguia, J. Ruths, Cusum and chi-squared attack detection of compromised sensors, in *2016 IEEE Conference on Control Applications (CCA)* (IEEE, 2016), pp. 474–480

Y. Nakahira, Y. Mo, Attack-resilient $\mathscr{H}_2$, $\mathscr{H}_\infty$, and ℓ_1 state estimator (2018), arXiv:1803.07053

M. Pajic, J. Weimer, N. Bezzo, P. Tabuada, O. Sokolsky, I. Lee, G. Pappas, Robustness of attack-resilient state estimators, in *ACM/IEEE Intl. Conference on Cyber-Physical Systems* (2014), pp. 163–174

M. Pajic, P. Tabuada, I. Lee, and G. Pappas, Attack-resilient state estimation in the presence of noise, in *IEEE Conference on Decision and Control* (2015), pp. 5827–5832

F. Pasqualetti, F. Dörfler, F. Bullo, Attack detection and identification in cyber-physical systems. IEEE Trans. Autom. Control **58**(11), 2715–2729 (2013)

R. Patton, R. Clark, P. Frank, *Fault Diagnosis in Dynamic Systems: Theory and Applications*, ser. Prentice-Hall International Series in Systems and Control Engineering (Prentice Hall, 1989)

V. Renganathan, N. Hashemi, J. Ruths, T. Summers, Higher-order moment-based anomaly detection. IEEE Control Syst. Lett. (2021)

G. Richards, Hackers versus slackers. Eng. Technol. **3**(19), 40–43 (2008)

J. Shamma, K. Tu, Set-valued observers and optimal disturbance rejection. IEEE Trans. Autom. Control **44**(2), 253–264 (1999)

J. Slay, M. Miller, Lessons learned from the Maroochy water breach, in *International Conference on Critical Infrastructure Protection* (Springer, 2007), pp. 73–82

S. Sundaram, C. Hadjicostis, Delayed observers for linear systems with unknown inputs. IEEE Trans. Autom. Control **52**(2), 334–339 (2007)

J. Weimer, S. Kar, K. Johansson, Distributed detection and isolation of topology attacks in power networks, in *Proceedings of the 1st International Conference on High Confidence Networked Systems*, ser. HiCoNS '12 (ACM, 2012), pp. 65–72

A. Wood, B. Wollenberg, G. Sheble, *Power Generation, Operation, and Control* (Wiley, 2013)

M. Yadegar, N. Meskin, W. Haddad, An output-feedback adaptive control architecture for mitigating actuator attacks in cyber-physical systems. Int. J. Adapt. Control Signal Process. **33**(6), 943–955 (2019)

S.Z. Yong, Simultaneous input and state set-valued observers with applications to attack-resilient estimation, in *Annual American Control Conference (ACC).* (IEEE, 2018), pp. 5167–5174

S.Z. Yong, M. Zhu, E. Frazzoli, Resilient state estimation against switching attacks on stochastic cyber-physical systems, in *Proceedings of the American Control Conference*, submitted (2015)

S.Z. Yong, M. Foo, E. Frazzoli, Robust and resilient estimation for cyber-physical systems under adversarial attacks, in *IEEE American Control Conference (ACC)* (2016a), pp. 308–315

S.Z. Yong, M. Zhu, E. Frazzoli, A unified filter for simultaneous input and state estimation of linear discrete-time stochastic systems. *Automatica*, vol. 63 (2016b), pp. 321–329. Extended version first appeared in September 2013 and is available from: http://arxiv.org/abs/1309.6627

S.Z. Yong, M. Zhu, E. Frazzoli, Simultaneous input and state estimation for linear time-varying continuous-time stochastic systems. IEEE Trans. Autom. Control **62**(5), 2531–2538 (2017)

S.Z. Yong, M. Zhu, E. Frazzoli, Switching and data injection attacks on stochastic cyber-physical systems: modeling, resilient estimation, and attack mitigation. ACM Trans. Cyber-Phys. Syst. **2**(2), 1–2 (2018)

S.Z. Yong, M. Zhu, E. Frazzoli, Simultaneous mode, input and state estimation for switched linear stochastic systems. Int. J. Robust Nonlinear Control **31**(2), 640–661 (2021)

K. Zetter, *Inside the Cunning, Unprecedented Hack of Ukraine's Power Grid* (Wired Magazine, 2016)

M. Zhu, S. Martínez, Stackelberg-game analysis of correlated attacks in cyber-physical systems, in *IEEE American Control Conference (ACC)* (2011), pp. 4063–4068

M. Zhu, S. Martínez, On distributed constrained formation control in operator-vehicle adversarial networks. Automatica **49**(12), 3571–3582 (2013)

Q. Zhu, T. Basar, Game-theoretic methods for robustness, security, and resilience of cyberphysical control systems: games-in-games principle for optimal cross-layer resilient control systems. IEEE Control Syst. **35**(1), 46–65 (2015)

Chapter 8
State and Attacks Estimation for Nonlinear Takagi–Sugeno Multiple Model Systems with Delayed Measurements

Souad Bezzaoucha Rebai, Holger Voos, and Mohamed Darouach

8.1 Introduction

The present work deals with state and cyber-attacks estimation for nonlinear Takagi–Sugeno systems with variable time-delay measurements. The use of the sector nonlinearity approach with the nonlinear Takagi–Sugeno systems allows us to extend the results to a wide variety of control process. Indeed, fuzzy control systems have been presented as an important tool to represent and implement human heuristic knowledge to control a system. This theory is based on a class of fuzzy models presented by the authors in Takagi and Sugeno (1985), which were designed to describe nonlinear systems as a collection of Linear Time-Invariant (LTI) models blended together with nonlinear functions, known as weighting functions. The Takagi–Sugeno (T–S) fuzzy structure, also called quasi-LPV (linear parameter variable) systems, offers an efficient representation of nonlinear behavior while relatively simple compared to general nonlinear models (Benzaouia and Hajaji 2014). In this contribution, we propose to represent the nonlinear system described by T–S models by an equivalent form extending the result presented in Bezzaoucha and Voos (2019) and Bezzaoucha Rebai and Voos (2019) for state and attacks estimation with delayed measurement. The objective is to obtain sufficient conditions in terms of $LMIs$ formulation for the observer design in order to ensure the asymptotic convergence of the estimation errors with an $\mathscr{L}_2$ attenuation constraint.

S. Bezzaoucha Rebai (✉)
EIGSI-La Rochelle, 26 Rue François de Vaux de Foletier, 17041 La Rochelle, France
e-mail: souad.bezzaoucha@eigsi.fr

H. Voos
Interdisciplinary Centre for Security, Reliability and Trust (SnT), Automatic Control Research Group, University of Luxembourg, Campus Kirchberg, 29 Avenue J.F Kennedy, L-1855 Luxembourg, Luxembourg
e-mail: holger.voos@uni.lu

M. Darouach
Research Center for Automatic Control of Nancy (CRAN), Université de Lorraine, IUT de Longwy, 186 rue de Lorraine, 54400 Cosnes et Romain, France
e-mail: mohamed.darouach@univ-lorraine.fr

M. Abbaszadeh and A. Zemouche (eds.), *Security and Resilience in Cyber-Physical Systems*, https://doi.org/10.1007/978-3-030-97166-3_8

The aim of this chapter is to tackle the state estimation of a nonlinear system subject to data deception attacks and variable time-delay measurements. Based on the same principle of own previous contributions (Gerard et al. 2018; Bezzaoucha Rebaiı et al. 2018) the malicious attacks can be modeled as adversary signals (i.e., like disturbances, unknown inputs, faults,...) introduced via the internal network by hackers and affecting the sensors and/or actuators data (Pajic et al. 2017; Teixeira et al. 2012). The isolation and reconstruction of these cyber-attacks can be seen from a control point of view as uncertain parameter problem.

Indeed, based on Bezzaoucha et al. (2013), we propose to use previously developed approach, applied for joint state and time-varying parameters estimation of Takagi–Sugeno models in order to reconstruct the state and cyber-attack signals for nonlinear LPV systems. In this book chapter, we will consider in addition the delayed measurement constraints.

The considered actuator/sensor attacks are modeled as time-varying parameters with multiplicative effect on the actuator input signal and sensor output signal, respectively. Based on the sector nonlinearity description, and using the convex property, the nonlinear model will be presented in a Linear Parameter-Varying (LPV) form, then an observer allowing both state and attack reconstruction is designed by solving an LMI optimization problem, exactly as detailed in Bezzaoucha and Voos (2019).

So far, to the best of our knowledge, there has been no delay-dependent method reported to study the observer-based H_∞ control for T–S fuzzy systems dealing with the state and attack reconstruction problem. Indeed, in general, practical problems, especially in Networked Control Systems (NCS) , the delayed measurement such as traffic flow in communication networks have to be considered, especially for stability reasons and measurement-based observer design. As it was developed in Orjuela et al. (2007) and Bezzaoucha et al. (2017), the considered approach provides an alternative and attractive path to deal with complex nonlinear systems and to obtain an equivalent representation by bounding the parameters and using the well-known sector nonlinearity transformation (SNT).

8.1.1 Contributions and Outline

Robust control and quadratic stabilization for linear systems with uncertain parameters have been considered in Shaked (2001). For fuzzy systems without uncertainties, Liu and Zhang in Liu and Zhang (2003) have proposed a new design method based on the H_∞ norm. However, their technique is based on a two-step approach which appears to be a drawback. Like in Bezzaoucha and Voos (2019), we proposed a method to simplify and to improve the existing design methods of robust fuzzy state observer design with disturbance attenuation for uncertain T–S fuzzy systems. The developed method gives not only the observer gains (for the state and the attacks) on a single-step analysis.

In practice, time delay often occurs in the transmission of information or material between different parts of a system. Transportation systems, communication systems, chemical systems, and power systems are example of time-delay systems. Also, it has been shown that the existence of time delay usually becomes the source of instability and deteriorates the performances of systems. Therefore, T–S fuzzy systems have been extended to deal with nonlinear systems with time-delay (Benzaouia and Hajaji 2014). The existing results of stabilization and stability criteria for this class of T–S fuzzy systems can be classified into two types: delay independent, which is applicable to delays of arbitrary sizes, and delay dependent, which includes information on the size of delays.

Although it is well known that delay-dependent results are less conservative than delay-independent ones, there are few delay-dependent results which study the problem of observer-based H_∞ control for T–S fuzzy systems with varying time delay. This motivates the research in this work to study this problem, i.e., the state and attacks reconstruction problem for nonlinear Takagi–Sugeno systems with delayed measurements. In this chapter, the asymptotic stabilization of uncertain (attacked) T–S observer systems with variable time-delay measurement is studied. Different from the methods currently found in the literature (Yue and Han 2005; Tian and Peng 2003), the proposed method does not need any transformation in the LKF (Lyapunov–Krasovskii functional), and thus avoids the restriction resulting from any used transformation. It improves the presented results in Bezzaoucha and Voos (2019) and Bezzaoucha et al. (2013) for two main aspects. The first one concerns the polytopic rewriting of the time-varying data deception attacks, and the second one is the time-delay measurement consideration and the delay-dependent stabilization conditions. Based on previous results, published in Bezzaoucha and Voos (2019), and on the sector nonlinearity approach, sufficient conditions in term of $LMIs$ formulation are given for the observer design. We will show that, despite the presence of cyber-attack (i.e., data deception attacks on both actuators and sensors) and the delayed measurements, the proposed observer is efficient and ensures the asymptotic convergence of the estimation errors with an $\mathscr{L}_2$ attenuation constraint.

8.1.2 Chapter Organization

The present contribution is organized as follows. After a brief introduction and a short overview of related works in Sect. 8.1, the problem statement is detailed in Sect. 8.2 by the presentation of the polytopic modeling of time-varying nonlinear systems and time-varying parameters (malicious attacks) with an LPV model of physical plant under data deception attacks. In Sect. 8.3, the main result/contribution of this work is given in terms of a general theorem for the observer design strategy and time-delay-dependent stability conditions. In Sect. 8.5, an illustrative example is given. From a basic nonlinear model of a biological wastewater treatment plant, the proposed approach is applied and illustrated with simulations. Conclusion will be given in the last section.

8.2 Problem Statement

The problem of state reconstruction in the presence of faults and attacks, also denoted as secure state estimation, has recently attracted considerable attention from the control community. The problem of reconstructing the state under actuator/sensor attacks is closely related to fault-detection and fault-tolerant state reconstruction. Based on the approach presented in previous works Bezzaoucha et al. (2013), Bezzaoucha et al. (2013) and adapted to the cyber-security problem, as presented in Bezzaoucha and Voos (2019) we address the design of observers that can accurately reconstruct the state and attacks of a cyber-physical system under actuator/sensor attacks with delayed measurements.

For that, we propose a simultaneous state and time-varying (attacks) observers for nonlinear systems in the presence of corrupted inputs and measurements, more specifically, the so-called false data injection attacks. In the spirit of a Luenberger observer, a state and attacks reconstruction algorithm is proposed based on the LMI approach and convex optimization problem. The second point of the problem statement will be about the variable time-delay measurements, which will be considered in the observer analysis, as shown in Orjuela et al. (2007).

8.2.1 *False Data Injection Attacks on Actuators/Sensors*

Based on results presented in Bezzaoucha and Voos (2019) and Orjuela et al. (2007), we assume that the attacker modifies the gain/s of the sensor and/or the actuator of the control system, which represent the injection of false information from sensors or controllers. This chapter is also dealing with a problem characterizing dynamical systems, which is the variable time-delay measurements. Mathematically speaking, explicit equations of both sensor and actuator signal attacks are derived and represented as time-varying multiplicative actuator/sensor faults/attacks. The Polytopic T–S approach is then used to reconstruct these signals in real time.

In this section, we assume that a malicious third party wants to compromise the integrity of the system. The attacker is assumed to have the following capabilities:

- He/she knows the system model, i.e., we assume that the hacker knows the system model and matrices.
- He/she can control the readings of the sensors and the actuators, i.e., modifies their values.
- The intrusions are represented as time-varying multiplicative actuator—sensor faults—attacks. The attacks signals are, of course, unknown, but bounded. Their min and max values are supposed to be known. Indeed, this assumption is not conservative since we suppose that if the boundaries are exceeded the attacks effect will be too obvious and easily detectable. Meaning, the hacker should respect the min and max values to a certain extent if he/she wants to remain undetectable.

- The nonlinear system is subject to time-variable delayed measurements. The time delay $\tau(t)$ is assumed perfectly known and satisfies the following conditions:

$$\begin{cases} 0 \leq \tau(t) \leq \tau \\ \dot{\tau}(t) \leq \gamma < 1. \end{cases} \tag{8.1}$$

8.2.2 *Polytopic Modeling of Time-Varying Nonlinear Systems with Delayed Measurements*

Let us consider the nonlinear system represented by the following equations:

$$\begin{cases} \dot{x}(t) = \sum_{i=1}^{r} \mu_i(x(t))(A_i x(t) + B_i(t)u(t)) \\ y(t) = C(t)x(t), \end{cases} \tag{8.2}$$

s.t. A_i, B_i, and $C(t)$ are constant matrices with appropriate dimensions.

With the time-varying matrices $B_i(t)$ and $C(t)$ defined by the following:

$$\begin{cases} B_i(t) = B_i + \sum_{j=1}^{n_{\theta_u}} \theta_j^u(t)\overline{B}_{ij} \\ C(t) = (I_m + F(t))C, \end{cases} \tag{8.3}$$

s.t. B_i, $\overline{B}_{ij}$ are constant matrices with appropriate dimensions and $\theta_j^u(t)$ time-varying unknown parameters and correspond to the multiplicative actuator attacks.

The matrix $F(t) \in \mathbb{R}^{m \times m}$ is defined by

$$F(t) = \text{diag}(\theta^y(t)), \tag{8.4}$$

s.t. $\text{diag}(\theta^y(t))$ corresponds to a diagonal matrix with the terms $\theta_j^y(t)$ (sensor attacks) on its diagonal.

The time-varying parameter vector $\theta(t), \theta(t) \in \mathbb{R}^n$ is defined by $\theta(t) = \begin{pmatrix} \theta^u(t) \\ \theta^y(t) \end{pmatrix}$ with $\theta^u(t) \in \mathbb{R}^{n_{\theta_u}}$ and $\theta^y(t) \in \mathbb{R}^{n_{\theta_y}}$ correspond, respectively, to the actuator and sensor attacks ($n = n_{\theta_u} + n_{\theta_y}$). $x(t) \in \mathbb{R}^{n_x}$, $y(t) \in \mathbb{R}^m$ and $u(t) \in \mathbb{R}^{n_u}$ correspond, respectively, to the system state, output, and control. The nonlinear system is modeled thanks to a polytopic representation with r sub-models. This representation may be obtained in a straightforward way by applying the Sector Nonlinearity Transformation (SNT). The interested readers can refer to Bezzaoucha et al. (2013) and Tanaka and Wang (2001) for more development details.

$F(t)$ may be expressed as

$$F(t) = \sum_{j=1}^{n_{\theta_y}} \theta_j^y(t) F_j, \tag{8.5}$$

with $n_{\theta_y} = m$, F_j are matrices of dimension $\mathbb{R}^{m \times m}$ and where the element of coordinate (j, j) is equal to 1 and 0 elsewhere. The coordinate j corresponds to the number of the attacked sensor. The terms $\theta_j^y(t)$ are time-varying unknown parameters and represent the multiplicative sensor attacks.

8.2.3 *Polytopic Modeling of Time-Varying Parameters (Malicious Attacks)*

As presented in Bezzaoucha and Voos (2019), the actuator data deception or false data injection is modeled thanks to the time-varying parameters $\theta_j^u(t)$. These attacks are of course unknown but bounded $\theta_j^u(t) \in [\theta_j^{2^u}, \theta_j^{1^u}]$, with known bounds. Applying the SNT transformation, each parameter $\theta_j^u(t)$ can always be expressed as

$$\theta_j^u(t) = \widetilde{\mu}_j^1(\theta_j^u(t))\theta_j^{1^u} + \widetilde{\mu}_j^2(\theta_j^u(t))\theta_j^{2^u}, \tag{8.6}$$

with

$$\widetilde{\mu}_j^1(\theta_j^u(t)) = \frac{\theta_j^u(t) - \theta_j^{2^u}}{\theta_j^{1^u} - \theta_j^{2^u}}, \widetilde{\mu}_j^2(\theta_j^u(t)) = \frac{\theta_j^{1^u} - \theta_j(t)}{\theta_j^{1^u} - \theta_j^{2^u}} \tag{8.7}$$

$$\widetilde{\mu}_j^1(\theta_j^u(t)) + \widetilde{\mu}_j^2(\theta_j^u(t)) = 1, \ \forall t.$$

Based on the same way, the sensor data deception or false data injection is modeled thanks to the time-varying parameters $\theta_j^y(t)$, such that

$$\theta_j^y(t) = \overline{\mu}_j^1(\theta_j^y(t))\theta_j^{1^y} + \overline{\mu}_j^2(\theta_j^y(t))\theta_j^{2^y} \tag{8.8}$$

with

$$\overline{\mu}_j^1(\theta_j^y(t)) = \frac{\theta_j^y(t) - \theta_j^{2^y}}{\theta_j^{1^y} - \theta_j^{2^y}}, \overline{\mu}_j^2(\theta_j^y(t)) = \frac{\theta_j^{1^y} - \theta_j(t)}{\theta_j^{1^y} - \theta_j^{2^y}} \tag{8.9}$$

$$\overline{\mu}_j^1(\theta_j^y(t)) + \overline{\mu}_j^2(\theta_j^y(t)) = 1, \ \forall t.$$

Replacing (8.6) and (8.8) into (8.3), we obtain

$$\begin{cases} B_i(t) = B_i + \sum_{j=1}^{n_{\theta_u}} \sum_{k=1}^{2} \widetilde{\mu}_j^k(\theta_j(t))\theta_j^{k^u}\overline{B}_{ij} \\ C(t) = \left(I + \sum_{j=1}^{n_{\theta_y}} \sum_{k=1}^{2} \overline{\mu}_j^k(\theta_j^y(t))\theta_j^{k^y} F_j \right) C. \end{cases} \tag{8.10}$$

8.2.4 LPV Model of Physical Plant Under Data Deception Attacks and Delayed Measurements

In order to have the same weighting functions for all the time-varying matrices $B_i(t)$ and write $C(t)$ as a simple polytopic matrix, exploiting the convex sum property of the weighting functions $\widetilde{\mu}_j(\theta_j^u(t))$ and $\overline{\mu}_j(\theta_j^y(t))$ of each parameter $\theta_j^u(t)$ and $\theta_j^y(t)$ (see Bezzaoucha et al. 2013 for computation details), (8.10) is written as

$$\begin{cases} B_i(t) = \sum_{j=1}^{n_{\theta_u}} \left[\left[(\widetilde{\mu}_j^1(\theta_j^u(t))\theta_j^{1^u} + \widetilde{\mu}_j^2(\theta_j^u(t))\theta_j^{2^u})\overline{B}_{ij}\right]\right] \times \\ \qquad \left[\prod_{\substack{k=1 \\ k\neq j}}^{n_{\theta_u}} \sum_{m=1}^{2} \widetilde{\mu}_k^m(\theta_k^u(t)) \right] + B_i \\ \qquad = B_i + \sum_{j=1}^{2^{n_{\theta_u}}} \widetilde{\mu_j}(\theta^u(t))\overline{\mathscr{B}}_{ij} \\ C(t) = \left(I + \sum_{j=1}^{2^{n_{\theta_y}}} \overline{\mu_j}(\theta^y(t))\overline{F}_j \right) C \end{cases} \tag{8.11}$$

with

$$\widetilde{\mu_j}(\theta^u(t)) = \prod_{k=1}^{n_{\theta_u}} \widetilde{\mu}_k^{\sigma_j^k}(\theta_k^u(t)),\ \overline{\mathscr{B}}_{ij} = \sum_{k=1}^{n_{\theta_u}} \theta_k^{u\sigma_j^k}\overline{B}_{ik} \tag{8.12}$$

and

$$\overline{\mu_j}(\theta^y(t)) = \prod_{k=1}^{n_{\theta_y}} \overline{\mu}_k^{\sigma_j^k}(\theta_k^y(t)),\ \overline{F}_j = \sum_{k=1}^{n_{\theta_y}} \theta_k^{y\sigma_j^k} F_j, \tag{8.13}$$

where the global weighting functions $\widetilde{\mu_j}(\theta^u(t))$ and $\overline{\mu_j}(\theta^y(t))$ satisfy the convex sum property. The index σ_j^k is either equal to 1 or 2 and indicates which partition of the kth parameter ($\widetilde{\mu_k}^1$ or $\widetilde{\mu_k}^2$, i.e., $\overline{\mu_k}^1$ or $\overline{\mu_k}^2$) is involved in the jth sub-model. The relation between the sub-model number j and the σ_j^k indices is given by the following equation:

$$j = 2^{n_{\theta u}-1}\sigma_j^1 + 2^{n_{\theta u}-2}\sigma_j^2 + \cdots + 2^0\sigma_j^{n_{\theta u}} - (2^1 + 2^2 + \cdots + 2^{n_{\theta u}-1}) \tag{8.14}$$

for the actuator, and in the same way for the sensor:

$$j = 2^{n_{\theta y}-1}\sigma_j^1 + 2^{n_{\theta y}-2}\sigma_j^2 + \cdots + 2^0\sigma_j^{n_{\theta y}} - (2^1 + 2^2 + \cdots + 2^{n_{\theta y}-1}). \tag{8.15}$$

Finally, using Eq. (8.11), the nonlinear LPV system (8.2) becomes

$$\begin{cases} \dot{x}(t) = \sum_{i=1}^{r}\sum_{j=1}^{2^{n_{\theta u}}} \mu_i(x(t))\widetilde{\mu_j}(\theta^u(t))(A_i x(t) + \mathscr{B}_{ij}u(t)) \\ y(t) = \sum_{k=1}^{2^{n_{\theta y}}} \overline{\mu_k}(\theta^y(t))\widetilde{C}_k x(t), \end{cases} \tag{8.16}$$

$$\mathscr{B}_{ij} = B_i + \overline{\mathscr{B}}_{ij},\ \widetilde{C}_k = C + \overline{F}_k C. \tag{8.17}$$

Now, if we consider some time-varying delay $\tau(t)$ in the output measurements, the nonlinear LPV system (8.16) becomes

$$\begin{cases} \dot{x}(t) = \sum_{i=1}^{r}\sum_{j=1}^{2^{n_{\theta u}}} \mu_i(x(t))\widetilde{\mu_j}(\theta^u(t))(A_i x(t) + \mathscr{B}_{ij}u(t)) \\ y(t) = \sum_{k=1}^{2^{n_{\theta y}}} \overline{\mu_k}(\theta^y(t-\tau(t)))\widetilde{C}_k x(t-\tau(t)). \end{cases} \tag{8.18}$$

8.3 Main Result: Observer Design

From the system equations (8.18), the aim of this chapter is to tackle the state and actuator/sensor data deception estimation of a nonlinear system subject to delayed measurements, and represented in a polytopic form. An $\mathscr{L}_2$ attenuation approach is applied in order to minimize the attacks effect on the state and malicious input estimation error.

The state and actuator/sensor data deception observer is given by the following equations:

$$\begin{cases} \dot{\hat{x}}(t) = \sum_{i=1}^{r}\sum_{j=1}^{2^{n_{\theta_u}}}\mu_i(\hat{x}(t))\widetilde{\mu_j}(\hat{\theta^u}(t)) \\ \qquad \big(A_i x(t) + \mathscr{B}_{ij}u(t) + L_{ij}(y(t) - \hat{y}(t))\big) \\ \dot{\hat{\theta^u}}(t) = \sum_{i=1}^{r}\sum_{j=1}^{2^{n_{\theta_u}}}\mu_i(\hat{x}(t))\widetilde{\mu_j}(\hat{\theta^u}(t)) \\ \qquad (K_{ij}^u(y(t) - \hat{y}(t)) - \alpha_{ij}^u\hat{\theta}^u(t)) \\ \dot{\hat{\theta^y}}(t) = \sum_{i=1}^{r}\sum_{k=1}^{2^{n_{\theta_y}}}\mu_i(\hat{x}(t))\overline{\mu_k}(\hat{\theta^y}(t-\tau(t))) \\ \qquad (K_{ik}^y(y(t) - \hat{y}(t)) - \alpha_{ik}^y\hat{\theta^y}(t)) \\ \hat{y}(t) = \sum_{k=1}^{2^{n_{\theta_y}}}\overline{\mu_k}(\hat{\theta^y}(t-\tau(t)))\widetilde{C}_k\hat{x}(t-\tau(t)), \end{cases} \tag{8.19}$$

where $L_{ij} \in \mathbb{R}^{n_x \times m}$, $K_{ij}^u \in \mathbb{R}^{n \times m}$, $\alpha_{ij}^u \in \mathbb{R}^{n \times n}$, $K_{ik}^y \in \mathbb{R}^{m \times m}$, and $\alpha_{ik}^y \in \mathbb{R}^{m \times m}$ are parameter matrices to be determined s.t. the estimated state and malicious input parameters converge to the real system state and attacks (i.e., the estimation errors for both state and malicious input parameters converge to zero).

Let us define the state and data deception estimation errors $e_x(t)$, $e_{\theta^u}(t)$ and $e_{\theta^y}(t)$ as

$$\begin{aligned} e_x(t) &= x(t) - \hat{x}(t) \\ e_{\theta^u}(t) &= \theta^u(t) - \hat{\theta^u}(t) \\ e_{\theta^y}(t) &= \theta^y(t) - \hat{\theta^y}(t). \end{aligned} \tag{8.20}$$

Based on the convex sum property of the weighting functions, from the results presented in Bezzaoucha et al. (2013) and in order to be able to calculate the estimation error dynamics, the system equations (8.16) are rewritten as follows:

$$\begin{cases} \dot{x}(t) = \sum_{i=1}^{r}\sum_{j=1}^{2^{n_{\theta_u}}}[\mu_i(\hat{x}(t))\widetilde{\mu_j}(\hat{\theta^u}(t))(A_i x(t) + \mathscr{B}_{ij}u(t)) + \\ \qquad \delta_{ij}(t)(A_i x(t) + \mathscr{B}_{ij}u(t))] \\ y(t) = \sum_{k=1}^{2^{n_{\theta_y}}}\Big[\overline{\mu_k}(\hat{\theta^y}(t-\tau(t)))\widetilde{C}_k x(t-\tau(t)) \\ \quad + \overline{\delta_k}(t-\tau(t))\widetilde{C}_k x(t-\tau(t))\Big], \end{cases} \tag{8.21}$$

where $\delta_{ij}(t)$ and $\overline{\delta_k}(t)$ are defined by the following equations:

$$\delta_{ij}(t) = \mu_i(x(t))\widetilde{\mu_j}(\theta^u(t)) - \mu_i(\hat{x}(t))\widetilde{\mu_j}(\hat{\theta^u}(t)) \tag{8.22}$$

$$\overline{\delta_k}(t-\tau(t)) = \overline{\mu_k}(\theta^y(t-\tau(t))) - \overline{\mu_k}(\hat{\theta^y}(t-\tau(t))) \tag{8.23}$$

and satisfy the inequalities:

$$-1 \leq \delta_{ij}(t) \leq 1, -1 \leq \overline{\delta_k}(t) \leq 1. \tag{8.24}$$

Equation (8.21) allows to deduce the state and data deception estimation error dynamics in a straightforward way, since the state and output are written now only depending on the weighting functions of the estimate $\mu_i(\hat{x}(t))$, $\widetilde{\mu_j}(\hat{\theta^u}(t))$, and $\overline{\mu_k}(\hat{\theta^y}(t))$.

Now, let us define the following matrices:

$$\Delta A(t) = \sum_{i=1}^{r} \sum_{j=1}^{2^{n_{\theta_u}}} \delta ij(t) A_i = \mathscr{A} \Sigma(t) E_A \tag{8.25}$$

$$\Delta B(t) = \sum_{i=1}^{r} \sum_{j=1}^{2^{n_{\theta_u}}} \delta_{ij}(t) \mathscr{B}_{ij} = \mathscr{B} \Sigma(t) E_B \tag{8.26}$$

$$\widetilde{C}(\nabla) = \left[\, \overline{\delta_1}(\nabla)\widetilde{C}_1 \ldots \overline{\delta_{2^{n_{\theta_y}}}}(\nabla)\widetilde{C}_{2^{n_{\theta_y}}} \,\right] \tag{8.27}$$

with

$$\mathscr{A} = \left[\underbrace{A_1 \ldots A_1}_{2^{n_{\theta_u}} times} \ldots \underbrace{A_r \ldots A_r}_{2^{n_{\theta_u}} times} \right] \tag{8.28}$$

$$\mathscr{B} = \left[\, \mathscr{B}_{11} \ldots \mathscr{B}_{r2^n} \,\right] \tag{8.29}$$

$$\Sigma(t) = \mathrm{diag}(\delta_{11}(t), \ldots, \delta_{r2^n}(t)) \tag{8.30}$$

$$E_A = \left[\, I_{n_x} \ldots I_{n_x} \,\right]^T, \; E_B = \left[\, I_{n_u} \ldots I_{n_u} \,\right]^T. \tag{8.31}$$

From (8.24) to (8.30), we have

$$\Sigma^T(t)\Sigma(t) \leq I. \tag{8.32}$$

By using (8.25)–(8.31) and the notation $\nabla = t - \tau(t)$, system (8.21) can be written as an uncertain system given by

$$\begin{cases} \dot{x}(t) = \displaystyle\sum_{i=1}^{r} \sum_{j=1}^{2^{n_{\theta_u}}} \mu_i(\hat{x}(t))\widetilde{\mu_j}(\hat{\theta^u}(t)) \\ \qquad ((A_i + \Delta A(t))x(t) + (\mathscr{B}_{ij} + \Delta B(t))u(t)) \\ y(t) = \displaystyle\sum_{k=1}^{2^{n_{\theta_y}}} \overline{\mu_k}(\hat{\theta^y}(\nabla))(\widetilde{C}_k + \widetilde{C}(\nabla))x(\nabla). \end{cases} \tag{8.33}$$

From Eqs. (8.33) and (8.20), the estimation error dynamics are then given by

$$
\begin{cases}
\dot{e}_x(t) = \sum_{i=1}^{r}\sum_{j=1}^{2^{n_{\theta u}}}\sum_{k=1}^{2^{n_{\theta y}}} \mu_i(\hat{x}(t))\widetilde{\mu_j}(\hat{\theta^u}(t))\overline{\mu_k}(\hat{\theta^y}(\nabla)) \\
\quad (A_i e_x(t) - L_{ij}\widetilde{C}_k e_x(\nabla) \\
\quad +\Delta A(t)x(t) - L_{ij}\widetilde{C}(\nabla)x(\nabla) + \Delta B(t)u(t)) \\
\dot{e}_{\theta^u}(t) = \sum_{i=1}^{r}\sum_{j=1}^{2^{n_{\theta u}}}\sum_{k=1}^{2^{n_{\theta y}}} \mu_i(\hat{x}(t))\widetilde{\mu_j}(\hat{\theta^u}(t))\overline{\mu_k}(\hat{\theta^y}(\nabla)) \\
\quad (-K_{ij}^u\widetilde{C}_k e_x(\nabla) - \alpha_{ij}^u e_{\theta^u}(t) \\
\quad -K_{ij}^u\widetilde{C}(\nabla)x(\nabla) + \alpha_{ij}^u\theta^u(t) + \dot{\theta^u}(t)) \\
\dot{e}_{\theta^y}(t) = \sum_{i=1}^{r}\sum_{k=1}^{2^{n_{\theta y}}} \mu_i(\hat{x}(t))\overline{\mu_k}(\hat{\theta^y}(\nabla)) \\
\quad (-K_{ik}^y\widetilde{C}_k e_x(\nabla) - \alpha_{ik}^y e_{\theta^y}(t) \\
\quad -K_{ik}^y\widetilde{C}(\nabla)x(\nabla) + \alpha_{ik}^y\theta^y(t) + \dot{\theta^y}(t)).
\end{cases}
\tag{8.34}
$$

Let us now consider the augmented vectors $e_a(t)$ and $\omega(t)$, such that

$$
e_a(t) = \begin{pmatrix} x(t) \\ e_x(t) \\ e_\theta^u(t) \\ e_\theta^y(t) \end{pmatrix}, \quad \omega(t) = \begin{pmatrix} \theta^u(t) \\ \theta^y(t) \\ \dot{\theta}^u(t) \\ \dot{\theta}^y(t) \\ u(t) \end{pmatrix}.
\tag{8.35}
$$

From (8.34) and (8.35), it follows that

$$
\begin{aligned}
\dot{e}_a(t) = &\sum_{i=1}^{r}\sum_{j=1}^{2^{n_{\theta u}}}\sum_{k=1}^{2^{n_{\theta y}}} \mu_i(\hat{x}(t))\widetilde{\mu_j}(\hat{\theta^u}(t))\overline{\mu_k}(\hat{\theta^y}(\nabla)) \\
&\big(\Phi_{ijk}(t)e_a(t) + \Psi_{ijk}(t)\omega(t) - R_{ijk}(\nabla)e_a(\nabla)\big)
\end{aligned}
\tag{8.36}
$$

with

$$
\Phi_{ijk}(t) = \begin{pmatrix} A_i & 0 & 0 & 0 \\ \Delta A(t) & A_i & 0 & 0 \\ 0 & 0 & -\alpha_{ij}^u & 0 \\ 0 & 0 & 0 & -\alpha_{ik}^y \end{pmatrix}
\tag{8.37}
$$

$$
\Psi_{ijk}(t) = \begin{pmatrix} 0 & 0 & 0 & 0 & \mathscr{B}_{ij} + \Delta B(t) \\ 0 & 0 & 0 & 0 & \Delta B(t) \\ \alpha_{ij}^u & 0 & I & 0 & 0 \\ 0 & \alpha_{ik}^y & 0 & I & 0 \end{pmatrix}
\tag{8.38}
$$

$$R_{ijk}(\nabla) = \begin{pmatrix} 0 & 0 & 0 & 0 \\ L_{ij}\widetilde{C}(\nabla) & L_{ij}\widetilde{C} & 0 & 0 \\ K^u_{ij}\widetilde{C}(\nabla) & K^u_{ij}\widetilde{C} & 0 & 0 \\ K^y_{ik}\widetilde{C}(\nabla) & K^y_{ik}\widetilde{C} & 0 & 0. \end{pmatrix}. \tag{8.39}$$

Now, the objective is to find the observer parameter matrices such that the transfer from $\omega(t)$ to $e_a(t)$ is minimized. This approach assumes that the disturbance, i.e., the external input $\omega(t)$ belongs to a set of norm bounded functions, i.e., is of finite energy. For the considered problem, knowing that the attacks do not appear all time (stealthy attacks), the assumption is realized.

Let us define the following Lyapunov–Krasovskii functional candidate Mondié and Kharitonov (2005):

$$V(t) = e_a^T(t)Pe_a(t) + \int_{-\tau(t)}^{0} e_a^T(t+\theta)e^{2\alpha\theta}Qe_a(t+\theta)d\theta, \tag{8.40}$$

where P and Q are symmetric, positive definite matrices. The convergence with L_2 attenuation is then guaranteed if the following conditions are satisfied:

$$V(t) > 0 \tag{8.41}$$

$$\dot{V}(t) + e_a^T(t)e_a(t) - \omega^T(t)\Gamma\omega(t) < -2\alpha V(t) \tag{8.42}$$

with

$$\Gamma = \text{diag}(\Gamma_l), \ \Gamma_l < \beta\, I, \ \text{for } l = 1, \ldots, 6. \tag{8.43}$$

An appropriate choice of Γ enables to attenuate the transfer from $\omega(t)$ to $e_a(t)$.

The time derivative of $V(t)$ along the trajectory of (8.36) is given by

$$\begin{array}{l} \dot{V}(t) = \dot{e_a}^T(t)Pe_a(t) + e_a(t)P\dot{e_a}^T(t) + e_a^T(t)Qe_at(t) \\ -(1-\dot{\tau}(t))e^{-2\alpha\tau(t)}e_a^T(\nabla)Qe_a(\nabla) \\ -2\alpha\int_{-\tau(t)}^{0} e_a^T(t+\theta)e^{2\alpha\theta}Qe_a(t+\theta)d\theta, \end{array} \tag{8.44}$$

which is upper bounded thanks to the time-delay condition (8.1) by

$$\begin{array}{l} \dot{V}(t) \leq \dot{e_a}^T(t)Pe_a(t) + e_a(t)P\dot{e_a}^T(t) + e_a^T(t)Qe_at(t) \\ -(1-\gamma)e^{-2\alpha\tau}e_a^T(\nabla)Qe_a(\nabla) \\ -2\alpha\int_{-\tau(t)}^{0} e_a^T(t+\theta)e^{2\alpha\theta}Qe_a(t+\theta)d\theta. \end{array} \tag{8.45}$$

By considering (8.36), we also have

$$\dot{V}(t) + e_a^T(t)e_a(t) - \omega^T(t)\Gamma\omega(t) = \sum_{i=1}^{r}\sum_{j=1}^{2^{n_{\theta_u}}}\sum_{k=1}^{2^{n_{\theta_y}}} \mu_i(\hat{x}(t))\widetilde{\mu_j}(\hat{\theta^u}(t))\overline{\mu_k}(\hat{\theta^y}(\nabla))$$

$$\begin{pmatrix} e_a(t) \\ \omega(t) \\ e_a(\nabla) \end{pmatrix}^T \left(\begin{array}{c|c|c} \Phi_{ij}^T(t)P + P\Phi_{ij}(t) + I & P\Psi_i(t) & -PR_{ijk}(\nabla) \\ \hline \Psi_i^T(t)P & -\Gamma & 0 \\ \hline * & * & -(1-\gamma)e^{2\alpha\tau}Q \end{array}\right) \begin{pmatrix} e_a(t) \\ \omega(t) \\ e_a(\nabla) \end{pmatrix}$$

$$-2\alpha \int_{-\tau(t)}^{0} e_a^T(t+\theta)e^{2\alpha\theta}Qe_a(t+\theta)d\theta \tag{8.46}$$

and

$$\begin{aligned} &\dot{V}(t) + e_a^T(t)e_a(t) - \omega^T(t)\Gamma\omega(t) + 2\alpha V(t) \leq \\ &\sum_{i=1}^{r}\sum_{j=1}^{2^{n_{\theta_u}}}\sum_{k=1}^{2^{n_{\theta_y}}} \mu_i(\hat{x}(t))\widetilde{\mu_j}(\hat{\theta^u}(t))\overline{\mu_k}(\hat{\theta^y}(\nabla)) \begin{pmatrix} e_a(t) \\ \omega(t) \\ e_a(\nabla) \end{pmatrix}^T \\ &\left[\left(\begin{array}{c|c|c} \Phi_{ij}^T(t)P + P\Phi_{ij}(t) + I & P\Psi_i(t) & -PR_{ijk}(\nabla) \\ \hline \Psi_i^T(t)P & -\Gamma & 0 \\ \hline * & * & -(1-\gamma)e^{2\alpha\tau}Q \end{array}\right)\right. \\ &\left. +2\alpha \left(\begin{array}{c|c|c} P & 0 & 0) \\ \hline 0 & 0 & 0 \\ \hline 0 & 0 & 0 \end{array}\right)\right] \begin{pmatrix} e_a(t) \\ \omega(t) \\ e_a(\nabla) \end{pmatrix}. \end{aligned} \tag{8.47}$$

The negativity of condition (8.47) due to the convex sum property of the weighting functions and the quadratic form of the vector $\begin{pmatrix} e_a(t) \\ \omega(t) \\ e_a(\nabla) \end{pmatrix}^T$ is therefore guaranteed if:

$$\left(\begin{array}{c|c|c} C_1 & P\Psi_i(t) & -PR_{ijk}(\nabla) \\ \hline \Psi_i^T(t)P & -\Gamma & 0 \\ \hline * & * & -(1-\gamma)e^{2\alpha\tau}Q \end{array}\right) < 0, \tag{8.48}$$

where $C_1 = (\Phi_{ij} + \alpha I)^T(t)P + P(\Phi_{ij}(t) + \alpha I) + I$. It is also important to highlight that the matrices $\widetilde{C}(\nabla)$ can be written as

$$\widetilde{C}(\nabla) = \sum_{l=1}^{2^{n_{\theta_y}}} \overline{\delta_l}(\nabla)\widetilde{C}_l. \tag{8.49}$$

From (8.49), and based on the convex sum property of $\overline{\delta_l}(t)$, the matrix inequalities (8.48) become

$$\sum_{l=1}^{2^{n_{\theta_y}}} \overline{\delta_l}(\nabla) \left(\begin{array}{c|c|c} C_1 & P\Psi_i(t) & -PR_{ijk} \\ \hline \Psi_i^T(t)P & -\Gamma & 0 \\ \hline * & * & -(1-\gamma)e^{2\alpha\tau}Q \end{array}\right) < 0, \tag{8.50}$$

where

$$R_{ijk} = \begin{pmatrix} 0 & 0 & 0 & 0 \\ L_{ij}\widetilde{C} & L_{ij}\widetilde{C} & 0 & 0 \\ K_{ij}^{u}\widetilde{C} & K_{ij}^{u}\widetilde{C} & 0 & 0 \\ K_{ik}^{y}\widetilde{C} & K_{ik}^{y}\widetilde{C} & 0 & 0 \end{pmatrix}, \tag{8.51}$$

which is equivalent to solve

$$\left(\begin{array}{c|c|c} C_1 & P\Psi_i(t) & -PR_{ijk} \\ \hline \Psi_i^T(t)P & -\Gamma & 0 \\ \hline * & * & -(1-\gamma)e^{2\alpha\tau}Q \end{array}\right) < 0. \tag{8.52}$$

The observer gains are then obtained by solving the above constraints with the sufficient condition inequality (8.52) for $i = 1, \ldots, r$, $j = 1, \ldots, 2^{n_{\theta_u}}$, $k = 1, \ldots, 2^{n_{\theta_y}}$, and $l = 1, \ldots, 2^{n_{\theta_y}}$.

The results may be summarized by the following theorem:

Theorem 8.1 *There exists a state and actuator/sensor date deception attack observer (8.19) for a nonlinear system (8.2) with delayed measurements and an $\mathcal{L}_2$ gain from $\omega(t)$ to $e_a(t)$ bounded by β ($\beta > 0$) if there exist positive symmetric matrices $P_1 = P_1^T > 0$, $P_2 = P_2^T > 0$, $P_3 = P_3^T > 0$, $P_4 = P_4^T > 0$ and $Q_1 = Q_1^T > 0$, $Q_2 = Q_2^T > 0$, $Q_3 = Q_3^T > 0$, $Q_4 = Q_4^T > 0$; positive matrices Γ_l, $l = 1, \ldots, 5$; matrices $\overline{\alpha}_{ij}^u$, $\overline{\alpha}_{ik}^y$, F_{ij}^u, F_{ik}^y, $\overline{R}_{ij}$; and scalars positive β, λ_A λ_{1B}, λ_{2B}, and α solution of the following optimization problem under LMI constraints (8.54) and (8.57) (see next page)*

$$\min_{\{P_1, P_2, P_3, \overline{R}_{ij}, F_{ij}^u, F_{ik}^y, \overline{\alpha}_{ij}^u, \overline{\alpha}_{ik}^y, \Gamma_l, \lambda_A, \lambda_{1B}, \lambda_{2B},\}} \beta, \tag{8.53}$$

for $i = 1, \ldots, r$, $j = 1, \ldots, 2^{n_{\theta_u}}$, $k = 1, \ldots, 2^{n_{\theta_y}}$, and $l = 1, \ldots, 2^{n_{\theta_y}}$, where the scalar α is called the delay rate.

$$\Gamma_l < \beta I \text{ for } l = 1, \ldots, 5 \tag{8.54}$$

with

$$\begin{aligned} Q_i^{11} &= P_1(A_i + \alpha I) + (A_i + \alpha I)^T P_1 + I_{n_x} \\ Q_5 &= -\Gamma_1 + \lambda_A E_A^T E_A \\ Q_8 &= -\Gamma_4 + \lambda_{1B} E_B^T E_B \\ Q_9 &= -\Gamma_5 + \lambda_{2B} E_B^T E_B \\ Q_{10} &= -(1-\gamma)e^{2\alpha\tau} Q_1 \\ Q_{11} &= -(1-\gamma)e^{2\alpha\tau} Q_2 \\ Q_{12} &= -(1-\gamma)e^{2\alpha\tau} Q_3 \\ Q_{13} &= -(1-\gamma)e^{2\alpha\tau} Q_4, \end{aligned} \tag{8.55}$$

where the observer gains are given by

$$\begin{cases} L_{ij} = P_2^{-1}\overline{R}_{ij} \\ K_{ij}^u = P_3^{-1}F_{ij}^u \\ K_{ik}^y = P_4^{-1}F_{ik}^y \\ \alpha_{ij}^u = P_3^{-1}\overline{\alpha}_{ij}^u \\ \alpha_{ik}^y = P_4^{-1}\overline{\alpha}_{ik}^y. \end{cases} \tag{8.56}$$

$$\begin{pmatrix} Q_i^{11} & 0 & 0 & 0 & 0 & 0 & 0 & 0 & P_1\mathscr{B}_{ij} & 0 & 0 & 0 & 0 & 0 & P_1\mathscr{B} & 0 \\ * & P_2A_i & 0 & 0 & 0 & 0 & 0 & 0 & 0 & -\overline{R}_{ij}\widetilde{C} & -\overline{R}_{ij}\widetilde{C} & 0 & 0 & P_2\mathscr{A} & 0 & P_2\mathscr{B} \\ * & * & -\overline{\alpha}_{ij}^u & 0 & \overline{\alpha}_{ij}^u & 0 & P_3 & 0 & 0 & F_{ij}^u\widetilde{C} & F_{ij}^u\widetilde{C} & 0 & 0 & 0 & 0 & 0 \\ * & * & * & -\overline{\alpha}_{ik}^y & 0 & \overline{\alpha}_{ik}^y & 0 & P_4 & 0 & F_{ik}^y\widetilde{C} & F_{ik}^y\widetilde{C} & 0 & 0 & 0 & 0 & 0 \\ * & * & * & * & Q_5 & 0 & 0 & 0 & 0 & 0 & 0 & 0 & 0 & 0 & 0 & 0 \\ * & * & * & * & * & -\Gamma_2 & 0 & 0 & 0 & 0 & 0 & 0 & 0 & 0 & 0 & 0 \\ * & * & * & * & * & * & -\Gamma_3 & 0 & 0 & 0 & 0 & 0 & 0 & 0 & 0 & 0 \\ * & * & * & * & * & * & * & Q_8 & 0 & 0 & 0 & 0 & 0 & 0 & 0 & 0 \\ * & * & * & * & * & * & * & * & Q_9 & 0 & 0 & 0 & 0 & 0 & 0 & 0 \\ * & * & * & * & * & * & & * & * & -Q_{10} & 0 & 0 & 0 & 0 & 0 & 0 \\ * & * & * & * & * & * & * & * & * & * & -Q_{11} & 0 & 0 & 0 & 0 & 0 \\ * & * & * & * & * & * & * & * & & * & * & -Q_{12} & 0 & 0 & 0 & 0 \\ * & * & * & * & * & * & * & * & * & * & * & * & -Q_{13} & 0 & 0 & 0 \\ * & * & * & * & * & * & * & * & * & * & * & * & * & -\lambda_A I & 0 & 0 \\ * & * & * & * & * & * & * & * & * & * & * & * & * & * & -\lambda_{1B} I & 0 \\ * & * & * & * & * & * & * & * & * & * & * & * & * & * & * & -\lambda_{2B} I \end{pmatrix} < 0. \tag{8.57}$$

Proof Based on condition (8.52), with (8.37) and the variable change (8.56), with the decomposition (8.25) and (8.26), properties (8.32), Schur's complement, and the following lemma:

Lemma 8.1 *Consider (Zhou and Khargonekar 1988) two matrices X and Y with appropriate dimensions, a time-varying matrix $\Delta(t)$ and a positive scalar ε. The following property is verified*

$$X^T \Delta^T(t) Y + Y^T \Delta(t) X \leq \varepsilon X^T X + \varepsilon^{-1} Y^T Y, \tag{8.58}$$

for $\Delta^T(t)\Delta(t) \leq I$

following the same development as the work presented in Bezzaoucha et al. (2013), Bezzaoucha et al. (2013), the Lyapunov stability with an $\mathscr{L}_2$ transfer from $\omega(t)$ to $e_a(t)$ is obtained by solving the optimization problem (8.53) under the LMI constraints (8.54) and (8.57), which ends the proof. □

8.4 Numerical Simulation

In the following, the proposed approach for state and attacks estimation is applied to a basic model of a biological wastewater treatment plant (Bezzaoucha et al. 2013).

The mathematical model is represented thanks to two state variables $x_1(t)$ and $x_2(t)$, corresponding to the biomass and substrate concentration, respectively, the input $u(t)$, which represents the dwell time in the treatment plant and the measured output which is the biomass concentration ($y(t) = x_1(t)$). The time delay that appears in the output of the system has the form $\tau(t) = 0.5 + 0.45\sin(0.5t)$. The upper bound of its derivative is then equal to $\gamma = 0.225$.

8.4.1 LPV Representation of The Process

First step, let us write the nonlinear system equations (8.59) in a polytopic form. As it was developed in Bezzaoucha et al. (2013), and under specific assumptions, some simplifications can be made and the nonlinear model may be given by

$$\begin{cases} \dot{x}_1(t) = \frac{ax_1(t)x_2(t)}{x_2(t)+b} - x_1(t)u(t) \\ \dot{x}_2(t) = -\frac{cax_1(t)x_2(t)}{x_2(t)+b} + (d - x_2(t))u(t), \end{cases} \tag{8.59}$$

where a, b, c, and d are known parameters.

From the system nonlinearities, applying the sector nonlinearity approach with the premise variables $\rho_1(t)$ and $\rho_2(t)$ chosen as follows:

$$\rho_1(t) = -u(t), \quad \rho_2(t) = \frac{ax_1(t)}{x_2(t) + b}. \tag{8.60}$$

From (8.59) to (8.60), the quasi-LPV system (8.61) is deduced as

$$\dot{x}(t) = \begin{pmatrix} \rho_1(t) & \rho_2(t) \\ 0 & -c\rho_2(t) + \rho_1(t) \end{pmatrix} x(t) + \begin{pmatrix} 0 \\ d \end{pmatrix} u(t). \tag{8.61}$$

Since an LPV representation is deduced in a compact set of the state space, the max and min values of the terms $\rho_1(t)$ and $\rho_2(t)$ may be calculated using the knowledge of the domain of variation of $u(t)$, i.e., $\rho_1(t) \in [-1, -0.2]$ and $\rho_2(t) \in [0.004, 15]$.

Applying the convex polytopic transformation, two partitions for each premise variable are defined as

$$\begin{cases} \rho_1(t) = \varrho_{11}(\rho_1)\rho_1^2 + \varrho_{12}(\rho_1)\rho_1^1 \\ \rho_2(t) = \varrho_{21}(\rho_2)\rho_2^2 + \varrho_{22}(\rho_2)\rho_2^1 \end{cases} \tag{8.62}$$

$$\text{with} \quad \varrho_{11}(\rho_1) = \frac{\rho_1(t) - \rho_1^2}{\rho_1^1 - \rho_1^2}, \; \varrho_{12}(\rho_1) = \frac{\rho_1^1 - \rho_1(t)}{\rho_1^1 - \rho_1^2}$$
$$\varrho_{21}(\rho_2) = \frac{\rho_2(t) - \rho_2^2}{\rho_2^1 - \rho_2^2}, \; \varrho_{22}(\rho_2) = \frac{\rho_2^1 - \rho_2(t)}{\rho_2^1 - \rho_2^2}, \tag{8.63}$$

where the scalars ρ_1^1, ρ_1^2, ρ_2^1, and ρ_2^2 are defined as

$$\begin{aligned} \rho_1^1 &= \max_u \rho_1(t), \ \rho_1^2 = \min_u \rho_1(t) \\ \rho_2^1 &= \max_x \rho_2(t), \ \rho_2^2 = \min_x \rho_2(t). \end{aligned} \tag{8.64}$$

The sub-models are defined by the sets (A_i, B_i, C) with $i = 1, 2, 3, 4$. Based on ρ_1 and ρ_2 definitions, all the B_i matrices are set to $B = \begin{bmatrix} 0 & d \end{bmatrix}^T$. The output matrix $C = \begin{bmatrix} 1 & 0 \end{bmatrix}$ and the matrices A_i are given by

$$A_1 = \begin{pmatrix} \rho_1^1 & \rho_2^1 \\ 0 & -c\rho_2^1 + \rho_1^1 \end{pmatrix}, \ A_2 = \begin{pmatrix} \rho_1^1 & \rho_2^2 \\ 0 & -c\rho_2^2 + \rho_1^1 \end{pmatrix}$$

$$A_3 = \begin{pmatrix} \rho_1^2 & \rho_2^1 \\ 0 & -c\rho_2^1 + \rho_1^2 \end{pmatrix}, \ A_4 = \begin{pmatrix} \rho_1^2 & \rho_2^2 \\ 0 & -c\rho_2^2 + \rho_1^2. \end{pmatrix}.$$

The weighting functions $\mu_i(t)$ are defined by the following equations:

$$\begin{aligned} \mu_1(t) &= \rho_{11}(\rho_1(t))\rho_{21}(\rho_2(t)), \mu_2(t) = \rho_{11}(\rho_1(t))\rho_{22}(\rho_2(t)) \\ \mu_3(t) &= \rho_{12}(\rho_1(t))\rho_{21}(\rho_2(t)), \mu_4(t) = \rho_{12}(\rho_1(t))\rho_{22}(\rho_2(t)). \end{aligned} \tag{8.65}$$

Since the polytopic representation is obtained in a compact set of the state space, maximum and minimum values that occur in $\rho_1(t)$ and $\rho_2(t)$ may be calculated using the knowledge of the domain of variation of $u(t)$: $\rho_1(t) \in [-1, -0.2]$ and $\rho_2(t) \in [0.004, 15]$.

8.4.2 *Date Deception Attacks Representation on The Actuator/Sensor*

Two types of data deception attacks are considered, i.e., attacks on actuators and sensors. It is assumed that, mathematically speaking, these attacks are modeled as bounded multiplicative actuator and sensor time-varying faults.

For the considered example, it is assumed that parameter d may be hacked. This actuator attack is represented by $d(t)$, such that

$$d(t) = d + \Delta d(t). \tag{8.66}$$

It can also be written as

$$d(t) = d + \theta^u(t)\overline{d}, \quad \theta^u(t) \in [\theta^{u2}, \theta^{u1}] \tag{8.67}$$

with $d = 2.5$, $\overline{d} = 2.1$ and $\theta^{u2} = -0.1958$, $\theta^{u1} = 0.1979$. Parameters a, b, and c have been identified and set to $a = 0.5$, $b = 0.07$, and $c = 0.7$.

Considering the attack on the actuator, the polytopic representation of the input matrix B is then given by two sub-models, such that

$$B_1 = B + \theta^{u1}\overline{B}, \quad B_2 = B + \theta^{u2}\overline{B}, \tag{8.68}$$

where it is defined by $\overline{B} := \begin{bmatrix} 0 \ \overline{d} \end{bmatrix}^T$. The weighting functions $\widetilde{\mu_j}(\theta^u(t))$ are defined as given in (8.7) and (8.12).

Now, for the sensor attack, it is assumed that a bounded multiplicative sensor fault $\theta^y(t)$ affects the output $y(t)$ such that

$$y(t) = (1 + \theta^y(t - \tau))x_1(t - \tau). \tag{8.69}$$

As previously explained, $\theta^y(t)$ can also be written as

$$\theta^y(t) = \overline{\mu}_1{}^1(\theta^y(t))\theta^{y1} + \overline{\mu}_1^2(\theta^y(t))\theta^{y2}, \ \theta^y(t) \in [\theta^{y2}, \theta^{y1}] \tag{8.70}$$

with $\theta^{y2} = 0.125$, $\theta^{y1} = 0.625$, $\overline{\mu}_1{}^1(\theta^y(t))$, and $\overline{\mu}_1^2(\theta^y(t))$ are defined by (8.9) and (8.13).

The polytopic form of the output is then given by

$$y(t) = \sum_{k=1}^{2} \overline{\mu}_k(\theta^y(t - \tau(t)))\widetilde{C}_k x(t - \tau(t)) \tag{8.71}$$

with $\widetilde{C}_1 = \begin{pmatrix} 1 + \theta^{y2} \ 0 \end{pmatrix}$, $\widetilde{C}_2 = \begin{pmatrix} 1 + \theta^{y1} \ 0 \end{pmatrix}$.

8.4.3 Simulation Results

From the considered example, with both attacks on the actuator/sensor, applying the proposed approach by solving Theorem 8.1, a simultaneous state and attacks observer is designed such that the system initial conditions are taken as $x(0) = \begin{pmatrix} 0.1 \ 1.5 \end{pmatrix}$ and $\hat{x}(0) = \begin{pmatrix} 0.09 \ 2.3 \end{pmatrix}$ for its observer. For both attacks, the initial conditions are set to zero, i.e., $\hat{\theta}^u(0) = 0$ and $\hat{\theta}^y(0) = 0$.

The state vector, its estimate as well as the data deception attacks with their estimates are depicted in Figs. 8.1, and 8.2, respectively. From the obtained plots, the efficiency of the proposed observer is highlighted; indeed, both system states and the time-varying multiplicative actuator/sensor attacks are well estimated.

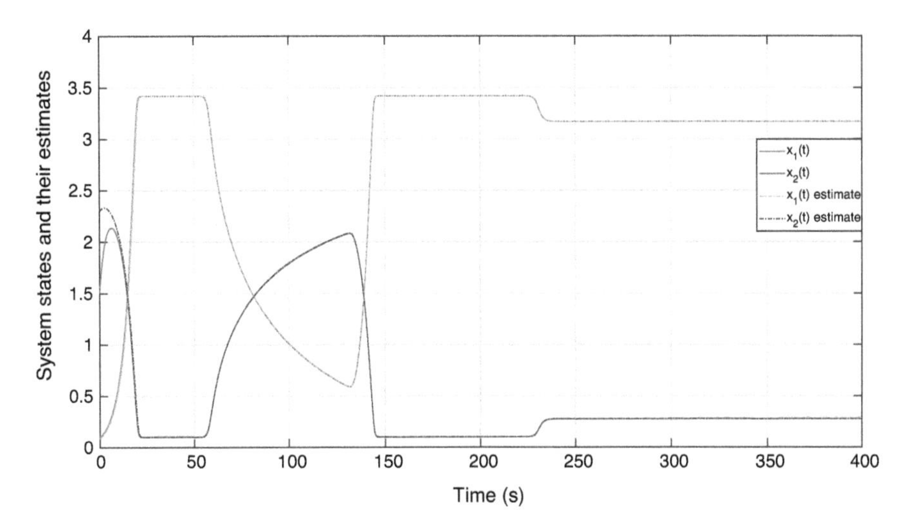

Fig. 8.1 System states and their estimates

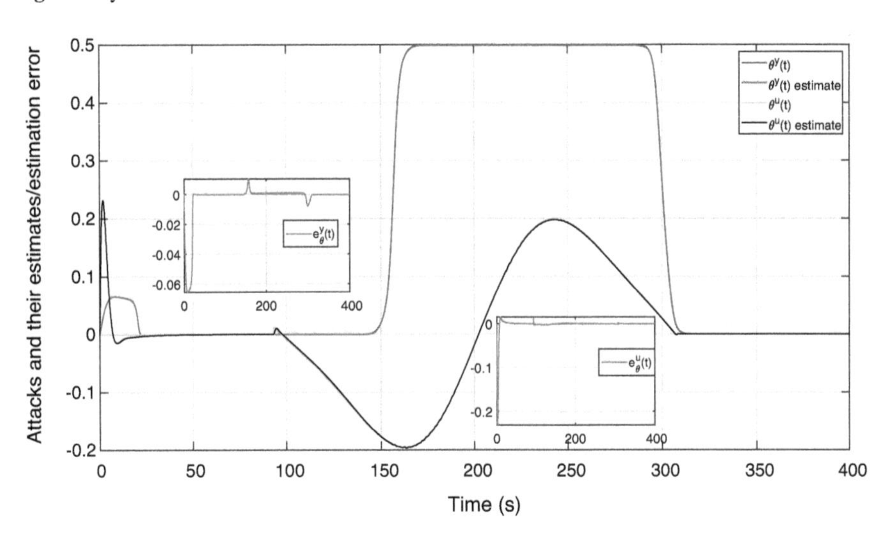

Fig. 8.2 Data deception attacks and their estimates

8.5 Conclusions

In the present book chapter, a polytopic approach was applied to cope with the system state and data deception attacks estimation and delayed measurements. Based on previous work, both attacks on actuator and sensor are modeled as multiplicative time-varying faults and written in a convex set, based only on their min and max bound. A simultaneous state and attack observer is designed by minimizing the $\mathscr{L}_2$

gain from the augmented input to the different estimation errors. The chosen application example is an activated sludge reactor with attacks represented by unknown time-varying parameters on the parameter d and the output. From the nonlinear equations of the system, an LPV model is derived. The proposed observer is designed and the obtained results illustrate its performance.

References

A. Benzaouia, A. El Hajaji, *Advanced Takagi-Sugeno Fuzzy Systems: Delay and Saturation* (Springer, 2014)

S. Bezzaoucha Rebai, H. Voos, Simultaneous state and false-data injection attacks reconstruction for nonlinear systems: an LPV approach, in *3rd International Conference on Automation, Control and Robots, Prague, Czech Republic* (2019)

S. Bezzaoucha Rebaiı, H. Voos, M. Darouach, Attack-tolerant control and observer-based trajectorytracking for cyber-physical systems. Eur. J. Control (2018). https://doi.org/10.1016/j.ejcon.2018.09.005

S. Bezzaoucha, B. Marx, D. Maquin, J. Ragot, State and multiplicative sensor fault estimation for nonlinear systems, in *2nd International Conference on Control and Fault-Tolerant Systems, Nice, France* (2013)

S. Bezzaoucha, H. Voos, Stability analysis of power networks under cyber-physical attacks: an LPV-descriptor approach, in *IEEE International Conference on Control, Decision and Information Technologies (CoDIT), Paris, France* (2019)

S. Bezzaoucha, H. Voos, M. Darouach, A contribution to cyber-security of networked control systems: an Event-based control approach, in *3rd International Conference on Event-Based Control, Communication and Signal Processing, Funchal, Madeira, Portugal* (2017)

S. Bezzaoucha, B. Marx, D. Maquin, J. Ragot, Nonlinear joint state and parameter estimation, Application to a wastewater treatment plant. Control Eng. Pract. **21**(10), 1377–1385 (2013)

B. Gerard, S. Bezzaoucha Rebai, H. Voos, M. Darouach, Cyber security, vulnerability analysis, of networked control system subject to false-data injection, in *The American Control Conference (Milwaukee* (Wisconsin, USA, 2018), p. 2018

X. Liu, G. Zhang, New approaches to H_∞ controller design based on fuzzy observers for T-S fuzzy systems via LMI. Automatica **39**, 1571–1582 (2003)

S. Mondié, V.L. Kharitonov, Exponential estimates for retarded time-delay systems: an LMI approach. IEEE Trans. Autom. Control **50**(2), 268–273 (2005)

R. Orjuela, B. Marx, J. Ragot, D. Maquin, A decoupled Multiple Model approach for state estimation of nonlinear systems subject to delayed measurements, in *3rd IFAC Workshop on Advanced Fuzzy and Neural Control (AFNC), Valenciennes, France* (2007)

M. Pajic, J. Weimer, N. Bezzo, O. Sokolsky, G.J. Pappas, I. Lee, Design and implementation of attack-resilient cyberphysical systems. IEEE Control Syst. Mag. **37**(2), 66–81 (2017)

U. Shaked, Improved LMI representations for the analysis and the design of continuous-time systems with polytopic type uncertainty. IEEE Trans. Autom. Control **46**, 652–656 (2001)

T. Takagi, M. Sugeno, Fuzzy identification of systems and its applications to modeling and control. IEEE Trans. Syst. Man Cybernet. **15**(1), 116–132 (1985)

K. Tanaka, H. Wang, *Fuzzy Control Systems Design and Analysis: A Linear Matrix Inequality Approach* (Wiley, New York, 2001)

A. Teixeira, D. Perez, H. Sandberg, K.H. Johansson, Attack models and scenarios for networked control systems, in *Proceedings of the 1st International Conference on High Confidence Networked Systems (HiCoNS '12), Beijing, China* (2012)

E. Tian, C. Peng, Delay-dependent stabilization analysis and synthesis for uncertain T-S fuzzy systems with time-varying delay. Fuzzy Sets Syst. **157**, 544–559 (2003)

D. Yue, Q.L. Han, Delayed feedback control of uncertain systems with time-varying input delay. Automatica **2**, 233–240 (2005)

K. Zhou, P.P. Khargonekar, Robust stabilization of linear systems with norm-bounded time-varying uncertainty. Syst. Control Lett. **10**(1), 17–20 (1988)

Chapter 9
Secure Estimation Under Model Uncertainty

Saurabh Sihag and Ali Tajer

9.1 Introduction

Cyber-physical systems are deployed in a variety of technical domains such as critical infrastructure, healthcare devices, and transportation. The rapid rise in their applications has exposed them to different vulnerabilities, threats, and attacks (Humayed et al. 2017). An abstract representation consisting of three main components: monitoring, communications, and computation and control, captures the fundamental aspects of cyber-physical systems. The monitoring component observes the environment and communicates with the computation and control component, which in turn processes the observations to form and communicate decisions. Each of these components could potentially be exploited or compromised, causing unexpected behaviors and compromised integrity and performance for the system.

The source of security threats to a cyber-physical system can broadly be categorized into three groups: an attacker with a malicious intent, functional failure of components in the system, and environmental threats such as natural disasters. While the impacts of operational failures of the system due to environmental threats or internal failures can be minimized by robust strategies (Hu et al. 2016), malicious attacks on cyber-physical systems intend to deceive the controller into making highly damaging decisions via well-crafted adversarial strategies. Therefore, specialized security measures are required to mitigate such attacks (Li et al. 2020).

Adversarial attacks that exploit the vulnerabilities of the inference and control algorithms deployed in the cyber-physical systems and potential defense strategies against them have been subjects of active research (Li et al. 2020; Fawzi et al. 2014;

S. Sihag
University of Pennsylvania, Philadelphia, PA 19104, USA
e-mail: saurabh.sihag@pennmedicine.upenn.edu

A. Tajer (✉)
Rensselaer Polytechnic Institute, Troy, NY 12108, USA
e-mail: tajer@ecse.rpi.edu

M. Abbaszadeh and A. Zemouche (eds.), *Security and Resilience in Cyber-Physical Systems*, https://doi.org/10.1007/978-3-030-97166-3_9

Ahmed et al. 2021). The taxonomy of the adversarial attacks on cyber-physical systems can be specified along three axes. The first axis pertains to the influence of the attack, where the attacker is capable of probing the algorithms for vulnerabilities. The attacker can further leverage these vulnerabilities to impose false decisions or outcomes in the system. The second axis pertains to the specificity of the attack, i.e., the attack can be either indiscriminate and affect all decisions made by the system, or targeted to impose false decisions only in specific scenarios. The third axis is related to the violation induced by the attack, where the attack can distort the integrity of the decisions made by the system in specific scenarios or overwhelm the system with malicious inputs, thus rendering it incapable of making any decision (for instance, through denial of service attacks).

In this chapter, we design a statistical inference framework for systems vulnerable to adversarial attacks. Statistical inference leverages the data sampled from a population to deduce its statistical properties. The commonly studied modes of statistical inference are broadly focused on discerning the statistical model of the population or estimating unknown, underlying parameters that characterize the statistical model of the population. Vulnerability to an attack induces uncertainties in the inference decisions, and therefore, must be accounted for in the design of inference algorithms that are resilient to adversarial attacks.

9.1.1 Overview and Contributions

We start by laying the context for the problem studied in this chapter. For this purpose, we consider the canonical parameter estimation problem in which the objective is to estimate a stochastic parameter X, which lies in a known set $\mathscr{X} \subseteq \mathbb{R}^p$, from the data samples $\mathbf{Y} \triangleq [Y_1, \ldots, Y_n]$, where the sample Y_r is distributed according to a statistical model with probability density function (pdf) P_X and lies in a known set $\mathscr{Y} \subseteq \mathbb{R}^m$. In practice, the dimension of the data points m could correspond to the number of data collecting entities in the system. Furthermore, the statistician *assumes* a prior data model for X and Y_r, determined through historical data. We denote the assumed underlying pdfs for X and $\mathbf{Y}$ by π and $f(\cdot \mid X)$, respectively, i.e.,

$$\mathbf{Y} \sim f(\cdot \mid X) , \qquad \text{with} \quad X \sim \pi . \tag{9.1}$$

For our analysis, we assume that the pdfs do not have any non-zero probability masses over lower-dimensional manifolds. The objective of the statistician is to formalize a reliable estimator

$$\hat{X}(\mathbf{Y}) : \mathscr{Y}^n \mapsto \mathscr{X} . \tag{9.2}$$

For elaborate discussions on the design of statistical estimators, we refer the readers to Poor (1998). In an adversarial environment, the attacker may launch an attack on different components of the data model defined in (9.1) to degrade the quality of $\hat{X}(\mathbf{Y})$. Next, we discuss two specific adversarial attack scenarios.

False data injection attacks: The purpose of false data injection attacks is to distort the data samples $\mathbf{Y}$ such that the data model deviates from (9.1) for at least a subset of coordinates in $\mathbf{Y}$.

Causative attacks: The purpose of a causative attack is to compromise the *process that underlies acquiring the statistical models* in (9.1). We emphasize that such an attack is different from false data injection attack because the effect of a causative attack is misleading the statistician about the true model $f(\cdot \mid X)$ that it assumes about the data. Such attacks are possible by compromising the historical (or training) data that is used for specifying a model for the data.

We remark that the nature of security vulnerabilities that inference algorithms are exposed to in causative attacks is fundamentally distinct from that of the data that faces false data injection attacks. Specifically, in the case of a false data injection attack, the information of the decision algorithm about the data model remains intact, while the data fed to the algorithm is anomalous. Therefore, when the sampled data is compromised, an inference algorithm produces decisions based on the true model for the data in the attack-free scenario, while the data that it receives and processes are compromised. On the other hand, when the historical data leveraged by the statistician to determine the true model are compromised, an inference algorithm functions based on an incorrect model for the data, in which case even un-compromised sampled data produces unreliable decisions. Both attack scenarios mentioned above force the inference algorithm to deviate from its optimal structure and, if not mitigated, may produce decisions that serve the adversary's purposes.

Depending on the specificity and the extent of an adversarial attack, e.g., the fraction of the observed data or training data that is compromised, the true model $f(\cdot \mid X)$ can be assumed to deviate to the space of alternative data models, which we denote by $\mathscr{F}$. The attack can be characterized by alterations in the statistical distributions of any number of the m coordinates of $\mathbf{Y}$. There are two major aspects of selecting $\mathscr{F}$ as a viable model space.

- An attack is effective in degrading the quality of estimation if the compromised model is sufficiently distinct from the model assumed by the statistician for designing the estimator. Hence, even though, in general, $\mathscr{F}$ can be thought of as any representation of possible kernels $f(\cdot \mid X)$ mapping $\mathscr{Y}$ to $\mathbb{R}^m$, only a subset of such mappings pertain to the set of effective attacks.
- There exists a tradeoff between the complexity of the model space and its expressiveness. Specifically, an overly expressive space can represent the possible compromised models with a more refined accuracy, albeit at the expense of more complex statistical inference rules.

We will discuss the specifics of the attack model in Sect. 9.2. Note that the potential adversarial presence induces a new dimension to the estimation problem in (9.2). Specifically, the optimal estimator design hinges on the knowledge of the true statistical model of the measurements $\mathbf{Y}$. However, detecting whether the data model has been compromised and discerning the true model, itself being an inference task, is never perfect. These observations imply an inherent coupling between the original estimation problem of interest and the introduced auxiliary problem due to potential adversarial behavior (i.e., detecting the presence of an attacker and isolating the true model). Therefore, the quality of the estimator is expected to degrade with respect to an attack-free setting due to uncertainties in the true model in the adversarial setting. Our objective is to characterize the fundamental interplay between the quality of discerning the true model and the degradation in the estimation quality.

9.1.2 Related Studies

The problem of secure inference is studied primarily in the context of sensor networks, where a subset of sensors may be corrupted by an attacker. The study in Wilson and Veeravalli (2016), in particular, considers the problem of secure estimation in a two-sensor network, in which one sensor is assumed to be secured , and the other sensor is vulnerable to attacks. According to the heuristic estimation design in this context, first, a decision is formed on the attacker's activity on the unsecured sensor . If it is deemed to be attacked, then the estimation design relies only on the secured sensor , and otherwise, it uses the data collected at both sensors. In contrast to Wilson and Veeravalli (2016), we consider a model with an arbitrary dimension of data, assume that all data coordinates are vulnerable to the attack, and characterize the optimal secure inference structure, which is distinct from being a detection -driven design studied in Wilson and Veeravalli (2016).

The adversarial setting considered in this chapter has similarities with the widely-investigated Byzantine attack models in sensor networks. In Byzantine attack models, the data corresponding to the compromised sensors is modified arbitrarily by the adversaries with an aim to degrade the inference quality. The impact of Byzantine attacks on the quality of inference and relevant mitigation strategies in sensor networks are discussed in Vempaty et al. (2013). Various detection-driven estimation strategies (i.e., when attack detection precedes and guides the estimation routine) for scenarios where the impacts of the Byzantine attacks on data are characterized by randomly flipped information bits, are discussed in Vempaty et al. (2013), Ebinger and Wolthusen (2009), Zhang et al. (2015), Zhang and Blum (2014). Furthermore, attack-resilient target localization strategies are studied in Vempaty et al. (2013, 2014), where the assumption is that the attacker adopts a fixed strategy that leads to maximum disruption in the inference. In these studies, however, an attacker can deviate from the worst-case attack strategy of incurring the maximum damage, and launch a less impactful but sustained attack, which may remain undetected. Finally, various strategies for isolating the compromised sensors in sensor networks are studied in Rawat et al. (2010), Soltanmohammadi et al. (2013), Vempaty et al. (2011). The emphasis of these studies is primarily detection of attacks or isolating the attacked sensors, whereas this chapter focuses on parameter estimation.

Secure estimation in linear *dynamical* systems that characterize cyber-physical systems has been actively studied in recent years (Fawzi et al. 2011, 2014; Yong et al. 2015; Pajic et al. 2014, 2015; Shoukry et al. 2017; Mishra et al. 2015). The studies with more relevance to the scope of this chapter include Fawzi et al. (2014), Mishra et al. (2015), and Pajic et al. (2014), which investigate robust estimation in dynamic systems. Specifically, a coding-theoretic interplay between the number of sensors compromised by an adversary and the guarantees on perfect system state recovery are characterized in Fawzi et al. (2014), a Kalman filter-based approach for identifying the most reliable set of sensors for inference is investigated in Mishra et al. (2015), and the design of estimators that is robust in the presence of dynamical model uncertainty is studied in Pajic et al. (2014). Furthermore, the degradation impact on estimation performance in a dynamical system consisting of a single sensor network is investigated from the adversary's perspective in Bai and Gupta (2014), where bounds on the degradation in estimation quality with the stealthiness of the attacker are characterized.

Secure estimation is also linked to robust estimation (Shen et al. 2014; Sayed 2001; Al-Sayed et al. 2017; Chen et al. 2017; Lin and Abur 2020; Zhao et al. 2016). These two problems share some aspects (e.g., data model uncertainty), but their inference tasks are distinct. Specifically, besides the estimation objective, both problems also face the problem of resolving uncertainties about the data model. The main distinction between secure estimation and robust estimation lies in their resolution of the model uncertainties, which results in significant differences in the formulation of the problems and the designs of the optimal decision rules. Specifically, in robust estimation , the emphasis is laid on forming the most reliable estimates, and as an intermediate step, the model uncertainty must also be resolved as a second inference task. Resolution of model uncertainties can be executed by a wide range of approaches, which include averaging out the effect of the model or forming an estimate of the model. The ultimate objective of robust estimation is optimizing the estimation quality, and it generally does not account for the quality of the decisions involved in resolving model uncertainty, i.e., model uncertainty resolution will be dictated by the decision rules optimized for producing the best estimates.

The aforementioned studies that study secure estimation, despite their discrepancies, conform to an underlying design principle, which decouples the estimation design from all other decisions involved (e.g., attack detection or attacked sensor isolation), and leads to either detection -driven estimators or estimation -driven detection routines. The sub-optimality of decoupling such intertwined estimation and detection problems into independent estimation and detection routines is well-investigated (Middleton and Esposito 1968; Zeitouni et al. 1992; Moustakides et al. 2012; Jajamovich et al. 2012). In contrast, in secure estimation, our focus is on the qualities of both decisions: estimating the desired parameter and detecting the unknown model. Hence, unlike robust estimation, we face *combined estimation and detection* decisions. The problem formulation is motivated by our recent work in Sihag and Tajer (2020), which emphasizes the natural coupling between the two inference tasks and requires that the optimal decisions are determined jointly.

9.2 Data Model and Definitions

Our focus is on the estimation problem in (9.2) and in this context, we discuss the data models under the attack-free and adversarial scenarios.

9.2.1 *Attack Model*

The objective is to form an optimal estimate $\hat{X}(\mathbf{Y})$ (under the general cost functions specified in Sect. 9.2.2) in the potential presence of an adversary. In the attack-free setting, the data is assumed to be generated according to a known model specified in (9.1). In an adversarial setting, an adversary, depending on its strength and desired impact, can launch an attack with the ultimate purpose of degrading the quality of the estimate of X. We assume that the adversary can corrupt the data model of *up to* $K \in \{1, \dots, m\}$ coordinates of $\mathbf{Y}$. Hence, for a given K, there exist $T = \sum_{i=1}^{K} \binom{m}{i}$ number of attack scenarios, each of which is associated with a distinct data model. To formalize this, we define $\mathscr{S} \triangleq \{S_1, \dots, S_T\}$ as the set of all possible attack scenarios, where $S_i \subseteq \{1, \dots, m\}$ describes the set of coordinates of $\mathbf{Y}$ the models of which are compromised under attack scenario $i \in \{1, \dots, T\}$.

Under the attack scenario $i \in \{1, \dots, T\}$, if $r \in S_i$, the data model deviates from f to a model in the space $\mathscr{F}_i$. Clearly, the attack can be effective if it encompasses sufficiently distinct models. For our analysis, we assume that $\mathscr{F}_i \triangleq \{f_i(\cdot \mid X)\}$, i.e., $\mathscr{F}_i$ consists of one alternative distribution. Based on this model, when the data models in the coordinates contained in S_i are compromised, the joint distribution changes from $f(\cdot \mid X)$ to $f_i(\cdot \mid X)$.

In practice, the resources and preferences of the attacker may determine the likelihood of an attack scenario. For instance, attacking one coordinate may be easier or more desirable as compared to others. To account for such likelihoods, we adopt a Bayesian framework in which we define ε_0 as the prior probability of an attack-free scenario and define ε_i as the prior probability of the event that the attacker compromises the data at coordinates specified by S_i. A block diagram of the attack model and the inferential goals is depicted in Fig. 9.1.

9.2.2 *Decision Cost Functions*

In the adversarial setting, the estimation decision is intertwined with the decision on the true model, and therefore, it constantly faces the uncertainty induced by the action or inaction of the adversary. A decoupled strategy of decisions for isolating the model and estimating the parameter under the isolated model does not generally guarantee optimal performance. In fact, there exist extensive studies on formalizing and analyzing such compound decisions, which generally aim to decouple the

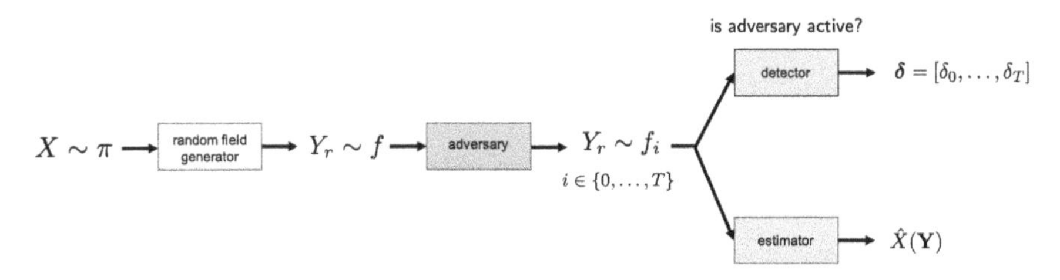

Fig. 9.1 The effect of the adversary on the data model, and the inferential decisions involved. Depending on the adversarial action, the data model may either deviate from f to one among the alternative data models $(\{f_i : i \in \{0, \ldots, T\}\})$ or retain the original data model (given by f_0)

inferential decisions. For instance, in Zeitouni et al. (1992), it is shown that the generalized likelihood ratio test (GLRT), which uses maximum likelihood estimates of unknown parameters in its decision rule, is not always optimal. In Moustakides et al. (2012) and Jajamovich et al. (2012), non-asymptotic frameworks for optimal joint detection and estimation are provided. Specifically, in Moustakides et al. (2012), a binary hypothesis testing problem is studied in a setting where one hypothesis is composite and consists of an unknown parameter to be estimated. In Jajamovich et al. (2012), the principles in Moustakides et al. (2012) are extended to a composite binary hypothesis testing problem in which both hypotheses correspond to composite models. We used similar principles as established in Moustakides et al. (2012) and Jajamovich et al. (2012) in our recent study on secure estimation in Sihag and Tajer (2020). We borrow the principles adopted in Sihag and Tajer (2020) to discuss secure estimation in the context of cyber-physical systems in this chapter. We next discuss the cost functions for true model detection and estimation quality.

9.2.2.1 Attack Detection Costs

Due to the existence of multiple attack scenarios, the true model detection problem can be formulated as the following $(T+1)$-composite hypothesis testing problem.

$$\begin{aligned} \mathsf{H}_0 &: \mathbf{Y} \sim f(\mathbf{Y} \mid X), \quad \text{with } X \sim \pi(X) \\ \mathsf{H}_i &: \mathbf{Y} \sim f_i(\mathbf{Y} \mid X), \quad \text{with } X \sim \pi(X)\,, \qquad \text{for } i \in \{1, \ldots, T\}, \end{aligned} \tag{9.3}$$

where H_0 is the hypothesis that represents the attack-free setting, and H_i is the hypothesis corresponding to an attack scenario where the attack is launched at the coordinates in $S_i \in \mathscr{S}$. For the convenience in notation, we denote the attack-free data model by $f_0(\cdot \mid X)$, i.e., $f_0(\cdot \mid X) = f(\cdot \mid X)$. To formalize relevant costs for the detection decisions, we define $\mathsf{D} \in \{\mathsf{H}_0, \ldots, \mathsf{H}_T\}$ as the decision on the hypothesis testing problem in (9.3), and $\mathsf{T} \in \{\mathsf{H}_0, \ldots, \mathsf{H}_T\}$ as the true hypothesis. The true hypothesis is discerned via a general *randomized* test $\boldsymbol{\delta}(\mathbf{Y}) \triangleq [\delta_0(\mathbf{Y}), \ldots, \delta_T(\mathbf{Y})]$, where $\delta_i(\mathbf{Y}) \in [0, 1]$ denotes the probability of deciding in favor of H_i. Clearly

$$\sum_{i=0}^{T} \delta_i(\mathbf{Y}) = 1. \tag{9.4}$$

Hence, the probability of forming a decision in favor of H_j while the true model is H_i is given by

$$\mathbb{P}(\mathsf{D}{=}\mathsf{H}_j\,|\,\mathsf{T}{=}\mathsf{H}_i) = \int_{\mathbf{Y}} \delta_j(\mathbf{Y}) f_i(\mathbf{Y})\,\mathrm{d}\mathbf{Y}. \tag{9.5}$$

We define P_{md} as the aggregate probability of error in identifying the true model when there exist compromised data coordinates due to attacker's activity, i.e.,

$$\begin{aligned}\mathsf{P}_{\mathsf{md}}(\boldsymbol{\delta}) &\triangleq \mathbb{P}(\mathsf{D} \neq \mathsf{T} \mid \mathsf{T} \neq \mathsf{H}_0) \\ &= \frac{1}{\mathbb{P}(\mathsf{T} \neq \mathsf{H}_0)} \sum_{i=1}^{T} \mathbb{P}(\mathsf{D} \neq \mathsf{H}_i \mid \mathsf{T} = \mathsf{H}_i)\mathbb{P}(\mathsf{T} = \mathsf{H}_i) \qquad (9.6)\\ &= \sum_{i=1}^{T} \frac{\varepsilon_i}{1-\varepsilon_0} \cdot \mathbb{P}(\mathsf{D} \neq \mathsf{H}_i \mid \mathsf{T} = \mathsf{H}_i). \qquad (9.7)\end{aligned}$$

Furthermore, we define P_{fa} as the aggregate probability of erroneously deciding that a set of coordinates is compromised while operating in an attack-free scenario. In this context, we have

$$\mathsf{P}_{\mathsf{fa}}(\boldsymbol{\delta}) \triangleq \mathbb{P}(\mathsf{D} \neq \mathsf{H}_0 \mid \mathsf{T} = \mathsf{H}_0) = \sum_{i=1}^{T} \mathbb{P}(\mathsf{D}{=}\mathsf{H}_i\,|\,\mathsf{T}{=}\mathsf{H}_0). \tag{9.8}$$

9.2.2.2 Secure Estimation Costs

In this subsection, we discuss the estimation cost functions that capture the quality of the estimate $\hat{X}(\mathbf{Y})$. For this purpose, we adopt a generic and *non-negative* cost function $\mathsf{C}(X, U(\mathbf{Y}))$ that quantifies the discrepancy between the ground truth X and a generic estimator $U(\mathbf{Y})$. Since the data models under different attack scenarios are distinct, we consider having possibly distinct estimators under each attack scenario. Therefore, we denote the estimate of X under model H_i by $\hat{X}_i(\mathbf{Y})$, and accordingly, we define

$$\hat{\mathbf{X}}(\mathbf{Y}) \triangleq [\hat{X}_0(\mathbf{Y}), \ldots, \hat{X}_T(\mathbf{Y})]. \tag{9.9}$$

Therefore, the estimation cost $\mathsf{C}(X, \hat{X}_i(\mathbf{Y}))$ is relevant only if the decision is H_i. Hence, for a generic estimator $U_i(\mathbf{Y})$ of X under model H_i, we define the *decision-specific average cost function* as

$$J_i(\delta_i, U_i(\mathbf{Y})) \triangleq \mathbb{E}_i[\mathsf{C}(X, U_i(\mathbf{Y})) \mid \mathsf{D} = \mathsf{H}_i] , \qquad \forall i \in \{0, \ldots, T\} \tag{9.10}$$

where the conditional expectation is with respect to X and $\mathbf{Y}$. Accordingly, we leverage (9.10) to define an aggregate average estimation cost according to

$$J(\boldsymbol{\delta}, \mathbf{U}) \triangleq \max_{i \in \{0,\ldots,T\}} J_i(\delta_i, U_i(\mathbf{Y})), \tag{9.11}$$

where we have $\mathbf{U} \triangleq [U_0(\mathbf{Y}), \ldots, U_T(\mathbf{Y})]$. Finally, in the attack-free scenario, corresponding to any generic estimator $V(\mathbf{Y})$, we define the average estimation according to

$$J_0(V) = \mathbb{E}[\mathsf{C}(X, V(\mathbf{Y}))], \tag{9.12}$$

where the expectation is with respect to X and $\mathbf{Y}$ under model f. Note that J_0 defined in (9.12) corresponds to the scenario in which the attack-free model f is the only possibility for the data model and is, therefore, fundamentally different from $J(\boldsymbol{\delta}, \mathbf{U})$ defined in (9.11). In the analysis, J_0 furnishes a baseline to assess the impact of potential adversarial action on the estimation quality.

9.3 Secure Parameter Estimation

In this section, we formalize the problem of secure estimation. There exists an inherent interplay between the quality of estimating X and the quality of isolation decision to identify the true model governing the data. On the one hand, detecting the adversary's attack model perfectly is not possible. At the same time, the estimation quality critically hinges on the successful isolation of the true data model. Therefore, an imperfection in the decision about the data model is expected to degrade the estimation quality with respect to the attack-free scenario. To quantify such an interplay as well as the degradation in estimation quality with respect to the attack-free scenario, we provide the following definition.

Definition 9.1 (*Estimation Degradation Factor*) For a given estimator V in the attack-free scenario, and a secure estimation framework specified by the rules $(\boldsymbol{\delta}, \mathbf{U})$ in the adversarial scenario, we define the estimation degradation factor (EDF) as

$$q(\boldsymbol{\delta}, \mathbf{U}, V) \triangleq \frac{J(\boldsymbol{\delta}, \mathbf{U})}{J_0(V)}. \tag{9.13}$$

Based on Definition 9.1, we define the performance region for secure estimation that encompasses all the pairs of estimation quality $q(\boldsymbol{\delta}, \mathbf{U}, V)$ and detection performance $\mathsf{P}_{\mathsf{md}}(\boldsymbol{\delta})$ over the space characterized by all possible decision rules $(\boldsymbol{\delta}, \mathbf{U}, V)$.

Definition 9.2 (*Performance Region*) We define the performance region as the region of all simultaneously achievable estimation quality $q(\boldsymbol{\delta}, \mathbf{U}, V)$ and detection performance $\mathsf{P}_{\mathsf{md}}(\boldsymbol{\delta})$.

Next, we leverage the definition of performance region to define the notion of (q, β)-security , which is instrumental for formalizing the secure estimation problem. For this purpose, we first note that the two estimation cost functions involved in the EDF $q(\boldsymbol{\delta}, \mathbf{U}, V)$ can be computed independently, and as a result, their attendant decision rules can be determined independently. For this purpose, we define V^* as the optimal estimator under the attack-free scenario, and J_0^* as the corresponding estimation cost, i.e.,

$$V^* \triangleq \arg\min_{V} J_0(V), \qquad \text{and} \qquad J_0^* \triangleq \min_{V} J_0(V). \tag{9.14}$$

Definition 9.3 (*(q, β)-security*) In the adversarial scenario, an estimation procedure specified by $(\boldsymbol{\delta}, \mathbf{U}, V^*)$ is called (q, β)-secure if the decision rules $(\boldsymbol{\delta}, \mathbf{U})$ yield the minimal EDF among all the decision rules corresponding to which the average rate of missing the attacks does not exceed $\beta \in (0, 1]$, i.e.,

$$q \triangleq \min_{\boldsymbol{\delta}, \mathbf{U}} q(\boldsymbol{\delta}, \mathbf{U}, V^*), \qquad \text{s.t.} \qquad \mathsf{P}_{\mathsf{md}}(\boldsymbol{\delta}) \leq \beta. \tag{9.15}$$

The performance region, and its boundary that specifies the interplay between q and β are illustrated figuratively in Fig. 9.2. Based on the definitions in this subsection, we aim to characterize the region of all simultaneously achievable values of $q(\boldsymbol{\delta}, \mathbf{U}, V^*)$ and $\mathsf{P}_{\mathsf{md}}(\boldsymbol{\delta})$ (represented by the dashed region in Fig. 9.2) and the (q, β)-secure decision rules that solve (9.15), and specify the boundary of the performance region (illustrated by a solid line as the boundary of the performance region in Fig. 9.2).

By noting that $q(\boldsymbol{\delta}, \mathbf{U}, V^*) = \frac{J(\boldsymbol{\delta}, \mathbf{U})}{J_0^*}$, where J_0^* is a constant, we formalize the problem of determining the performance region and the (q, β)-secure decision rules as

$$\mathcal{Q}(\beta) \triangleq \begin{cases} \min_{\boldsymbol{\delta}, \mathbf{U}} & J(\boldsymbol{\delta}, \mathbf{U}) \\ \text{s.t.} & \mathsf{P}_{\mathsf{md}}(\boldsymbol{\delta}) \leq \beta \end{cases}. \tag{9.16}$$

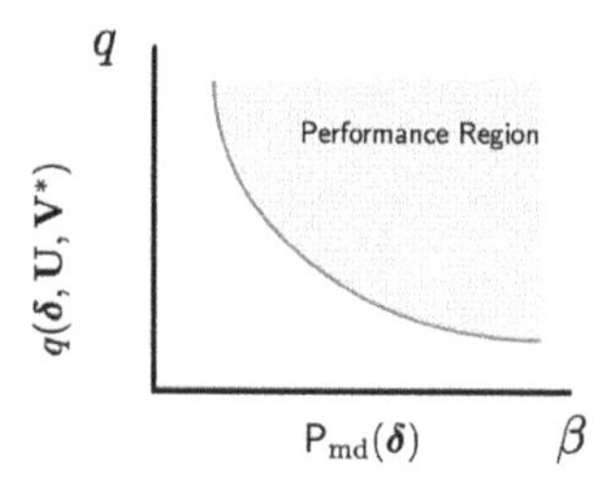

Fig. 9.2 Performance region

We note that although $\mathscr{Q}(\beta)$ ensures that the likelihood of missing an attack is confined below β, it is insensitive to the rate of the false alarms, that is, the rate of erroneously declaring an attack when there is no attack. If it is also desirable to control the rate of false alarms, we can further extend the notion of (q, β)-security as follows.

Definition 9.4 An estimation procedure is (q, α, β)-secure if it is (q, β)-secure and the likelihood of false alarms does not exceed $\alpha \in (0, 1]$.

The (q, α, β)-secure decisions are determined by the optimal decision rules that form the solution to

$$\mathscr{P}(\alpha, \beta) = \begin{cases} \min_{\boldsymbol{\delta}, \mathbf{U}} & J(\boldsymbol{\delta}, \mathbf{U}) \\ \text{s.t.} & \mathsf{P}_{\mathsf{md}}(\boldsymbol{\delta}) \leq \beta \\ & \mathsf{P}_{\mathsf{fa}}(\boldsymbol{\delta}) \leq \alpha \end{cases}. \tag{9.17}$$

Remark 9.1 It is straightforward to verify that $\mathscr{Q}(\beta) = \mathscr{P}(1, \beta)$.

Remark 9.2 (*Feasibility*) The Neyman–Pearson theory (Poor 1998) dictates that the probabilities $\mathsf{P}_{\mathsf{md}}(\boldsymbol{\delta})$ and $\mathsf{P}_{\mathsf{fa}}(\boldsymbol{\delta})$ cannot be made arbitrarily small simultaneously. Specifically, for any given α, there exists a smallest feasible value for β, denoted by $\beta^*(\alpha)$.

We provide the optimal solution to problems $\mathscr{P}(\alpha, \beta)$ and $\mathscr{Q}(\beta)$ in closed-forms in Sect. 9.4.

9.4 Secure Parameter Estimation: Optimal Decision Rules

In this section, we characterize an optimal solution to the general problem $\mathscr{P}(\alpha, \beta)$ to determine the designs for the estimators $\{\hat{X}_i(\mathbf{Y}) : i \in \{0, \ldots, T\}\}$ and the detectors $\{\delta_i(\mathbf{Y}) : i \in \{0, \ldots, T\}\}$. We first leverage the expansions of the error probability terms $\mathsf{P}_{\mathsf{md}}(\boldsymbol{\delta})$ and $\mathsf{P}_{\mathsf{fa}}(\boldsymbol{\delta})$ in terms of the data models and decision rules. Based on (9.5) and (9.6), we have

$$\mathsf{P}_{\mathsf{md}}(\boldsymbol{\delta}) = \sum_{i=1}^{T} \frac{\varepsilon_i}{1 - \varepsilon_0} \sum_{\substack{j=0 \\ j \neq i}}^{T} \int_{\mathbf{Y}} \delta_j(\mathbf{Y}) f_i(\mathbf{Y}) \, \mathrm{d}\mathbf{Y}. \tag{9.18}$$

Similarly, by noting (9.5) and based on (9.8), we have

$$\mathsf{P}_{\mathsf{fa}}(\boldsymbol{\delta}) = \sum_{i=1}^{T} \int_{\mathbf{Y}} \delta_i(\mathbf{Y}) f_0(\mathbf{Y}) \, \mathrm{d}\mathbf{Y}. \tag{9.19}$$

By using the expansions in (9.18) and (9.19), the equivalent problem to (9.17) is given by

$$\mathscr{P}(\alpha,\beta)=\begin{cases}\min_{(\boldsymbol{\delta},U)} & J(\boldsymbol{\delta},\mathbf{U})\\ \text{s.t.} & \sum_{i=1}^{T}\frac{\varepsilon_i}{1-\varepsilon_0}\sum_{\substack{j=0\\ j\neq i}}^{T}\int_{\mathbf{Y}}\delta_j(\mathbf{Y})f_i(\mathbf{Y})\,\mathrm{d}\mathbf{Y}\leq\beta\\ & \sum_{i=1}^{T}\int_{\mathbf{Y}}\delta_i(\mathbf{Y})f_0(\mathbf{Y})\,\mathrm{d}\mathbf{Y}\leq\alpha\end{cases}. \tag{9.20}$$

Note that the estimators $\{U_i(\mathbf{Y}) : i \in \{0,\ldots,T\}\}$ are restricted to the utility function $J(\boldsymbol{\delta},\mathbf{U})$, which allows us to decouple the problem $\mathscr{P}(\alpha,\beta)$ into two sub-problems, formalized next.

Theorem 9.1 *The optimal secure estimators of X under different models, i.e., $\hat{\mathbf{X}} = [\hat{X}_0,\ldots,\hat{X}_T]$ are the solutions to*

$$\hat{\mathbf{X}} = \arg\min_{\mathbf{U}} J(\boldsymbol{\delta},\mathbf{U}). \tag{9.21}$$

Furthermore, the solution of $\mathscr{P}(\alpha,\beta)$, and subsequently the design of the attack detectors, can be found by equivalently solving

$$\mathscr{P}(\alpha,\beta)=\begin{cases}\min_{\boldsymbol{\delta}} & J(\boldsymbol{\delta},\hat{\mathbf{X}})\\ s.t. & \sum_{i=1}^{T}\frac{\varepsilon_i}{1-\varepsilon_0}\sum_{\substack{j=0\\ j\neq i}}^{T}\int_{\mathbf{Y}}\delta_j(\mathbf{Y})f_i(\mathbf{Y})\,\mathrm{d}\mathbf{Y}\leq\beta\\ & \sum_{i=1}^{T}\int_{\mathbf{Y}}\delta_i(\mathbf{Y})f_0(\mathbf{Y})\,\mathrm{d}\mathbf{Y}\leq\alpha\end{cases}. \tag{9.22}$$

By leveraging the design in (9.21) and the decoupled structure of the problem $\mathscr{P}(\alpha,\beta)$ in (9.22), in the following theorem, we discuss optimal designs for the estimators in the secure estimation problem.

Theorem 9.2 ((q,α,β)-secure Estimators) *For the optimal secure estimators $\hat{\mathbf{X}}$, we have:*

1. *The minimizer of the estimation cost $J_i(\delta_i, U_i(\mathbf{Y}))$, i.e., the estimation cost function under model H_i, is given by*

$$U_i^*(\mathbf{Y}) \triangleq \arg\inf_{U_i(\mathbf{Y})} \mathsf{C}_{\mathrm{p},i}(U_i(\mathbf{Y}) \mid \mathbf{Y}), \tag{9.23}$$

where $\mathsf{C}_{\mathrm{p},i}(U(\mathbf{Y}) \mid \mathbf{Y})$ is the average posterior cost function denoted by

$$\mathsf{C}_{\mathrm{p},i}(U(\mathbf{Y}) \mid \mathbf{Y}) \triangleq \mathbb{E}_i\left[\mathsf{C}(X, U(\mathbf{Y})) \mid \mathbf{Y}\right], \tag{9.24}$$

where the conditional expectation in (9.24) *is with respect to X under model H_i.*

2. *The optimal estimator* $\hat{\mathbf{X}} = [\hat{X}_0, \ldots, \hat{X}_T]$, *specified in* (9.21), *is given by*

$$\hat{X}_i(\mathbf{Y}) = U_i^*(\mathbf{Y}). \tag{9.25}$$

3. *The cost function* $J(\boldsymbol{\delta}, \hat{\mathbf{X}})$ *is given by*

$$J(\boldsymbol{\delta}, \hat{\mathbf{X}}) = \max_{i\in\{0,\ldots,T\}} \left\{ \frac{\int_{\mathbf{Y}} \delta_i(\mathbf{Y}) \mathsf{C}_{\mathrm{p},i}^*(\mathbf{Y}) f_i(\mathbf{Y}) d\mathbf{Y}}{\int_{\mathbf{Y}} \delta_i(\mathbf{Y}) f_i(\mathbf{Y}) d\mathbf{Y}} \right\}, \tag{9.26}$$

where we have defined

$$\mathsf{C}_{\mathrm{p},i}^*(\mathbf{Y}) \triangleq \inf_{U_i(\mathbf{Y})} \mathsf{C}_{\mathrm{p},i}(U_i(\mathbf{Y}) \mid \mathbf{Y}). \tag{9.27}$$

Proof See Appendix 1. ■

We next discuss the application of decision rules in Theorem 9.2 in a specific example. Specifically, in the next corollary, we discuss the closed-forms of these decision rules when the distributions $\{f_i(\cdot \mid X) : i \in \{0, \ldots, T\}\}$ are Gaussian.

Corollary 9.1 ((q, α, β)-secure Estimators in Gaussian Models) *When the data models are Gaussian, i.e.,*

$$f_i(\cdot \mid X) \sim \mathcal{N}(\theta_i, X), \qquad \textit{for} \;\; \theta_i \in \mathbb{R} \tag{9.28}$$

such that the mean values are distinct, and

$$X \sim \mathscr{X}^{-1}(\zeta, \phi), \tag{9.29}$$

where $\mathscr{X}^{-1}(\zeta, \phi)$ *denotes the inverse chi-squared distribution with parameters* ζ *and* ϕ, *such that* $\zeta + n > 4$, *and the cost* $\mathsf{C}(X, U(\mathbf{Y}))$ *is the mean squared error, given by*

$$\mathsf{C}(X, U(\mathbf{Y})) = \|X - U(Y)\|^2, \tag{9.30}$$

for the optimal secure estimators $\hat{\mathbf{X}}$, *we have:*

1. *The minimizer of the estimation cost* $J(\delta_i, U_i(\mathbf{Y}))$, *i.e., the estimation cost function under model* H_i, *is given by*

$$U_i^*(\mathbf{Y}) = \frac{\zeta\phi + \sum_{r=1}^{n} \|Y_r - \theta_i\|_2^2}{\zeta + n - 2}. \tag{9.31}$$

2. *The optimal estimator* $\hat{\mathbf{X}} = [\hat{X}_0, \ldots, \hat{X}_T]$, *specified in* (9.21), *is given by*

$$\hat{X}_i(\mathbf{Y}) = U_i^*(\mathbf{Y}). \tag{9.32}$$

3. *The cost function* $J(\boldsymbol{\delta}, \hat{\mathbf{X}})$ *is given by*

$$J(\boldsymbol{\delta}, \hat{\mathbf{X}}) = \max_{i \in \{0,\ldots,T\}} \left\{ \frac{\displaystyle\int_{\mathbf{Y}} \delta_i(\mathbf{Y}) \mathsf{C}^*_{\mathrm{p},i}(\mathbf{Y}) f_i(\mathbf{Y}) d\mathbf{Y}}{\displaystyle\int_{\mathbf{Y}} \delta_i(\mathbf{Y}) f_i(\mathbf{Y}) d\mathbf{Y}} \right\}, \tag{9.33}$$

where we have

$$\mathsf{C}^*_{\mathrm{p},i}(\mathbf{Y}) = \frac{2(\zeta\phi + \sum_{r=1}^{n} \|Y_r - \theta_1\|^2)^2}{(\zeta_i + n - 2)^2(\zeta + n - 4)}. \tag{9.34}$$

Next, given the optimal estimators $\hat{\mathbf{X}}$, we provide the optimal detection rules in the next theorem. We note that the decision rules depend on the metrics computed based on the optimal estimation costs, establishing the coupling of estimation and true model detection decisions. We show that by using the solution of the specific auxiliary convex problem in a variational form in the next theorem, we can solve $\mathscr{P}(\alpha, \beta)$ in (9.22).

Theorem 9.3 *For any arbitrary* $u \in \mathbb{R}_+$, *we have* $\mathscr{P}(\alpha, \beta) \leq u$ *if and only if* $\mathbb{R}(\alpha, \beta, u) \leq 0$, *where we have defined*

$$\mathbb{R}(\alpha, \beta, u) = \begin{cases} \min_{\delta} & \eta \\ \textit{s.t.} & \displaystyle\int_{\mathbf{Y}} \delta_i(\mathbf{Y}) f_i(\mathbf{Y}) [\mathsf{C}^*_{\mathrm{p},i}(\mathbf{Y}) - u] \, \mathrm{d}\mathbf{Y} \leq \eta, \quad \forall\, i \in \{0, \ldots, T\} \\ & \displaystyle\sum_{i=1}^{T} \frac{\varepsilon_i}{1 - \varepsilon_0} \sum_{\substack{j=0 \\ j \neq i}}^{T} \int_{\mathbf{Y}} \delta_j(\mathbf{Y}) f_i(\mathbf{Y}) \, \mathrm{d}\mathbf{Y} \ \leq \beta + \eta \\ & \displaystyle\sum_{i=1}^{T} \int_{\mathbf{Y}} \delta_i(\mathbf{Y}) f_0(\mathbf{Y}) \, \mathrm{d}\mathbf{Y} \leq \alpha + \eta \end{cases}. \tag{9.35}$$

Furthermore, $\mathbb{R}(\alpha, \beta, u)$ *is convex, and* $\mathbb{R}(\alpha, \beta, u) = 0$ *has a unique solution in* u, *which we denote by* u^*.

Proof See Appendix 2. ∎

The point u^* plays a pivotal role in the structure of optimal detection decision rules. We define the constants $\{\ell_i : i \in \{0, \ldots, T+2\}\}$ as the dual variables in the Lagrange function for the convex problem $\mathbb{R}(\alpha, \beta, u^*)$. Given u^* and $\{\ell_i : i \in$

$\{0, \ldots, T+2\}\}$, we can characterize the optimal detection rules in closed-forms, as specified in the following theorem.

Theorem 9.4 ((q, α, β)-secure Detection Rules) *The optimal decision rules for isolating the compromised coordinates are given by*

$$\delta_i(\mathbf{Y}) = \begin{cases} 1, & \text{if } i = i^* \\ 0, & \text{if } i \neq i^* \end{cases}, \tag{9.36}$$

where we have defined

$$i^* \triangleq \underset{i \in \{0,\ldots,T\}}{\operatorname{argmin}} \ A_i \ . \tag{9.37}$$

Constants $\{A_0, \ldots, A_T\}$ *are specified by the data models,* u^**, and its associated Langrangian multipliers* $\{\ell_i : i \in \{0, \ldots, T+2\}\}$*. Specifically, we have*

$$A_0 \triangleq \ell_0 f_0(\mathbf{Y})[\mathsf{C}^*_{\mathrm{p},0}(\mathbf{Y}) - u^*] + \ell_{T+1} \sum_{i=1}^{T} \frac{\varepsilon_i}{1-\varepsilon_0} f_i(\mathbf{Y}), \tag{9.38}$$

and for $i \in \{1, \ldots, T\}$*, we have*

$$A_i \triangleq \ell_i f_i(\mathbf{Y})[\mathsf{C}^*_{\mathrm{p},i}(\mathbf{Y}) - u^*] + \ell_{T+1} \sum_{j=1, j\neq i}^{T} \frac{\varepsilon_j}{1-\varepsilon_0} f_j(\mathbf{Y}) + \ell_{T+2} f_0(\mathbf{Y}). \tag{9.39}$$

Proof See Appendix 3. ■

In the next corollary, we discuss the application of these decision rules when the distributions $\{f_i(\cdot \mid X) : i \in \{0, \ldots, T\}\}$ are all Gaussian.

Corollary 9.2 ((q, α, β)-secure Detection Rules in Gaussian Models) *When the data models* $\{f_i(\cdot \mid X) : i \in \{0, \ldots, T\}\}$ *have the following Gaussian distributions*

$$f_i(\cdot \mid X) \sim \mathcal{N}(\theta_i, X) \ , \quad \textit{for} \ \ \theta_i \in \mathbb{R} \tag{9.40}$$

where the mean values are distinct, and

$$X \sim \mathcal{X}^{-1}(\zeta, \phi), \tag{9.41}$$

the optimal decision rules for isolating the compromised coordinates are given by

$$\delta_i(\mathbf{Y}) = \begin{cases} 1, & \text{if } i = i^* \\ 0, & \text{if } i \neq i^* \end{cases}, \tag{9.42}$$

where we have defined

$$i^* \triangleq \underset{i\in\{0,\dots,T\}}{\text{argmin}}\ A_i. \tag{9.43}$$

Constants $\{A_0,\dots,A_T\}$ *are specified by the data models,* u^**, and its associated Langrangian multipliers* $\{\ell_i : i \in \{0,\dots,T+2\}\}$*. Specifically, we have*

$$A_0 \triangleq \ell_0 f_0(\mathbf{Y})(\mathsf{C}^*_{\text{p},0}(\mathbf{Y}) - u^*) + \ell_{T+1}\sum_{i=1}^{T}\frac{\varepsilon_i}{1-\varepsilon_0} f_i(\mathbf{Y}), \tag{9.44}$$

and for $i \in \{1,\dots,T\}$*, we have*

$$A_i \triangleq \ell_i f_i(\mathbf{Y})(\mathsf{C}^*_{\text{p},i}(\mathbf{Y}) - u^*) + \ell_{T+1}\sum_{\substack{j=1\\ j\neq i}}^{T}\frac{\varepsilon_j}{1-\varepsilon_0} f_j(\mathbf{Y}) + \ell_{T+2} f_0(\mathbf{Y}). \tag{9.45}$$

When the cost function $\mathsf{C}(X, U(\mathbf{Y}))$ *is the mean squared error cost, and* $\mathsf{C}^*_{\text{p},i}(\mathbf{Y})$ *is evaluated using* (9.34)*, we obtain*

$$f_i(\mathbf{Y}) = \frac{(\zeta\phi)^{\frac{\zeta}{2}}}{\pi^{\frac{n}{2}}\Gamma(\zeta/2)} \cdot \frac{\Gamma(\zeta+n)/2}{(\zeta\phi + \sum_{r=1}^{n}\|Y_r - \theta_i\|^2)^{\frac{\zeta+n}{2}}}. \tag{9.46}$$

Figure 9.3 illustrates the performance region and the corresponding (q,β)-security curve for the case $T=1$, $n=1$, $\theta_0=0$, $\theta_1=2$, $\zeta=4$, and $\phi=1$. The (q,β)-security curve in Fig. 9.3 depicts the tradeoff between the quality of the true model detection and the degradation in the estimation quality. Note that this tradeoff is inherently due to secure estimation problem formulation. Essentially, the design of the problem $\mathscr{P}(\alpha,\beta)$ as specified in (9.17) enables the trade of the quality of detection in favor of improving the estimation cost.

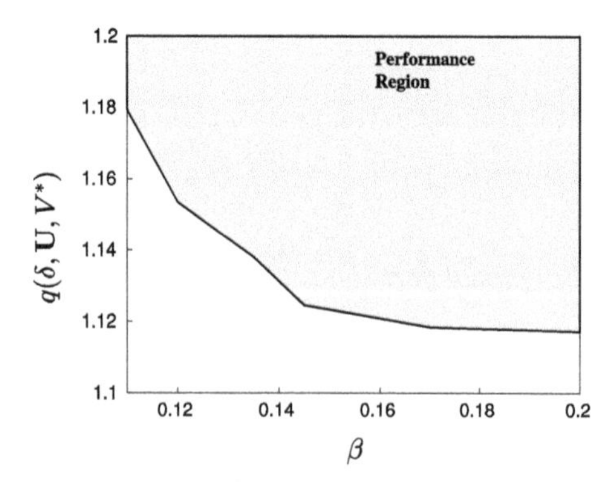

Fig. 9.3 Performance region for the Gaussian data model

We provide Algorithm 9.1, which summarizes all the steps for solving $\mathscr{P}(\alpha, \beta)$ for any feasible pair of α and β, and it encapsulates the decision rules specified by the theorems in this section and the detailed steps of specifying the parameters involved in characterizing the decision rules.

Algorithm 9.1 – Solving $\mathscr{P}(\alpha, \beta)$

Input: α and β and evaluate $\beta^*(\alpha)$
if $\beta < \beta^*(\alpha)$ **then**
  $\mathscr{P}(\alpha, \beta)$ not feasible for given choice of α and β;
  break;
else
  Initialize $u_0 = 0, u_1$;
  Evaluate optimal posterior estimation costs in (9.27);
  repeat
    $\hat{u} \leftarrow (u_0 + u_1)/2$;
    for *every* $\hat{\boldsymbol{\ell}} \succcurlyeq 0$ *in the discretized space* $\|\hat{\boldsymbol{\ell}}\|_1 = 1$ **do**
      Compute $\boldsymbol{\delta}$ from Theorem 9.4;
      Compute $M(\hat{\boldsymbol{\ell}}) \triangleq \mathbb{R}(\alpha, \beta, \hat{u})$;
      if $\min_{\hat{\boldsymbol{\ell}}} M(\hat{\boldsymbol{\ell}}) \leq 0$ **then**
        $u_1 \leftarrow \hat{u}, \quad \boldsymbol{\ell} \leftarrow \hat{\boldsymbol{\ell}}$;
      else
        $u_0 \leftarrow \hat{u}$;
  until $u_1 - u_0 \leq \varepsilon$, *for* ε *sufficiently small*;
  $\mathscr{P}(\alpha, \beta) \leftarrow u^* = u_1$;
  Output: Decision rules $\boldsymbol{\delta}$

9.5 Case Studies: Secure Estimation in Sensor Networks

We evaluate the secure estimation framework using the example of a two-sensor network with a fusion center (FC). Each sensor collects a stream of data consisting of n samples. Sensor $i \in \{1, 2\}$ collects n measurements, denoted by $\mathbf{Y}_i = [Y_1^i, \dots, Y_n^i]$, where each sample $Y_j^i \in \mathbb{R}$ in an attack-free scenario follows the model

$$Y_j^i = h^i X + N_j^i, \tag{9.47}$$

where h^i models the channel connecting sensor i to the FC and N_j^i accounts for the additive channel noise. Different noise terms are assumed to be independent and identically distributed (i.i.d.) generated according to a known distribution. We will consider two adversarial scenarios that impact the data model in (9.47) and evaluate the optimal performance.

9.5.1 Case 1: One Sensor Vulnerable to Causative Attacks

We first consider an adversarial setting in which the data model from only one sensor (sensor 1) is vulnerable to an adversarial attack while the other sensor (sensor 2) is secured. Under this setting, we clearly have only one attack scenario, i.e., $T = 1$ and $S_1 = \{1\}$. Accordingly, we have $\varepsilon_0 + \varepsilon_1 = 1$. Under the attack-free scenario, the noise terms N_j^i are distributed according to $\mathcal{N}(0, \sigma_n^2)$, i.e.,

$$Y_j^i \mid X \sim \mathcal{N}(h^i X, \sigma_n^2). \tag{9.48}$$

When sensor 1 is compromised, the actual conditional distribution of $Y_j^1|X$ is distinct from the above distribution. The inference objective under such a setting, in principle, becomes similar to the adversarial setting of Wilson and Veeravalli (2016), which focuses on data injection attack. Hence, for comparison with the performance of the secure estimation framework with that of Wilson and Veeravalli (2016), we assume that the conditional distribution of $Y_j^1|X$ when sensor 1 is under attack is $\mathcal{N}(h^i X, \sigma_n^2) * \mathsf{Unif}[a, b]$, where $a, b \in \mathbb{R}$ are fixed constants and $*$ denotes convolution. Therefore, the composite hypothesis test for estimating X and discerning the model in (9.3) simplifies to a binary test with the prior probabilities ε_0 and ε_1.

$$\begin{array}{ll} \mathsf{H}_0 : \mathbf{Y} \sim f_0(\mathbf{Y} \mid X), & \text{with } X \sim \mathcal{N}(0, \sigma^2) \\ \mathsf{H}_1 : \mathbf{Y} \sim f_1(\mathbf{Y} \mid X), & \text{with } X \sim \mathcal{N}(0, \sigma^2). \end{array} \tag{9.49}$$

Figure 9.4 shows the variations of the estimation quality, captured by q, versus the miss-detection rate β, where it is observed that the estimation quality improves monotonically with an increase in β, and it reaches its maximum quality as β approaches 1. This observation is in line with the analytic implications of the formulations of the secure parameter estimation problem in (9.16) and (9.17). A similar setting is studied in Wilson and Veeravalli (2016), where the attack is induced additively into the data of sensor 1 and can be any real number. This setting can be studied in the context of adversarial attacks where the attacker compromises the data by adding a uniformly distributed disturbance. Figure 9.4 also shows the comparison of the estimation quality of the secure estimation framework in this chapter, with that from the methodology in Wilson and Veeravalli (2016). In Wilson and Veeravalli (2016), the estimator is designed to obtain the most robust estimate corresponding to an optimal false alarm probability α^*, which, in turn, fixes the miss-detection error probability. Therefore, the framework in Wilson and Veeravalli (2016) does not provide the flexibility to change the miss-detection rate β.

The results presented in Fig. 9.4 correspond to $\sigma = 3$, $\sigma_n = 1$, $h^1 = 1$, $h^2 = 4$, $a = -40$, $b = 40$. The upper bound on P_{fa} is set to $\alpha^* = 0.1$, where α^* is obtained using the methodology in Wilson and Veeravalli (2016).

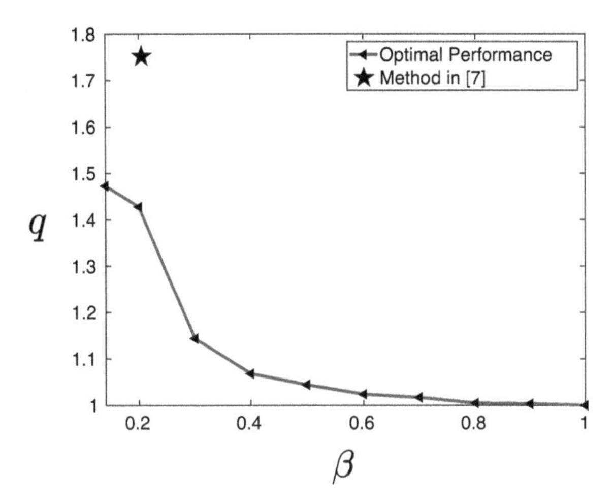

Fig. 9.4 q versus β for fixed $\alpha^* = 0.1$

9.5.2 *Case 2: Both Sensors Vulnerable to Adversarial Attacks*

We consider the same model for X, and in this setting, we assume that both sensors are vulnerable to attack. The attacker can compromise the data of at most one sensor. Under this setting, we have $T = 2$, $S_1 = \{1\}$, and $S_2 = \{2\}$. Therefore, in the adversarial setting, the following hypothesis model forms the basis of the secure estimation problem

$$\begin{array}{ll} \mathsf{H}_0 : \mathbf{Y} \sim f_0(\mathbf{Y} \mid X), & \text{with } X \sim \pi(X) \\ \mathsf{H}_1 : \mathbf{Y} \sim f_1(\mathbf{Y} \mid X), & \text{with } X \sim \pi(X) \\ \mathsf{H}_2 : \mathbf{Y} \sim f_2(\mathbf{Y} \mid X), & \text{with } X \sim \pi(X), \end{array} \tag{9.50}$$

where H_0 is the attack-free setting and H_i corresponds to sensor i being compromised. Since the sensor with higher gain h^i is expected to provide a better estimate, we explore a scenario in which the sensor with the higher gain is more likely to be attacked. Hence, we select the parameters $h^1 = 1$, and $h^2 = 2$, and set the probabilities $(\varepsilon_0, \varepsilon_1, \varepsilon_2) = (0.2, 0.2, 0.6)$. We assume the distribution of X to be $\mathsf{Unif}[-2, 2]$. We assume that Y_j^i, for $i \in \{1, 2\}$, given X, is distributed according to $\mathcal{N}(h^i X, 1)$ in the attack-free setting. When sensor i is compromised, we assume that Y_j^i, for $i \in \{1, 2\}$, given X, follows the distribution $\mathcal{N}(h^i X, 5)$.

Figure 9.5 shows the performance region illustrated in Fig. 9.2, which corresponds to the variations of q with β for three different values of α. The region spanned by the plots between q and β for different values of α is the feasible region of operation and allows the FC to adjust the emphasis on either the estimation or detection decisions. As expected, the estimation quality improves monotonically as α and β increase.

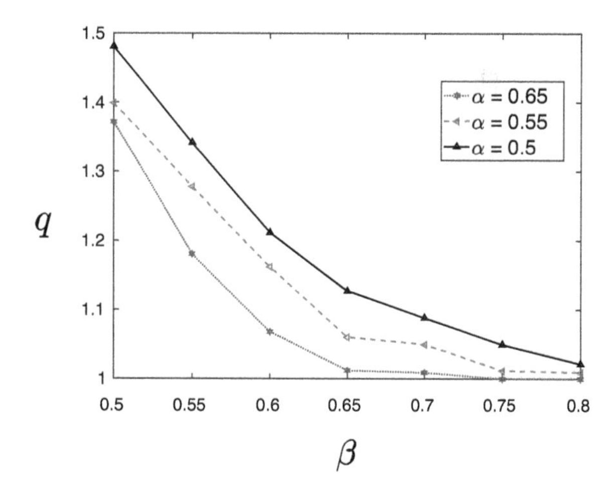

Fig. 9.5 q versus β for different values of α

9.6 Conclusions

We have formalized and analyzed the problem of secure estimation under adversarial attacks on the data model. The possible presence of adversaries results in uncertainty in the statistical model of the data. This further leads the estimation algorithm to exhibit degraded performance compared to the attack-free setting. We have characterized closed-form optimal decision rules that provide the optimal estimation quality (minimum estimation cost) while controlling for the error in detecting the attack and isolating the true model of the data. Our analysis has shown that the design of optimal estimators is intertwined with that of the detection rules to determine the true model of the data. Based on this, we have provided the optimal decision rules that combine the estimation quality with detection power. This allows the decision-maker to place any desired emphasis on the estimation and detection routines involved to study the tradeoff between the two.

Appendix 1

We start the proof of Theorem 9.2 by defining the cost function $J_i(\delta_i, U_i)$ and analyzing a lower bound on it. Our analysis will show that the lower bound on the $J_i(\delta_i, U_i)$ is achieved for the choice of estimator in (9.53). From (9.10), we have

$$J_i(\delta_i, U_i) = \mathbb{E}\left[\mathsf{C}(X, U_i(\mathbf{Y})) \,|\, \mathsf{D}{=}\mathsf{H}_i\right]$$
$$= \frac{\int_{\mathbf{Y}} \int_{\mathbf{X}} \delta_i(\mathbf{Y}) \mathsf{C}(X, U_i(\mathbf{Y})) f_i(\mathbf{Y} \,|\, X) \pi(X) \mathrm{d}X \mathrm{d}\mathbf{Y}}{\int_{\mathbf{Y}} \delta_i(\mathbf{Y}) f_i(\mathbf{Y}) \mathrm{d}\mathbf{Y}} .$$

By leveraging the definition of $\mathsf{C}_{\mathrm{p},i}(U_i(\mathbf{Y}) \mid \mathbf{Y})$ from (9.24), we have

$$J_i(\delta_i, U_i) = \frac{\int_{\mathbf{Y}} \delta_i(\mathbf{Y}) \mathsf{C}_{\mathrm{p},i}(U_i(\mathbf{Y}) \mid \mathbf{Y}) f_i(\mathbf{Y}) \mathrm{d}\mathbf{Y}}{\int_{\mathbf{Y}} \delta_i(\mathbf{Y}) f_i(\mathbf{Y}) \mathrm{d}\mathbf{Y}}$$
$$\geq \frac{\int_{\mathbf{Y}} \delta_i(\mathbf{Y}) \inf_{U_i(\mathbf{Y})} \mathsf{C}_{\mathrm{p},i}(U_i(\mathbf{Y}) \mid \mathbf{Y}) f_i(\mathbf{Y}) \mathrm{d}\mathbf{Y}}{\int_{\mathbf{Y}} \delta_i(\mathbf{Y}) f_i(\mathbf{Y}) \mathrm{d}\mathbf{Y}} , \tag{9.51}$$

which implies that

$$J_i(\delta_i, U_i) \geq \frac{\int_{\mathbf{Y}} \delta_i(\mathbf{Y}) \mathsf{C}^*_{\mathrm{p},i}(\mathbf{Y}) f_i(\mathbf{Y}) \mathrm{d}\mathbf{Y}}{\int_{\mathbf{Y}} \delta_i(\mathbf{Y}) f_i(\mathbf{Y}) \mathrm{d}\mathbf{Y}} . \tag{9.52}$$

Using the definition of $\hat{X}_i(\mathbf{Y})$ in (9.23), the above lower bound is achieved when the estimator $U_i(\mathbf{Y})$ is selected to be

$$\hat{X}_i(\mathbf{Y}) = \arg \inf_{U_i(\mathbf{Y})} \mathsf{C}_{\mathrm{p},i}(U_i(\mathbf{Y}) \mid \mathbf{Y}), \tag{9.53}$$

which proves that the estimator in (9.23) is the optimal estimator for minimizing the cost $J_i(\delta_i, U_i)$. The corresponding minimum average estimation cost is

$$J_i(\delta_i, \hat{X}_i) = \frac{\int_{\mathbf{Y}} \delta_i(\mathbf{Y}) \mathsf{C}^*_{\mathrm{p},i}(\mathbf{Y}) f_i(\mathbf{Y}) \mathrm{d}\mathbf{Y}}{\int_{\mathbf{Y}} \delta_i(\mathbf{Y}) f_i(\mathbf{Y}) \mathrm{d}\mathbf{Y}} . \tag{9.54}$$

Next, we prove that

$$\max_i \min_{\mathbf{U}} \{J_i(\delta_i, U_i)\} \equiv \min_{\mathbf{U}} \max_i \{J_i(\delta_i, U_i)\} . \tag{9.55}$$

Recall from (9.11), the estimation cost $J(\boldsymbol{\delta}, \mathbf{U})$ is defined as

$$J(\boldsymbol{\delta}, \mathbf{U}) = \max_{i} \{J_i(\delta_i, U_i)\}. \quad (9.56)$$

We define $\mathscr{C}(\boldsymbol{\Omega}, \boldsymbol{\delta}, \mathbf{U})$ as a convex function of $J_i(\delta_i, U_i), i \in \{0, \ldots, T\}$, given by

$$\mathscr{C}(\boldsymbol{\Omega}, \boldsymbol{\delta}, \mathbf{U}) \triangleq \sum_{i=0}^{T} \Omega_i J_i(\delta_i, U_i), \quad (9.57)$$

where $\boldsymbol{\Omega} = [\Omega_0, \ldots, \Omega_T]$, and Ω_i satisfy

$$\sum_{i=0}^{T} \Omega_i = 1 \ , \ \text{and } \Omega_i \in [0, 1]. \quad (9.58)$$

$J(\boldsymbol{\delta}, \mathbf{U})$ can be represented as a function of $\mathscr{C}(\boldsymbol{\Omega}, \boldsymbol{\delta}, \mathbf{U})$ in the following form

$$J(\boldsymbol{\delta}, \mathbf{U}) = \max_{\boldsymbol{\Omega}} \mathscr{C}(\boldsymbol{\Omega}, \boldsymbol{\delta}, \mathbf{U}).$$

Let $\boldsymbol{\Omega}^* = \{\Omega_j^* : j = 0, \ldots, T\}$ be defined as

$$\boldsymbol{\Omega}^* \triangleq \arg\max_{\boldsymbol{\Omega}} \mathscr{C}(\boldsymbol{\Omega}, \boldsymbol{\delta}, \mathbf{U}),$$

where $\Omega_j^* = 1$ if

$$j = \arg\max_{i} \{J_i(\delta_i, U_i)\}. \quad (9.59)$$

From (9.53) and (9.54), we observe that

$$\begin{aligned} \max_{\boldsymbol{\Omega}} \min_{\mathbf{U}} \mathscr{C}(\boldsymbol{\Omega}, \boldsymbol{\delta}, \mathbf{U}) &= \max_{\boldsymbol{\Omega}} \mathscr{C}(\boldsymbol{\Omega}, \boldsymbol{\delta}, \hat{\mathbf{X}}) \\ &\geq \min_{\mathbf{U}} \max_{\boldsymbol{\Omega}} \mathscr{C}(\boldsymbol{\Omega}, \boldsymbol{\delta}, \mathbf{U}). \end{aligned} \quad (9.60)$$

Also, we have

$$\max_{\boldsymbol{\Omega}} \mathscr{C}(\boldsymbol{\Omega}, \boldsymbol{\delta}, \mathbf{U}) \geq \max_{\boldsymbol{\Omega}} \min_{\mathbf{U}} \mathscr{C}(\boldsymbol{\Omega}, \boldsymbol{\delta}, \mathbf{U}), \quad (9.61)$$

which implies that

$$\min_{\mathbf{U}} \max_{\boldsymbol{\Omega}} \mathscr{C}(\boldsymbol{\Omega}, \boldsymbol{\delta}, \mathbf{U}) \geq \max_{\boldsymbol{\Omega}} \min_{\mathbf{U}} \mathscr{C}(\boldsymbol{\Omega}, \boldsymbol{\delta}, \mathbf{U}). \quad (9.62)$$

From (9.60) and (9.62), it is easily concluded that

$$\max_{\boldsymbol{\Omega}} \min_{\mathbf{U}} \mathscr{C}(\boldsymbol{\Omega}, \boldsymbol{\delta}, \mathbf{U}) = \min_{\mathbf{U}} \max_{\boldsymbol{\Omega}} \mathscr{C}(\boldsymbol{\Omega}, \boldsymbol{\delta}, \mathbf{U}), \quad (9.63)$$

which completes the proof for (9.55). Using the results in (9.55) and (9.54), the cost function $J(\boldsymbol{\delta}, \hat{\mathbf{X}})$ is given by

$$\begin{aligned} J(\boldsymbol{\delta}, \hat{\mathbf{X}}) &= \min_{\mathbf{U}} \max_{i} \{J_i(\delta_i, U_i)\} \\ &= \max_{i} \min_{\mathbf{U}} \{J_i(\delta_i, U_i)\} \\ &= \max_{i} \left\{J_i(\delta_i, \hat{X}_i)\right\} \qquad (9.64) \\ &= \max_{i} \left\{ \frac{\int_{\mathbf{Y}} \delta_i(\mathbf{Y}) \mathbf{C}^*_{\mathrm{p},i}(\mathbf{Y}) f_i(\mathbf{Y}) \mathrm{d}\mathbf{Y}}{\int_{\mathbf{Y}} \delta_i(\mathbf{Y}) f_i(\mathbf{Y}) \mathrm{d}\mathbf{Y}} \right\}. \qquad (9.65) \end{aligned}$$

Appendix 2

The function $J_i(\delta_i, U_i)$ is a quasi-convex function. The weighted maximum function preserves the quasi-convexity and therefore, $J_i(\delta_i, \hat{X}_i)$ is a quasi-convex function from its definition in (9.26). This allows us to find the solution by solving an equivalent feasibility problem given below (Boyd and Vandenberghe 2004). Specifically, for $u \in \mathbb{R}_+$, it is observed that

$$J(\boldsymbol{\delta}, \hat{\mathbf{X}}) \leq u \equiv \int_{\mathbf{Y}} \delta_i(\mathbf{Y}) f_i(\mathbf{Y})(\mathbf{C}^*_{\mathrm{p},i}(\mathbf{Y}) - u) \mathrm{d}\mathbf{Y} \leq 0, \text{ for } i \in \{0, \ldots, T\}. \qquad (9.66)$$

Hence, the feasibility problem equivalent to (9.22) is given by

$$\mathscr{P}(\alpha, \beta) = \begin{cases} \min_{\boldsymbol{\delta}} & u \\ \text{s.t.} & \int_{\mathbf{Y}} \delta_i(\mathbf{Y}) f_i(\mathbf{Y})(\mathbf{C}^*_{\mathrm{p},i}(\mathbf{Y}) - u) \mathrm{d}\mathbf{Y} \leq 0, \quad \forall i \in \{0, \ldots, T\} \\ & \sum_{j=1}^{T} \sum_{i=0, i \neq j}^{T} \frac{\varepsilon_j}{1-\varepsilon_0} \int_{\mathbf{Y}} \delta_i(\mathbf{Y}) f_j(\mathbf{Y}) \mathrm{d}\mathbf{Y} \leq \beta \\ & \sum_{i=1}^{T} \int_{\mathbf{Y}} \delta_i(\mathbf{Y}) f_0(\mathbf{Y}) \mathrm{d}\mathbf{Y} \leq \alpha \end{cases} . \qquad (9.67)$$

The above problem is feasible if $\mathscr{P}(\alpha, \beta) \leq u$, where $\mathscr{P}(\alpha, \beta)$ is the lowest value of u for which the problem is feasible and all constraints are satisfied. Given an interval $[u_0, u_1]$ containing $\mathscr{P}(\alpha, \beta)$, the detection rule $\boldsymbol{\delta}$ and the estimation cost $\mathscr{P}(\alpha, \beta)$ are determined by a bi-section search between u_0 and u_1 iteratively, solving the feasibility problem in each iteration. We define an auxiliary convex optimization problem that allows us to solve the feasibility problem

$$\mathbb{R}(\alpha, \beta, u) = \begin{cases} \min_{\delta} & \eta \\ \text{s.t.} & \int_{\mathbf{Y}} \delta_i(\mathbf{Y}) f_i(\mathbf{Y})(\mathsf{C}^*_{\mathrm{p},i}(\mathbf{Y}) - u)\mathrm{d}\mathbf{Y} \le \eta, \quad \forall i \in \{0, \ldots, T\} \\ & \sum_{j=1}^{T} \sum_{i=0, i\neq j}^{T} \frac{\varepsilon_j}{1-\varepsilon_0} \int_{\mathbf{Y}} \delta_i(\mathbf{Y}) f_j(\mathbf{Y})\mathrm{d}\mathbf{Y} \ \le \beta + \eta \\ & \sum_{i=1}^{T} \int_{\mathbf{Y}} \delta_i(\mathbf{Y}) f_0(\mathbf{Y})\mathrm{d}\mathbf{Y} \le \alpha + \eta \end{cases} . \tag{9.68}$$

Algorithm 9.2 summarizes the steps for determining $\mathscr{P}(\alpha, \beta)$.

Algorithm 9.2 Bi-section Search

Input: Initialize u_0, u_1

repeat

- $\hat{u} \leftarrow (u_0 + u_1)/2$;
- Solve $\mathbb{R}(\alpha, \beta, \hat{u})$;
- **if** $\mathscr{J}(\alpha, \beta, \hat{u}) \le 0$ **then**
  - $u_1 \leftarrow \hat{u}$;
- **else**
  - $u_0 \leftarrow \hat{u}$;

until $u_1 - u_0 \le \varepsilon$, *for* ε *sufficiently small*;

Output: $\mathscr{P}(\alpha, \beta) \leftarrow u_1$

Appendix 3

To solve the problem in (9.68), a Lagrangian function is constructed according to

$$\begin{aligned} \mathscr{Q}(\boldsymbol{\delta}, \eta, \boldsymbol{\ell}) \triangleq & \left(1 - \sum_{i=0}^{T+2} \ell_i\right) \eta \\ & + \sum_{i=0}^{T} \ell_i \int_{\mathbf{Y}} \delta_i(\mathbf{Y}) f_i(\mathbf{Y})(\mathsf{C}^*_{\mathrm{p},i}(\mathbf{Y}) - u)\mathrm{d}\mathbf{Y} \\ & + \ell_{T+1} \sum_{j=1}^{T} \sum_{i=0, i\neq j}^{T} \frac{\varepsilon_j}{1-\varepsilon_0} \int_{\mathbf{Y}} \delta_i(\mathbf{Y}) f_j(\mathbf{Y})\mathrm{d}\mathbf{Y} \ - \ell_{T+1}\beta \\ & + \ell_{T+2} \sum_{i=1}^{T} \int_{\mathbf{Y}} \delta_i(\mathbf{Y}) f_0(\mathbf{Y})\mathrm{d}\mathbf{Y} - \ell_{T+2}\alpha, \end{aligned}$$

where $\boldsymbol{\ell} \triangleq \left[\ell_0, \ldots, \ell_{T+2}\right]$ are the non-negative Lagrangian multipliers selected to satisfy the constraints in (9.22), such that

$$\sum_{i=0}^{T+2} \ell_i = 1. \tag{9.69}$$

The Lagrangian dual function is given by

$$\begin{aligned} d(\boldsymbol{\ell}) &\triangleq \min_{\boldsymbol{\delta},\boldsymbol{\eta}} \mathscr{Q}(\boldsymbol{\delta}, \eta, \boldsymbol{\ell}) \\ &= \min_{\boldsymbol{\delta}} \left(\sum_{i=0}^{T} \int_{\mathbf{Y}} \delta_i(\mathbf{Y}) A_i \mathrm{d}\mathbf{Y} \right) - \ell_{T+1}\beta - \ell_{T+2}\alpha, \end{aligned} \tag{9.70}$$

where

$$A_0 \triangleq \ell_0 f_0(\mathbf{Y})[\mathsf{C}^*_{\mathrm{p},0}(\mathbf{Y}) - u] + \ell_{T+1} \sum_{i=1}^{T} \frac{\varepsilon_i}{1-\varepsilon_0} f_i(\mathbf{Y}), \tag{9.71}$$

and for $i \in \{1, \ldots, T\}$

$$A_i \triangleq \ell_i f_i(\mathbf{Y})[\mathsf{C}^*_{\mathrm{p},i}(\mathbf{Y}) - u] + \ell_{T+1} \sum_{j=1, j\neq i}^{T} \frac{\varepsilon_j}{1-\varepsilon_0} f_j(\mathbf{Y}) + \ell_{T+2} f_0(\mathbf{Y}). \tag{9.72}$$

Therefore, the optimum detection rules that minimize $d(\boldsymbol{\ell})$ are given by:

$$\delta_i(\mathbf{Y}) = \begin{cases} 1, & \text{if} \quad i = i^* \\ 0, & \text{if} \quad i \neq i^* \end{cases}, \tag{9.73}$$

where $i^* = \operatorname{argmin}_{i\in\{0,\ldots,T\}} A_i$. Hence, the proof is concluded.

References

C.M. Ahmed, M.A. Umer, B.S.S.B. Liyakkathali, M.T. Jilani, J. Zhou, Machine learning for cps security: applications, challenges and recommendations, in *Machine Intelligence and Big Data Analytics for Cybersecurity Applications* (Springer, 2021), pp. 397–421

S. Al-Sayed, A.M. Zoubir, A.H. Sayed, Robust distributed estimation by networked agents. IEEE Trans. Signal Process. **65**(15), 3909–3921 (2017)

C.Z. Bai, V. Gupta, On Kalman filtering in the presence of a compromised sensor: fundamental performance bounds, in *Proceedings of American Control Conference*, Portland, OR (2014), pp. 3029–3034

S. Boyd, L. Vandenberghe, *Convex Optimization* (Cambridge University Press, 2004)

K. Chen, V. Gupta, Y. Huang, Minimum variance unbiased estimation in the presence of an adversary, in *Proceedings of Conference on Decision and Control (CDC)* (2017), pp. 151–156

P. Ebinger, S.D. Wolthusen, Efficient state estimation and Byzantine behavior identification in tactical MANETs, in *Proceedings of IEEE Military Communications Conference*, Boston, MA (2009)

H. Fawzi, P. Tabuada, S. Diggavi, Secure state-estimation for dynamical systems under active adversaries, in *Proceedings of Allerton Conference on Communication, Control, and Computing*, Monticello, IL (2011), pp. 337–344

H. Fawzi, P. Tabuada, S. Diggavi, Secure estimation and control for cyber-physical systems under adversarial attacks. IEEE Transactions on Automatic Control **59**(6), 1454–1467 (2014)

H. Fawzi, P. Tabuada, S. Diggavi, Secure estimation and control for cyber-physical systems under adversarial attacks. IEEE Trans. Autom. Control **59**(6), 1454–1467 (2014)

F. Hu, Y. Lu, A.V. Vasilakos, Q. Hao, R. Ma, Y. Patil, T. Zhang, J. Lu, X. Li, N.N. Xiong, Robust cyber-physical systems: Concept, models, and implementation. Future generation computer systems **56**, 449–475 (2016)

A. Humayed, J. Lin, F. Li, B. Luo, Cyber-physical systems security'a survey. IEEE Internet of Things Journal **4**(6), 1802–1831 (2017)

G.H. Jajamovich, A. Tajer, X. Wang, Minimax-optimal hypothesis testing with estimation-dependent costs. IEEE Transactions on Signal Processing **60**(12), 6151–6165 (2012)

J. Li, Y. Liu, T. Chen, Z. Xiao, Z. Li, J. Wang, Adversarial attacks and defenses on cyber-physical systems: a survey. IEEE Internet Things J. **7**(6), 5103–5115 (2020)

Y. Lin, A. Abur, Robust state estimation against measurement and network parameter errors, IEEE Trans. Power Syst. 33, (5), pp. 4751–4759 2020

D. Middleton, R. Esposito, Simultaneous optimum detection and estimation of signals in noise. IEEE Trans. Inf. Theory **14**(3), 434–444 (1968)

S. Mishra, Y. Shoukry, N. Karamchandani, S. Diggavi, P. Tabuada, Secure state estimation: optimal guarantees against sensor attacks in the presence of noise, in *Proceedings of IEEE International Symposium on Information Theory*, Hong Kong, China (2015), pp. 2929–2933

G.V. Moustakides, G.H. Jajamovich, A. Tajer, X. Wang, Joint detection and estimation: Optimum tests and applications. IEEE Transactions on Information Theory **58**(7), 4215–4229 (2012)

M. Pajic, J. Weimer, N. Bezzo, P. Tabuada, O. Sokolsky, I. Lee, G.J. Pappas, Robustness of attack-resilient state estimators, in *Proceedings of IEEE International Conference on Cyber-Physical Systems* (2014), pp. 163–174

M. Pajic, P. Tabuada, I. Lee, G.J. Pappas, Attack-resilient state estimation in the presence of noise, in *Proceedings of IEEE Conference on Decision and Control*, Osaka, Japan (2015), pp. 5827–5832

H.V. Poor, *An Introduction to Signal Detection and Estimation*, 2nd edn. (Springer-Verlag, New York, 1998)

A.S. Rawat, P. Anand, H. Chen, P.K. Varshney, Countering Byzantine attacks in cognitive radio networks, in *Proceedings of IEEE International Conference on Acoustics, Speech and Signal Processing*, Dallas, TX (2010), pp. 3098–3101

A.H. Sayed, A framework for state-space estimation with uncertain models. IEEE Trans. Autom. Control **46**(7), 998–1013 (2001)

X. Shen, P.K. Varshney, Y. Zhu, Robust distributed maximum likelihood estimation with dependent quantized data. Automatica **50**(1), 169–174 (Jan. 2014)

Y. Shoukry, P. Nuzzo, A. Puggelli, A.L. Sangiovanni-Vincentelli, S.A. Seshia, P. Tabuada, Secure state estimation for cyber physical systems under sensor attacks: a satisfiability modulo theory approach. IEEE Transactions on Automatic Control **62**(10), 4917–4932 (2017)

S. Sihag, A. Tajer, Secure estimation under causative attacks. IEEE Transactions on Information Theory **66**(8), 5145–5166 (2020)

E. Soltanmohammadi, M. Orooji, M. Naraghi-Pour, Decentralized hypothesis testing in wireless sensor networks in the presence of misbehaving nodes. IEEE Transactions on Information Forensics and Security **8**(1), 205–215 (2013)

A. Vempaty, K. Agrawal, P. Varshney, H. Chen, Adaptive learning of Byzantines' behavior in cooperative spectrum sensing, in *Proceedings of IEEE Wireless Communications and Networking Conference*, Cancun, Mexico (2011), pp. 1310–1315
A. Vempaty, L. Tong, P.K. Varshney, Distributed inference with Byzantine data: state-of-the-art review on data falsification attacks. IEEE Signal Process. Mag. **30**(5), 65–75 (2013)
A. Vempaty, O. Ozdemir, K. Agrawal, H. Chen, P.K. Varshney, Localization in wireless sensor networks: Byzantines and mitigation techniques. IEEE Transactions on Signal Processing **61**(6), 1495–1508 (2013)
A. Vempaty, Y.S. Han, P.K. Varshney, Target localization in wireless sensor networks using error correcting codes. IEEE Transactions on Information Theory **60**(1), 697–712 (2014)
C. Wilson, V.V. Veeravalli, MMSE estimation in a sensor network in the presence of an adversary, in *Proceedings of IEEE International Symposium on Information Theory*, Barcelona, Spain (2016), pp. 2479–2483
S.Z. Yong, M. Zhu, E. Frazzoli, Resilient state estimation against switching attacks on stochastic cyber-physical systems, in *Proceedings of IEEE Conference on Decision and Control*, Osaka, Japan (2015), pp. 5162–5169
O. Zeitouni, J. Ziv, N. Merhav, When is the generalized likelihood ratio test optimal? IEEE Transactions on Information Theory **38**(5), 1597–1602 (1992)
J. Zhang, R.S. Blum, Distributed estimation in the presence of attacks for large scale sensor networks, in *Proceedings of Conference on Information Sciences and Systems*, Princeton, NJ (2014)
J. Zhang, R.S. Blum, X. Lu, D. Conus, Asymptotically optimum distributed estimation in the presence of attacks. IEEE Transactions on Signal Processing **63**(5), 1086–1101 (2015)
J. Zhao, M. Netto, L. Mili, A robust iterated extended kalman filter for power system dynamic state estimation. IEEE Transactions on Power Systems **32**(4), 3205–3216 (2016)

Chapter 10
Resilient Control of Nonlinear Cyber-Physical Systems: Higher-Order Sliding Mode Differentiation and Sparse Recovery-Based Approaches

Shamila Nateghi, Yuri Shtessel, Christopher Edwards, and Jean-Pierre Barbot

10.1 Introduction

Cyber-physical system security including information security, protection of CPS from being attacked and detection in adversarial environments have been considered in the literature (Pasqualetti et al. 2013; Jafarnia-Jahromi et al. 2012; Antsaklis 2014; Nekouei et al. 2018; Cardenas et al. 2008). Cryptography and Randomization are the two main approaches to protect a CPS against disclosure attacks: Cryptography is an approach to prevent third parties or the public from reading private messages by defining some protocols (Chen et al. 2016; Diffie and Hellman 1976). Randomization is a defensive strategy to confuse the potential attacker about deterministic rules and information of the system (Farokhi et al. 2017).

However, another challenge is to ensure that the CPS can continue functioning properly if a cyber-attack has happened. If the defense strategy just relies on detection, then the system's performance still degrades, and the threat of the same attack recur-

S. Nateghi (✉) · Y. Shtessel
University of Alabama in Huntsville, Huntsville, AL, USA
e-mail: sb0086@uah.edu

Y. Shtessel
e-mail: shtessy@uah.edu

C. Edwards
The University of Exeter, Exeter, UK
e-mail: C.Edwards@exeter.ac.uk

J.-P. Barbot
QUARTZ Laboratory, ENSEA, Cergy-Pontoise, France
e-mail: barbot@ensea.fr

M. Abbaszadeh and A. Zemouche (eds.), *Security and Resilience in Cyber-Physical Systems*, https://doi.org/10.1007/978-3-030-97166-3_10

ring is not diminished. In addition, in the interval between the onset of the attack and detection, the system could experience significant damage (Jafarnia-Jahromi et al. 2012). A good example of such a scenario is the Stuxnet (Chen 2010). The Maroochy attack happened because of the lack of detection and resilience mechanisms as well (Slay and Miller 2007). In RQ-170, the absence of resilience control caused the system to be unable to defend itself against the spoofing attack (Hartmann and Steup 2013).

It is suggested in Dibaji et al. 2019 that information security mechanisms must be complemented by specially designed resilient control systems until the system is restored to normal operation. The focus of this chapter is on the reconstruction of the cyber-attack as a step to provide resilient control for a CPS.

The control/estimation algorithms are proposed in the literature for recovering CPS performance online if an attacker penetrates the information security mechanisms. A game-theoretic approach that provides resilience consists of trying to minimize the damage that an attacker can apply to the system or maximize the price of attacking a system. For example, a zero-sum stochastic differential game between a defender and an attacker is used to find an optimal control design to provide system security in Zhu and Başar (2011). Event-triggered control schemes instead of time-triggered schemes, which are based on how frequent the attacks occur, are an appropriate strategy to increase the resilience of CPS (Heemels et al. 2012). Event-triggered control is especially used to mitigate the effect of a disruption attack (Cetinkaya et al. 2016). Mean Subsequence Reduced as a resilient control approach ignores suspicious values and computes the control input at every moment (LeBlanc et al. 2013; Dibaji et al. 2017). In trust-based approaches, a function of trust value between the nodes of the system is defined since some of the nodes of the system may be untrustworthy (Ahmed et al. 2015). In Fawzi et al. (2014), authors found the number of attacks that can be tolerated so that the state of the system can still be exactly recovered. They designed a secure local control loop to improve the resilience of the system. In Jin et al. (2017), new adaptive control architectures that can foil malicious sensors and actuator attacks are developed for linear CPS without reconstructing the attacks, by means of feedback control only.

The mentioned approaches suffer some limitations including: I. It is assumed that the maximum number of malicious sensors in the network is known and bounded. Once the number of attacked sensors exceeds the upper bound, the proposed secure estimation or resilient control schemes fail to work. II. Only specific types of malicious actions acting on the cyber layer are considered. III. Only special structures of the cyber-physical system are considered.

On the other hand, the Sliding Mode Control and Higher-Order Sliding Mode Control (SMC/HOSM) and observation/differentiation techniques can handle systems of arbitrary relative degree perturbed by bounded attacks of arbitrary shape. The Sliding Mode Observers/differentiators (SMO/D) are capable of estimating the system states and reconstruct the bounded attacks asymptotically or in finite time (Fridman et al. 2007; Utkin 1992; Shtessel et al. 2014; Fridman et al. 2008; Levant 2003; Nateghi and Shtessel 2018; Nateghi et al. 2020a, 2018a, b) while addressing the outlined challenges.

Detection and observation of a scalar attack by a SMO has been accomplished for a linearized differential-algebraic model of an electric power network when plant and sensor attacks do not occur simultaneously (Wu et al. 2018). An adaptive SMO is designed coupled with a parameter estimator and a robust differentiator for detection and reconstruction of attacks in linear cyber-physical systems in Huang et al. (2018) when state and sensor attacks do not happen simultaneously. In Nateghi et al. (2020b, 2021), fixed-gain and adaptive-gain SMO are proposed for the online reconstruction of sensor attacks. Especially, dynamic filters that address the attack propagation dynamics are employed for attack reconstruction. A probabilistic risk mitigation model for cyber-attacks against Phasor Measurement Unit (PMU) networks is presented in Mousavian et al. (2014), where a risk mitigation technique determines whether a certain PMU should be kept connected to the network or removed while minimizing the maximum threat level for all connected PMUs. In Taha et al. (2016), the sliding mode-based observation algorithm is used to reconstruct the attacks asymptotically. This reconstruction is approximate only since pseudo-inverse techniques are used. In the above mentioned studies, which use a Sliding Mode approach for resilient control of CPSs, they all consider linear CPS and have their specific limitations.

In this chapter, online cyber-attack reconstruction for nonlinear CPSs is investigated. Two complement cases are considered: (I) When the number of sensors is less than the number of potential sparse attacks. A sparse signal recovery (SR) algorithm with a finite time convergence property (Yu et al. 2017) is used to reconstruct the attacks and presented in Sect. 10.3. (II) when the number of sensors is equal or greater than the number of potential attacks. A certain number of sensors are assumed to be protected from cyber-attacks. A higher-order sliding mode observer/differentiator (Fridman et al. 2008) is applied to estimate the states and reconstruct the attacks provided in Sect. 10.4. The proposed algorithm ensures finite-time state estimation of observable variables and asymptotic estimation of the unobservable variables for the case when the system has asymptotically stable internal dynamics. In order to maintain the CPS closed-loop dynamics to be the same as those prior to the attacks, it is proposed to clean the corrupted measurements, as soon as the attacks are reconstructed, thus preventing the attack propagation to the CPS through feedback control. Actuator attacks are also cleaned from the reconstructed actuator attacks. The effectiveness of the proposed algorithms in Sects. 10.3 and 10.4 to estimate the states and reconstruct the attacks are tested on the attacked US WECC power network system.

10.2 Mathematical Modeling

Consider the following nonlinear CPS which is completely observable and asymptotically stable affected by attack

$$\dot{x} = f_1(t) + B_1(x)(u + d_u(t)), \tag{10.1}$$

where $x \in R^n$ presents the state vector of CPS, $f_1(x) \in R^n$ is a smooth vector field, $y \in R^p$ denotes the sensor measurement vector, and $u \in R^{q_1}$ is the control signal. The $d_u \in R^{q_1}$ and $d_y \in R^{q_2}$ are the actuator and sensor attack, respectively. The vector $C_x \in R^p$ is the output smooth vector field, $B_1(x) \in R^{n\times q_1}$ and $D_1 \in R^{p\times q_2}$ denote the attack/fault distribution matrices.

The output feedback control signal u is a function of sensor measurement y which can be corrupted by the sensor attacks. This is

$$u(y) = \gamma(C(x) + d_y) = \gamma(x + D_1 d_y). \tag{10.2}$$

Replacing control signal u in CPS (10.1) to find the closed-loop CPS model gives

$$\begin{aligned}\dot{x} &= f_1(t) + B_1(x)(\gamma(x, d_y), d_u(t)) = f_1(t) + B_1(x)(\gamma(x, d_y) + B_1(x)d_u(t)\\ y &= C(x) + D_1 d_y(t).\end{aligned} \tag{10.3}$$

Assume that u can be written as

$$\gamma(x, d_y) = \gamma_1(x) + \gamma_2(d_y), \tag{10.4}$$

then, the closed-loop CPS (10.3) is given as

$$\begin{aligned}\dot{x} &= f_1(t) + B_1(x)(\gamma(x, d_y), d_u(t))\\ &= f_1(t) + B_1(x)\gamma_1(x) + B_1(x)\gamma_2(d_y) + B_1(x)d_u(t)\\ y &= C(x) + D_1 d_y(t).\end{aligned} \tag{10.5}$$

Therefore, the CPS (10.1) after applying control signal u is presented as

$$\begin{aligned}\dot{x} &= f(t) + B_1(x)d_x(t)\\ y &= C(x) + D_1 d_y(t),\end{aligned} \tag{10.6}$$

where

$$\begin{aligned}f(x) &= f_1(x) + B_1(x)\gamma_1(x)\\ d_x(t) &= \gamma_2(d_y) + d_u(t),\end{aligned} \tag{10.7}$$

where $d_x(t)$ represents the plant/state attack.

Define the attack signal $d(t) \in R^q$ where $q = q_1 + q_2$ as

$$d = \begin{bmatrix} d_x \\ d_y \end{bmatrix}, \tag{10.8}$$

where $d_x \in R^{q_1}$ and $d_y \in R^{q_2}$, and

$$\begin{aligned} B(x) &= \left[B_1(x)\ 0_1\right] \\ D &= \left[0_2\ D_1\right], \end{aligned} \tag{10.9}$$

where $B_1(x) \in R^{n\times q_1}$, $D_1 \in R^{p\times(q-q_1)}$, $0_1 \in R^{n\times(q-q_1)}$, $0_2 \in R^{p\times q_1}$. Then, the closed-loop CPS (10.6) is rewritten as

$$\begin{aligned} \dot{x} &= f(x) + B(x)d(t) \\ y &= C(x) + Dd(t). \end{aligned} \tag{10.10}$$

10.2.1 Problem Statement

The problem is two-fold
1. Develop an observation algorithm that reconstructs online the state $x \in R^n$ and attack signal $d(t) \in R^q$ in CPS (10.10) so that

$$\begin{aligned} \hat{x}(t) &\rightarrow x(t) \\ \hat{d}(t) &\rightarrow d(t). \end{aligned} \tag{10.11}$$

2. Develop an observation algorithm that reconstructs online the state $x \in R^n$, the plant attack signal $d_x(t) \in R^{q_1}$, and sensor attack signal $d_y(t) \in R^{q_2}$ in CPS (10.6) as shown in the table below so that

$$\begin{aligned} \hat{x}(t) &\rightarrow x(t) \\ \hat{d}_x(t) &\rightarrow d_x(t) \\ \hat{d}_y(t) &\rightarrow d_y(t) \end{aligned} \tag{10.12}$$

as time increases.

Attack plan	$d_u(t) \neq 0$	$d_y(t) \neq 0$	Access to all sensors	Need to know the system model
Stealth attack		√		
Deception attack	√			
Replay attack	√	√	√	
Covert attack	√	√		√
False data injection attack		√		√

Remark 10.1 As soon as the sensor attack $d_y(t)$ and the state attack $d_x(t)$ are estimated/reconstructed the measurement $y = C(x) + D_1 d_y(t)$ could be cleaned as

$$y_{clean} = y - D_1\hat{d}_y(t) = C(\hat{x}) + D_1(d_y(t) - \hat{d}_y(t)) \rightarrow y_{clean} = C(\hat{x}). \tag{10.13}$$

Next, the clean measurement y_{clean} can be used in the feedback control of CPS. This allows blocking the propagation of the sensor attack to the dynamics of CPS through the feedback control. The modified actuator commands are also cleaned from estimated actuator attacks, i.e., the actuator attack $d_u(t)$ can be estimated/reconstructed from (10.7) as $\hat{d}_u(t) = \hat{d}_y(t) - \gamma_2(\hat{d}_y)$, and the system (10.5) dynamics converge to

$$\dot{x} = f_1(x) + B_1(x)(u + d_u(t) - \hat{d}_u(t)) \rightarrow \dot{x} = f_1(t) + B_1(x)u \tag{10.14}$$

as time increases.

In this chapter, attack reconstruction is divided to two cases: when the number of potential attacks is (I) greater or equal, and (II) less than the number of sensors. In the following two sections, the mentioned cases are investigated.

10.3 Preliminary: Sparse Recovering Algorithm

The problem of recovering an unknown input signal from measurements is well known, as a left invertibility problem, as seen in Sain and Massey (1969), Barbot et al. (2009), but this problem was only treated in the case where the number of measurements is equal or greater than the number of unknown inputs. The left invertibility problem in the case of fewer measurements than unknown inputs has no solution or more exactly has an infinity of solutions.

Note that the input signals can be considered sparse or compressive for transmission. The compressive sensing theory could be a proper candidate to deal with these constraints. Sparse recovery algorithm is used to address this problem. The problem is to find the exact recovery under sparse assumption denoted for the sake of simplicity as "Sparse Recovery", i.e., finding a concise representation of a signal which is described as

$$\kappa = \Theta(s + \varepsilon), \tag{10.15}$$

where $s \in R^N$ are the unknown inputs with no more than j non-zero entries, $\kappa \in R^M$ are the measurements, ε is a measurement noise, and $\Theta \in R^M \times N$ is a matrix where $M < N$.

Assumption 10.1 The matrix Θ satisfies the Restricted Isometry Property (RIP) condition of j-order with constant $\zeta_j \in (0, 1)$ (ζ_j is as small as possible for computational reasons).

Note that the condition of RIP in compressive sensing is an essential requirement that ensures the recovery of sparse signal vectors. RIP property provides the necessary and sufficient requirements for the compressive sensing matrix; however, it is not robust enough for consideration under the noise.

Assumption 10.1 implies that for any j sparse of signal s, i.e., vectors with at most j non-zero elements, the following condition is verified

$$(1 - \zeta_s)\|s\|_2^2 \leq \|\Theta s\|_2^2 \leq (1 + \zeta_s)\|s\|_2^2. \tag{10.16}$$

Consider Γ as the index set of non-zero elements of Θ, then (10.16) is equivalent to Yu et al. (2017), Candes and Tao (2005)

$$1 - \zeta_s \leq eig(\Theta_\Gamma^T \Theta_\Gamma) \leq 1 + \zeta_s, \tag{10.17}$$

where Θ_Γ is the sub-matrix of Θ with active nodes. The problem of SR is often cast as an optimization problem that minimizes a cost function constructed by leveraging the observation error term and the sparsity inducing term (Yu et al. 2017), i.e.,

$$s^* = arg \min_{s \in R^N} \frac{1}{2}\|\kappa - \Theta s\|_1^2 + \lambda \Lambda(s), \tag{10.18}$$

where the sparsity term $\Lambda(s)$ can be replaced by $\Lambda(s) = \|s\|_1 \equiv \sum_i |s_i|$ as long as the RIP conditions hold. The $\lambda > 0$ in (10.18) is the balancing parameter and s^* is the critical point, i.e., the solution of (10.15).

For sparse vectors s with j-sparsity, where j must be equal or smaller than $\frac{M-1}{2}$, solution to the SR problem is unique and coincides with the critical point of (10.15) when the RIP condition for Θ with order $2j$ is verified (Yu et al. 2017). Under the sparse Assumption 10.1 of s and fulfilling j-RIP condition of matrix, the estimate of the sparse signal s as proposed in Yu et al. (2017) is

$$\begin{aligned} &\mu \dot{\nu}(t) = -\lceil \nu(t) + (\Theta^T \Theta - I_{N \times N})a(t) - \Theta^T \kappa \rfloor^\beta \\ &\hat{s} = a(t), \end{aligned} \tag{10.19}$$

where $\nu \in R^N$ is the state vector, $\hat{s}(t)$ represents the estimate of the sparse signal s of (10.15), and $\mu > 0$ is a time-constant determined by the physical properties of the implementing system. Note that $\lceil . \rfloor = |.|^\beta sign(.)$ and $a(t) = H_\lambda(\nu)$, where $H_\lambda(.)$ is a continuous soft thresholding function and is defined as

$$H_\lambda(\nu) = max(|\nu| - \lambda, 0)sgn(\nu), \tag{10.20}$$

where $\lambda > 0$ is chosen with respect to the noise and the minimum absolute value of the non-zero terms.

Under Assumption 10.1 the state ν of (10.19) converges in finite time to its equilibrium point ν^*, and $\hat{s}(t)$ in (10.19) converges in finite time to s^* of (10.18).

10.4 Attack Reconstruction When the Number of Potential Attacks is Greater Than the Number of Sensors

The nonlinear CPS in (10.10) is considered when the number of potential attacks is greater than the number of sensors, i.e.,

$$\begin{aligned} \dot{x} &= f(x) + B(x)d(t) \\ y &= C(x) + Dd(t) \qquad where \quad q > p. \end{aligned} \tag{10.21}$$

Assumption 10.2 It is assumed that the attack vector is sparse, meaning that numerous attacks are possible, but the attacks are not coordinated, and only few non-zero attacks happen at the same time, i.e., the index set of non-zero attacks is presented as

$$\begin{aligned} &\Phi_\Gamma = \{k_1, k_2, \ldots, k_j\}, \quad j < q \quad where \\ &2j + 1 \leq p. \end{aligned} \tag{10.22}$$

The objective is to reconstruct online the time-varying attack sparse vector based on the sensor measurement in CPS (10.21).

10.4.1 System Transformation

Feeding the sensor measurements under attack, y, of the CPS (10.21) to the input of the low-pass filter that facilitates filtering out the possible measurement noise gives Nateghi et al. (2018b)

$$\dot{z} = \frac{1}{\tau}(-z + C(x) + D(x)d(t)), \tag{10.23}$$

whose output $z \in R^p$, is available. Then, the CPS in (10.21) is rewritten as

$$\begin{aligned} \dot{\xi} &= \eta(\xi) + \Omega d(t) \\ \psi &= C\xi, \end{aligned} \tag{10.24}$$

where $\psi \in R^p$, and

$$\begin{aligned} &\xi = \begin{bmatrix} z \\ x \end{bmatrix}_{(p+n)\times 1}, \quad \eta(\xi) = \begin{bmatrix} -\frac{1}{\tau}I & 0 \\ 0 & 0 \end{bmatrix}\begin{bmatrix} z \\ x \end{bmatrix} + \begin{bmatrix} \frac{1}{\tau}C(x) \\ f(x) \end{bmatrix} \\ &\Omega = \begin{bmatrix} \frac{1}{\tau}B(x) \\ B(x) \end{bmatrix} = [\Omega_1, \Omega_2, \ldots, \Omega_q]_{(p+n)\times q} \\ &C = [C_1, C_2, \ldots, C_{p+n}] = [I_{p\times p} \quad 0_{p\times n}]. \end{aligned} \tag{10.25}$$

Assumption 10.3 The transformed CPS (10.25) is assumed to have a vector relative degree $r = \{r_1, r_2, \ldots, r_p\}$, i.e.,

$$\begin{aligned} &\Gamma_{\Omega_j}\Gamma_{\eta}^{\lambda}\psi_i(\xi) = 0 \quad \forall j = 1, \ldots, q \quad \forall \lambda < r_i - 1 \quad \forall i = 1, \ldots, p \\ &\Gamma_{\Omega_j}\Gamma_{\eta}^{r_i-1}\psi_i(\xi) \neq 0 \quad for \quad at \quad least \quad one \quad 1 \leq j \leq q. \end{aligned} \tag{10.26}$$

Assumption 10.4 The distribution $\Gamma = span\{b_1, b_2, \ldots, b_q\}$ is involutive, where b_i is the ith column of matrix B in (10.21). This means that no new direction is generated by the Lie bracket of the distribution vector fields. This ensures that the zero dynamics (when exist) can be rewritten independently of the unknown input.

Assumption 10.5 Here it is assumed that there are no zero dynamics in system (10.24), i.e., total relative degree equal to the system's (10.10) order: $n = r_1 + r_2 + \cdots + r_p$.

Assuming that the Assumptions (10.4) and (10.5) are satisfied, then input–output dynamics of system (10.24) are presented as Fridman et al. (2008)

$$\dot{\Upsilon}_i = \begin{bmatrix} 0\;1\;0\ldots0 \\ 0\;0\;1\ldots0 \\ \vdots\;\vdots\;\vdots\ldots\vdots \\ 0\;0\;0\;\;0\;\;0 \end{bmatrix}\Upsilon_i + \begin{bmatrix} 0 \\ 0 \\ \vdots \\ L_f^{r_i}\psi_i(\xi) \end{bmatrix} + \begin{bmatrix} 0 \\ 0 \\ \vdots \\ \sum_{j=1}^{q} L_{\Omega_j}L_f^{r_i-1}\psi_i(\xi)d_i \end{bmatrix}, \tag{10.27}$$

where

$$\Upsilon_i = \begin{bmatrix} \Upsilon_1^i(\xi) \\ \Upsilon_2^i(\xi) \\ \vdots \\ \Upsilon_{r_i}^i(\xi) \end{bmatrix} = \begin{bmatrix} \psi_i(\xi) \\ L\psi_i(\xi) \\ \vdots \\ L_f^{r_i-1}\psi_i(\xi) \end{bmatrix} \quad for \quad i = 1, \ldots, p, \tag{10.28}$$

where $\psi_i(\xi)$ is the ith entry of vector $\psi(\xi)$. Each of system output ψ_i at its own relative degree r_i, satisfies following equation (Fridman et al. 2008)

$$\dot{\Upsilon}_{r_i}^i(\xi) = L_f^{r_i}\psi_i(\xi) + \sum_{j=1}^{\alpha} L_{\Omega_j}L_f^{r_i-1}\psi_i d_i \quad i = 1, \ldots, p. \tag{10.29}$$

Therefore, system (10.24) can be rewritten as the following algebraic equation

$$Z_p = F(\xi)d(t), \tag{10.30}$$

where

$$Z_p = \begin{bmatrix} \dot{\Upsilon}_{r_1}^1 \\ \vdots \\ \dot{\Upsilon}_{r_p}^p \end{bmatrix} - \begin{bmatrix} L_f^{r_1}\psi_1(\xi) \\ \vdots \\ L_f^{r_p}\psi_p(\xi) \end{bmatrix}, \tag{10.31}$$

where $Z_p \in R^p$, $F(\xi) \in R^{p\times q}$, and

$$F(\xi) = \begin{bmatrix} L_{\Omega_1}L_f^{r_1-1}\psi_1 & L_{\Omega_2}L_f^{r_1-1}\psi_1 & \dots & L_{\Omega_q}L_f^{r_1-1}\psi_1 \\ L_{\Omega_1}L_f^{r_2-1}\psi_2 & L_{\Omega_2}L_f^{r_2-1}\psi_2 & \dots & L_{\Omega_q}L_f^{r_2-1}\psi_2 \\ \vdots & & & \vdots \\ L_{\Omega_1}L_f^{r_p-1}\psi_p & L_{\Omega_2}L_f^{r_p-1}\psi_p & \dots & L_{\Omega_q}L_f^{r_p-1}\psi_p \end{bmatrix}. \tag{10.32}$$

Remark 10.2 The derivative $\dot{\Upsilon}_{r_1}^1, \dots, \dot{\Upsilon}_{r_p}^p$ are computed exactly in finite time using higher-order sliding mode differentiators (Fridman et al. 2008; Levant 2003). The details about the HOSMC differentiation algorithms and their parametric tuning can be found in Fridman et al. (2008), Levant (2003).

10.4.2 Attack Reconstruction

Assumption 10.6 The matrix $F(\xi)$ in (10.30)–(10.32) is assumed to satisfy the RIP condition as in Assumption 10.1.

The attack in (10.30) is reconstructed using the SR Algorithm as

$$\begin{aligned} \mu\dot{v}(t) &= -\lceil v(t) + (F(\xi)^T F(\xi) - I_{N\times N})a(t) - F(\xi)^T Z_p \rfloor^\beta \\ \hat{d} &= a(t), \end{aligned} \tag{10.33}$$

where $\hat{d}(t)$ represents the estimate of the sparse signal $d(t)$ of (10.30).

Under Assumption 10.6, the $\hat{d}(t)$ in (10.33) converges in finite time to $d(t)$ of (10.30) (Yu et al. 2017).

10.5 Attack Reconstruction When the Number of Sensors is Greater Than the Number of Potential Sensor Attacks

Consider the nonlinear CPS model under the state and sensor attack in (10.10) when the number of sensors is greater than the number of sensor attacks, that is

$$\begin{aligned} \dot{x} &= f(x) + B_1(x)d_x(t) \\ y &= C(x) + D_1 d_y(t) \qquad where \quad p > q - q_1, \end{aligned} \tag{10.34}$$

where $y \in R^p$, $d_x(t) \in R^{q_1}$ and $d_y(t) \in R^{q-q_1}$. Since there are more sensors than potential sensor attacks in CPS (10.34), there exists a nonsingular output transformation $M \in R^{R\times R}$ so that

$$\bar{y} = M^{-1}y = M^{-1}C(x) + M^{-1}D_1 d_y, \tag{10.35}$$

where the matrix M is selected to satisfy the condition

$$M^{-1}D = \begin{bmatrix} 0_3 \\ D_2 \end{bmatrix}, \tag{10.36}$$

where $0_3 \in R^{p_1 \times (q-q_1)}$, $D_2 \in R^{(p-p_1)\times(q-q_1)}$, and $p - p_1 \leq q - q_1$. The transformed sensor measurement vector in (10.35) is partitioned as

$$\bar{y} = \begin{bmatrix} \bar{y}_1 \\ \bar{y}_2 \end{bmatrix}, \tag{10.37}$$

where $\bar{y}_1 \in R_1^p$ and $\bar{y}_2 \in R^{p-p_1}$.

Next, CPS (10.34) is presented in a partitioned format in accordance with (10.37) as

$$\begin{aligned} \dot{x} &= f(x) + B_1(x)d_x(t) \\ \bar{y}_1 &= C_1(x) \\ \bar{y}_2 &= C_2(x) + D_2 d_y(t). \end{aligned} \tag{10.38}$$

$C_1 \in R^{p_1}$ and $C_2 \in R^{p-p_1}$.

Remark 10.3 The virtual measurement $\bar{y}_1$ in (10.38) is not affected by the attack corruption signal and can be classified as a protected measurement.

Assumption 10.7 The number of protected measurements is equal or greater than the number of plant attacks, i.e.,

$$q_1 \leq p_1. \tag{10.39}$$

Remark 10.4 Equation (10.39) gives that the number of unprotected measurements is equal or less than the number of attacks that may corrupt the measurements, i.e.,

$$p - p_1 \leq q - q_1. \tag{10.40}$$

The considered problem is: given the nonlinear CPS dynamics in Eq. (10.38) with virtual protected $\bar{y}_1 \in R_1^p$ and $\bar{y}_2 \in R^{p-p_1}$ unprotected sensors, and attack signals $d_x \in R^{q_1}$ on the plant and $d_y \in R^{q-q_1}$ on the sensors (sensor corruption signals), reconstruct the attack signals. The attack reconstruction is to be accomplished in two steps:

Step 1: The plant state $x(t)$ and the attack $d_x(t)$ vectors are estimated by applying the HOSM observer, described in the next section, with respect to the protected output $\bar{y}_1$ only, so that

$$\hat{x}(t) \to x(t), \quad \hat{d}_x(t) \to d_x(t) \tag{10.41}$$

in finite time, where $\hat{x}(t)$ and $\hat{d}_x(t)$ are the estimation of CPS states and the reconstruction of plant attack, respectively.

Step 2: Given the state $\hat{x}(t)$, which is estimated online, the unprotected sensor attack d_y is then estimated by applying the SR algorithm described in Sect. 10.3.

10.5.1 State Attack Reconstruction

Consider the part of CPS (10.38) associated with the virtual measurements protected from the attacks

$$\begin{aligned} \dot{x} &= f(x) + B_1(x)d_x(t) \\ \bar{y}_1 &= C_1(x). \end{aligned} \tag{10.42}$$

Note that only q_1 out of p_1 virtual protected measurements are employed, and that the other $p_1 - q_1$ virtual protected measurements can be used at the second step of the proposed algorithm. The aforementioned modifications are addressed by defining $\bar{y}_1$ and B_1 in (10.42) as $\bar{y}_1 = [\bar{y}_{11}, \dots, \bar{y}_{1q_1}]^T$, $B_1 = [b_1, b_2, \dots, b_{q_1}] \in R^{n \times q_1}$, where $b_i \in R^n, \forall i = 1, 2, \dots, q_1$ are smooth vector fields defined on an open $\Omega \subset R^n$. The problem is to estimate the states of nonlinear CPS (10.42) with unknown input, and reconstruct the state attack vector $d_x(t)$.

Assume that the CPS in (10.42) has the vector relative degree $r = \{r_1, r_2, \dots, r_{q_1}\}$ as it is defined in Assumption 10.3.

Assumption 10.8 The matrix

$$L(x) = \begin{bmatrix} L_{b_1}(L_f^{r_1-1}\bar{y}_1) & L_{b_2}(L_f^{r_1-1}\bar{y}_1) & \dots & L_{b_{q_1}}(L_f^{r_1-1}\bar{y}_1) \\ L_{b_1}(L_f^{r_2-1}\bar{y}_2) & L_{b_2}(L_f^{r_2-1}\bar{y}_2) & \dots & L_{b_{q_1}}(L_f^{r_2-1}\bar{y}_2) \\ \vdots & & & \vdots \\ L_{b_1}(L_f^{r_{q_1}-1}\bar{y}_{q_1}) & L_{b_2}(L_f^{r_{q_1}-1}\bar{y}_{q_1}) & \dots & L_{b_{q_1}}(L_f^{r_{q_1}-1}\bar{y}_{q_1}) \end{bmatrix} \tag{10.43}$$

is full rank.

If the CPS in (10.42) satisfies Assumptions (10.4) and (10.8), then the CPS given by Eq. (41) with the involutive distribution $\Gamma = span\{b_1, b_2, \dots, b_{q_1}\}$ and total relative degree $r = \sum_{i=1}^{q_1} r_i \leq n$ can be rewritten as Fridman et al. (2008)

$$\begin{aligned} \dot{\delta}_i &= \begin{bmatrix} 0\,1\,0 \dots 0 \\ 0\,0\,1 \dots 0 \\ \vdots\;\vdots\;\vdots \dots \vdots \\ 0\,0\,0\;\;0\;\;0 \end{bmatrix} \delta_i + \begin{bmatrix} 0 \\ 0 \\ \vdots \\ L_f^{r_i}\bar{y}_{1_i}(x) \end{bmatrix} + \begin{bmatrix} 0 \\ 0 \\ \vdots \\ \sum_{j=1}^{m} L_{b_j} L_f^{r_i-1}\bar{y}_{1_i}(x) d_x(t) \end{bmatrix} \\ \forall i &= 1, \dots, q_1 \\ \dot{\gamma} &= g(\delta, \gamma), \end{aligned} \tag{10.44}$$

where

$$\delta = \begin{bmatrix} \delta_1 \\ \delta_2 \\ \vdots \\ \delta_{q_1} \end{bmatrix}, \quad \delta_i = \begin{bmatrix} \delta_{i_1} \\ \delta_{i_2} \\ \vdots \\ \delta_{i_{r_1}} \end{bmatrix} = \begin{bmatrix} \eta_{i_1}(x) \\ \eta_{i_2}(x) \\ \vdots \\ \eta_{i_{r_1}}(x) \end{bmatrix} = \begin{bmatrix} \bar{y_1}_i(x) \\ L_f \bar{y_1}_i(x) \\ \vdots \\ L_f^{r_i-1} \bar{y_1}_i(x) \end{bmatrix} \in R^{r_i} \quad \forall i = 1, \ldots, q_1$$
$$\gamma = \begin{bmatrix} \gamma_1 \\ \gamma_2 \\ \vdots \\ \gamma_{n-r} \end{bmatrix} = \begin{bmatrix} \eta_{r+1}(x) \\ \eta_{r+2}(x) \\ \vdots \\ \eta_n(x) \end{bmatrix}. \tag{10.45}$$

Assumption 10.9 The norm-bounded solution of the internal dynamics (10.44) $\dot{\gamma} = g(\delta, \gamma)$ is assumed to be locally asymptotically stable (Fridman et al. 2008) as it is mentioned in (A3).
The variables $\eta_{r+1}(x), \eta_{r+2}(x), \ldots, \eta_n(x)$ are defined to satisfy

$$L_{b_j} \eta_i(x) = 0 \quad \forall i = r+1, \ldots, n, \quad \forall j = 1, \ldots, q_1, \tag{10.46}$$

if Assumption 10.4 is satisfied, then it is always possible to find $n - r$ functions $\eta_{r+1}(x), \eta_{r+2}(x), \ldots, \eta_n(x)$ such that

$$\Psi(x) = col\{\eta_{11}(x), \ldots, \eta_{1r_1}(x), \eta_{q_1 1}(x), \ldots, \eta_{q_1 r_{q_1}}(x), \eta_{r+1}(x), \ldots, \eta_n(x)\} \in R^n. \tag{10.47}$$

is a local diffeomorphism in a neighborhood of any point $x \in \bar{\Omega} \subset \Omega \subset R^n$, which means that

$$x = \Psi^{-1}(x)(\delta, \gamma). \tag{10.48}$$

To estimate the derivatives $\delta_{ij}, \forall i = 1, \ldots, q_1, \forall j = 1, \ldots, r_i$ of the outputs y_i in finite time, higher-order sliding mode differentiators (Levant 2003) are used

$$\begin{aligned}
&\dot{z}_0^i = v_0^i, \quad v_0^i = -\lambda_0^i |z_0^i - y_i(t)|^{(r_i/(r_i+1))} sign(z_0^i - y_i(t)) + z_1^i \\
&\dot{z}_1^i = v_1^i, \quad v_1^i = -\lambda_1^i |z_1^i - v_0^i|^{((r_i-1)/r_i)} sign(z_1^i - v_0^i) + z_2^i \\
&\vdots \\
&\dot{z}_{r_i-1}^i = v_{r_i-1}^i, \quad v_{r_i-1}^i = -\lambda_{r_i-1}^i |z_{r_i-1}^i - v_{r_i-2}^i|^{(1/2)} sign(z_{r_i-1}^i - v_{r_i-2}^i) + z_{r_i}^i \\
&\dot{z}_{r_i}^i = -\lambda_{r_i}^i sign(z_{r_i}^i - v_{r_i-1}^i),
\end{aligned} \tag{10.49}$$

for $i = 1, \ldots, q_1$.

By construction

$$\begin{aligned}
&\hat{\delta}_1^1 = \hat{\eta}_1^1(x) = z_0^1, \quad \ldots, \quad \hat{\delta}_{r_1}^1 = \hat{\eta}_{r_1}^1(x) = z_{r_1-1}^1, \quad \hat{\dot{\delta}}_1^1 = \hat{\dot{\eta}}_{r_1}^1(x) = z_{r_1}^1 \\
&\vdots \\
&\hat{\delta}_1^{q_1} = \hat{\eta}_1^{q_1} = z_0^{q_1}, \quad \ldots, \quad \hat{\delta}_{r_{q_1}}^{q_1} = \hat{\eta}_{r_{q_1}}^{q_1} = z_{r_{q_1}-1}^{q_1}, \quad \hat{\dot{\delta}}_{r_{q_1}}^{q_1} = \hat{\dot{\eta}}_{r_{q_1}}^{q_1} = z_{r_{q_1}}^1.
\end{aligned} \tag{10.50}$$

Therefore, the following exact estimates are available in finite time

$$\hat{\delta}_i = \begin{bmatrix} \hat{\delta}_{i1} \\ \hat{\delta}_{i2} \\ \vdots \\ \hat{\delta}_{ir_1} \end{bmatrix} = \begin{bmatrix} \hat{\eta}_{i1}(\hat{x}) \\ \hat{\eta}_{i2}(\hat{x}) \\ \vdots \\ \hat{\eta}_{ir_1}(\hat{x}) \end{bmatrix} \in R^{r_i} \quad \forall i = 1, \ldots, q_1 \quad \hat{\delta} = \begin{bmatrix} \hat{\delta}^1 \\ \hat{\delta}^2 \\ \vdots \\ \hat{\delta}^{q_1} \end{bmatrix} \in R^{r_t}. \tag{10.51}$$

Integrating the second equation in (10.44) and replacing δ by $\hat{\delta}$, the internal dynamics is given as

$$\dot{\hat{\gamma}} = g(\hat{\gamma}, \hat{\delta}), \tag{10.52}$$

and with some initial condition from the stability domain of the internal dynamics, a asymptotic estimate $\hat{\gamma}$ can be obtained locally as

$$\hat{\gamma} = \begin{bmatrix} \hat{\gamma}_1 \\ \hat{\gamma}_2 \\ \vdots \\ \hat{\gamma}_{n-r} \end{bmatrix} = \begin{bmatrix} \hat{\eta}_{r+1}(x) \\ \hat{\eta}_{r+2}(x) \\ \vdots \\ \hat{\eta}_n(x) \end{bmatrix}. \tag{10.53}$$

Therefore, the asymptotic estimate for the mapping (10.49) is identified as

$$\Psi(\hat{x}) = col\{\hat{\eta}_{11}(\hat{x}), \ldots, \eta_{1r_1}(\hat{x}), \ldots, \eta_{q_1 r_{q_1}}(\hat{x}), \hat{\eta}_{r+1}(\hat{x}), \hat{\eta}_n(\hat{x})\}. \tag{10.54}$$

The asymptotic estimate $\hat{x}$ of the state vector x of CPS (10.42) can be easily identified via (10.51) and (10.53) as

$$\hat{x} = \Psi^{-1}(\hat{\delta}, \hat{\gamma}). \tag{10.55}$$

An asymptotic estimate $\hat{d}_x(t)$ of the cyber state attack $d_x(t)$ in (10.42) can be identified as Nateghi et al. (2018a)

$$\hat{d}_x(t) = L^{-1}(\Psi^{-1}(\hat{\delta}, \hat{\gamma})) \left[\begin{bmatrix} \hat{\dot{\delta}}_{1r_1} \\ \hat{\dot{\delta}}_{2r_2} \\ \vdots \\ \hat{\dot{\delta}}_{qr_q} \end{bmatrix} - \begin{bmatrix} L_f^{r_1} y_1(\Psi^{-1}(\hat{\delta}, \hat{\gamma})) \\ L_f^{r_2} y_2(\Psi^{-1}(\hat{\delta}, \hat{\gamma})) \\ \vdots \\ L_f^{r_q} y_q(\Psi^{-1}(\hat{\delta}, \hat{\gamma})) \end{bmatrix} \right], \tag{10.56}$$

where $L^{-1}(\Psi^{-1}(\hat{\delta}, \hat{\gamma})) = \sum_{j=1}^{q} L_{b_j} L_f^{r_i-1} \bar{y}_{1_i}(x)$.

10.5.2 Sensor Attacks Reconstruction

After the state vector $x(t)$ and the plant attack $d_x(t)$ of CPS (10.34) are reconstructed in (10.55) and (10.56), then the sensor attacks $d_y(t)$ can be reconstructed as the following discussion: Consider the attacked part of system (10.38) as

$$\begin{aligned} \dot{x} &= f(x) + B_1(x)d_x(t) \\ \bar{y}_2 &= C_2(x) + D_2 d_y(t), \end{aligned} \tag{10.57}$$

where $y_2 \in R^{p-q_1}$, $D_2 \in R^{(p-q_1)\times(q-q_1)}$, $d_y(t) \in R^{q-q_1}$.

Two cases that cover all possible situations are considered to reconstruct the sensor attack $d_y(t)$.

Case 1: If the number of sensor attacks and the number of corrupted sensors is the same, i.e., $p - q_1 = q - q_1$, and D_2 is invertible, then using $\hat{x}$ estimated by the SMO in (10.55), there is a unique solution for estimation of sensor attack as Nateghi et al. (2018a)

$$\hat{d}_y(t) = D_2^{-1}(y_2 - C_2(\hat{x})). \tag{10.58}$$

Case 2: If the number of sensor attacks is greater than the number of corrupted sensors, i.e., $p - q_1 < q - q_1$ and the following assumption is verified for sensor attack d_y.

Assumption 10.10 It is assumed that the sensor attack vector $d_y \in R^{q-q_1}$ is sparse, meaning that there is only a small number of non-zero sensor attacks at any point in time.

Assumption 10.11 Matrix D_2 satisfies the RIP condition in Assumption 10.1.

Under Assumptions (10.10) and (10.11), then the attack vector $d(t)$ in (10.57) is reconstructed using the SR algorithm presented in Sect. 10.3 as

$$\hat{d}_y(t) = a(t), \tag{10.59}$$

where $v \in R^q$ is the state vector, $\hat{d}_y(t)$ represents the estimate of the sparse signal $d_y(t)$, and $\mu > 0$ is a time-constant determined by the physical properties of the implementing system. The sensor attack estimation in (10.59) converges in finite time to sensor attack $d_y(t)$ in CPS (10.34) (Yu et al. 2017).

10.6 Case Study: Cyber Attack Reconstruction in the US Western Electricity Coordinating Council Power System

In a real-world electrical power network, only small groups of generator rotor angles and rates are directly measured, and typical attacks aim at injecting disturbance signals that mainly affect the sensor-less generators (Wu et al. 2018). The CPS that motivates the results presented in this section is the US WECC power system (Scholtz 2004; Pasqualetti et al. 2015) under attack with three generators and six buses. The proposed approaches in Sects. 10.4 and 10.5 are applied to the linearized model of the US WECC, to estimate the states and reconstruct the attacks affected the considered WECC.

10.6.1 Mathematical Model of Electrical Power Network

The descriptor (Differential Algebraic Equations (DAE)) swing mathematical model is adopted to describe the electromechanical behavior of the considered electrical power networks (Taha et al. 2016; Yu et al. 2017). The DAE swing mathematical model for a power network stabilized by a linear output feedback controller is given by Yu et al. (2017):

$$\begin{bmatrix} I & 0 & 0 \\ 0 & M_g & 0 \\ 0 & 0 & 0 \end{bmatrix} \begin{bmatrix} \dot{\delta} \\ \dot{\omega} \\ \dot{\theta} \end{bmatrix} = - \begin{bmatrix} 0 & -I & 0 \\ L^{\theta}_{g,g} & E_g & L^{\theta}_{g,l} \\ L^{\theta}_{l,g} & 0 & L^{\theta}_{l,l} \end{bmatrix} \begin{bmatrix} \delta \\ \omega \\ \theta \end{bmatrix} + \begin{bmatrix} 0 \\ B_{\omega} \\ B_{\theta} \end{bmatrix} d(t) + \begin{bmatrix} 0 \\ P_{\omega} \\ P_{\theta} \end{bmatrix} \quad (10.60)$$
$$y = Cx + Dd(t),$$

where $x = \begin{bmatrix} \delta^T & \omega^T & \theta^T \end{bmatrix} \mathrm{T}^T$ is the vector of states of the system, $\delta \in R^a$, $\omega \in R^a$ and $\theta \in R^b$ are vectors of the phase angles of the source measured in rad, generator speed deviations from synchronous measured in rad/s, and the bus angles measured in rad, respectively. The index a is the number of generators, and b is the number of buses in the electrical system. The vector $y \in R^p$ is the sensor measurement vector, the vector $d \in R^q$ is the attack vector, and $B \in R^{(2a+b)\times q}$, $D \in R^{p\times q}$ are the attack distribution matrices; P_{ω}, P_{θ} are known changes in the mechanical input power to the generators or real power demand at the loads. The matrices E_g, $M_g \in R^{a\times a}$ are diagonal matrices whose non-zero entries consist of the damping coefficients and the normalized inertias of the generators, respectively. Finally, the matrices $L^{\theta}_{g,g}$, $L^{\theta}_{g,l}$, $L^{\theta}_{l,g}$ $L^{\theta}_{l,l}$ form the following symmetric susceptance matrix

$$L^{\theta} = \begin{bmatrix} L^{\theta}_{g,g} & L^{\theta}_{g,l} \\ L^{\theta}_{l,g} & L^{\theta}_{l,l} \end{bmatrix} \quad (10.61)$$

that is the Laplacian associated with the susceptance-weighted graph.

Assumption 10.12 The matrix $L^{\theta}_{l,l}$ is nonsingular (such an assumption usually holds in practical electric power systems).

Note that the following terms that appear in the electric power network model (59)

$$\begin{bmatrix} 0 \\ B_{\omega} \\ B_{\theta} \end{bmatrix} d(t) + \begin{bmatrix} 0 \\ P_{\omega} \\ P_{\theta} \end{bmatrix} \tag{10.62}$$

are due to the output feedback control that processes the output corrupted by the attack signal.

10.6.2 Transformation of DAE to ODE

Assuming (A10) holds, then the variable θ can be expressed as

$$\theta = (R^{\theta}_{l,l})^{-1}(-R^{\theta}_{l,g}\delta + P_{\theta} + B_{\theta}d) \tag{10.63}$$

substituting (10.63) into (10.60) gives

$$\begin{aligned} \begin{bmatrix} \dot{\delta} \\ \dot{\omega} \end{bmatrix} &= \begin{bmatrix} \phi_{\delta}(\delta, \omega) \\ \phi_{\omega}(\delta, \omega) \end{bmatrix} + \begin{bmatrix} 0 \\ P_{\theta\omega} \end{bmatrix} + \begin{bmatrix} 0 \\ B_{\theta\omega} \end{bmatrix} d(t) \\ y &= C \begin{bmatrix} \delta \\ \omega \end{bmatrix} + Dd(t), \end{aligned} \tag{10.64}$$

where

$$\begin{aligned} \begin{bmatrix} \phi_{\delta}(\delta, \omega) \\ \phi_{\omega}(\delta, \omega) \end{bmatrix} &= \begin{bmatrix} 0 & I_{p\times p} \\ M_g^{-1}(-R^{\theta}_{g,g} + R^{\theta}_{g,l}(R^{\theta}_{l,l})^{-1}R^{\theta}_{l,g}) & -M_g^{-1}E_g \end{bmatrix} \begin{bmatrix} \delta \\ \omega \end{bmatrix} \\ P_{\theta\omega} = M_g^{-1}(P_{\omega} - R^{\theta}_{g,l}&(R^{\theta}_{l,l})^{-1}P_{\theta}),\ B_{\theta\omega} = M_g^{-1}(B_{\omega} - R^{\theta}_{g,l}(R^{\theta}_{l,l})^{-1}B_{\theta}). \end{aligned} \tag{10.65}$$

10.6.3 Parameterization of Mathematical Model of Western Electricity Coordinating Council Power System

The electrical power network considered here is a classical nine-bus configuration adopted from Scholtz (2004), Pasqualetti et al. (2015). It consists of 3 generators $\{g_1, g_2, g_3\}$ and 6 load buses $\{b_1, \ldots, b_6\}$. Therefore, we have $\omega = \begin{bmatrix} \omega_1 & \omega_2 & \omega_3 \end{bmatrix} \mathsf{T}^T \in R^3$, $\delta = \begin{bmatrix} \delta_1 & \delta_2 & \delta_3 \end{bmatrix} \mathsf{T}^T \in R^3$, and $\theta \in R^6$.

The matrices $E_g, M_g \in R^{a\times a}$ are given as

$$M_g = \begin{bmatrix} 0.125 & 0 & 0 \\ 0 & 0.034 & 0 \\ 0 & 0 & 0.016 \end{bmatrix}, E_g = \begin{bmatrix} 0.125 & 0 & 0 \\ 0 & 0.068 & 0 \\ 0 & 0 & 0.048 \end{bmatrix}. \tag{10.66}$$

The symmetric susceptance matrix L^θ including $L^\theta_{g,g} \in R^{3\times3}$, $L^\theta_{g,l} \in R^{3\times6}$, $L^\theta_{l,g} \in R^{6\times3}$, $L^\theta_{l,l} \in R^{6\times6}$ is equal to

$$L^\theta = \begin{bmatrix} 0.058 & 0 & 0 & -0.058 & 0 & 0 & 0 & 0 & 0 \\ 0 & 0.063 & 0 & 0 & -0.063 & 0 & 0 & 0 & 0 \\ 0 & 0 & 0.059 & 0 & 0 & 0.059 & 0 & 0 & 0 \\ -0.058 & 0 & 0 & 0.0265 & 0 & 0 & -0.085 & -0.092 & 0 \\ 0 & -0.063 & 0 & 0 & 0.296 & 0 & -0.161 & 0 & -0.072 \\ 0 & 0 & -0.059 & 0 & 0 & 0.330 & 0 & -0.170 & -0.101 \\ 0 & 0 & 0 & -0.085 & -0.161 & 0 & 0.246 & 0 & 0 \\ 0 & 0 & 0 & -0.092 & 0 & -0.170 & 0 & 0.262 & 0 \\ 0 & 0 & 0 & 0 & -0.072 & -0.101 & 0 & 0 & 0.173 \end{bmatrix}. \tag{10.67}$$

The inputs P_ω and P_θ are defined as

$$P_\omega = \left[0.716\ 1.62\ 0.85\right] \mathrm{T}^T, P_\theta = \left[0\ -1.25\ 0.94\ 0\ -1\ 0\right] \mathrm{T}^T. \tag{10.68}$$

10.6.4 *Reconstruction of Attacks via Sparse Recovery Algorithm: The Number of Potential Attacks is Greater Than the Number of Sensors*

Consider the WECC power system (10.60) under attack signal $d = \left[d_x^T\ d_y^T\right]^T \in R^{18}$ where $d_x \in R^{12}$, and $d_y \in R^6$ are the attacks of the plant and sensors, respectively. The attacks d_x,d_y are further decoupled as follows:

$$d_1 = \begin{bmatrix} d^\delta_{x(3\times1)} \\ d^\omega_{x(3\times1)} \\ d^\theta_{x(6\times1)} \end{bmatrix}, d_2 = \begin{bmatrix} d^\delta_{y(3\times1)} \\ d^\omega_{y(3\times1)} \end{bmatrix}, \tag{10.69}$$

where d_x^δ, d_x^ω, d_x^θ are attacks on δ, ω, θ, and d_y^δ, d_y^ω are attacks on measurements of δ and ω, respectively. It is considered that

$$\begin{aligned}
B_\delta &\in R^{3\times 18} = \begin{bmatrix} I_{3\times 3} & 0_{3\times 15} \end{bmatrix} \\
B_\omega &\in R^{3\times 18} = \begin{bmatrix} 0_{3\times 3} & I_{3\times 3} & 0_{3\times 12} \end{bmatrix} \\
B_\theta &\in R^{6\times 18} = \begin{bmatrix} 0_{6\times 6} & I_{6\times 6} & 0_{6\times 6} \end{bmatrix} \\
D_\delta &\in R^{3\times 18} = \begin{bmatrix} 0_{3\times 12} & I_{3\times 3} & 0_{3\times 3} \end{bmatrix} \\
D_\omega &\in R^{3\times 18} = \begin{bmatrix} 0_{3\times 15} & I_{3\times 3} \end{bmatrix}.
\end{aligned} \tag{10.70}$$

The corrupted sensor measurements $y = \begin{bmatrix} \delta \\ \omega \end{bmatrix} \in R^6$ are fed to the low-pass filter (10.23) and the new variable ξ is defined as

$$\xi = \begin{bmatrix} z \\ y \end{bmatrix} \in R^{12}, \tag{10.71}$$

where $z = \begin{bmatrix} z_{1_{3\times 1}} \\ z_{2_{3\times 1}} \end{bmatrix} \in R^6$ is the output of LPF.

Then, the WECC (10.60) with the LPF (10.23)–(10.25) is presented as

$$\begin{aligned}
\dot{\xi} &= \begin{bmatrix} \frac{-1}{\tau} & 0 & \frac{1}{\tau} & 0 \\ 0 & \frac{-1}{\tau} & 0 & \frac{1}{\tau} \\ 0 & 0 & 0 & 1 \\ 0 & 0 & M_g^{-1}(-P^\theta_{g,g} + P^\theta_{g,l}(R^\theta_{l,l})^{-1}R^\theta_{l,g}) & -M_g^{-1}E_g \end{bmatrix} \times \xi + \begin{bmatrix} \frac{1}{\tau}D_\delta \\ \frac{1}{\tau}D_\omega \\ B_\delta \\ B_{\delta\omega} \end{bmatrix} d + \\
&\begin{bmatrix} 0 \\ 0 \\ 0 \\ -M_g^{-1}P^\theta_{g,l} + P^{\theta -1}_{l,l}P_\theta + M_g^{-1}P_\omega) \end{bmatrix} \\
\psi &= \begin{bmatrix} I_{6,6} & 0_{6,6} \end{bmatrix} \xi.
\end{aligned} \tag{10.72}$$

Considering $\psi = \begin{bmatrix} \psi_1 & \psi_2 \end{bmatrix}^T$ where $\psi_{1_{(3\times 1)}} = z_{1_{(3\times 1)}}$, $\psi_{2_{(3\times 1)}} = z_{2_{(3\times 1)}}$, then

$$\dot{z}_1 = \frac{1}{\tau}(-z_1 + \delta + d_2^\delta), \dot{z}_2 = \frac{1}{\tau}(-z_2 + \omega + d_2^\omega). \tag{10.73}$$

To verify if the (10.73) satisfies the RIP condition in Assumption 10.1, (10.17), the Eq. (10.73) is rewritten in a format of (10.15) as Nateghi et al. (2018b)

$$\begin{bmatrix} \dot{z}_1 + \frac{1}{\tau}z_1 - \frac{1}{\tau}\delta \\ \dot{z}_2 + \frac{1}{\tau}z_2 - \frac{1}{\tau}\omega \end{bmatrix} = \begin{bmatrix} 0_{3\times 3} & 0_{3\times 3} & 0_{3\times 6} & (\frac{1}{\tau})I_{3\times 3} & 0_{3\times 3} \\ 0_{3\times 3} & 0_{3\times 3} & 0_{3\times 6} & 0_{3\times 3} & (\frac{1}{\tau})I_{3\times 3} \end{bmatrix} \begin{bmatrix} d_1^\delta \\ d_1^\omega \\ d_1^\theta \\ d_2^\delta \\ d_2^\omega \end{bmatrix}. \tag{10.74}$$

Apparently, $F(\xi)$ in (10.74) doesn't satisfy the RIP condition (10.17), therefore, another differentiation of $\dot{z}_1$, $\dot{z}_2$ is required:

$$\ddot{z}_1 = \frac{1}{\tau}(-\dot{z}_1 + \dot{\delta} + \dot{d}_2^{\delta}), \ddot{z}_2 = \frac{1}{\tau}(-\dot{z}_2 + \dot{\omega} + \dot{d}_2^{\omega}). \tag{10.75}$$

Taking into account the output filter dynamics (10.23), and bearing in mind that

$$\dot{\delta} = \omega + B_{\delta}d = (\tau\dot{z}_2 + z_2 - d_2^{\omega}) + B_{\delta}d \tag{10.76}$$

and

$$\begin{aligned} \dot{\omega} &= \phi_{21}\delta + \phi_{22}\omega + P_{\theta\omega} + B_{\theta\omega}d(t) \\ &= \phi_{21}(\tau\dot{z}_1 + z_1 - d_2^{\delta}) + \phi_{22}(\tau\dot{z}_2 + z_2 - d_2^{\omega}) + P_{\theta\omega} + B_{\theta\omega}d(t), \end{aligned} \tag{10.77}$$

where $B_{\theta\omega}d(t) = M_g^{-1}d_{g,l}^{\omega} - M_g^{-1}p_{g,l}^{\theta}(p_{l,l}^{\theta})^{-1}d_1^{\theta}$
then (10.75) is rewritten as

$$\tilde{Z} = \tilde{F}\tilde{d} \tag{10.78}$$

where

$$\tilde{Z}_m = \begin{bmatrix} \ddot{z}_1 + \frac{1}{\tau}\dot{z}_1 - \dot{z}_2 - \frac{1}{\tau}z_2 \\ \ddot{z}_2 + \frac{1}{\tau}\dot{z}_2 - \phi_{21}\dot{z}_1 - \frac{1}{\tau}\phi_{21}z_1 - \phi_{22}\dot{z}_2 - \frac{1}{\tau}\phi_{22}z_2 - \frac{1}{\tau}P_{\theta\omega} \end{bmatrix} \tag{10.79}$$

$$\tilde{F} = \begin{bmatrix} \frac{1}{\tau} & 0 & 0 & 0 & -\frac{1}{\tau} & \frac{1}{\tau} & 0 \\ 0 & \frac{M_g^{-1}}{\tau} & \frac{M_g^{-1}P_{g,l}^{\theta}(P_{l,l}^{\theta})^{-1}}{\tau} & \frac{-\phi_{21}}{\tau} & \frac{-\phi_{22}}{\tau} & 0 & \frac{1}{\tau} \end{bmatrix} \tag{10.80}$$

$$\tilde{d}_{24\times 1} = \left[(d_1^{\delta})^T \; (d_1^{\omega})^T \; (d_1^{\theta})^T \; (d_2^{\delta})^T \; (d_2^{\omega})^T \; (\dot{d}_2^{\delta})^T \; (\dot{d}_2^{\omega})^T\right]^T. \tag{10.81}$$

Now, $\tilde{F}$ in (10.80) satisfies the RIP condition (10.17), therefore, the SR algorithm can be applied to (10.78).

Remark 10.5 The derivatives $\ddot{z}_1$, $\ddot{z}_2$, $\dot{z}_1$ and $\dot{z}_2$ that appear in the entries of the virtual measurement vector $\tilde{Z}_m$ are obtained using HOSM differentiators (Fridman et al. 2008).

Assumption 10.13 The sensor attack signals d_2^{δ} and d_2^{ω} are assumed to be slow with respect to system (10.17) dynamics. In other words, it is assumed $\dot{d}_s^{\delta} \approx 0$ and $d_s^{\omega} \approx 0$ (Nateghi et al. 2018b).

Assumption 10.14 The attacks are assumed to be not coordinated, and only two out of possible 18 attacks of following attack signal

$$d_{18\times 1} = \left[(d_1^\delta)^T \ (d_1^\omega)^T \ (d_1^\theta)^T \ (d_2^\delta)^T \ (d_2^\omega)^T \right]^T, \tag{10.82}$$

are assumed to happen (it is not known which ones), the other 16 unknown attacks are assumed non-existent. These two attacks are recovered using the SR algorithm described in Sect. 3 applied to filtered WECC power system (10.72).

10.6.4.1 Simulation Results

The simulation results have been obtained via MATLAB.

Simulation Experiment 1 *Two constant attacks* $(d_1^\omega)_2 = -1$ *which is the second entry of* d_1^ω, *and* $(d_2^\omega)_1 = 1$ *affect the filtered WECC power system (10.72) at the time* $t = 0.4$ *s, and* $\tau = 0.01$. *The SR algorithm was used to recover the attacks. The results of the simulations are shown in Fig. 10.1. The simulated two non-zero attacks, which are shown by dash line and dot line, are accurately recovered in finite time, while the estimated values of other zero attacks, which are shown by solid lines, converge to zero in finite time. In Figs. 10.1, 10.2 and 10.3, Attack1 and Attack2 are used to describe the real attack signals and* $d_1 - d_{18}$ *display the reconstructed plant and sensor attacks.*

Simulation Experiment 2 *Two time-varying attacks,* $(d_1^\omega)_1 = sin(\pi t)$ *and* $(d_1^\omega)_2 = sin(\pi t)$ *affect the filtered WECC power system (10.60) at the time* $t = 0.4$ *s. The simulated two time-varying non-zero attacks are accurately recovered in finite time, which are illustrated by dash line and dot line, while the estimated values of other 16 zero attacks appear to converge to zero in finite time. The solid lines illustrate them.*

Simulation Experiment 3 *Two non-zero attacks are generated and affected the filtered WECC power system (10.60) at the time* $t = 0.4$ *s, the plant attack is time varying* $(d_1^\omega)_2 = sin(\pi t)$, *and sensor attack is constant* $(d_2^\omega)_1 = -1$. *The simulation result in Fig. 10.3 shows 2 non-zero and 16 zero attacks were accurately recovered in finite time.*

The Simulation results in Figs. 10.1, 10.2 and 10.3 show that SR algorithm can reconstruct the time-varying sparse attack signal in finite time.

10.6.5 Reconstruction of Attacks and Estimation of States: The Number of Sensors is Greater Than the Number of Potential Sensor Attacks

In this section, we investigate the WECC power system (10.60) as a nonlinear system when we have more sensors rather than potential sensor attacks, i.e., there are 6 sensor

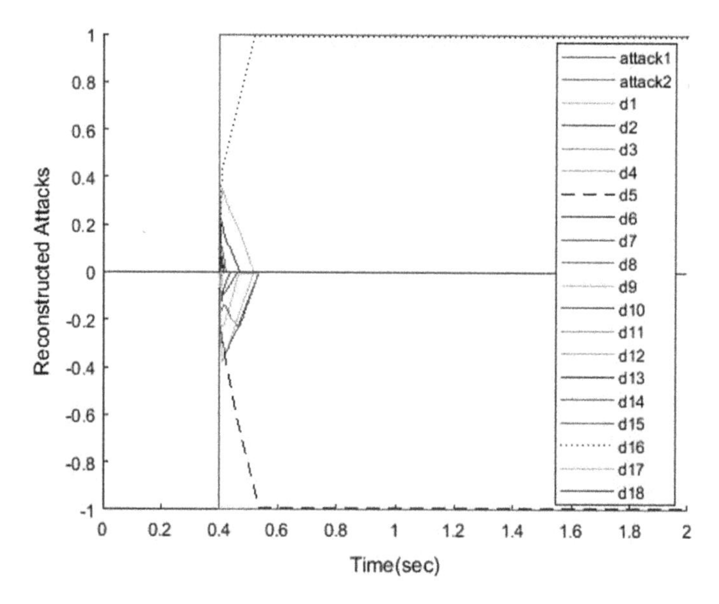

Fig. 10.1 Reconstruction of Two Constant Plant Attack and Sensor Attack in a Sparse Attack Signal, ©2018 IEEE. Reprinted, with permission, from Nateghi et al. (2018b)

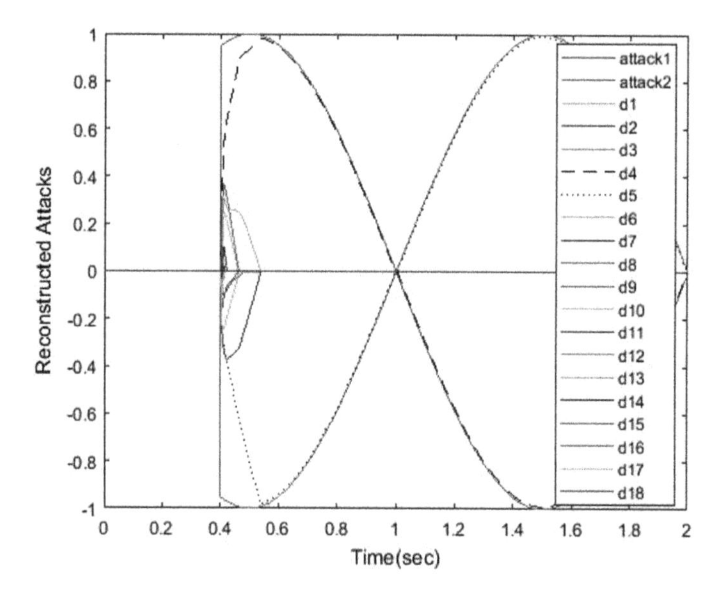

Fig. 10.2 Reconstruction of Two Time Varying Plant Attack in a Sparse Attack Signal, ©2018 IEEE. Reprinted, with permission, from Nateghi et al. (2018b)

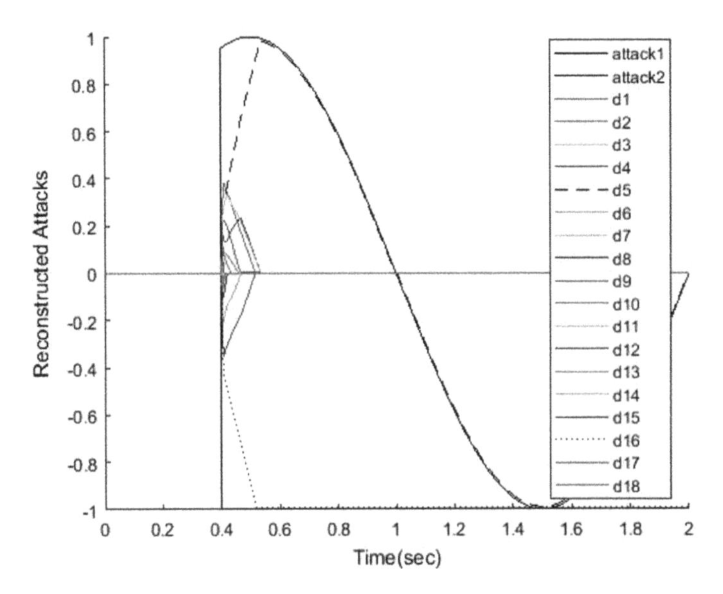

Fig. 10.3 Reconstruction of Time Varying Plant Attack and Constant Sensor Attack in a Sparse Attack Signal, ©2018 IEEE. Reprinted, with permission, from Nateghi et al. (2018b)

measurements and 3 plant attacks. The matrices B and D in (10.60) are defined in such a way that plant attack d_x and sensor attack d_y can be written separately as follows:

$$\begin{bmatrix} I & 0 & 0 \\ 0 & M_g & 0 \\ 0 & 0 & 0 \end{bmatrix} \begin{bmatrix} \dot{\delta} \\ \dot{\omega} \\ \dot{\theta} \end{bmatrix} = - \begin{bmatrix} 0 & -I & 0 \\ R^{\theta}_{g,g} & E_g & R^{\theta}_{g,l} \\ R^{\theta}_{l,g} & 0 & R^{\theta}_{l,l} \end{bmatrix} \begin{bmatrix} \delta \\ \omega \\ \theta \end{bmatrix} + \begin{bmatrix} 0 \\ I \\ 0 \end{bmatrix} d_x(t) + \begin{bmatrix} 0 \\ P_{\omega} \\ P_{\theta} \end{bmatrix}$$
$$y = \begin{bmatrix} C_{\delta} & 0 \\ 0 & C_{\omega} \end{bmatrix} \begin{bmatrix} \delta \\ \omega \end{bmatrix} + \begin{bmatrix} D_{\delta} \\ D_{\omega} \end{bmatrix} d_y(t), \tag{10.83}$$

where

$$C_{\delta} = I_3 \ , C_{\omega} = I_3 \ , D_{\delta} = 0_{3\times 6} \ , D_{\omega} \in R^{3\times 6} = \begin{bmatrix} 0 & 1 & 2 & 0 & 1 & 1 \\ 1 & 0 & 0 & 2 & 1 & 0 \\ 0 & 0 & 1 & 0 & 1 & 0 \end{bmatrix}. \tag{10.84}$$

The WECC power system (10.84) can be rewritten as

$$\begin{bmatrix} \dot{\delta} \\ \dot{\omega} \end{bmatrix} = \begin{bmatrix} \omega \\ M_g^{-1}(-R^{\theta}_{g,g} + R^{\theta}_{g,l}(R^{\theta}_{l,l})^{-1} R^{\theta}_{l,g})\delta - M_g^{-1} E_g \omega + P_{\theta\omega} \end{bmatrix} + \bar{B} d_x(t)$$
$$\begin{bmatrix} y_1 \\ y_2 \end{bmatrix} = \begin{bmatrix} \bar{C}_{\delta} \\ \bar{C}_{\omega} \end{bmatrix} \begin{bmatrix} \delta \\ \omega \end{bmatrix} + \begin{bmatrix} 0 \\ D_{\omega} \end{bmatrix} d_y(t) \tag{10.85}$$

where

$$\begin{aligned} P_{\theta\omega} &= M_g^{-1}(P_\omega - L_{g,l}^\theta (L_{l,l}^\theta)^{-1} P_\theta) \\ B_{\theta\omega} &= M_g^{-1}(B_\omega - L_{g,l}^\theta (L_{l,l}^\theta)^{-1} B_\theta) \\ \bar{C}_\delta &= \begin{bmatrix} I_3 \; 0_3 \end{bmatrix}, \quad \bar{C}_\omega = \begin{bmatrix} 0_3 \; I_3 \end{bmatrix}, \quad \bar{B} = \begin{bmatrix} 0_3 \\ M_g^{-1} \end{bmatrix}. \end{aligned} \tag{10.86}$$

Remark 10.6 It can be verified that D_ω satisfies the RIP condition defined in (10.16).

Suppose that the following three plant attacks (Nateghi et al. 2018a)

$$d_x = \begin{bmatrix} d_{x1} \\ d_{x2} \\ d_{x3} \end{bmatrix} = (t-10) \begin{bmatrix} sin(0.5t) \\ 0.5cos(0.5t) \\ 0.5sin(0.5t) + 0.5cos(0.5t) \end{bmatrix} \tag{10.87}$$

and the time-varying sensor attack

$$d_y = 1(t-10). \begin{bmatrix} 0 \; 0 \; 0 \; 0.5cos(0.5t) \; 0 \; 0 \end{bmatrix} \tag{10.88}$$

affect system (10.83) at $t = 10$ s.

The states $\hat{\delta}$, $\hat{\omega}$ and plant attacks $d_x(t)$ in (10.83) are reconstructed by using HOSM observer. Then, the estimated $\hat{\omega}$ is used in to give

$$y_2 - \hat{\omega} = D_\omega d_y(t). \tag{10.89}$$

The SR algorithm described in Sect. 10.3 can be applied to reconstruct the sparse $d_y(t)$ in WECC power system (10.89), where only one out of six potential attacks $d_{y1} \ldots d_{y6}$ is non-zero.

10.6.5.1 Simulation Results

The MATLAB software is used to simulate the system. The simulated plant attacks d_{x1}, d_{x2}, d_{x3} and sensor attack $d_{y1} \ldots d_{y6}$ are accurately recovered in finite time and are shown in Figs. 10.4 and 10.5, respectively. Reconstructed attacks are used for cleaning the corrupted plant input and measurements. Figures 10.6 and 10.7 compare the corrupted measurements with the measurements when the system is not under attack, and with the compensated measurements after being attacked.

Therefore, simulation results illustrate that compensated measurements converge to the measurements without attack in finite time. As a result, actual measurements are recovered from corrupted ones in finite time by using the HOSM observer and SR algorithm.

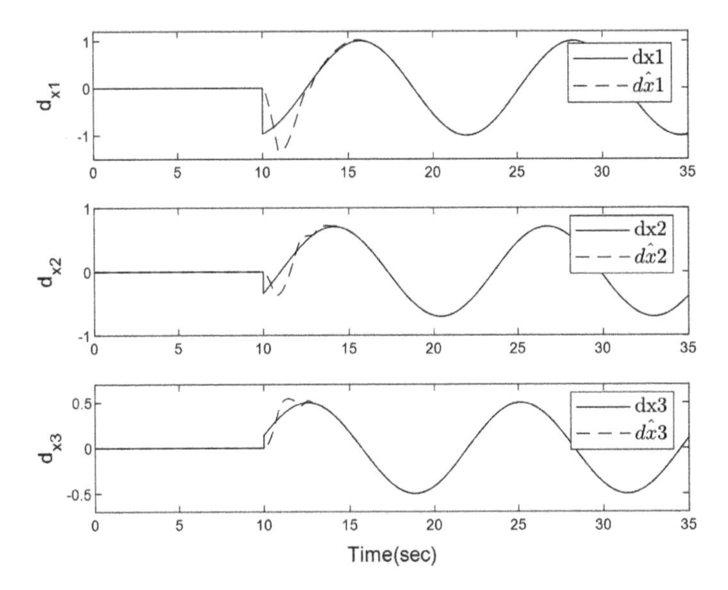

Fig. 10.4 Plant Attack $d_{x_1}, d_{x_2}, d_{x_3}$ Compare with its Reconstruction $\hat{d}_{x_1}, \hat{d}_{x_2}, \hat{d}_{x_3}$, ©2018 IEEE. Reprinted, with permission, from Nateghi et al. (2018a)

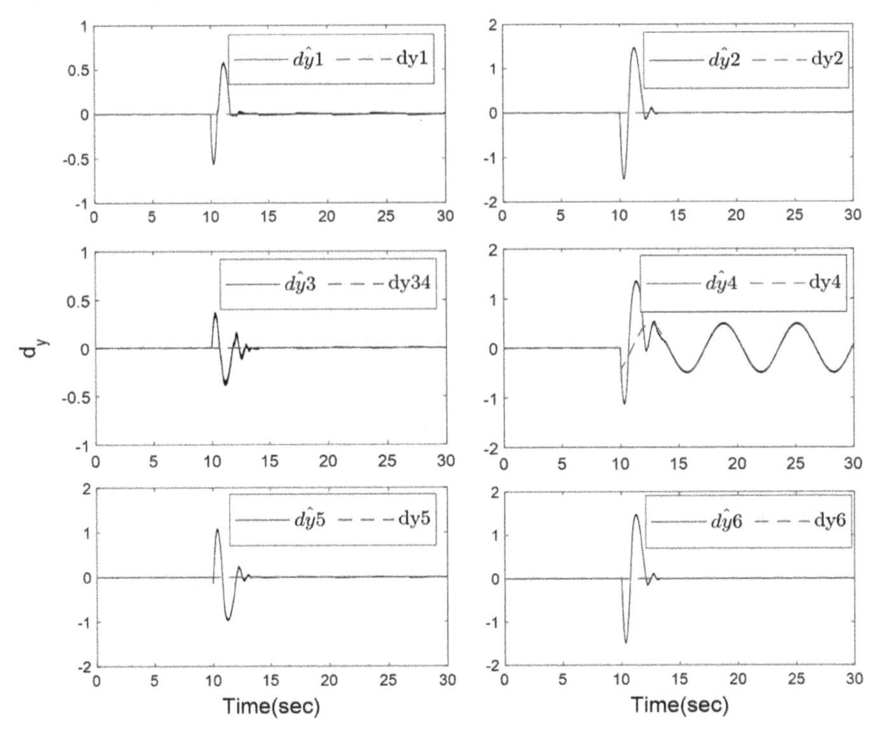

Fig. 10.5 Sensor Attack d_y Reconstruction, ©2018 IEEE. Reprinted, with permission, from Nateghi et al. (2018a)

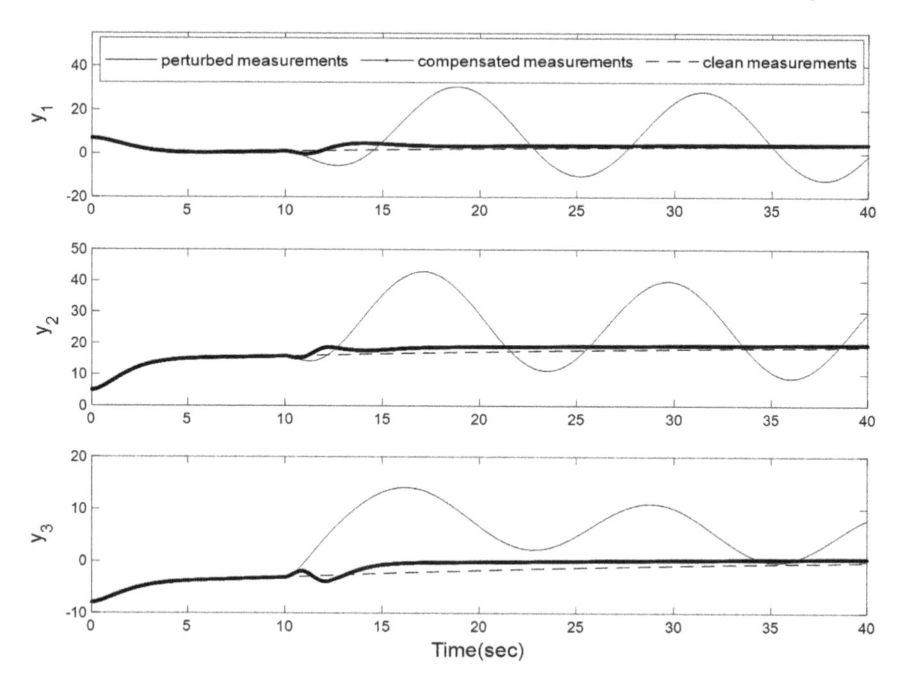

Fig. 10.6 Corrupted WECC Power System Sensor Measurements y_1, y_2, y_3 Compared with the Compensated Measurements and to the Measurements without Attacks, ©2018 IEEE. Reprinted, with permission, from Nateghi et al. (2018a)

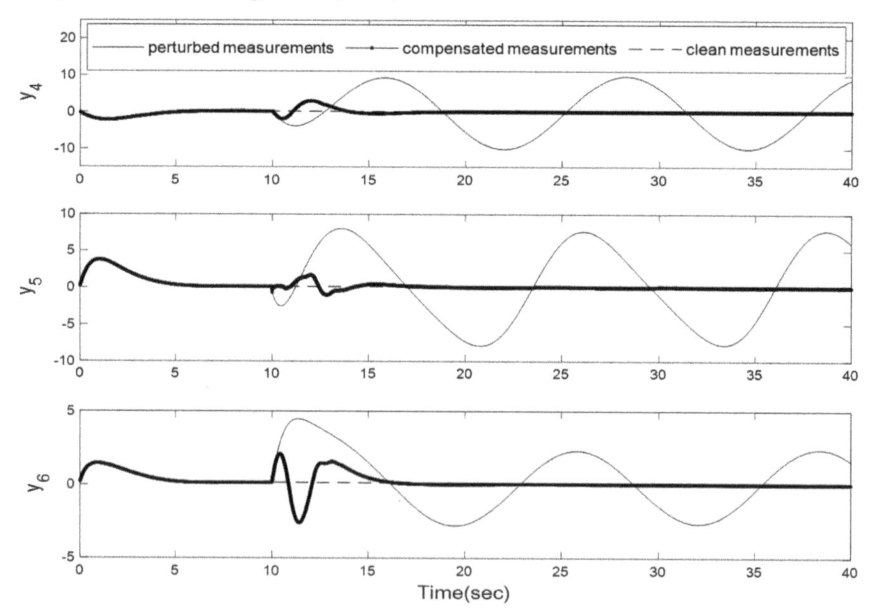

Fig. 10.7 Corrupted WECC Power System Sensor Measurements y_4, y_5, y_6 Compared with the Compensated Measurements and to the Measurements without Attacks (Nateghi et al. 2018a)

10.7 Conclusions

In this chapter, considering the nonlinear cyber-physical systems under deception attacks and sparse sensor attacks, two complimentary cases are investigated. In the first case, when the number of potential attacks is greater than the number of sensor measurements, attacks are reconstructed using higher-order sliding mode differentiation techniques in concert with the SR algorithm, when only several unknown attacks out of all possible attacks are non-zero. In the second case, when the number of sensor measurements is equal or greater than the number of potential sensor attacks, the states of the system and the state attacks are reconstructed online using a HOSM observer. A SR algorithm is used to reconstruct the stealth sensor attacks to the unprotected sensors. The effectiveness of the proposed algorithms to estimate the states and reconstruct the attacks are tested on the US WECC power network system. The simulation results confirm that the attacks degrade the performance of CPS under attack and imply that cleaning the measurements from the reconstructed attacks before using them in the feedback control can elevate CPS performance close to the one without attack.

References

A. Ahmed, K.A. Bakar, M.I. Channa, K. Haseeb, A.W. Khan, A survey on trust based detection and isolation of malicious nodes in ad-hoc and sensor networks. Front. Comput. Sci. **9**(2), 280–296 (2015)

P. Antsaklis, Goals and challenges in cyber-physical systems research editorial of the editor in chief. IEEE Trans. Autom. Control **59**(12), 3117–3119 (2014)

J.-P. Barbot, D. Boutat, T. Floquet, An observation algorithm for nonlinear systems with unknown inputs. Automatica **45**(8), 1970–1974 (2009)

E. Candes, T. Tao, Decoding by linear programming. IEEE Trans. Inf. Theory **51**(12), 4203–4215 (2005)

A.A. Cardenas, S. Amin, S. Sastry, Secure control: towards survivable cyber-physical systems, in *2008 The 28th International Conference on Distributed Computing Systems Workshops* (IEEE, 2008), pp. 495–500

A. Cetinkaya, H. Ishii, T. Hayakawa, Networked control under random and malicious packet losses. IEEE Trans. Autom. Control **62**(5), 2434–2449 (2016)

T.M. Chen, Stuxnet, the real start of cyber warfare?[editor's note]. IEEE Netw. **24**(6), 2–3 (2010)

S. Chen, M. Ma, Z. Luo, An authentication scheme with identity-based cryptography for m2m security in cyber-physical systems. Secur. Commun. Netw. **9**(10), 1146–1157 (2016)

S.M. Dibaji, H. Ishii, R. Tempo, Resilient randomized quantized consensus. IEEE Trans. Autom. Control **63**(8), 2508–2522 (2017)

S.M. Dibaji, M. Pirani, D.B. Flamholz, A.M. Annaswamy, K.H. Johansson, A. Chakrabortty, A systems and control perspective of cps security. Annu. Rev. Control **47**, 394–411 (2019)

W. Diffie, M. Hellman, New directions in cryptography. IEEE Trans. Inf. Theory **22**(6), 644–654 (1976)

F. Farokhi, I. Shames, N. Batterham, Secure and private control using semi-homomorphic encryption. Control Eng. Pract. **67**, 13–20 (2017)

H. Fawzi, P. Tabuada, S. Diggavi, Secure estimation and control for cyber-physical systems under adversarial attacks. IEEE Trans. Autom. Control **59**(6), 1454–1467 (2014)

L. Fridman, A. Levant, J. Davila, Observation of linear systems with unknown inputs via high-order sliding-modes. Int. J. Syst. Sci. **38**(10), 773–791 (2007)

L. Fridman, Y. Shtessel, C. Edwards, X.-G. Yan, Higher-order sliding-mode observer for state estimation and input reconstruction in nonlinear systems. Int. J. Robust Nonlinear Control IFAC-Affiliated J. **18**(4–5), 399–412 (2008)

K. Hartmann, C. Steup, The vulnerability of uavs to cyber attacks-an approach to the risk assessment, in *5th International Conference on Cyber Conflict (CYCON 2013)* (IEEE, 2013), pp. 1–23

W.P. Heemels, K.H. Johansson, P. Tabuada, An introduction to event-triggered and self-triggered control, in *IEEE 51st Ieee Conference on Decision and Control (CDC)* (IEEE, 2012), pp. 3270–3285

X. Huang, D. Zhai, J. Dong, Adaptive integral sliding-mode control strategy of data-driven cyber-physical systems against a class of actuator attacks. IET Control Theory Appl. **12**(10), 1440–1447 (2018)

A. Jafarnia-Jahromi, A. Broumandan, J. Nielsen, G. Lachapelle, Gps vulnerability to spoofing threats and a review of antispoofing techniques. Int. J. Navig. Obs. **2012** (2012)

X. Jin, W.M. Haddad, T. Yucelen, An adaptive control architecture for mitigating sensor and actuator attacks in cyber-physical systems. IEEE Trans. Autom. Control **62**(11), 6058–6064 (2017)

H.J. LeBlanc, H. Zhang, X. Koutsoukos, S. Sundaram, Resilient asymptotic consensus in robust networks. IEEE J. Sel. Areas Commun. **31**(4), 766–781 (2013)

A. Levant, Higher-order sliding modes, differentiation and output-feedback control. Int. J. Control **76**(9–10), 924–941 (2003)

S. Mousavian, J. Valenzuela, J. Wang, A probabilistic risk mitigation model for cyber-attacks to pmu networks. IEEE Trans. Power Syst. **30**(1), 156–165 (2014)

S. Nateghi, Y. Shtessel, Robust stabilization of linear differential inclusion using adaptive sliding mode control. Annu. Am. Control Conf. (ACC) **2018**, 5327–5331 (2018)

S. Nateghi, Y. Shtessel, J.-P. Barbot, C. Edwards, Cyber attack reconstruction of nonlinear systems via higher-order sliding-mode observer and sparse recovery algorithm. IEEE Conf. Decis. Control (CDC) **2018**, 5963–5968 (2018a)

S. Nateghi, Y. Shtessel, J.-P. Barbot, G. Zheng, L. Yu, Cyber-attack reconstruction via sliding mode differentiation and sparse recovery algorithm: electrical power networks application, in *2018 15th International Workshop on Variable Structure Systems (VSS)* (2018b), pp. 285–290

S. Nateghi, Y. Shtessel, R. Rajesh, S.S. Das, Control of nonlinear cyber-physical systems under attack using higher order sliding mode observer, in *2020 IEEE Conference on Control Technology and Applications (CCTA)* (IEEE, 2020a), pp. 1–6

S. Nateghi, Y. Shtessel, C. Edwards, Cyber-attacks and faults reconstruction using finite time convergent observation algorithms: electric power network application. J. Frankl. Inst. **357**(1), 179–205 (2020b)

S. Nateghi, Y. Shtessel, C. Edwards, Resilient control of cyber-physical systems under sensor and actuator attacks driven by adaptive sliding mode observer. Int. J. Robust Nonlinear Control (2021)

E. Nekouei, M. Skoglund, K.H. Johansson, Privacy of information sharing schemes in a cloud-based multi-sensor estimation problem, in *2018 Annual American Control Conference (ACC)*. (IEEE, 2018), pp. 998–1002

F. Pasqualetti, F. Dorfler, F. Bullo, Attack detection and identification in cyber-physical systems. IEEE Trans. Autom. Control **58**(11), 2715–2729 (2013)

F. Pasqualetti, F. Dorfler, F. Bullo, Control-theoretic methods for cyberphysical security: geometric principles for optimal cross-layer resilient control systems. IEEE Control Syst. Mag. **35**(1), 110–127 (2015)

M. Sain, J. Massey, Invertibility of linear time-invariant dynamical systems. IEEE Trans. Autom. Control **14**(2), 141–149 (1969)

E. Scholtz, Observer-based monitors and distributed wave controllers for electromechanical disturbances in power systems, Ph.D. dissertation, Massachusetts Institute of Technology (2004)

Y. Shtessel, C. Edwards, L. Fridman, A. Levant et al., *Sliding Mode Control and Observation*, vol. 10 (Springer, 2014)

J. Slay, M. Miller, Lessons learned from the maroochy water breach, in *International Conference on Critical Infrastructure Protection* (Springer, 2007), pp. 73–82
A.F. Taha, J. Qi, J. Wang, J.H. Panchal, Risk mitigation for dynamic state estimation against cyber attacks and unknown inputs. IEEE Trans. Smart Grid **9**(2), 886–899 (2016)
V.I. Utkin, Manipulator control system, in *Sliding Modes in Control and Optimization* (Springer, 1992), pp. 239–249
C. Wu, Z. Hu, J. Liu, L. Wu, Secure estimation for cyber-physical systems via sliding mode. IEEE Trans. Cybern. **48**(12), 3420–3431 (2018)
L. Yu, G. Zheng, J.-P. Barbot, Dynamical sparse recovery with finite-time convergence. IEEE Trans. Signal Process. **65**(23), 6146–6157 (2017)
Q. Zhu, T. Başar, Robust and resilient control design for cyber-physical systems with an application to power systems, in *2011 50th IEEE Conference on Decision and Control and European Control Conference* (IEEE, 2011), pp. 4066–4071

Chapter 11
Resilient Cooperative Control of Input Constrained Networked Cyber-Physical Systems

Junjie Fu, Guanghui Wen, Yongjun Xu, Ali Zemouche, and Fan Zhang

11.1 Introduction

With the rapid development of sensing, communication, and computing technology, intensive research attention has been devoted to the coordination control of NCPS in recent years. NCPS can be used to model a large class of complex networked infrastructures where the network layer is closely intertwined with the physical layer. Different Cyber-Physical systems in the network communicate with each other and the interaction has a direct impact on the operation of the physical plants in the local system. Therefore, they are more complex than the traditional networked systems as constraints from the physical processes have to be taken into consideration when designing cooperative control strategies. Potential applications range across wide areas such as mobile sensor networks, unmanned aerial vehicles, and small satellite groups (Olfati-Saber and Murray 2004; Beard et al. 2001). In distributed control of NCPS, the objective is to design distributed control laws that use only local information such that some global control tasks can be completed. Common coordination tasks include consensus, coordinated tracking, flocking, swarming, and

J. Fu (✉) · G. Wen
School of Mathematics, Southeast University, Nanjing 210096, People's Republic of China
e-mail: fujunjie@seu.edu.cn

Y. Xu
Institute of Computing Technology, Chinese Academy of Sciences, Beijing 100190, China
e-mail: xyj@ict.ac.cn

A. Zemouche
CRAN CNR-UMR 7039, IUT Henri Poincaré de Longwy, Université de Lorraine, Cosnes-et-Romain, France
e-mail: ali.zemouche@univ-lorraine.fr

F. Zhang
School of Aeronautics and Astronautics (Shenzhen), Sun Yat-sen University, Guangzhou, People's Republic of China
e-mail: zhangfan6@mail.sysu.edu.cn

M. Abbaszadeh and A. Zemouche (eds.), *Security and Resilience in Cyber-Physical Systems*, https://doi.org/10.1007/978-3-030-97166-3_11

so forth (Murray 2007; Cao et al. 2013). In the coordinated tracking problem, there generally exists a leader agent which determines the final desired trajectory of the network and the follower agents need to track such a trajectory asymptotically. This control paradigm is especially suitable for the distributed control of systems such as smart grids and mobile vehicle networks (Bidram et al. 2013; Sheikholeslam and Desoer 1992). Various results have been obtained for this problem regarding different agent dynamics and communication topologies (Hong et al. 2006; Chen and Song 2014; Wen et al. 2014; Li et al. 2014; Zhang et al. 2017; Vanli et al. 2017).

When implementing the coordination controllers on real systems, practical limitations of the agent dynamics have to be considered. One common limitation is the input saturation effect resulted from the finite actuation power of physical systems. It may lead to serious performance degradation or even instability if not properly handled (Hu and Lin 2001; Zaccarian and Teel 2011). Therefore, designing coordination controllers for NCPS subject to input saturation has great importance. In Li et al. (2011) and Du et al. (2013), input saturated consensus for first-order integrators was studied under directed communication graphs. In Ren (2008), global bounded consensus algorithms for double-integrator dynamics were designed. In Abdessameud and Tayebi (2010), consensus strategies accounting for actuator saturations and the lack of velocity measurements were designed for a group of agents with double-integrator dynamics based on auxiliary systems. In Meng et al. (2013), input saturated global coordinated tracking problem was investigated for NCPS with, respectively, neutrally stable dynamics and double-integrator dynamics subject to detail-balanced directed graphs. Global input saturated consensus problem for discrete-time neutrally stable and double-integrator NCPS was studied in Yang et al. (2014). A multi-hop relay-based distributed controller was proposed in Zhao and Lin (2016) to achieve global consensus tracking for asymptotically null controllable with bounded control (ANCBC) linear NCPS under detail-balanced directed graphs. Bounded observer-based control strategies were proposed in Meng and Lin (2013) and Fu and Wang (2014) to ensure finite-time coordinated tracking for both low- and high-order uncertain integrator NCPS under general directed communication graphs. Apart from these results on global coordinated tracking, semi-global coordination of ANCBC linear NCPS with input saturation has been investigated in Song et al. (2016), Su et al. (2013), Zhao and Lin (2015), and Wang et al. (2018) using low-gain control approach. Specifically, semi-global consensus of ANCBC linear NCPS with input saturation using relative output feedback was investigated in Fan et al. (2015). Low- and high-gain control approaches were employed in Wang et al. (2017a) to achieve global consensus tracking for ANCBC linear NCPS with input saturation under directed switching graphs. Robust global coordinated tracking for ANCBC linear NCPS with input saturation and input-additive uncertainties was achieved in Wang et al. (2017b) where the communication graphs among the followers were assumed to be undirected.

Note that for the global consensus of second- or high-order NCPS with input saturation, a common assumption in the existing results is that the communication graphs are undirected or special directed graphs (e.g., detail-balanced directed graphs) (Ren 2008; Abdessameud and Tayebi 2010; Meng et al. 2013; Yang et al.

2014; Zhao and Lin 2016; Wang et al. 2017b). Furthermore, uncertain dynamics and input disturbances are usually not well dealt with in existing works (Zhao and Lin 2016; Wang et al. 2017a). The observer-based controllers in Meng and Lin (2013) and Fu and Wang (2014) can be applied to high-order NCPS with uncertain dynamics under general directed graphs. However, to implement the control strategies, both the relative state measurements and transmission of internal observer states are needed. In this work, we aim to design distributed coordinated tracking controllers for input saturated high-order NCPS under general directed communication graphs with reduced communication. New kinds of nonlinear distributed controllers using only local measurement information are proposed to achieve coordinated tracking. Sending internal states using digital communication between neighboring agents is avoided which may both simplify the agent design and reduce the energy consumption of the network. Application to the platoon control of autonomous vehicles is used to illustrate the effectiveness of the proposed control strategies.

Most of the existing consensus results are asymptotic algorithms which means the coordination objective can be achieved as time goes to infinity. In many cases, finite-time convergence is preferable due to the mission requirement. Furthermore, finite-time controller also enjoys the benefits of faster convergence speed and more robustness to uncertainties and disturbances (Bhat and Bernstein 2000). The finite-time consensus problem for first-order integrator systems was studied in Cortés (2006) and Xiao et al. (2009). Homogeneous system theory was employed in Wang and Hong (2008) to develop a class of continuous finite-time consensus controllers for second-order systems under undirected communication graphs. A robust finite-time consensus tracking controller was developed in Khoo et al. (2009) using terminal sliding mode control techniques. In Cao et al. (2010) and Meng and Lin (2013), finite-time formation tracking problems for first-order and second-order integrator systems with directed switching communication graphs were solved by designing decentralized finite-time sliding mode estimators. In Meng et al. (2010) and Du et al. (2011), finite-time attitude consensus algorithms were proposed for nonlinear spacecraft models. Observer-based finite-time consensus tracking problem for high-order integrator systems with bounded external disturbances was studied in Fu and Wang (2014).

However, existing finite-time consensus control designs have rarely considered input saturation. In Wang and Hong (2008) and Fu et al. (2018), bounded finite-time consensus controllers were proposed for second-order integrator systems under undirected communication graphs. Sliding mode control-based controllers were proposed in Fu et al. (2019) for input constrained second-order NCPS with directed communication graphs. Considering that many practical NCPS have high-order dynamics and the directed communication graphs are more general due to the presence of link failures or communication constraints, designing finite-time consensus controller for high-order NCPS with general directed communication graphs has great importance. Notably, the observer-based controllers proposed in Fu and Wang (2014) were able to achieve finite-time consensus tracking of high-order integrator systems subject to input saturation. However, explicit communication of observer states was required to implement the controller. In this chapter, a switching control strategy is proposed which achieves robust finite-time consensus control of high-order NCPS with input saturation under general directed communication graphs using only local relative measurement information.

The main contribution of this chapter is on the design of input constrained distributed consensus tracking strategies for high-order triangular form NCPS subject to general directed communication graphs. Both asymptotic and finite-time consensus problems have been studied. The controllers have the feature that sliding mode control techniques have been employed and only relative state or output measurement is needed to implement the control strategies. As a result, they are resilient to both the control input constraints, the unknown external disturbances and the possible digital communication restraints.

Organization: The contents of this chapter can be concluded and summarized as follows. The notations and some preliminaries will be given next. In Sect. 11.2, new classes of input constrained consensus tracking controllers for high-order NCPS are proposed. In Sect. 11.3, the finite-time input constrained consensus tracking problem for high-order NCPS is studied. In Sect. 11.4, Simulation examples are provided. Finally, conclusions are provided in Sect. 11.5.

11.1.1 Notation

Here, we introduce the notations that will be used throughout this chapter. $\mathbf{1_N}$ is a vector of all 1s. $\|x\|_1$ and $\|x\|_\infty$ represent the 1-norm and infinity-norm of a vector $x = [x_1, \ldots, x_n]^T$, respectively. $\text{diag}\{x_1, \ldots, x_n\}$ is a diagonal matrix composed of the elements $x_1, \ldots, x_n$. $\text{sgn}(x) = [\text{sgn}(x_1), \ldots, \text{sgn}(x_n)]^T$ denotes the signum function. $\text{sig}(x)^\alpha = \text{sgn}(x)|x|^\alpha$. $\|A\|_\infty$ is the induced infinity-norm of a square matrix A. $\lambda_{\min}(A)$ denotes the smallest eigenvalue when all the eigenvalues of A are real.

11.1.2 Preliminaries on Algebraic Graph Theory

A directed graph $\mathscr{G} = (\mathscr{V}(\mathscr{G}), \mathscr{E}(\mathscr{G}))$ can be used to represent the communication relation among the agents where $\mathscr{V}(\mathscr{G}) = \{e_0, e_1, \ldots, e_N\}$ is the vertex set and $\mathscr{E}(\mathscr{G}) \subset \mathscr{V}(\mathscr{G}) \times \mathscr{V}(\mathscr{G})$ is the edge set. Agent i is represented by vertex e_i in $\mathscr{V}(\mathscr{G})$ and an edge (e_i, e_j) represents the information flow from agent j to agent i. The set of neighbors of node e_i is denoted by $\mathscr{N}_i = \{j : (e_i, e_j) \in \mathscr{E}(\mathscr{G})\}$. A directed path $\mathscr{P}$ in $\mathscr{G}$ from e_{i_0} to e_{i_k} is a sequence of distinct vertices $\{e_{i_0}, \ldots, e_{i_k}\}$ where $(e_{i_{j-1}}, e_{i_j}) \in \mathscr{E}(\mathscr{G})$ for $j = 1, \ldots, k$. Node e_j is reachable from e_i if there exists a path from e_i to e_j. Graph $\mathscr{G}$ is strongly connected if there exists a path between any two ordered vertices and contains a spanning tree if there exists a vertex, named as root, which is reachable from all the other vertices in the graph. An induced subgraph $\mathscr{G}_s$ of $\mathscr{G}$ is a graph such that $\mathscr{V}(\mathscr{G}_s) \subset \mathscr{V}(\mathscr{G})$ and for any $e_i, e_j \in \mathscr{V}(\mathscr{G}_s)$, $(e_i, e_j) \in \mathscr{E}(\mathscr{G}_s)$ if and only if $(e_i, e_j) \in \mathscr{E}(\mathscr{G})$. In this chapter, the vertex set $\mathscr{V}(\mathscr{G}_s) = \{e_1, \ldots, e_N\}$

of the subgraph $\mathscr{G}_s$ is used to represent the follower agents. The adjacency matrix $\mathscr{A} = [a_{ij}]$ associated with $\mathscr{G}$ is defined as $a_{ii} = 0$ and $a_{ij} > 0$ if $(e_i, e_j) \in \mathscr{E}(\mathscr{G})$ where $i \neq j$. The Laplacian matrix of $\mathscr{G}$ is defined as $\mathscr{L} = [l_{ij}]$ where $l_{ii} = \sum_{j \neq i} a_{ij}$ and $l_{ij} = -a_{ij}$ where $i \neq j$. The follower agents have no influence over the leader; therefore, $a_{0i} = 0$, $i = 1, \ldots, N$. Moreover, the communication relation between the leader and the followers is indicated by a_{i0}, $i = 1, \ldots, N$ where $a_{i0} > 0$ means that follower i directly has access to the information of the leader and $a_{i0} = 0$ otherwise. Let $\mathscr{L}_s \in \mathbb{R}^{N \times N}$ denote the Laplacian matrix associated with the subgraph $\mathscr{G}_s$ and $\mathscr{A}_0 = \mathrm{diag}\{a_{10}, \ldots, a_{N0}\} \in \mathbb{R}^{N \times N}$. A matrix $H = [h_{ij}] := \mathscr{L}_s + \mathscr{A}_0 \in \mathbb{R}^{N \times N}$ is defined for further analysis.

Lemma 11.1 (Zhang et al. 2015) *Suppose that the graph $\mathscr{G}$ contains a directed spanning tree with the leader as the root, then H is invertible. Moreover, let*

$$
\begin{aligned}
r &= [r_1, \ldots, r_N]^T = (H^{-1})^T \mathbf{1}_N, \\
R &= \mathrm{diag}\{r_1, \ldots, r_N\}, \\
W &= RH + H^T R,
\end{aligned}
\tag{11.1}
$$

then both the diagonal matrix R and the symmetric matrix W are positive definite.

11.1.3 Preliminaries on Finite-Time Stability

Consider the system

$$
\dot{x} = f(t, x), \; f(t, 0) = 0, \; x(0) = x_0, \; x \in \mathbb{R}^n, \tag{11.2}
$$

where $f : \mathbb{R}_{\geq 0} \times U_0 \to \mathbb{R}^n$ is piecewise continuous on an open neighborhood U_0 of the origin.

Definition 11.1 (Hong et al. 2002) The equilibrium point $x = 0$ of (11.2) is locally finite-time stable if it is Lyapunov stable and locally finite-time convergent in U_0. If $U_0 = \mathbb{R}^n$, then the origin is globally finite-time stable.

Definition 11.2 (Hong et al. 2002) Let $f(x) = [f_1(x), \ldots, f_n(x)]^T$ be a continuous vector field. $f(x)$ is said to be homogeneous of degree k with respect to $(r_1, r_2, \ldots, r_n) \in \mathbb{R}^n_+$ if for any given $\alpha > 0$ it holds $f_i(\alpha^{r_1} x_1, \alpha^{r_2} x_2, \ldots, \alpha^{r_n} x_n) = \alpha^{k + r_i} f_i(x)$, $i = 1, \ldots, n$.

Lemma 11.2 (Hong et al. 2002) *Consider the system*

$$
\dot{x} = f(x), \; f(0) = 0, \; x \in \mathbb{R}^n, \tag{11.3}
$$

where $f(x)$ is a continuous homogeneous vector field of degree $k < 0$ with respect to $(r_1, r_2, \ldots, r_n)$. Assume $x = 0$ is an asymptotically stable equilibrium of the system.

Then, $x = 0$ is a locally finite-time stable equilibrium of the system (11.3). Moreover, if the stable equilibrium $x = 0$ is globally asymptotically stable, then $x = 0$ is a globally finite-time stable equilibrium of (11.3).

The following lemma is useful which is based on the finite-time robust exact differentiators proposed in Levant (2003).

Lemma 11.3 *For any integer $n \geq 1$, $L > 0$, let*

$$\begin{aligned}
&\dot{z}_1(t) = w_1(t),\\
&w_1(t) = -\lambda_n L^{1/n} |z_1(t)|^{(n-1)/n} \operatorname{sgn}(z_1(t)) + z_2(t),\\
&\dot{z}_k(t) = w_k(t),\\
&w_k(t) = -\lambda_{n-k} L^{1/(n-k)} |z_k(t) - w_{k-1}(t)|^{(n-k-1)/(n-k)} \cdot\\
&\ \operatorname{sgn}(z_k(t) - w_{k-1}(t)) + z_{k+1}(t), k = 2, \ldots, n-1,\\
&\dot{z}_n(t) = v(t) - \lambda_1 L \operatorname{sgn}(z_n(t) - w_{n-1}(t)),
\end{aligned} \tag{11.4}$$

where $v(t)$ is any bounded signal satisfying $|v(t)| \leq L$. Then, there exist positive parameters λ_i, $i = 1, \ldots, n$ such that $z_i(t), i = 1, \ldots, n$ converge to zero after a finite time.

Remark 11.1 The positive parameters $\lambda_i, i = 1, \ldots, n$ can be determined in advance for given n. For $n \leq 6$, a possible set of choice is given in Levant (2005) as $\lambda_1 = 1.1$, $\lambda_2 = 1.5$, $\lambda_3 = 2$, $\lambda_4 = 3$, $\lambda_5 = 5$, and $\lambda_6 = 8$. The convergence speed generally increases with increasing design parameters.

For a signal $\sigma(t) \in \mathbb{R}$ which satisfies the condition $\left|\sigma^{(n)}\right| \leq L$, a uniform finite-time exact differentiator is proposed in Angulo et al. (2013).

Lemma 11.4 *The $(n-1)$-th-order differentiator*

$$\begin{aligned}
&\dot{z}_i = -\lambda_i \theta sig(z_1 - \sigma)^{\frac{n-i}{n}} - \eta_i (1-\theta) sig(z_1 - \sigma)^{\frac{n+\beta i}{n}} + z_2, z_{i+1}\\
&i = 1, \ldots, n-1,\\
&\dot{z}_n = -\lambda_n \theta \operatorname{sgn}(z_1 - \sigma) - \eta_n (1-\theta) sig(z_1 - \sigma)^{1+\beta}
\end{aligned}$$

is uniformly finite-time exact when its parameters are selected as follows:

- *$\{\lambda_i, i = 1, \ldots, n\}$ are selected based on the bound of the perturbation L using the formulas for the HOSM differentiator (Levant 2003);*
- *$\beta > 0$ is chosen small enough and $\{\eta_i, i = 1, \ldots, n\}$ are selected such that the polynomial $p^n + \eta_n p^{n-1} + \cdots + \eta_2 p + \eta_1$ is Hurwitz;*
- *the function $\theta : [0, \infty) \rightarrow \{0, 1\}$ is selected as*

$$\theta(t) = \begin{cases} 0 & \text{if } t \leq T_l, \\ 1 & \text{otherwise,} \end{cases}$$

with some arbitrarily chosen $T_l > 0$.

Furthermore, there exists a $T_u > T_l$ such that for any initial estimation z_i, it holds $z_i = \sigma^{(i-1)}$ where $i = 2, \ldots, n$.

11.2 Input Constrained Robust Consensus Tracking for High-Order NCPS

In this section, we propose new classes of consensus tracking controllers for high-order NCPS with input saturation constraints. Note that it is a non-trivial control task as simple control strategies generally cannot achieve global convergence of high-order NCPS with a globally bounded control input. First, the case of a static leader is considered. Then, the case of a dynamic leader with an unknown control input is studied. By using high-order finite-time convergent observers, consensus tracking controllers using only relative output information are also designed to further reduce the sensing requirement of the system. Different from the results in Meng and Lin (2013) and Fu and Wang (2014) where observer states must be transmitted among neighboring agents, the proposed controllers only need relative measurement information and avoid additional information transmission. Considering that communication usually takes a large part of the overall energy consumption of the agents, this is an advantage of the proposed controllers.

11.2.1 Problem Formulation

Consider a leader–follower network where the followers have the following high-order dynamics

$$\begin{aligned} \dot{x}_{ij} &= x_{i(j+1)}, \quad j = 1, \ldots, n-1, \\ \dot{x}_{in} &= u_i + d_i, \quad\; i = 1, \ldots, N, \end{aligned} \tag{11.5}$$

and the leader agent is modeled by

$$\begin{aligned} \dot{x}_{0j} &= x_{0(j+1)}, \quad j = 1, \ldots, n-1, \\ \dot{x}_{0n} &= u_0, \end{aligned} \tag{11.6}$$

where $x_i = [x_{i1}, \ldots, x_{in}]^T \in \mathbb{R}^n$, $i = 0, \ldots, N$ are the state vectors of the agents, d_i are the external disturbances which satisfy $|d_i| \leq \delta$, and $u_i \in \mathbb{R}$, $i = 1, \ldots, N$ are the control inputs of the followers. Suppose that the input saturation constraint requires that $|u_i| \leq u_m$ where $u_m > 0$ is a positive constant. The leader's input satisfies $|u_0(t)| \leq \rho$ where ρ is a positive constant.

Remark 11.2 The considered systems (11.5) and (11.6) include first-order and second-order dynamical systems studied in Li et al. (2011), Du et al. (2013), Ren (2008), Abdessameud and Tayebi (2010), Meng et al. (2013), and Yang et al. (2014)

as special cases. Furthermore, they can represent many practical high-order NCPS that can be put into their form after performing feedback linearization as shown in Khoo et al. (2014). An example is the platoon control of autonomous vehicles which is presented in the simulation section.

The following general communication graph is considered.

Assumption 11.1 The communication graph $\mathscr{G}$ of the leader–follower network is a general directed graph that contains a spanning tree with the leader as the root.

Note that, in order to track the trajectory of the leader precisely as the time approaches infinity, the followers' control inputs u_i have to dominate the effects of u_0 and d_i as the tracking errors approach zero. Therefore, the following assumption is necessary to achieve robust coordinated tracking with a dynamic leader.

Assumption 11.2 The input saturation level of the follower agents u_m, the upper bound of the leader's control input ρ, and the upper bound of the external disturbances δ satisfy the relation $u_m > \delta + \rho$.

In this work, we want to design input saturated distributed controllers which achieve robust global coordinated tracking for the high-order multi-agent system (11.5) and (11.6) using only local measurement information. The control objective is formally defined as follows:

Definition 11.3 (*Input saturated coordinated tracking*) Design a distributed controller for each follower $i = 1, \ldots, N$ in (11.5) which uses only local measurement information and satisfies $|u_i| \leq u_m$ such that for any initial condition, it holds $x_i(t) - x_0(t) \to 0$ as $t \to \infty$.

11.2.2 Input Constrained Robust Consensus Tracking with a Static Leader

First, we study the case when $x_{01} =$ const, that is, the leader's position is fixed. Then, it holds that $x_{0j} = 0$, $j = 2, \ldots, n$ and $u_0 = 0$. A distributed controller which uses only local state measurement and relative state measurement is proposed. Since there exists external disturbance in (11.5), we design the controller based on integral sliding mode control method by considering first the undisturbed case, that is, $d_i = 0$. Then, the leader–follower system becomes

$$\begin{aligned} \dot{x}_{ij} &= x_{i(j+1)}, \quad j = 1, \ldots, n-1, \\ \dot{x}_{in} &= \bar{u}_i, \qquad i = 1, \ldots, N. \end{aligned} \tag{11.7}$$

Definition 11.4 Given two positive constants L, M with $L \leq M$, a function $\sigma : \mathbb{R} \to \mathbb{R}$ is said to be a linear saturation for (L, M) if it is a continuous, nondecreasing function satisfying

- $s\sigma(s) > 0$ for all $s \neq 0$;
- $\sigma(s) = s$ when $|s| \leq L$;
- $|\sigma(s)| \leq M$ for all $s \in \mathbb{R}$.

A simple linear saturation function example is $\sigma(s) = s$ when $|s| \leq M$ and $\sigma(s) = Msgn(s)$ when $|s| > M$. Using the linear coordinate transformation $y_i = Tx_i$ where

$$y_{i(n-j)} = \sum_{k=0}^{j} \binom{j}{k} x_{i(n-k)}, \quad \binom{j}{k} = \frac{j!}{k!(j-k)!},$$

one can transform (11.7) and (11.6) into

$$\dot{y}_{ij} = y_{i(j+1)} + \cdots + y_{in} + \bar{u}_i, \, j = 1, \ldots, n-1, \quad \dot{y}_{in} = \bar{u}_i,$$

where $i = 0, 1, \ldots, N$ and $\bar{u}_0 = u_0$.

For each follower $i = 1, \ldots, N$, consider the following distributed controller:

$$\bar{u}_i = -\sigma_n\Big(y_{in} + \sigma_{n-1}\Big(y_{i(n-1)} + \cdots + \sigma_2\Big(y_{i2} + \sigma_1\Big(\sum_{j=0}^{N} a_{ij}\big(y_{i1} - y_{j1}\big)\Big)\Big)\cdots\Big)\Big), \tag{11.8}$$

where $\{\sigma_j\}$ are linear saturations for (L_j, M_j), $j = 1, \ldots, n$. Note that only local state measurement and relative state measurement are needed in (11.8).

Theorem 11.1 *Suppose that Assumptions 11.1 and 11.2 hold. Global coordinated tracking for (11.7) and (11.6) is achieved with (11.8) if*

$$M_j < \frac{1}{2} L_{j+1}, \quad j = 1, \ldots, n-1. \tag{11.9}$$

Furthermore, it holds $|\bar{u}_i| \leq M_n$.

Proof For state y_{in} of follower i, consider the Lyapunov function $V_{in} = y_{in}^2$. The derivative of V_{in} is given by

$$\dot{V}_{in} = -2y_{in}\sigma_n\Big(y_{in} + \sigma_{n-1}\Big(y_{i(n-1)} + \cdots + \sigma_1\Big(\sum_{j=0}^{N} a_{ij}\big(y_{i1} - y_{j1}\big)\Big)\cdots\Big)\Big). \tag{11.10}$$

Since $M_{n-1} < \frac{1}{2}L_n$, we see that $\dot{V}_{in} < 0$ for all $y_{in} > \frac{1}{2}L_n$. Therefore, it will hold $y_{in} \leq \frac{1}{2}L_n$ in finite time. Now consider the evolution of the state $y_{i(n-1)}$. Note that when $y_{in} \leq \frac{1}{2}L_n$ we have

$$\Big|y_{in} + \sigma_{n-1}\Big(y_{i(n-1)} + \cdots + \sigma_1\Big(\sum_{j=0}^{N} a_{ij}\big(y_{i1} - y_{j1}\big)\Big)\cdots\Big)\Big| \leq \frac{1}{2}L_n + M_{n-1} \leq L_n.$$

Consequently, σ_n operates in its linear region. Then the evolution of $y_{i(n-1)}$ is given by

$$\dot{y}_{i(n-1)} = -\sigma_{n-1}\Big(y_{i(n-1)} + \cdots + \sigma_1\Big(\sum_{j=0}^{N} a_{ij}\big(y_{i1} - y_{j1}\big)\Big)\cdots\Big).$$

Following the same argument as for y_{in}, we can show that $y_{i(n-1)}$ satisfies $|y_{i(n-1)}| \leq \frac{1}{2}L_{n-1}$ in finite time. Continuing this procedure, it can be shown that after some finite time the argument of every function σ_i, $i = 2, \ldots, n$ has entered the region where the function is linear. Therefore, after this finite time, we have $\dot{y}_{i1} = -\sigma_1\left(\sum_{j=0}^{N} a_{ij}\left(y_{i1} - y_{j1}\right)\right)$, where $i = 1, \ldots, N$. Let $\tilde{y}_{i1} = y_{i1} - y_{01}$, then we have $\dot{\tilde{y}}_{i1} = -\sigma_1\left(\sum_{j=0}^{N} a_{ij}\left(\tilde{y}_{i1} - \tilde{y}_{j1}\right)\right)$. Let $\tilde{y}^1 = [\tilde{y}_{11}, \ldots, \tilde{y}_{N1}]^T$ and $\xi = H\tilde{y}^1$, then $\dot{\xi} = -H\sigma_1(\xi)$, where $\sigma_1(\xi) = [\sigma_1(\xi_1), \ldots, \sigma_1(\xi_N)]^T$. Consider the Lyapunov function

$$V = \sum_{i=1}^{N} r_i \int_0^{\xi_i} \sigma_1(s)ds,$$

where r_i, $i = 1, \ldots, N$ are defined in Lemma 11.1. From the properties of the saturation functions given in Definition 11.4, it is easy to show that the Lyapunov function V is positive definite in ξ. Furthermore, it holds that

$$\dot{V} = -\sigma_1^T(\xi) RH\sigma_1(\xi) \leq -\frac{\lambda_{\min}(W)}{2}\sigma_1^T(\xi)\sigma_1(\xi).$$

Therefore, $\xi \to 0$ as $t \to \infty$. Since H is invertible, we have $\tilde{y}^1 \to 0$ as $t \to \infty$. Let $\tilde{x}_{ij} = x_{ij} - x_{0j}$, it holds $\tilde{y}_{i1} = \sum_{k=0}^{n-1}\binom{n-1}{k}\tilde{x}_{i(n-k)} \to 0$. Noting (11.7), it leads to $\tilde{x}_{ij} \to 0, i = 1, \ldots, N, j = 1, \ldots, n$ which means $x_i(t) - x_0(t) \to 0, t \to \infty$. Therefore, global coordinated tracking is achieved for (11.7) and (11.6). ■

Based on the controller (11.8), we consider the following integral sliding mode control-based controller to handle the effect of disturbances:

$$s_i = x_{in} - \int \bar{u}_i dt, \quad u_i = \bar{u}_i - k\,\mathrm{sgn}(s_i), \tag{11.11}$$

where $k > 0$.

Theorem 11.2 *Let Assumptions 11.1 and 11.2 hold. Global input saturated coordinated tracking for (11.5) and (11.6) is achieved with (11.11) if $k > \delta$ and*

$$M_j < \frac{1}{2}L_{j+1}, \quad j = 1, \ldots, n-1, \quad M_n \leq u_m - k. \tag{11.12}$$

Proof With the controller (11.11), we have

$$\dot{s}_i = u_i + d_i - \bar{u}_i = -k\,\mathrm{sgn}(s_i) + d_i,$$

where $i = 1, 2, \ldots, N$. Then under the condition $k > \delta$, we have that after finite time it holds $s_i = 0$. On the sliding surface, the closed-loop system evolves according to

$$\begin{aligned}\dot{x}_{ij} &= x_{i(j+1)}, \quad j = 1, \ldots, n-1, \\ \dot{x}_{in} &= \bar{u}_i, \quad \quad i = 1, \ldots, N.\end{aligned} \tag{11.13}$$

It follows from Theorem 11.1 that $x_i(t) - x_0(t) \to 0, t \to \infty$. Noting that $|u_i| \leq M_n + k$, it holds $|u_i| \leq u_m$. By definition, robust global input saturated coordinated tracking for (11.5) and (11.6) is achieved. ■

Remark 11.3 For each follower agent $i = 1, 2, \ldots, N$, the proposed controller (11.11) only depends on local state measurement $x_{i2}, \ldots, x_{in}$ and relative state measurement $\sum_{j=0}^{N} a_{ij}(x_{i1} - x_{j1}), \ldots, \sum_{j=0}^{N} a_{ij}(x_{in} - x_{jn})$. Furthermore, no global information about the communication graph is needed in the controller design.

11.2.3 Input Constrained Robust Consensus Tracking with a Dynamic Leader

Next, we focus on the case when the leader has unknown control input. Distributed coordinated tracking controllers are proposed based on sliding mode observers to estimate the unknown terms involving neighbors' inputs and external disturbances.

Let $\tilde{x}_{ij} = x_{ij} - x_{0j}$, $i = 1, 2, \ldots, N$, $j = 1, 2 \ldots, n$, and $e_{ij} = \sum_{k=0}^{N} a_{ik}(\tilde{x}_{ij} - \tilde{x}_{kj})$, then from (11.5) and (11.6), we have

$$\begin{aligned}\dot{e}_{ij} &= e_{i(j+1)}, \quad j = 1, 2, \ldots, n-1, \\ \dot{e}_{in} &= \sum_{j=1}^{N} a_{ij}(u_i + d_i - u_j - d_j) + a_{i0}(u_i + d_i - u_0) \\ &= h_i u_i - \sum_{j=0}^{N} a_{ij} u_j + h_i d_i - \sum_{j=1}^{N} a_{ij} d_j,\end{aligned} \tag{11.14}$$

where $h_i = \sum_{j=0}^{N} a_{ij}$.

Lemma 11.5 *Suppose that Assumption 11.1 holds. The coordinated tracking problem for the leader–follower network (11.5) and (11.6) is solved if $e^{ij}, i = 1, 2, \ldots, N$, $j = 1, 2 \ldots, n$ converge to zero as $t \to \infty$.*

Proof Let $\tilde{x}^k = [\tilde{x}_{1k}, \ldots, \tilde{x}_{Nk}]^T$ and $e^k = [e_{1k}, \ldots, e_{Nk}]^T, k = 1, 2, \ldots, n$, then we have $e^k = H\tilde{x}^k$. Since H is of full rank from Lemma 11.1, it holds that $\tilde{x}^k$ converge to zero if $e^k, k = 1, \ldots, n$ converge to zero as $t \to \infty$. Therefore, coordinated tracking problem for the leader–follower network (11.5) and (11.6) is solved. ■

Denote $\gamma_i = -\sum_{j=0}^{N} a_{ij} u_j + h_i d_i - \sum_{j=1}^{N} a_{ij} d_j$. Suppose that we have designed a local observer $\hat{\gamma}_i$ for each follower such that for some $T_1 > 0$ it holds $|\tilde{\gamma}_i(t)| = |\hat{\gamma}_i(t) - \gamma_i(t)| \leq \gamma_\tau, t \geq T_1$ where γ_τ is a positive constant which can be made arbitrarily small. Then, the following distributed controller is considered:

$$u_i = \frac{\beta_i - \hat{\gamma}_i}{h_i}, \tag{11.15}$$

where β_i is to be designed. Let $\beta = [\beta_1, \ldots, \beta_N]^T$, $\tilde{\gamma} = [\tilde{\gamma}_1, \ldots, \tilde{\gamma}_N]^T$, $u = [u_1, \ldots, u_N]^T$, and $d = [d_1, \ldots, d_N]^T$. The control input (11.15) satisfies $H(u - \mathbf{1}_N u_0 + d) = \beta - \tilde{\gamma}$. If $\|\beta(t)\|_\infty \leq \frac{u_m - \rho - \delta}{\|H^{-1}\|_\infty} - \gamma_\tau$, then we will have

$$\begin{aligned}\|u(t)\|_\infty &= \left\| H^{-1}(\beta(t) - \tilde{\gamma}(t)) + \mathbf{1}_N u_0(t) - d(t) \right\|_\infty \\ &\leq \left\| H^{-1} \right\|_\infty (\|\beta(t)\|_\infty + \gamma_\tau) + \rho + \delta \\ &\leq u_m.\end{aligned} \tag{11.16}$$

Moreover, with the controller (11.15), the closed-loop system (11.14) becomes

$$\begin{aligned}\dot{e}_{ij} &= e_{i(j+1)},\ j = 1, 2, \ldots, n-1 \\ \dot{e}_{in} &= \beta_i - \tilde{\gamma}_i.\end{aligned} \tag{11.17}$$

Thus, if we design β_i such that e_{ij} converge to zero as $t \to \infty$ and satisfy $|\beta_i(t)| \leq \frac{u_m - \rho - \delta}{\|H^{-1}\|_\infty} - \gamma_\tau$, then input saturated coordinated tracking is achieved. The following design of β is considered which is shown to satisfy this property in Lemma 11.6:

$$\begin{aligned}&\beta_i = \bar{\beta}_i - k_s \operatorname{sgn}(s_i), \quad s_i = e_{in} - \int \bar{\beta}_i(t) dt, \\ &\bar{\beta}_i(t) = -\sigma_n(\bar{e}_{in} + \sigma_{n-1}(\bar{e}_{i(n-1)} + \cdots + \sigma_1(\bar{e}_{i1}) \cdots)),\end{aligned} \tag{11.18}$$

where

$$\bar{e}_{i(n-j)} = \sum\nolimits_{k=0}^{j} \binom{j}{k} e_{i(n-k)}, \quad \binom{j}{k} = \frac{j!}{k!(j-k)!},$$

and $L_j, M_j > 0$, $j = 1, 2, \ldots, n$, $k_s > 0$ are design parameters.

Lemma 11.6 *Consider the closed-loop system (11.17) with the control input β_i given in (11.18). If*

$$k_s > \gamma_\tau, \quad M_n + k_s \leq \frac{u_m - \rho - \delta}{\left\| H^{-1} \right\|_\infty} - \gamma_\tau, \tag{11.19}$$

then, e_{ij}, $j = 1, 2, \ldots, n$ will converge to zero as $t \to \infty$ and $|\beta_i(t)| \leq \frac{u_m - \rho - \delta}{\|H^{-1}\|_\infty} - \gamma_\tau$.

Proof Consider the sliding mode variable s_i. From (11.17) to (11.18), it holds $\dot{s}_i = \beta_i - \tilde{\gamma}_i - \bar{\beta}_i = -k_s \operatorname{sgn}(s_i) - \tilde{\gamma}_i$. Consider the Lyapunov function $V_s = (1/2)s_i^2$. It follows that $\dot{V}_s = s_i \dot{s}_i \leq -(k_s - \gamma_\tau)\,|s_i|$. Therefore, under the condition $k_s > \gamma_\tau$, s_i will reach zero in finite time. On the sliding surface, the closed-loop system evolves according to

$$\dot{e}_{ij} = e_{i(j+1)}, \quad \dot{e}_{in} = -\sigma_n(\bar{e}_{in} + \sigma_{n-1}(\bar{e}_{i(n-1)} + \cdots + \sigma_1(\bar{e}_{i1}) \cdots)).$$

Following similar steps as in the proof of Theorem 11.1, it is easy to show that e_{ij} will converge to zero asymptotically. Furthermore, since $|\bar{\beta}_i(t)| \leq M_n$, under the condition $M_n + k_s \leq \frac{u_m-\rho-\delta}{\|H^{-1}\|_\infty} - \gamma_\tau$, it holds that $|\beta_i(t)| \leq \frac{u_m-\rho-\delta}{\|H^{-1}\|_\infty} - \gamma_\tau$. ■

Next, we consider the construction of desired $\hat{\gamma}_i$ for each follower. From (11.14), we have $\dot{e}_{in} = h_i u_i + \gamma_i$. The following auxiliary observer is proposed for each follower $i = 1, \ldots, N$:

$$\dot{\hat{e}}_{in} = h_i u_i - k_i \text{sgn}(\hat{e}_{in} - e_{in}), \tag{11.20}$$

where $k_i > 0$ are the design parameters. The effectiveness of the observer is shown in the following Lemma.

Lemma 11.7 *Consider the observer (11.20), if $|u_i(t)| \leq u_m$, $i = 1, 2, \ldots, N$ and the observer parameters k_i, $i = 1, \ldots, N$ satisfy*

$$k_i > h_i u_m + 2h_i\delta, \tag{11.21}$$

then the sliding surface $s_i = \hat{e}_{in} - e_{in} = 0$ will be reached in finite time, and in the sliding mode, it holds $\gamma_i = [-k_i \text{sgn}(\hat{e}_{in} - e_{in})]_{eq}$ where $[-k_i \text{sgn}(\hat{e}_{in} - e_{in})]_{eq}$ denotes the equivalent control of the switching term $-k_i \text{sgn}(\hat{e}_{in} - e_{in})$.

Proof Let $s_i = \hat{e}_{in} - e_{in}$, then we have $\dot{s}_i = -k_i \text{sgn}(s_i) - \gamma_i$. Under the conditions $|u_i(t)| \leq u_m$, $|d_i(t)| \leq \delta$, and Assumption 11.2, it holds $|\gamma_i(t)| \leq h_i u_m + 2h_i\delta$. Therefore, under the condition (11.21), the sliding surface $s_i = 0$ will be reached in finite time. Moreover, during the sliding mode, the equivalent control (Utkin 1992) of the discontinuous term $-k_i \text{sgn}(\hat{e}_{in} - e_{in})$ can be determined from $\dot{s}_i = 0$ which leads to $\gamma_i = [-k_i \text{sgn}(\hat{e}_{in} - e_{in})]_{eq}$. ■

According to the results in Utkin and Poznyak (2013), $[-k_i \text{sgn}(\hat{e}_{in} - e_{in})]_{eq}$ can be obtained by passing the discontinuous term through a low-pass filter and filtering out the high-frequency component. Consider the following low-pass filter:

$$\dot{\hat{\gamma}}_i = -\frac{\hat{\gamma}_i}{\tau} - \frac{k_i \text{sgn}(\hat{e}_{in} - e_{in})}{\tau}, \quad \hat{\gamma}_i(0) = 0, \tag{11.22}$$

with a small time constant $\tau > 0$. Then, the output $\hat{\gamma}_i$ is an estimate of $[-k_i \text{sgn}(\hat{e}_{in} - e_{in})]_{eq}$ and satisfies $\left|\hat{\gamma}_i - \left[-k_i \text{ sgn}(\hat{e}_{in} - e_{in})\right]_{eq}\right| \underset{\tau\to 0}{\to} 0$. Therefore, for sufficiently small $\tau > 0$, we have $|\tilde{\gamma}_i(t)| = |\hat{\gamma}_i(t) - \gamma_i(t)| \leq \gamma_\tau$ where $\gamma_\tau \to 0$ as $\tau \to 0$.

Combining (11.20), (11.22), (11.18), and (11.15), we propose the following distributed control input:

$$
\begin{aligned}
u_i &= \mathrm{sat}_{u_m}\left(\tfrac{\beta_i-\hat{\gamma}_i}{h_i}\right),\\
\beta_i &= \bar{\beta}_i - k_s\,\mathrm{sgn}(s_i),\quad s_i = e_{in} - \textstyle\int \bar{\beta}_i(t)dt,\\
\bar{\beta}_i(t) &= -\sigma_n(\bar{e}_{in} + \sigma_{n-1}(\bar{e}_{i(n-1)} + \cdots + \sigma_1(\bar{e}_{i1})\cdots)),\\
\dot{\hat{e}}_{in} &= h_i u_i - k_i \mathrm{sgn}(\hat{e}_{in} - e_{in}),\\
\dot{\hat{\gamma}}_i &= -\tfrac{\hat{\gamma}_i}{\tau} - \tfrac{k_i \mathrm{sgn}(\hat{e}_{in}-e_{in})}{\tau},
\end{aligned}
\tag{11.23}
$$

where $i = 1, \ldots, N$, M_j, L_j, $j = 1, 2, \ldots, n$, $k_s > 0$, $k_i > 0$, and $\tau > 0$ satisfy the conditions (11.9), (11.19), and (11.21).

Theorem 11.3 *Suppose that Assumptions 11.1 and 11.2 hold. The robust global input saturated coordinated tracking problem for (11.5) and (11.6) is solved by the distributed control input (11.23) with the controller parameters*

$$
\begin{aligned}
M_j &< \frac{1}{2}L_{j+1},\quad j = 1, \ldots, n-1,\\
M_n + k_s + \gamma_\tau &\le \frac{u_m - \rho - \delta}{\left\|H^{-1}\right\|_\infty},\\
\gamma_\tau &< k_s,\quad k_i > h_i u_m + 2h_i\delta.
\end{aligned}
\tag{11.24}
$$

Furthermore, since $\gamma_\tau \to 0$ as $\tau \to 0$, there always exists a sufficiently small $\tau > 0$ such that there exist control parameters M_j, L_j, $j = 1, 2, \ldots, n$, $k_s > 0$, $k_i > 0$ that satisfy condition (11.24).

Proof From Lemma 11.7, it follows that with the proposed observer

$$
\begin{aligned}
\dot{\hat{e}}_{in} &= h_i u_i - k_i\,\mathrm{sgn}(\hat{e}_{in} - e_{in}),\\
\dot{\hat{\gamma}}_i &= -\frac{\hat{\gamma}_i}{\tau} - \frac{k_i\,\mathrm{sgn}(\hat{e}_{in} - e_{in})}{\tau},
\end{aligned}
$$

there exists a $T_1 > 0$ such that $|\tilde{\gamma}_i(t)| = |\hat{\gamma}_i(t) - \gamma_i(t)| \le \gamma_\tau$ for $t \ge T_1$. Considering the facts that the control input u_i is bounded by $|u_i| \le u_m$ and the disturbance $|d_i| \le \delta$, it is easy to obtain that all the closed-loop signals are bounded for $t \in [0, T_1]$. For $t \ge T_1$, we have that $u_i = \frac{\beta_i - \hat{\gamma}_i}{h_i}$. It follows from Lemma 11.6 that $e_{ij}, i = 1, 2, \ldots, N$, $j = 1, 2, \ldots, n$ converge to zero as $t \to \infty$. Furthermore, we have $|u_i| \le u_m$. Therefore, robust global input saturated coordinated tracking for (11.5) and (11.6) is achieved. ■

Remark 11.4 Under some circumstances, each follower can estimate the upper bound of $\|H^{-1}\|_\infty$ using only local information. Then, the controller (11.23) can be implemented in a fully distributed fashion without knowing the global communication graph. One such example is given in the simulation section where we consider the platoon control of autonomous vehicles.

Remark 11.5 The convergence speed of the proposed distributed controller (11.23) can be adjusted by tuning the control parameters M_j, L_j, $j = 1, 2, \ldots, n$, $k_s > 0$, $k_i > 0$, and $\tau > 0$.

11.2.4 Output-Based Input Constrained Robust Consensus Tracking

In this section, we consider the case when only local output measurement x_{i1} and/or relative output measurement e_{i1} are available. Note that in this case the proposed controllers cannot be directly used since for controller (11.11) we need $x_{i2}, \ldots, x_{in}$ and $e_{i2}, \ldots, e_{in}$ and for controller (11.23) we need $e_{i2}, \ldots, e_{in}$. Therefore, to design the distributed controllers with only output measurement, we focus on developing observers for the local states and relative state information.

The following finite-time observer of local state information is considered for each follower i:

$$\begin{aligned}
\dot{\eta}_{i1} &= w_{i1}, \\
w_{i1} &= -\lambda_n L^{1/n} \left|\eta_{i1} - x_{i1}\right|^{(n-1)/n} \operatorname{sgn}(\eta_{i1} - x_{i1}) + \eta_{i2}, \\
\dot{\eta}_{ik} &= w_{ik}, \\
w_{ik} &= -\lambda_{n-k} L^{1/(n-k)} \left|\eta_{ik} - w_{i(k-1)}\right|^{(n-k-1)/(n-k)} \cdot \\
&\quad \operatorname{sgn}(\eta_{ik} - w_{i(k-1)}) + \eta_{i(k+1)}, k = 2, \ldots, n-1, \\
\dot{\eta}_{in} &= u_i - \lambda_1 L \operatorname{sgn}(\eta_{in} - w_{i(n-1)}),
\end{aligned} \tag{11.25}$$

where $L \geq \delta$ and the parameters λ_i are determined according to Remark 11.1.

Let $\tilde{x}_{ij} = \eta_{ij} - x_{ij}$, $j = 1, 2, \ldots, n$, from (11.5) and (11.25), it follows that

$$\begin{aligned}
\dot{\tilde{x}}_{i1}(t) &= \tilde{w}_{i1}(t), \\
\tilde{w}_{i1}(t) &= -\lambda_n L^{1/n} \left|\tilde{x}_{i1}(t)\right|^{(n-1)/n} \operatorname{sgn}(\tilde{x}_{i1}(t)) + \tilde{x}_{i2}(t), \\
\dot{\tilde{x}}_{ik}(t) &= \tilde{w}_{ik}(t), \\
\tilde{w}_{ik}(t) &= -\lambda_{n-k} L^{1/(n-k)} \left|\tilde{x}_{ik}(t) - \tilde{w}_{i(k-1)}(t)\right|^{(n-k-1)/(n-k)} \cdot \\
&\quad \operatorname{sgn}(\tilde{x}_{ik}(t) - \tilde{w}_{i(k-1)}(t)) + \tilde{x}_{i(k+1)}(t), k = 2, \ldots, n-1, \\
\dot{\tilde{x}}_{in}(t) &= -d_i(t) - \lambda_1 L \operatorname{sgn}(\tilde{x}_{in}(t) - \tilde{w}_{i(n-1)}(t)).
\end{aligned} \tag{11.26}$$

Noting that $|d_i| \leq \delta$ and $L \geq \delta$, we have $\eta_{ij} = x_{ij}$, $j = 1, 2 \ldots, n$ after a finite time according to Lemma 11.3.

To design the finite-time observer of relative state information for each follower i, note that from (11.14) it follows

$$\begin{aligned}
\dot{e}_{ij} &= e_{i(j+1)}, j = 1, 2, \ldots, n-1, \\
\dot{e}_{in} &= h_i u_i + \gamma_i,
\end{aligned} \tag{11.27}$$

where $i = 1, \ldots, N$ and $|\gamma_i| \leq h_i u_m + 2h_i\delta$. Consider the following distributed observer for each follower i:

$$
\begin{aligned}
\dot{\zeta}_{i1} &= w_{i1}, \\
w_{i1} &= -\lambda_n L^{1/n} \left|\zeta_{i1} - e_{i1}\right|^{(n-1)/n} \operatorname{sgn}(\zeta_{i1} - e_{i1}) + \zeta_{i2}, \\
\dot{\zeta}_{ik} &= w_{ik}, \\
w_{ik} &= -\lambda_{n-k} L^{1/(n-k)} \left|\zeta_{ik} - w_{i(k-1)}\right|^{(n-k-1)/(n-k)} \cdot \\
&\operatorname{sgn}(\zeta_{ik} - w_{i(k-1)}) + \zeta_{i(k+1)}, k = 2, \ldots, n-1, \\
\dot{\zeta}_{in} &= h_i u_i - \lambda_1 L \operatorname{sgn}(\zeta_{in} - w_{i(n-1)}),
\end{aligned}
\tag{11.28}
$$

where $L \geq h_i u_m + 2h_i\delta$ and the parameters λ_i are determined according to Remark 11.1. It can be similarly obtained from Lemma 11.3 that there exists a finite time $T_3 > 0$ such that for $t \geq T_3$, we have $\zeta_{ij} = e_{ij}$, $j = 1, 2, \ldots, n$ and $\gamma_i = [-\lambda_1 L \operatorname{sgn}(\tilde{e}_{in}(t) - \tilde{w}_{n-1}(t))]_{eq}$.

As in the previous section, we can obtain the equivalent control of $-\lambda_1 L \operatorname{sgn}(\tilde{e}_{in}(t) - \tilde{w}_{i(n-1)}(t))$ using a low-pass filter

$$
\dot{\hat{\gamma}}_i = -\frac{\hat{\gamma}_i}{\tau} - \frac{-\lambda_1 L \operatorname{sgn}(\tilde{e}_{in}(t) - \tilde{w}_{i(n-1)}(t))}{\tau}, \tag{11.29}
$$

where $\tau > 0$ is a small time constant. For sufficiently small $\tau > 0$, we have $|\tilde{\gamma}_i(t)| = |\hat{\gamma}_i(t) - \gamma_i(t)| \leq \gamma_\tau$ where $\gamma_\tau \to 0$ as $\tau \to 0$.

With the finite-time local state observer (11.25) and the finite-time relative state observer (11.28), we can construct the distributed controller (11.11) using only local and relative output measurement. Note that the trajectory of the closed-loop system is bounded in any finite-time interval since the control inputs of the agents are bounded. Therefore, the separation principle is trivially satisfied with the proposed finite-time convergent observers since after the convergence of the observers, the controller with only output measurements will reduce to the state feedback controller. Then, the global asymptotic convergence of the tracking errors can be easily obtained from the analysis in the previous sections.

Similarly, with the relative state observer (11.28) and the equivalent control filter (11.29), we can implement controller (11.23) using only relative output measurements. The convergence of the closed-loop system can be concluded following a similar argument as given above.

Remark 11.6 The finite-time convergent observers employed in this section have advantages over other types of commonly used observers such as high-gain ones due to the ease of theoretical analysis and practical implementation.

11.3 Input Constrained Robust Finite-Time Consensus Tracking for High-Order NCPS

Many practical applications of NCPS may require finite convergence time of the consensus tracking task. In this section, we propose a switching control strategy which combines a globally bounded asymptotic consensus tracking controller with a local finite-time convergent consensus controller to achieve robust global finite-time convergent consensus tracking of high-order NCPS subject to input saturation constraints. Both the cases of relative state measurement and relative output measurement are considered. There also requires no exchange of control inputs or internal states between neighboring agents in the controller design.

11.3.1 Problem Formulation

Consider a network of high-order integrator systems with input disturbances

$$\begin{aligned} \dot{x}_{ij} &= x_{i(j+1)}, \quad j = 1, 2, \ldots, n-1, \\ \dot{x}_{in} &= u_i + d_i, \\ y_i &= x_{i1}, \quad i = 1, \ldots, N, \end{aligned} \tag{11.30}$$

with an active leader modeled by

$$\begin{aligned} \dot{x}_{0j} &= x_{0(j+1)}, \quad j = 1, 2, \ldots, n-1, \\ \dot{x}_{0n} &= u_0, \\ y_0 &= x_{01}, \end{aligned} \tag{11.31}$$

where $x_i = [x_{i1}, \ldots, x_{in}]^T \in \mathbb{R}^n$, $i = 0, \ldots, N$ are the state vectors of the agents, $y_i \in \mathbb{R}$ are the outputs, d_i are the external disturbances which satisfy $|d_i| \leq \delta$, and $u_i \in \mathbb{R}$, $i = 1, \ldots, N$ are the control inputs of the followers. The input u_0 of the leader is assumed to be bounded and satisfies $|u_0| \leq C$.

Suppose that the actuators of the agents can only provide control inputs satisfying $|u_i| \leq u_{\max}$, $i = 1, 2, \ldots, N$ where $u_{\max}$ is a known positive constant. Then, the input saturated finite-time consensus tracking problem for (11.30) and (11.31) is to design distributed control input u_i satisfying $|u_i(t)| \leq u_{\max}$ for each follower which uses only local information from their neighbors such that for any initial condition, there exists a time $T > 0$ such that for any $t \geq T$

$$x_i(t) = x_0(t), \quad i = 1, \ldots, N.$$

Note that, to achieve precise consensus tracking, the control input u_i needs to provide the control input u_0 after consensus is achieved. Therefore, a necessary condition of the input saturation bound is that $C < u_{\max}$.

11.3.2 Input Constrained Robust Finite-Time Consensus Tracking with Relative State Measurements

In this section, we solve the input saturated finite-time consensus tracking problem for (11.30) and (11.31) when relative state information is available. The distributed controller is designed based on globally bounded finite-time stabilizing controllers for single high-order integrators. We will first design such a controller using a switching strategy between an asymptotic stabilization controller and a finite-time convergent controller for a single high-order integrator. After that, we propose the distributed consensus tracking controller for the multi-agent system (11.30) and (11.31) employing sliding mode control ideas.

For a single high-order integrator system described by

$$\begin{aligned} \dot{q}_i &= q_{i+1}, \quad i = 1, 2, \ldots, n-1, \\ \dot{q}_n &= u, \end{aligned} \tag{11.32}$$

where $q = [q_1, \ldots, q_n]^T \in \mathbb{R}^n$ is the system state and u is the control input, a globally bounded asymptotic stabilizing controller is proposed in Ding and Zheng (2015) as follows.

Definition 11.5 Define a series of polynomials as follows:

$$\begin{aligned} p_1(s) &= s + k_1 \\ p_2(s) &= s^2 + k_2 s + k_2 k_1 \\ p_3(s) &= s^3 + k_3 s^2 + k_3 k_2 s + k_3 k_2 k_1 \\ &\vdots \\ p_n(s) &= s^n + k_n s^{n-1} + k_n k_{n-1} s^{n-2} + \cdots + k_n \cdots k_1. \end{aligned}$$

If $p_i(s), i = 1, \ldots, n$ are stable polynomials, then we call $p(s) = p_n(s)$ a $\mathscr{P}$-stable polynomial.

Denote $\bar{\varepsilon}_i = [\varepsilon_1, \ldots, \varepsilon_i]^T$, $\varepsilon_i > 0, i = 1, \ldots, n$ and

$$\begin{aligned} a_1(\bar{\varepsilon}_1) &= 0 \\ a_2(\bar{\varepsilon}_2) &= \varepsilon_2 + k_1 \varepsilon_1 + k_0 a_1(\bar{\varepsilon}_1) \\ a_3(\bar{\varepsilon}_3) &= \varepsilon_3 + k_2 \varepsilon_2 + k_1 a_2(\bar{\varepsilon}_2) \\ &\vdots \\ a_n(\bar{\varepsilon}_n) &= \varepsilon_n + k_{n-1} \varepsilon_{n-1} + k_{n-2} a_{n-1}(\bar{\varepsilon}_{n-1}) \end{aligned}$$

with $k_i > 0, i = 0, 1, \ldots, n$. Let

$$\sigma_\varepsilon(x) = \begin{cases} \varepsilon \mathrm{sgn}(x), & for\ |x| > \varepsilon, \\ x, & for\ |x| \le \varepsilon, \end{cases}$$

where $\varepsilon > 0$.

Lemma 11.8 *If* $p(s) = s^n + k_n s^{n-1} + k_n k_{n-1} s^{n-2} + \cdots + k_n \cdots k_1$ *is a* $\mathscr{P}$*-stable polynomial, then system (11.32) can be globally asymptotically stabilized by the following nested-saturation-based controller* $u = \alpha_1(q)$ *where*

$$\alpha_1(q) = -k_n \sigma_{\varepsilon_n}(q_n + k_{n-1}\sigma_{\varepsilon_{n-1}}(q_{n-1} + \cdots + k_1 \sigma_{\varepsilon_1}(q_1))\cdots) \tag{11.33}$$

with

$$\begin{aligned} &k_i \varepsilon_i > \varepsilon_{i+1} + k_{i-1} a_i(\bar{\varepsilon}_1), i = 1, \ldots, n-1, \\ &u_m \geq k_n \varepsilon_n > k_{n-1} a_n(\bar{\varepsilon}_n) \end{aligned}$$

and it holds $|u| \leq u_m$.

A local finite-time stabilizing feedback controller for single high-order integrators was proposed in Bhat and Bernstein (2005) based on homogeneity theory.

Lemma 11.9 *Let the positive constants* $c_1, \ldots, c_n$ *be such that polynomial* $p^n + c_n p^{n-1} + \cdots + c_2 p + c_1$ *is Hurwitz. There is* $\gamma \in (0, 1)$ *such that, for every* $\nu \in (1 - \gamma, 1)$*, system (11.32) is stabilized at the origin in finite time under the feedback* $u = \alpha_2(q)$ *where*

$$\alpha_2(q) = -c_1 \operatorname{sgn}(q_1)\, |q_1|^{\nu_1} - \cdots - c_n \operatorname{sgn}(q_n)\, |q_n|^{\nu_n} \tag{11.34}$$

where the standard notation sgn$(\cdot)$ *denotes the signum function and* $\nu_1, \ldots, \nu_n$ *satisfy*

$$\nu_{i-1} = \frac{\nu_i \nu_{i+1}}{2\nu_{i+1} - \nu_i}, \qquad i = 2, \ldots, n \tag{11.35}$$

with $\nu_{n+1} = 1$ *and* $\nu_n = \nu$.

Based on the controllers (11.33) and (11.34), a globally bounded finite-time convergent controller is proposed for (11.32) as follows. Suppose that ρ is such that for $\forall q(0) \in \mathscr{Q} = \{q | \|q\|_2 \leq \rho\}$, $\|\alpha_2(t)\|_\infty \leq u_m$. That is, as long as q enters the region $\mathscr{Q}$, then the state converges to zero in finite time while the control input satisfies $\|\alpha_2(t)\|_\infty \leq u_m$. Then, the following switching controller is considered:

$$u(q(t)) = \alpha(q(t)) = \begin{cases} \alpha_1(q(t)), t \leq \min\{t | \|q(t)\|_2 \leq \rho\} \\ \alpha_2(q(t)), t > \min\{t | \|q(t)\|_2 \leq \rho\}. \end{cases} \tag{11.36}$$

Note that only a single switching is needed for any initial condition. Since $\alpha_1(q(t))$ is a globally asymptotically convergent controller, the switching time is finite. Furthermore, after the switching, the state converges to zero in finite time while the controller always satisfies the control input constraints $|u(t)| \leq u_m$.

Then, we design the distributed finite-time consensus tracking controller with input saturation. For each follower, define the consensus tracking errors

$$e_{ik} = \sum_{j=1}^{N} a_{ij}(x_{ik} - x_{jk}) + b_i(x_{ik} - x_{0k}), k = 1, \ldots, n, \tag{11.37}$$

where $b_i = a_{i0}$, $i = 1, \ldots, N$. Similar to the proof of Lemma 11.5, it can be shown that the finite-time consensus tracking problem for (11.30) and (11.31) is solved if there exists a finite time $T > 0$ such that for $t \geq T$, we have $e_{ij} = 0, i = 1, \ldots, N, j = 1, \ldots, n$.

From (11.37), (11.30), and (11.31), we have that

$$\begin{aligned} \dot{e}_{ij} &= e_{i(j+1)}, \quad j = 1, 2, \ldots, n-1, \\ \dot{e}_{in} &= h_i u_i - \sum_{j=1}^{N} a_{ij} u_j - b_i u_0 + h_i d_i - \sum_{j=1}^{N} a_{ij} d_j, \end{aligned} \tag{11.38}$$

where $h_i = \sum_{j=1}^{N} a_{ij} + b_i, i = 1, \ldots, N$.

For the following high-order integrator system,

$$\begin{aligned} \dot{e}_{ij} &= e_{i(j+1)}, \quad j = 1, \ldots, n-1, \\ \dot{e}_{in} &= \alpha(e_i), \end{aligned} \tag{11.39}$$

where $\alpha(\cdot)$ is defined in (11.36), we know that it is finite-time stable where $e_i = [e_{i1}, \ldots, e_{in}]^T$.

To stabilize the system (11.38) in finite time, we design the control input u_i based on (11.39) using integral sliding mode control techniques. Define the sliding variable $s_i \in \mathbb{R}, i = 1, \ldots, N$ as

$$s_i = e_{in} + e_{auxi}, \quad \dot{e}_{auxi} = -\alpha(e_i) \tag{11.40}$$

with $e_{auxi}(0) = -e_{in}(0)$. Then we have the following lemma.

Lemma 11.10 *Suppose the communication graph $\mathcal{G}$ contains a spanning tree with the leader as the root. If the sliding variables s_i, $i = 1, \ldots, N$ defined as in (11.40) are kept at 0, then the finite-time consensus tracking problem for (11.30) and (11.31) is solved along the sliding surfaces.*

Proof Note that on the sliding surface, the dynamics of the closed-loop system can be determined from $\dot{s} = 0$ as

$$\begin{aligned} \dot{e}_{ij} &= e_{i(j+1)}, \quad j = 1, \ldots, n-1 \\ \dot{e}_{in} &= \alpha(e_i). \end{aligned}$$

Then it follows that $[e_{i1}, \ldots, e_{in}]$ will converge to zero in finite time which means the consensus tracking problem for (11.30) and (11.31) is solved. ■

Next, we design the control inputs such that the sliding surfaces are kept at 0. From (11.40) and (11.38), we have that

$$\dot{s}_i = \sum_{j=1}^{N} a_{ij}(u_i - u_j) + b_i(u_i - u_0) + h_i d_i - \sum_{j=1}^{N} a_{ij} d_j - \alpha(e_i), \quad (11.41)$$

where $i = 1, \ldots, N$. The following control input is considered:

$$u_i(t) = -l\mathrm{sgn}(s_i), \quad (11.42)$$

where $l > 0$ is the design parameter. The main result is presented below.

Theorem 11.4 *Suppose the communication graph $\mathscr{G}$ contains a spanning tree with the leader as the root, then the finite-time consensus tracking problem for (11.30) and (11.31) is solved with the distributed controller (11.42) if*

$$\frac{2r_{\max}(u_m + 2h_m\delta)}{\lambda_{\min}(W)} + C < l \le u_{\max},$$

where $r_{\max} = \max\{r_1, \ldots, r_N\}$, $h_m = \max_{i=1,\ldots,N}\{h_i\}$, W is defined in Lemma 11.1, u_m is the design parameter of $\alpha(\cdot)$, and C is the upper bound of the leader's input.

Proof Let $s = [s_1, \ldots, s_N]^T$, from (11.41), we have $\dot{s} = -H\left(l\mathrm{sgn}(s) + \bar{u}_0\right) + \tau - \bar{\alpha}$ where $\bar{u}_0 = [u_0, \ldots, u_0]^T$, $\tau = [h_1 d_1 - \sum_{j=1}^{N} a_{1j} d_j, \ldots, h_N d_N - \sum_{j=1}^{N} a_{Nj} d_j]^T$, and $\bar{\alpha} = [\alpha(e_1), \ldots, \alpha(e_N)]^T$. Note that it holds $\|\tau\|_\infty \le 2h_m\delta$ and $\|\bar{\alpha}\|_\infty \le u_m$. From Lemma 11.1, we have that there exists $r = [r_1, \ldots, r_N]^T$ such that both R and W are positive definite. Consider the Lyapunov function $V = \sum_{i=1}^{N} r_i |s_i|$. The derivative of V satisfies

$$\begin{aligned}\dot{V} &\le \sum_{i=1}^{N} r_i \mathrm{sgn}(s_i)\Big[\sum_{j=1}^{N} a_{ij}\left(-C\mathrm{sgn}(s_i) + C\mathrm{sgn}(s_j)\right) + b_i\left(-C\mathrm{sgn}(s_i) - u_0\right)\Big] \\ &\quad - (l - C)\,\mathrm{sgn}^T(s) RH \mathrm{sgn}(s) + r_{\max}(u_m + 2h_m\delta)\|\mathrm{sgn}(s)\|_1 \\ &\le -\left[\tfrac{l-C}{2}\lambda_{\min}(W) - r_{\max}(u_m + 2h_m\delta)\right]\|\mathrm{sgn}(s)\|_1.\end{aligned}$$

Therefore, under the condition $\frac{2r_{\max}(u_m+2h_m\delta)}{\lambda_{\min}(W)} + C < l$, we have that $s_i(t) = 0$, $i = 1, 2\ldots, N$. Then from Lemma 11.10, finite-time consensus tracking of (11.30) and (11.31) is achieved. ■

Remark 11.7 From Theorem 11.4, a sufficient condition for the existence of the controller gain l is $\frac{4r_{\max}h_m\delta}{\lambda_{\min}(W)} + C < u_{\max}$. That is, the input saturation bound has a lower bound $\frac{4r_{\max}h_m\delta}{\lambda_{\min}(W)} + C$ which is related to the bound of the external disturbance, the communication graph, and the leader's control input.

Remark 11.8 With the sliding variable design (11.40), the consensus tracking problem of the high-order NCPS is transformed into a stabilization problem for a first-order system which facilitates the distributed controller design. Furthermore, the convergence rate can be easily tuned by properly choosing the design parameters.

11.3.3 Input Constrained Robust Finite-Time Consensus Tracking with Relative Output Measurements

In this section, we consider the case when only the relative output measurement is available for each agent. Note that the distributed controller (11.42) cannot be directly implemented in this situation since the agents no longer have access to $e_{i2}, \ldots, e_{in}, i = 1, \ldots, N$ which are needed in the construction of s_i. Note, however, with the control input (11.42), the closed-loop system takes the form of

$$\begin{aligned} \dot{e}_{ij} &= e_{i(j+1)}, \quad j = 1, 2, \ldots, n-1, \\ \dot{e}_{in} &= -h_i l \operatorname{sgn}(s_i) + \sum\nolimits_{j=1}^{N} a_{ij} l \operatorname{sgn} s_j - b_i u_0 + h_i d_i - \sum\nolimits_{j=1}^{N} a_{ij} d_j. \end{aligned} \tag{11.43}$$

Thus, it is easy to see that under controller (11.42) we have $\left|e_{i1}^{(n)}\right| = |\dot{e}_{in}| \le 2h_i(l + \delta) + b_i C$. Therefore, an $(n-1)$-th-order uniform finite-time exact differentiator can be designed according to Sect. 11.1.3 as

$$\begin{aligned} \dot{z}_{ij} =& -\lambda_i \theta sig(z_{i1} - e_{i1})^{\frac{n-i}{n}} - \eta_i (1-\theta) sig(z_{i1} - e_{i1})^{\frac{n+\beta i}{n}} \\ &+ z_{i(j+1)}, \, j = 1, \ldots, n-1, \\ \dot{z}_{in} =& -\lambda_n \theta \operatorname{sgn}(z_{i1} - e_{i1}) - \eta_n (1-\theta) sig(z_{i1} - e_{i1})^{1+\beta}, \end{aligned} \tag{11.44}$$

where $i = 1, \ldots, N$. The uniform convergence time is designed as $T_u > 0$. Then we have that for any initial estimation, after $t \ge T_u$, it holds $z_{i2} = e_{i2}, \ldots, z_{in} = e_{in}$ for $i = 1, \ldots, N$.

Based on the above reasoning, when only relative output information is available for each agent, we modify the distributed controller (11.42) into

$$u_i = -l \operatorname{sgn}(s_i), \quad s_i = z_{in} + e_{auxi}, \quad \dot{e}_{auxi} = -\hat{\alpha}(\hat{e}_i), \tag{11.45}$$

where $e_{auxi}(T_u) = -z_{in}(T_u)$, $\hat{e}_i = [e_{i1}, z_{i2}, \ldots, z_{in}]$,

$$\hat{\alpha}(q(t)) = \begin{cases} \alpha_1(q(t)), t \le \min\{t | t \ge T_u, \|q(t)\|_2 \le \rho\}, \\ \alpha_2(q(t)), t > \min\{t | t \ge T_u, \|q(t)\|_2 \le \rho\} \end{cases}$$

and $z_{ij}, i = 1, \ldots, N, j = 2, \ldots, n$ are the outputs of the uniform finite-time differentiator (11.44). The main result in this section is summarized in the following theorem.

Theorem 11.5 *Suppose the communication graph $\mathcal{G}$ defined on the $N+1$ agents is a directed graph which contains a spanning tree with the leader as the root. With the uniform finite-time differentiator (11.44) and the distributed control input (11.45), the finite-time consensus tracking problem for (11.30) and (11.31) is solved if*

$$\frac{2r_{\max}(u_m + 2h_m\delta)}{\lambda_{\min}(W)} + C < l \le u_{\max}, \tag{11.46}$$

where $r_{\max} = \max\{r_1, \ldots, r_N\}$, $h_m = \max_{i=1,\ldots,N}\{h_i\}$, *$W$ is defined in Lemma 11.1,* u_m *is the design parameter of* $\alpha(\cdot)$*, and C is the upper bound of the leader's input.*

Proof We divide the convergence process into two phases. The first phase is the differentiator convergence phase $[0, T_u]$. From (11.45), the closed-loop system takes the form of

$$\begin{aligned} \dot{e}_{ij} &= e_{i(j+1)}, \quad j = 1, 2, \ldots, n-1, \\ \dot{e}_{in} &= -h_i l\,\mathrm{sgn}(\hat{s}_i) - \sum_{j=1}^{N} a_{ij}u_j - b_i u_0 + h_i d_i - \sum_{j=1}^{N} a_{ij}d_j. \end{aligned} \tag{11.47}$$

Consider the closed-loop system (11.47) on the time interval $[0, T_u]$. Take the Lyapunov function $V_i = \frac{1}{2}e_{i1}^2 + \frac{1}{2}e_{i2}^2 + \cdots + \frac{1}{2}e_{in}^2$, then we have

$$\begin{aligned} \dot{V}_i &\le e_{i1}e_{i2} + e_{i2}e_{i3} + \cdots + e_{i(n-1)}e_{in} + \left[h_i(l + 2\delta) + b_i C + \sum_{j=1}^{N} a_{ij}l\right]|e_{in}| \\ &\le \frac{1}{2}|e_{i1}|^2 + |e_{i2}|^2 + \cdots + \left|e_{i(n-1)}\right|^2 + \frac{1}{2}|e_{in}|^2 + \left[h_i(l + 2\delta) + b_i C + \sum_{j=1}^{N} a_{ij}l\right]|e_{in}| \\ &\le K_{i1}V_i + K_{i2}\sqrt{V_i}, \end{aligned}$$

where

$$K_{i1} = 2, \quad K_{i2} = \sqrt{2}\left[h_i(l + 2\delta) + b_i C + \sum_{j=1}^{N} a_{ij}l\right].$$

Thus, it follows $V_i(t) \le \left(e^{(K_{i1}/2)t}\sqrt{V_i(0)} + K_{i2}/K_{i1}(e^{(K_{i1}/2)t} - 1)\right)^2$, that is, the state of the closed-loop system will not escape in finite time. For $t \ge T_u$, it holds $z_{i2} = e_{i2}, \ldots, z_{in} = e_{in}$, $i = 1, \ldots, N$ and the controller (11.45) becomes the same as the state feedback controller (11.42). Then following the same process as in the proof of Theorem 11.4, we conclude that the finite-time consensus tracking problem for (11.30) and (11.31) is solved under the condition (11.46). ■

Remark 11.9 It is easy to see that the controller (11.45) is bounded. Furthermore, the proposed control strategy (11.44) and (11.45) only requires relative output measurement. No exchange of control inputs or internal states between neighboring agents is needed in the controller design. The convergence rate can be easily tuned by properly choosing the differentiator parameters and the controller parameters.

11.4 Numerical Examples

In this section, several simulation examples are provided to illustrate the performance of the previously designed controllers for high-order NCPS subject to input saturation constraints.

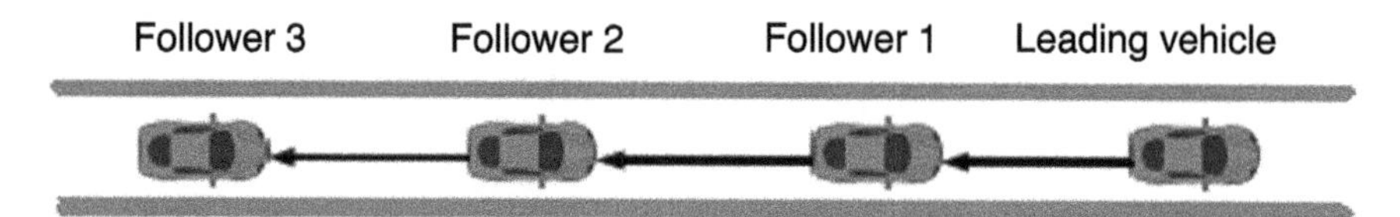

Fig. 11.1 Illustration of the autonomous vehicle platoon

11.4.1 Input Constrained Robust Consensus Tracking for High-Order NCPS

In this section, the proposed control strategies are applied to the longitudinal control of a platoon of autonomous vehicles. In this problem, there exists a fleet of autonomous vehicles moving along the highway with a leading vehicle which determines the desired speed (cf. Fig. 11.1). The objective is to design control input for each follower vehicle such that they maintain the speed of the leader while keeping desired relative distances between the vehicles. The vehicle platoon may have the advantages of reducing traffic load, the chance of collision, and also fuel consumption (Chiu et al. 1977; Shladover 1989). Assume that each of the follower vehicle is only equipped with sensors such as laser or radar to measure relative position with respect to its immediate preceding vehicle. The interactive relation is then shown in Fig. 11.1. First, we model the dynamics of each vehicle. Under the assumption of horizontal road surface and negligible wind disturbance, each vehicle can be modeled as follows (Sheikholeslam and Desoer 1992, 1993):

$$\begin{aligned} m_i\dot{v}_i &= m_i\xi_i - K_{di}v_i^2 - d_{mi}, \\ \dot{\xi}_i &= -\frac{\xi_i}{\tau_i(v_i)} + \frac{u_i}{m_i\tau_i(v_i)}, \end{aligned} \tag{11.48}$$

where m_i is the mass, v_i is the velocity, $m_i\xi_i$ represents the engine force applied to the i-th vehicle, K_{di} denotes the aerodynamic drag coefficient for the i-th vehicle, and d_{mi} is mechanical drag. $\tau_i(v_i)$ denotes the i-th vehicle's engine time constant, and u_i represents the throttle input to the vehicle's engine.

Let x_{i1} denote the position of the i-th vehicle, x_{i2} the velocity, and x_{i3} the acceleration. Then, the dynamics of the i-th vehicle can be determined as

$$\dot{x}_{i1} = x_{i2}, \quad \dot{x}_{i2} = x_{i3}, \quad \dot{x}_{i3} = a_i(v_i)u_i + b_i(v_i, \dot{v}_i), \tag{11.49}$$

where

$$a_i(v_i) = \frac{1}{m_i\tau_i(v_i)}, \quad b_i(v_i, \dot{v}_i) = -2\frac{K_{di}}{m_i}v_i\dot{v}_i - \frac{1}{\tau_i(v_i)}\left[\dot{v}_i + \frac{K_{di}}{m_i}v_i^2 + \frac{d_{mi}}{m_i}\right].$$

With the input transformation

$$u_i = \frac{1}{\hat{a}(v_i)}[q_i - \hat{b}(v_i, \dot{v}_i)], \tag{11.50}$$

where $\hat{a}$ and $\hat{b}$ are obtained with the nominal parameters $\hat{m}_i$, $\hat{\tau}_i$, $\hat{K}_{di}$, and $\hat{d}_{mi}$, it can be obtained

$$\dot{x}_{i1} = x_{i2}, \quad \dot{x}_{i2} = x_{i3}, \quad \dot{x}_{i3} = q_i + d_i, \tag{11.51}$$

where $d_i = (\frac{a}{\hat{a}} - 1)q_i - \frac{a\hat{b} - \hat{a}b}{\hat{a}}$. The control objective of the platoon can be expressed as follows:

$$x_{i1} - x_{(i-1)1} - L_d \to 0, x_{i2} - x_{02} \to 0, i = 1, \ldots, N, t \to \infty,$$

where L_d is the desired separation between the vehicles. By denoting $\bar{x}_{i1} = x_{i1} + i * L_d, i = 0, 1, \ldots, N$, it can be seen that coordinated tracking of $\bar{x}_{i1}$ with the leader's position $\bar{x}_{01}$ means $x_{i1} - x_{(i-1)1} = L_d$. Therefore, the platoon control problem can be transformed into the coordinated tracking problem with the dynamics

$$\dot{\bar{x}}_{i1} = x_{i2}, \quad \dot{x}_{i2} = x_{i3}, \quad \dot{x}_{i3} = q_i + d_i \tag{11.52}$$

and the leader

$$\dot{x}_{01} = x_{02}, \quad \dot{x}_{02} = x_{03}, \quad \dot{x}_{03} = q_0. \tag{11.53}$$

Suppose that the finite actuation power of each vehicle's engine requires that $|q_i| \leq q_m$. The bound on the disturbances d_i can also be estimated in practical situations. Then, the proposed control strategies can be used to design the control input q_i while the original control input u_i can be obtained from the input transformation (11.50).

It is easy to see that the communication graph of the platoon satisfies Assumption 11.1. Suppose that the Laplacian matrix is taken as $a_{11} = N_{\max}, a_{ii} = N_{\max}, a_{(i-1)i} = -N_{\max}, i = 2, \ldots, N$ where the number $N_{\max}$ is the upper bound of the scale of the platoon. Then it can be verified that $\|H^{-1}\|_\infty \leq 1$ for any $N \leq N_{\max}$. In this case, the proposed controller (11.23) is fully distributed which only needs the relative position measurement between the neighboring two vehicles.

For the simulation, consider a platoon of four follower vehicles. The desired separation between the vehicles is set to 10 m under the nominal velocity 25 m/s. Suppose that at the beginning of the platoon maneuvering, all the vehicles are moving at a constant velocity larger than 25 m/s and with the separations larger than 10 m. Specifically, set the initial position of the leader at 0, and the initial positions of the followers are $[-30, -60, -90, -120]$. The initial velocity is set as 28 m/s. To make the platoon forming more challenging, suppose that the velocity of the leader is time varying as $x_{02} = 28 - 3\sin(0.2t)$. Then, $q_0 = 0.12\sin(0.2t)$ which leads to $\rho = 0.12$. The parameters of the vehicles are $m_1 = 1300\,\text{kg}, \tau_1 = 0.16, K_{d1} = 0.3, m_2 = 1400\,\text{kg}, \tau_2 = 0.22, K_{d2} =$

0.35, $m_3 = 1200\,\text{kg}$, $\tau_3 = 0.18$, $K_{d3} = 0.2$, $m_4 = 1350\,\text{kg}$, $\tau_4 = 0.24$, $K_{d4} = 0.45$ as in Seshagiri and Khalil (1989). The nominal parameters of the vehicles are chosen the same for each vehicle as $m = 1300\,\text{kg}$, $\tau = 0.2$, $K_d = 0.3$. The disturbances are assumed to be $d_{m1} = 100 + 10\sin(0.2t)$, $d_{m2} = 100 + 15\sin(0.3t)$, $d_{m3} = 100 + 10\cos(0.5t)$, $d_{m4} = 100 + 20\sin(0.1t)$ with the nominal value set as $d_m = 100$. Then, the upper bound of the disturbances can be taken as $\delta = 0.3$.

Suppose the Laplacian matrix is taken with $N_{\max} = 4$. Then we have

$$H = \begin{bmatrix} 4 & 0 & 0 & 0 \\ -4 & 4 & 0 & 0 \\ 0 & -4 & 4 & 0 \\ 0 & 0 & -4 & 4 \end{bmatrix},$$

which leads to $\|H\|_\infty^{-1} = 1$. It is assumed that the engine power limit leads to input saturation $q_m = 4$. With the above setting, the distributed controller (11.23) which uses only relative position measurement between the neighboring vehicles is designed with the parameters $M_3 = L_3 = 3.4$, $M_2 = L_2 = 1.6$, $M_1 = L_1 = 0.6$, $k_s = 0.1$, $\lambda_1 = 1.1$, $\lambda_2 = 1.5$, $\lambda_3 = 2$, and $\tau = 0.1$, $k_i = 30$ determined according to Theorem 11.3. The initial conditions of the observers are set to zero. The simulation results are given in Figs. 11.2, 11.3, 11.4, and 11.5.

It can be observed that the desired separations between the vehicles are achieved successfully. Furthermore, the control inputs q_i are bounded by $q_m = 4$ during the whole maneuver.

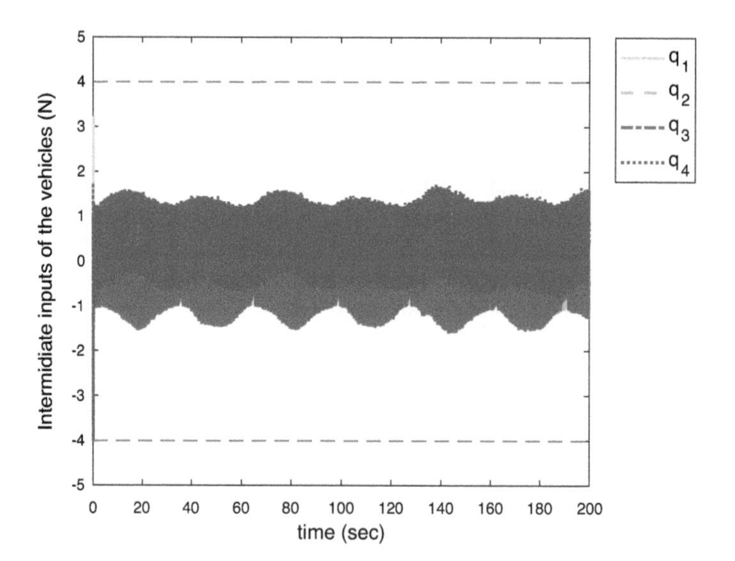

Fig. 11.2 Intermediate inputs of the vehicles

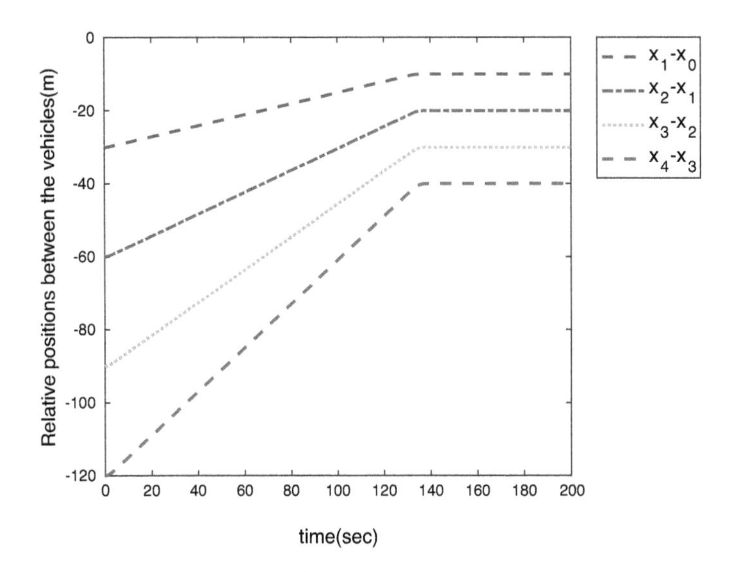

Fig. 11.3 Relative positions between the vehicles

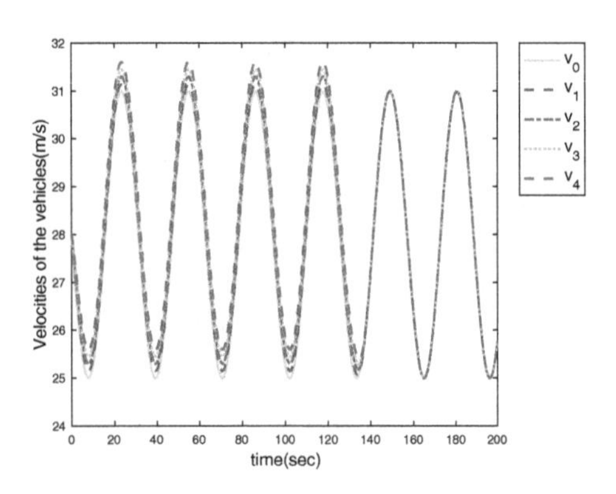

Fig. 11.4 Velocities of the vehicles

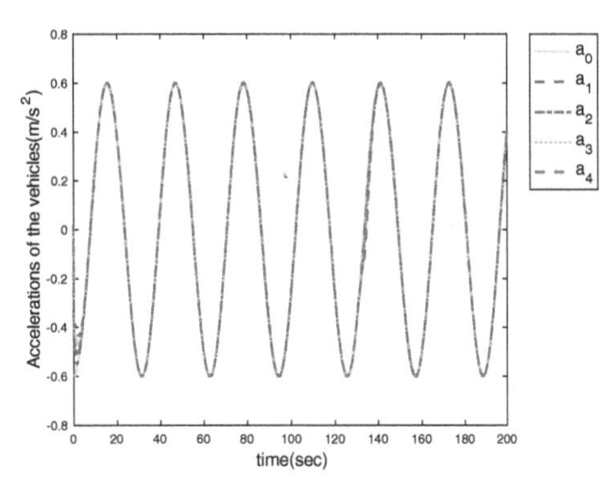

Fig. 11.5 Accelerations of the vehicles

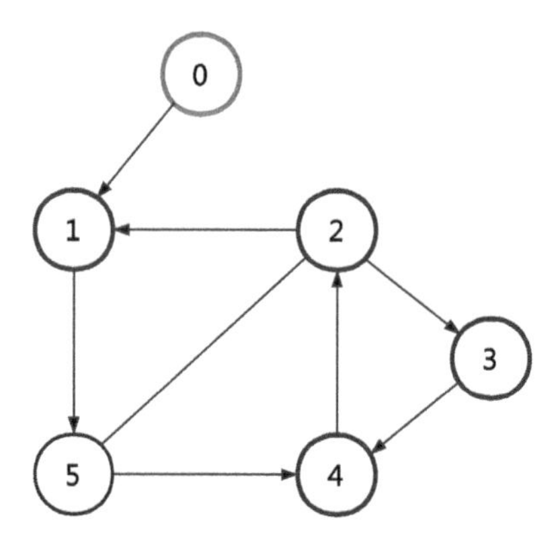

Fig. 11.6 Communication graphs

11.4.2 Input Constrained Robust Finite-Time Consensus Tracking for High-Order NCPS

In this section, we demonstrate the performance of the proposed robust finite-time convergent consensus controllers for high-order NCPS subject to input saturation constraints. Consider a leader–follower network of five followers and one leader with third-order dynamics as shown in (11.30) and (11.31) where the external disturbance $d_i = (0.5 + 0.5 * i)\sin(0.2 * i * t)$.

The communication graph $\mathscr{G}$ is a directed graph as shown in Fig. 11.6. It is easy to see that $\mathscr{G}$ contains a spanning tree with the leader as the root. The element a_{ij} $i, j = 1, 2, \ldots, 5$ of the adjacent matrix equals to 1 or 0 where $a_{ij} = 1$ if there is information flow from agent j to agent i and $a_{ij} = 0$ otherwise. The leader's initial condition is set as [3; 0; 0] and the control input is chosen such that $|u_0| \le 3$.

We consider the case when only relative output measurements are available. The uniform finite-time exact observer is designed with the parameters $\lambda_1 = 10, \lambda_2 = 25$, $\lambda_3 = 40$, $\eta_1 = 1$, $\eta_2 = 3$, $\eta_3 = 3$, $T_l = 0.1$, and $T_u = 15$ with the initial states setting to zero. The tracking controller (11.45) is implemented for each follower with parameters $k_1 = 1/6, k_2 = 2/3, k_3 = 4, \varepsilon_1 = 8, \varepsilon_2 = 5/4, \varepsilon_3 = 1/4, c_1 = 1, c_2 = 3$, $c_3 = 3$, $\nu_1 = 7/10$, $\nu_2 = 7/9$, $\nu_3 = 7/8$, $\rho = 0.1$, and $l = 10$ which are determined from Theorem 11.5. The followers are assumed to start moving from rest with initial positions chosen randomly in the interval $[-10, 10]$. The tracking results of the followers are shown in Fig. 11.7. We see that the finite-time consensus tracking problem is solved with the proposed control strategy. Furthermore, the mode switching of the controllers $\alpha(\cdot)$ for each follower in this case is shown in Fig. 11.8.

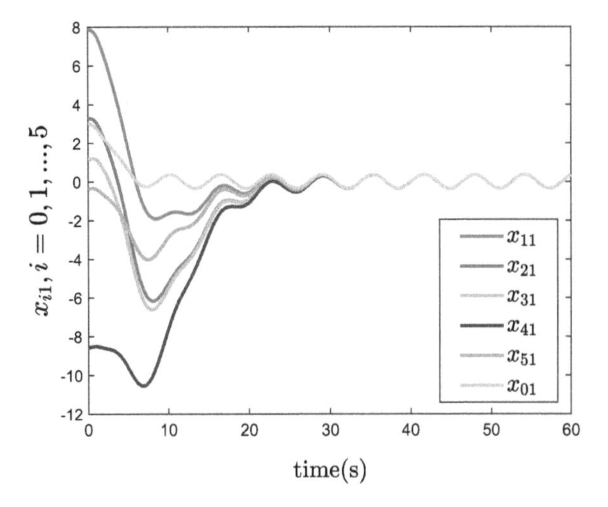

Fig. 11.7 Finite-time consensus tracking with relative output measurements

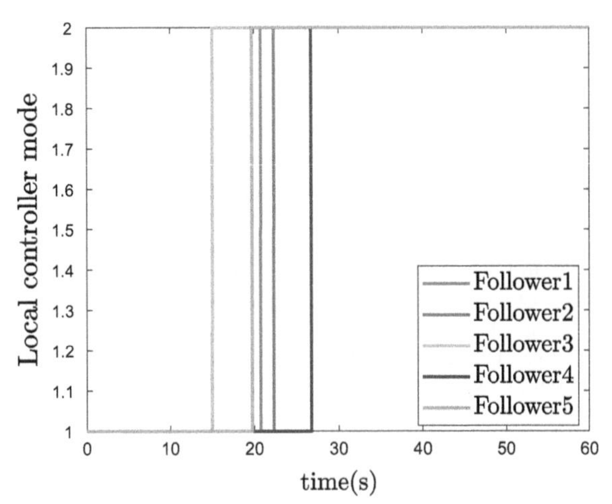

Fig. 11.8 Local controller mode 1: $\alpha(\cdot) = \alpha_1(\cdot)$. Local controller mode 2: $\alpha(\cdot) = \alpha_2(\cdot)$

11.5 Conclusions

In this chapter, the consensus tracking problem for high-order NCPS subject to input saturation constraints is studied and several effective controller design methods have been proposed which achieve asymptotic and finite-time convergence, respectively. In Sect. 11.2, the robust global coordinated tracking problem has been studied for a class of high-order NCPS with general directed communication graphs subject to input saturation. Some new kinds of nonlinear distributed controllers have been proposed which achieve global coordinated tracking with only local and relative measurements. Digital communication between neighboring agents has been avoided. Both the cases with static and dynamic leaders have been considered. In Sect. 11.3, a switching control strategy is proposed to realize finite-time consensus control of high-order NCPS. The sliding mode control method is employed to guarantee that the sliding variables remain to be zero using the distributed control input. Application in the platoon control of autonomous vehicles has been presented to illustrate the effectiveness of the proposed controllers. Some limitations of the obtained results include that they are developed only for systems with matched disturbances and only consensus tracking task is considered for the NCPS. Therefore, future work includes considering more general system dynamics and richer classes of coordination tasks such as containment control and coordinated searching in the presence of input saturation.

Acknowledgements This work is supported by the National Nature Science Foundation of China through Grant No. 61703094, the Natural Science Foundation of Jiangsu Province of China through Grant No. BK20170695, and the Fundamental Research Funds for the Central Universities of China. A. Zemouche would like to thank the ANR agency for the partial support of this work via the project ArtISMo ANR-20-CE48-0015.

References

A. Abdessameud, A. Tayebi, On consensus algorithms for double-integrator dynamics without velocity measurement and with input constraints. Syst. Control Lett. **59**(12), 812–821 (2010)
M.T. Angulo, J.A. Moreno, L. Fridman, Robust exact uniformly convergent arbitrary order differentiator. Automatica **49**(8), 2489–2495 (2013)
R.W. Beard, J. Lawton, F.Y. Hadaegh, A coordination architecture for spacecraft formation control. Control Syst. Technol. IEEE Trans. **9**(6), 777–790 (2001)
S.P. Bhat, D.S. Bernstein, Finite-time stability of continuous autonomous systems. SIAM J. Control Optim. **38**(3), 751–766 (2000)
S.P. Bhat, D.S. Bernstein, Geometric homogeneity with applications to finite-time stability. Math. Control Signals Syst. **17**(2), 101–127 (2005)
A. Bidram, A. Davoudi, F.L. Lewis, J.M. Guerrero, Distributed cooperative secondary control of microgrids using feedback linearization. IEEE Trans. Power Syst. **28**(3), 3462–3470 (2013)
Y. Cao, W. Ren, Z. Meng, Decentralized finite-time sliding mode estimators and their applications in decentralized finite-time formation tracking. Syst. Control Lett. **59**(9), 522–529 (2010)
Y. Cao, W. Yu, W. Ren, G. Chen, An overview of recent progress in the study of distributed multi-agent coordination. IEEE Trans. Ind. Inf. **9**(1), 427–438 (2013)
G. Chen, Y.D. Song, Cooperative tracking control of nonlinear multiagent systems using self-structuring neural networks. IEEE Trans. Neural Netw. Learn. Syst. **25**(8), 1496–1507 (2014)
H.Y. Chiu Jr., G.B.S., Jr, S.J.B., Vehicle-follower control with variable-gains for short headway automated guideway transit systems. J. Dyn. Syst. Meas. Control **99**(3), 183–189 (1977)
J. Cortés, Finite-time convergent gradient flows with applications to network consensus. Automatica **42**(11), 1993–2000 (2006)
S. Ding, W.X. Zheng, Robust control of multiple integrators subject to input saturation and disturbance. Int. J. Control **88**(4), 844–856 (2015)
H. Du, S. Li, C. Qian, Finite-time attitude tracking control of spacecraft with application to attitude synchronization. IEEE Trans. Autom. Control **56**(11), 2711–2717 (2011)
H. Du, S. Li, S. Ding, Bounded consensus algorithms for multi-agent systems in directed networks. Asian J. Control **15**(1), 282–291 (2013)
M.C. Fan, H.T. Zhang, Z. Lin, Distributed semi-global consensus with relative output feedback and input saturation under directed switching networks. IEEE Trans. Circuits Syst. II Express Briefs **62**(8), 796–800 (2015)

J. Fu, J. Wang, Observer-based finite-time coordinated tracking for high-order integrator systems with matched uncertainties under directed communication graphs, in *11th IEEE International Conference on Control & Automation (ICCA)* (IEEE, 2014), pp. 880–885

J. Fu, G. Wen, W. Yu, Z. Ding, Finite-time consensus for second-order multi-agent systems with input saturation. IEEE Trans. Circuits Syst. II Express Briefs **65**(11), 1758–1762 (2018)

J. Fu, Q. Wang, J. Wang, Robust finite-time consensus tracking for second-order multi-agent systems with input saturation under general directed communication graphs. Int. J. Control **92**(8), 1785–1795 (2019)

Y. Hong, Y. Xu, J. Huang, Finite-time control for robot manipulators. Syst. Control Lett. **46**(4), 243–253 (2002)

Y. Hong, J. Hu, L. Gao, Tracking control for multi-agent consensus with an active leader and variable topology. Automatica **42**(7), 1177–1182 (2006)

T. Hu, Z. Lin, *Control Systems with Actuator Saturation: Analysis and Design* (Springer Science & Business Media, 2001)

S. Khoo, L. Xie, Z. Man, Robust finite-time consensus tracking algorithm for multirobot systems. IEEE/ASME Trans. Mechatron. **14**(2), 219–228 (2009)

S. Khoo, L. Xie, S. Zhao, Z. Man, Multi-surface sliding control for fast finite-time leader-follower consensus with high order siso uncertain nonlinear agents. Int. J. Robust Nonlinear Control **24**(16), 2388–2404 (2014)

A. Levant, Higher-order sliding modes, differentiation and output-feedback control. Int. J. Control **76**(9–10), 924–941 (2003)

A. Levant, Quasi-continuous high-order sliding-mode controllers. IEEE Trans. Autom. Control **50**(11), 1812–1816 (2005)

H. Li, X. Liao, T. Huang, W. Zhu, Y. Liu, Second-order global consensus in multiagent networks with random directional link failure. IEEE Trans. Neural Netw. Learn. Syst. **26**(3), 565–575 (2014)

Y. Li, J. Xiang, W. Wei, Consensus problems for linear time-invariant multi-agent systems with saturation constraints. Iet Control Theory Appl. **5**(6), 823–829 (2011)

Z. Meng, Z. Lin, On distributed finite-time observer design and finite-time coordinated tracking of multiple double integrator systems via local interactions. Int. J. Robust Nonlinear Control **24**(16), 2473–2489 (2013)

Z. Meng, W. Ren, Z. You, Distributed finite-time attitude containment control for multiple rigid bodies. Automatica **46**(12), 2092–2099 (2010)

Z. Meng, Z. Zhao, Z. Lin, On global leader-following consensus of identical linear dynamic systems subject to actuator saturation. Syst. Control Lett. **62**(2), 132–142 (2013)

R.M. Murray, Recent research in cooperative control of multivehicle systems. J. Dyn. Syst. Meas. Control **129**(5), 571 (2007)

R. Olfati-Saber, R. Murray, Consensus problems in networks of agents with switching topology and time-delays. IEEE Trans. Autom. Control **49**(9), 1520–1533 (2004)

W. Ren, On consensus algorithms for double-integrator dynamics. IEEE Trans. Autom. Control **53**(6), 1503–1509 (2008)

S. Seshagiri, H.K. Khalil, Longitudinal adaptive control of a platoon of vehicles, in *Proceedings of the American Control Conference, 1999*, vol. 5 (1989), pp. 3681–3685

S. Sheikholeslam, C.A. Desoer, A system level study of the longitudinal control of a platoon of vehicles. J. Dyn. Syst. Meas. Control **114**(2), 286–292 (1992)

S. Sheikholeslam, C.A. Desoer, Longitudinal control of a platoon of vehicles with no communication of lead vehicle information: a system level study. IEEE Trans. Veh. Technol. **42**(4), 546–554 (1993)

S.E. Shladover, Longitudinal control of automotive vehicles in close-formation platoons. J. Dyn. Syst. Meas. Control **113**(2), 231–241 (1989)

Q. Song, F. Liu, H. Su, A.V. Vasilakos, Semi-global and global containment control of multi-agent systems with second-order dynamics and input saturation. Int. J. Robust Nonlinear Control **26**(16), 3460–3480 (2016)

H. Su, M.Z.Q. Chen, J. Lam, Z. Lin, Semi-global leader-following consensus of linear multi-agent systems with input saturation via low gain feedback. IEEE Trans. Circuits Syst. I Regul. Pap. **60**(7), 1881–1889 (2013)

V.I. Utkin, *Slides Modes in Control and Optimization* (Springer, NY, Berlin, 1992)

V.I. Utkin, A.S. Poznyak, Adaptive sliding mode control with application to super-twist algorithm: equivalent control method. Automatica **49**(1), 39–47 (2013)

N.D. Vanli, M.O. Sayin, I. Delibalta, S.S. Kozat, Sequential nonlinear learning for distributed multiagent systems via extreme learning machines. IEEE Trans. Neural Netw. Learn. Syst. **28**(3), 546–558 (2017)

B. Wang, J. Wang, B. Zhang, X. Li, Global cooperative control framework for multiagent systems subject to actuator saturation with industrial applications. IEEE Trans. Syst. Man Cybern. Syst. **47**(7), 1270–1283 (2017a)

Q. Wang, C. Sun, X. Xin, Robust consensus tracking of linear multiagent systems with input saturation and input-additive uncertainties. Int. J. Robust Nonlinear Control **27**(14), 2393–2409 (2017b)

X. Wang, Y. Hong, Finite-time consensus for multi-agent networks with second-order agent dynamics, in *IFAC World Congress* (2008), pp. 15,185–15,190

X. Wang, H. Su, M.Z.Q. Chen, X. Wang, Observer-based robust coordinated control of multiagent systems with input saturation. IEEE Trans. Neural Netw. Learn. Syst. **29**(5), 1933–1946 (2018)

G. Wen, Z. Duan, G. Chen, W. Yu, Consensus tracking of multi-agent systems with lipschitz-type node dynamics and switching topologies. IEEE Trans. Circuits Syst. I Regul. Pap. **61**(2), 499–511 (2014)

F. Xiao, L. Wang, J. Chen, Y. Gao, Finite-time formation control for multi-agent systems. Automatica **45**(11), 2605–2611 (2009)

T. Yang, Z. Meng, D.V. Dimarogonas, K.H. Johansson, Global consensus for discrete-time multi-agent systems with input saturation constraints. Automatica **50**(2), 499–506 (2014)

L. Zaccarian, A.R. Teel, Modern Anti-windup Synthesis: Control Augmentation for Actuator Saturation (Princeton University Press, 2011)

H. Zhang, Z. Li, Z. Qu, F.L. Lewis, On constructing lyapunov functions for multi-agent systems. Automatica **58**, 39–42 (2015)

H. Zhang, H. Liang, Z. Wang, T. Feng, Optimal output regulation for heterogeneous multiagent systems via adaptive dynamic programming. IEEE Trans. Neural Netw. Learn. Syst. **28**(1), 18–29 (2017)

Z. Zhao, Z. Lin, Semi-global leader-following consensus of multiple linear systems with position and rate limited actuators. Int. J. Robust Nonlinear Control **25**(13), 2083–2100 (2015)

Z. Zhao, Z. Lin, Global leader-following consensus of a group of general linear systems using bounded controls. Automatica **68**, 294–304 (2016)

Chapter 12
Optimal Subsystem Decomposition and Resilient Distributed State Estimation for Wastewater Treatment Plants

Langwen Zhang, Miaomiao Xie, Wei Xie, and Bohui Wang

12.1 Introduction

The wastewater treatment plant (WWTP) is an important step in water recycling (Qu et al. 2013). WWTP is usually composed of several interconnected operation units. State estimation is a process of constructing system state based on output measurements and system model. State estimation is important for WWTP since many related states in WWTP cannot be measured or affected by significant noise. Cyber-physical systems (CPS) integrates communication network, engineering, computing, and physical process components and uses the network to realize the interaction between computing processes and physical processes (Zhang et al. 2021) and operate physical entities in a remote, reliable, real-time, secure (Wang et al. 2021), and cooperative way (Ding et al. 2020). When the distributed state estimation of sewage treatment system is carried out, in order to improve the efficiency, two or more computer equipment are often used for calculation and processing. That is, each computer processes a subsystem state estimation. Among the subsystems, the state information needs to be exchanged through the communication network. Therefore, WWTP can be regarded as an information physical system in which physical pro-

L. Zhang (✉) · M. Xie · W. Xie
College of Automation Science and Technology, South China University of Technology, Guangzhou, China
e-mail: aulwzhang@scut.edu.cn

M. Xie
e-mail: 1503312925@qq.com

W. Xie
e-mail: weixie@scut.edu.cn

B. Wang
School of Electrical and Electronic Engineering, Nanyang Technological University, Singapore, Singapore
e-mail: wang31aa@126.com

M. Abbaszadeh and A. Zemouche (eds.), *Security and Resilience in Cyber-Physical Systems*, https://doi.org/10.1007/978-3-030-97166-3_12

cesses and networks are connected (Anter et al. 2020; Wei et al. 2020). In Barbu et al. (2011), deterministic and stochastic observers were developed. A univariate statistical technique was proposed in Baklouti et al. (2018) to enhance the monitoring of WWTP and the state estimation of two-time scale nonlinear systems was considered in Kiss et al. (2011). In Busch et al. (2013), a synthesis method based on optimization for the design and estimation of sensor networks was proposed and was applied to the WWTP system. Extended Kalman filter (EKF) and unscented Kalman filter (UKF) were used to estimate the unmeasurable states in WWTP system in Wahab et al. (2012). In Yin and Liu (2018), EKF and the moving horizon estimator (MHE) estimators were proposed based on model reduction for improved computational efficiency. Also, there are some researches on distributed state estimation.

WWTP is considered critical infrastructure and their resiliency is vital. The resilience against natural disasters (storm and rain) as opposed to cyber-incidents is critical. Performing distributed state estimation is one of the effective ways to improve the resiliency of the system, compared to a centralized scheme applied to the whole system. There are some existing results on the distributed state estimation method for the WWTP system, i.e., in Zeng et al. (2016), distributed EKF (DEKF) was applied to WWTP system, and distributed MHE was studied in Yin et al. (2018). However, the existing distribution control and estimation usually assume that the system decomposition is available. A systematic approach to decompose the large-scale system into subsystems has not yet received enough attention and is crucial for distributed state estimation (Dunbar 2007). When applying the distributed state estimation, a good subsystem decomposition with weak inter-subsystem interactions can improve the resiliency of the system.

In Heo et al. (2015), there are some important results about the decomposition algorithm of distributed control system. In Yin et al. (2016), a subsystem decomposition method for distributed estimation was proposed and applied to WWTP system. In Yin and Liu (2019), the existing community discovery algorithm was extended to the common framework of distributed state estimation and control. However, the above subsystem partition based on community discovery algorithm only considers the correlation degree of the system and ignores the connection strength between different variables. In Zhang et al. (2019), a method of subsystem partition based on weighted edge group detection was proposed and was applied in distributed state estimation. However, the weighted community discovery algorithm for distributed control has not been investigated. Also, there are little results about subsystem decomposition method for WWTP system. Thus, community structure detection is used to decompose the WWTP into smaller groups, such that the intra-connection within each group is made much stronger than the interaction among different groups. Subsystem models that are appropriate for distributed state estimation are configured based on the variables assigned to the groups.

In this work, an optimal subsystem decomposition method is investigated for complex cyber-physical systems for the purpose of improving the resiliency under the distributed state estimation. The main contributions lie in the following: (1) a subsystem decomposition method based on community structure discovery algorithm is proposed for the WWTP system; (2) to deal with the natural disasters and the unreliable communication networks, a resilient distributed framework is proposed with information compensation strategy; (3) comparative study is carried out for WWTP system to show that the subsystem decomposition and resilient distributed state estimation scheme improves the resiliency of the system, compared to a centralized scheme applied to the whole system.

12.2 Model Description of Wastewater Treatment Plants

In this work, optimal subsystem decomposition and distributed EKF methods are designed. The theoretical results will be validated in a benchmark WWTP system (Alex et al. 2008). The benchmark WWTP system model will be reviewed in this part and the motivation for decomposing the WWTP system will be derived. The plant layout is shown in Fig. 12.1. In this process, the five activated sludge reactors are composed of two sections: (1) The anoxic section: reactor 1 and reactor 2, where the bacteria convert nitrate into nitrogen (i.e., denitrification biological reactions). (2) The aerated section: reactor 3, reactor 4, and reactor 5, where the bacteria oxidize ammonium to nitrate (i.e., nitrification reactions).

For each reactor, the following variables ($k = 1$ to 5) are defined: flow rate: Q_k; concentration: $Z_{as,k}$; the volume of anoxic section: $V_{as,1} = V_{as,2} = 1000\,\text{m}^3$; the volume of aerobic section: $V_{as,3} = V_{as,4} = V_{as,5} = 1333\,\text{m}^3$; reaction rate: r_i. In this model, the general equation of mass balance of bioreactor (two anoxic reactors and three aerobic reactors) follows from Alex et al. (2008).

For reactor 1 ($k = 1$),

$$\frac{dZ_{as,1}}{dt} = \frac{1}{V_{as,1}}(Q_{int}Z_{int} + Q_r Z_r + Q_i Z_i + r_{Z,1}V_{as,1} - Q_1 Z_{as,1}), \tag{12.1}$$

$$Q_1 = Q_{int} + Q_r + Q_i. \tag{12.2}$$

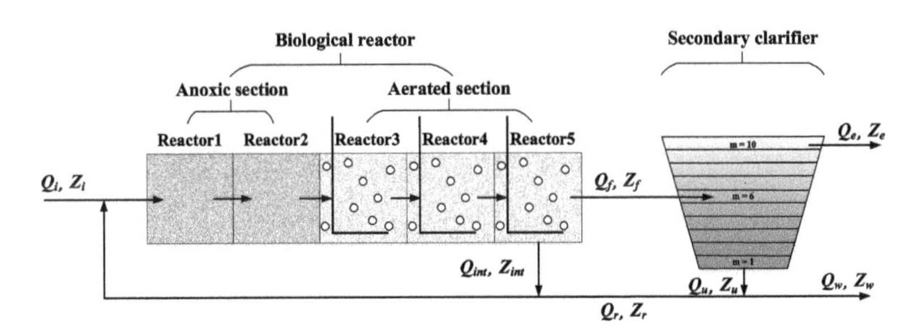

Fig. 12.1 General overview of the BSM1 plant

For reactor 2–5 ($k = 2 - 5$),

$$\frac{dZ_{as,k}}{dt} = \frac{1}{V_{as,k}}(Q_{k-1}Z_{as,k-1} + r_{Z,k}V_{as,k} - Q_kZ_{as,k}), \tag{12.3}$$

$$Q_k = Q_{k-1}. \tag{12.4}$$

Special case for oxygen ($S_{O,k}$):

$$\frac{dS_{O,as,k}}{dt} = \frac{1}{V_{as,k}}(Q_{k-1}S_{O,as,k-1} + r_{Z,k}V_{as,k} + (K_{La})_kV_{as,k}(S_O^* - S_{O,as,k}) - Q_kS_{O,as,k}), \tag{12.5}$$

where the saturation concentration of oxygen is $S_O^* = 8g \cdot m^3$, and r_k denotes the conversion rate of different compounds in the reactor; the detailed calculation of $r_{Z,k}$ can be found in Alex et al. (2008). The flow rate of the reaction process in Fig. 12.1 satisfies the following:

$$\begin{aligned} Z_{int} &= Z_{as,5}, \\ Z_f &= Z_{as,5}, \\ Z_w &= Z_r, \\ Q_f &= Q_e + Q_r + Q_w = Q_e + Q_u. \end{aligned} \tag{12.6}$$

The solid flux caused by gravity is $J_s = v_s(X_{sc})X_{sc}$, where X_{sc} denotes the total sludge concentration (i.e., including X_I, X_S, $X_{B,H}$, $X_{B,A}$, X_P, and X_{ND}). The double exponential settlement velocity function is selected:

$$v_s(X_{sc}) = max\left[0, min\left\{v_0^{'}, v_0\left(e^{-r_h(X_{sc}-X_{min})} - e^{-r_p(X_{sc}-X_{min})}\right)\right\}\right], \tag{12.7}$$

where $X_{min} = f_{ns}X_f$, X_f is the total solids concentration from the bioreactor. The upward velocity (v_{up}) and the downward velocity (v_{dn}) are calculated as follows:

$$\begin{aligned} v_{up} &= \frac{Q_u}{A} = \frac{Q_r+Q_w}{A}, \\ v_{dn} &= \frac{Q_e}{A}. \end{aligned} \tag{12.8}$$

According to these symbols, the mass balance of sludge is written as follows:

$$m = 1: \frac{dX_{sc,m}}{dt} = \frac{v_{dn}(X_{sc,m+1} - X_{sc,m}) + min(J_{s,m}, J_{sc,m+1})}{z_m} \tag{12.9a}$$

$$m = 2 - 5: \frac{dX_{sc,m}}{dt} = \frac{v_{dn}(X_{sc,m+1} - X_{sc,m}) + min(J_{s,m}, J_{sc,m+1}) - min(J_{s,m}, J_{sc,m-1})}{z_m} \tag{12.9b}$$

$$m = 6: \frac{dX_{sc,m}}{dt} = \frac{\frac{Q_fX_f}{A} + J_{sc,m+1} - (v_{up} + v_{dn})X_{sc,m} - min(J_{s,m}, J_{sc,m-1})}{z_m} \tag{12.9c}$$

$$m = 7 - 9: \frac{dX_{sc,m}}{dt} = \frac{v_{up}(X_{sc,m-1} - X_{sc,m}) + J_{sc,m+1} - J_{s,m}}{z_m} \quad (12.9d)$$

$$m = 10: \frac{dX_{sc,m}}{dt} = \frac{v_{up}(X_{sc,m-1} - X_{sc,m}) - J_{s,m}}{z_m}, \quad (12.9e)$$

where the critical concentration is $X_t = 3000g/m^3$, and the detailed calculation of $J_{sc,m}$ can be found in Alex et al. (2008).

For soluble components (i.e., S_I, S_S, S_O, S_{NO}, S_{NH}, S_{ND}, and S_{ALK}), each layer represents the volume of complete mixing, and the concentration of soluble components is calculated accordingly.

$$m = 1 - 5: \frac{dZ_{sc,m}}{dt} = \frac{v_{dn}(Z_{sc,m+1} - Z_{sc,m})}{z_m} \quad (12.10a)$$

$$m = 6: \frac{dZ_{sc,m}}{dt} = \frac{\frac{Q_f X_f}{A} - (v_{up} + v_{dn})Z_{sc,m}}{z_m} \quad (12.10b)$$

$$m = 7 - 10: \frac{dZ_{sc,m}}{dt} = \frac{v_{up}(Z_{sc,m-1} - Z_{sc,m})}{z_m}, \quad (12.10c)$$

where the concentration in the recycle and waste stream is equal to that in the first layer (bottom layer), that is, $Z_u = Z_{sc,1}$.

According to the concentration in compartment 5 of the activated sludge reactor, the sludge concentration can be calculated directly as follows:

$$X_f = \frac{1}{fr_{COD-SS}}(X_{S,as,5} + X_{P,as,5} + X_{I,as,5} + X_{B,H,as,5} + X_{B,A,as,5}), \quad (12.11)$$

where the conversion coefficient fr_{COD-SS} from COD to SS is equal to 4/3. The same principle applies $X - u$ (in the underflow of the secondary sedimentation tank) and $X - e$ (at the outlet of the secondary sedimentation tank).

Then, Eqs. (12.1)–(12.11) form the WWTP system model. Typically, there are 145 states in the system and it is difficult to design centralized controller or estimator for the WWTP system. Thus, distributed control/estimation method is necessary. To do this, we have to (1) decompose the whole system into subsystems and (2) design distributed controller/estimator.

12.3 Subsystem Decomposition

In this section, we will present a subsystem decomposition method for WWTP system. The existing decomposition method usually ignores the weights on the edges. We will present a weighted directed graph-based subsystem decomposition method for large-scale system. The whole system will be represented by a weighted directed graph, and the community structure detection method is derived for decomposition.

The dynamic model of WWTP system can be formulated as follows:

$$\dot{x}(t) = f(x(t), u(t)) \quad (12.12a)$$

$$y(t) = h(x(t)), \quad (12.12b)$$

where $x \in \mathbb{R}^{n_x}$, $u \in \mathbb{R}^{n_y}$, and $y \in \mathbb{R}^{n_y}$, respectively, represent the state vector of the system, the vector of inputs, and the vector of measured outputs, and f and h are two vector fields describing the dynamics of the nonlinear system and the output relation, respectively. The objective is to decompose system (12.12) into subsystems with the form:

$$\dot{x}_i(t) = f_i(x_i(t), X_i(t), u_i(t), U_i(t)) \quad (12.13a)$$

$$y_i(t) = h_i(x_i(t)), \quad (12.13b)$$

where $i = 1, \ldots, p$, with p being the number of subsystems, $x_i \in \mathbb{R}^{n_{x_i}}$ denotes the state vector of the ith subsystem, $u_i \in \mathbb{R}^{n_{u_i}}$ denotes the input vector of the ith subsystem, $y_i \in \mathbb{R}^{n_{y_i}}$ is the output vector of the ith subsystem, and X_i and U_i are the vector that comprises the states and input of all the subsystems that affect the dynamics of subsystem i directly.

The weighted subsystem decomposition method is extended from our early work (Zhang et al. 2019) to simultaneous state estimation and control (see Fig. 12.2). In the proposed approach, system (12.12) is characterized by a weighted directed graph. The graph characterizes the connectivity between the state, input, and measured output variables. When the entire system is not observable or not controllable, we have to adjust the system structure to get a observable or controllable system. For the observation, we can add more sensors to make sure that the observability matrix is full rank. Also, we can add more control variables to guarantee that the controllability matrix is full rank.

Specifically, the weighted directed graph is created based on the methods for generating unweighted directed graphs described. All the state, input, and measured output variables are considered as vertices of a graph, which are connected through directed edges. Let $f_i, i = 1, \cdots, n_x$, denote the ith element of the vector field f, and $h_j, j = 1, \ldots, n_y$, denote the jth element of the vector field h. In addition, let us denote $x_i, i = 1, \ldots, n_x$, as the ith element of x, denote $u_k, k = 1, \ldots, n_u$, as the jth element of u, and denote $y_j, j = 1, \ldots, n_y$, as the jth element of y. The edges in a directed graph are constructed based on the following rules:

- State-to-state edge: there is a unidirectional edge from a state vertex x_i to another state vertex x_l, if $\frac{\partial f_l(x,u)}{\partial x_i} \neq 0, l, i = 1, \ldots, n_x$.
- State-to-input edge: there is a unidirectional edge from an output vertex y_j to a state vertex x_l, if $\frac{\partial f_i(x,u)}{\partial u_k} \neq 0, i = 1, \ldots, n_x, k = 1, \ldots, n_u$.
- State-to-output edge: there is a bidirectional edge from an output vertex y_j to a state vertex x_l, if $\frac{\partial h_j(x)}{\partial x_l} \neq 0, l = 1, \ldots, n_x, j = 1, \ldots, n_y$.

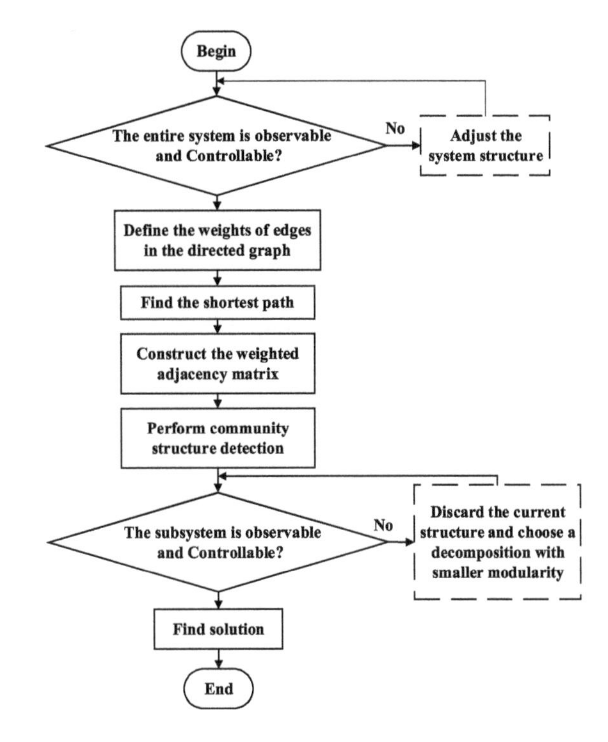

Fig. 12.2 The flowchart of the optimal subsystem decomposition

Then, the weighted directed graphs are constructed with $\mathscr{G} = (V, E)$, where $V = \{x_i, y_i, u_i\}$ is set of the vertices and E is the set of the edges. The edges are constructed as follows:

$$S(u_k, x_i) = \frac{\partial f_i(x, u)}{\partial u_k}\Big|_{(x_s, u_s)} \tag{12.14a}$$

$$S(x_i, x_l) = \frac{\partial f_l(x, u)}{\partial x_i}\Big|_{(x_s, u_s)} \tag{12.14b}$$

$$S(x_l, y_j) = \frac{\partial h_j(x)}{\partial x_l}\Big|_{x=x_s}, \tag{12.14c}$$

where $S(x_i, x_l)$ is the sensitivity for a state-to-state pair (x_i, x_j) and $S(x_l, y_j)$ is the sensitivity for a state-to-output pair (x_l, y_j). A sensitivity matrix is constructed as follows:

$$S = \begin{bmatrix} \overline{A} & \overline{B} & \overline{C}^T \\ 0_{n_u \times n_x} & 0_{n_u \times n_u} & 0_{n_u \times n_y} \\ \overline{C} & 0_{n_y \times n_u} & 0_{n_y \times n_y} \end{bmatrix}_{n_x + n_u + n_y}, \tag{12.15}$$

where $\bar{A}$, $\bar{B}$, and $\bar{C}$ are obtained by taking the Jacobian of system (12.12) at (x_s, u_s), respectively, as follows:

$$\overline{A}=\begin{bmatrix}\frac{\partial f_1}{\partial x_1} & \cdots & \frac{\partial f_1}{\partial x_{nx}}\\ \vdots & \ddots & \vdots\\ \frac{\partial f_{nx}}{\partial x_1} & \cdots & \frac{\partial f_{nx}}{\partial x_{nx}}\end{bmatrix}_{(x_s,u_s)},\ \overline{B}=\begin{bmatrix}\frac{\partial f_1}{\partial u_1} & \cdots & \frac{\partial f_1}{\partial u_{nu}}\\ \vdots & \ddots & \vdots\\ \frac{\partial f_{nx}}{\partial u_1} & \cdots & \frac{\partial f_{nx}}{\partial u_{nu}}\end{bmatrix}_{(x_s,u_s)},\ \overline{C}=\begin{bmatrix}\frac{\partial h_1}{\partial x_1} & \cdots & \frac{\partial h_1}{\partial x_{ny}}\\ \vdots & \ddots & \vdots\\ \frac{\partial h_{ny}}{\partial x_1} & \cdots & \frac{\partial h_{ny}}{\partial x_{ny}}\end{bmatrix}_{x=x_s}. \tag{12.16}$$

The weights of the edges are defined as follows:

- Weight of state-to-state edge:

$$w(x_i,x_l)=\begin{cases}\frac{1}{|S(x_i,x_l)|}, & \text{if } \frac{\partial f_l(x,u)}{\partial x_i}\Big|_{(x_s,u_s)}\neq 0,\ l,i=1,\ldots,n_x\\ \infty, & \text{otherwise.}\end{cases} \tag{12.17}$$

- Weight of input-to-state edge:

$$w(u_k,x_i)=\begin{cases}\frac{1}{|S(u_k,x_i)|}, & \text{if } \frac{\partial f_i(x,u)}{\partial u_k}\Big|_{(x_s,u_s)}\neq 0,\ k=1,\ldots,n_u, i=1,\ldots,n_x\\ \infty, & \text{otherwise.}\end{cases} \tag{12.18}$$

- Weight of state-to-output edge:

$$w(x_l,y_j)=\begin{cases}\frac{1}{|S(x_l,y_j)|}, & \text{if } \frac{\partial h_j(x)}{\partial x_l}\Big|_{x_s}\neq 0,\ l=1,\ldots,n_x,\ j=1,\ldots,n_y\\ \infty, & \text{otherwise.}\end{cases} \tag{12.19}$$

The shortest paths can be identified for constructing the adjacency matrix. The lengths of the path $L_{il}(P_{il})$ from x_i to x_l and x_l to an output vertex y_j have been given in Zhang et al. (2019). Additionally, we need to take the connection from u_k to a state vertex x_i and path from a input vertex u_k to an output vertex y_j into account.

The length of the path $L_{ki}(P_{ki})$ from a input vertex u_k to a state vertex x_i is given as follows:

$$L_{ki}(P_{ki})=w\left(u_k,x_1^{(k,i)}\right)+\cdots+w\left(x_{N(P_{ki})}^{(k,i)},x_i\right). \tag{12.20}$$

The corresponding shortest path $\underline{d}(u_k,x_i)$ is calculated as follows:

$$\begin{aligned}\underline{d}(u_k,x_i)&=\min_{P_{ki}\in\mathbb{P}_{ki}} L_{ki}(P_{ki})\\ &=\min_{P_{ki}\in\mathbb{P}_{ki}}\left\{\frac{1}{\left|S\left(u_k,x_1^{(k,i)}\right)\right|^{\alpha}}+\cdots+\frac{1}{\left|S\left(x_{N(P_{ki})}^{(k,i)},x_i\right)\right|^{\alpha}}\right\},\end{aligned} \tag{12.21}$$

where $k=1,\ldots,n_u, i=1,\ldots,n_x$, and $\mathbb{P}_{ki}$ and P_{ki} represent the set of all the paths and one of path from a input vertex u_k to a state vertex x_i, respectively.

The length of the path $L_{kj}(P_{kj})$ from a input vertex u_k to an output vertex y_j is given as follows:

$$L_{kj}(P_{kj}) = w\left(u_k, x_1^{(k,j)}\right) + \cdots + w\left(x_{N(P_{kj})}^{(k,j)}, y_j\right). \tag{12.22}$$

The corresponding shortest path $\underline{d}(u_k, y_j)$ is calculated as follows:

$$\begin{aligned}\underline{d}(u_k, y_j) &= \min_{P_{kj} \in \mathbb{P}_{kj}} L_{kj}(P_{kj}) \\ &= \min_{P_{kj} \in \mathbb{P}_{kj}} \left\{ \frac{1}{\left|S\left(u_k, x_1^{(k,j)}\right)\right|^{\alpha}} + \cdots + \frac{1}{\left|S\left(x_{N(P_{kj})}^{(k,j)}, y_j\right)\right|^{\alpha}} \right\},\end{aligned} \tag{12.23}$$

where $k = 1, \ldots, n_u$, $j = 1, \ldots, n_y$, and $\mathbb{P}_{kj}$ and P_{kj} represent the set of all the paths and one of path from a input vertex u_k to an output vertex y_j, respectively.

The length of the path $L_{jm}(P_{jm})$ from an output vertex y_j to another output vertex y_m is given as follows:

$$L_{jm}(P_{jm}) = w\left(y_j, x_1^{(j,m)}\right) + \cdots + w\left(x_{N(P_{jm})}^{(j,m)}, y_m\right). \tag{12.24}$$

The corresponding shortest path $\underline{d}(y_j, y_m)$ is calculated as follows:

$$\begin{aligned}\underline{d}(y_j, y_m) &= \min_{P_{jm} \in \mathbb{P}_{jm}} L_{jm}(P_{jm}) \\ &= \min_{P_{jm} \in \mathbb{P}_{jm}} \left\{ \frac{1}{\left|S\left(y_j, x_1^{(j,m)}\right)\right|^{\alpha}} + \cdots + \frac{1}{\left|S\left(x_{N(P_{jm})}^{(j,m)}, y_m\right)\right|^{\alpha}} \right\},\end{aligned} \tag{12.25}$$

where $j, m = 1, \ldots, n_y$, and $\mathbb{P}_{jm}$ and P_{jm} represent the set of all the paths and one of path from an output vertex y_j to another output vertex y_m, respectively.

An adjacency matrix $A_w \in \mathbb{R}^{n_a \times n_a}$ involving all the vertices is then constructed based on $\underline{d}(x_i, x_l)$, $\underline{d}(x_l, y_j)$, $\underline{d}(u_k, x_i)$, $\underline{d}(u_k, y_j)$, and $\underline{d}(y_j, y_m)$, $i, l = 1, \ldots, n_x$, $k = 1, \ldots, n_u$, $j, m = 1, \ldots, n_y$:

$$\overline{A}_w = \begin{bmatrix} \overline{A}_{w,11} & \overline{A}_{w,12} & \overline{A}_{w,31}^T \\ 0_{n_u \times n_x} & 0_{n_u \times n_u} & 0_{n_u \times n_y} \\ \overline{A}_{w,31} & \overline{A}_{w,32} & \overline{A}_{w,33} \end{bmatrix}_{n_a \times n_a}, \tag{12.26}$$

where $\overline{A}_{w,11}$, $\overline{A}_{w,12}$, $\overline{A}_{w,31}$, $\overline{A}_{w,32}$, and $\overline{A}_{w,33}$ are constructed as follows:

$$\overline{A}_{w,11} = \begin{bmatrix} \frac{1}{\underline{d}(x_1,x_1)} & \cdots & \frac{1}{\underline{d}(x_1,x_{n_x})} \\ \vdots & \ddots & \vdots \\ \frac{1}{\underline{d}(x_{n_x},x_1)} & \cdots & \frac{1}{\underline{d}(x_{n_x},x_{n_x})} \end{bmatrix}, \overline{A}_{w,12} = \begin{bmatrix} \frac{1}{\underline{d}(u_1,x_1)} & \cdots & \frac{1}{\underline{d}(u_1,x_{n_x})} \\ \vdots & \ddots & \vdots \\ \frac{1}{\underline{d}(u_{n_u},x_1)} & \cdots & \frac{1}{\underline{d}(u_{n_u},x_{n_x})} \end{bmatrix},$$

$$\overline{A}_{w,31} = \begin{bmatrix} \frac{1}{\underline{d}(x_1,y_1)} & \cdots & \frac{1}{\underline{d}(x_1,y_{n_y})} \\ \vdots & \ddots & \vdots \\ \frac{1}{\underline{d}(x_{n_x},y_1)} & \cdots & \frac{1}{\underline{d}(x_{n_x},y_{n_y})} \end{bmatrix}, \overline{A}_{w,32} = \begin{bmatrix} \frac{1}{\underline{d}(u_1,y_1)} & \cdots & \frac{1}{\underline{d}(u_1,y_{n_y})} \\ \vdots & \ddots & \vdots \\ \frac{1}{\underline{d}(u_{n_u},y_1)} & \cdots & \frac{1}{\underline{d}(u_{n_u},y_{n_y})} \end{bmatrix}, \quad (12.27)$$

$$\overline{A}_{w,33} = \begin{bmatrix} \frac{1}{\underline{d}(y_1,y_1)} & \cdots & \frac{1}{\underline{d}(y_1,y_{n_y})} \\ \vdots & \ddots & \vdots \\ \frac{1}{\underline{d}(y_{n_y},y_1)} & \cdots & \frac{1}{\underline{d}(y_{n_y},y_{n_y})} \end{bmatrix}.$$

The weighted adjacency matrix A_w is constructed following (Zhang et al. 2019). The problem of subsystem decomposition is equivalent to performing community structure detection by finding a higher value of modularity Q. Community structure detection (Zhang et al. 2019) is used to decompose the network into smaller groups, such that the intra-connection within each group is made much stronger than the interaction among different groups. Subsystem models that are appropriate for distributed state estimation are configured based on the variables assigned to the groups. To this end, we have constructed the subsystem model for distributed state estimation. In the following section, we will present a resilient distributed estimator for WWTP system.

12.4 Resilient Distributed State Estimator Design

In distributed state estimation design, the distributed operation of each subsystem is usually carried out by multiple physical devices and the subsystem information needs to be exchanged. The physical equipment operation is supported by the communication network, and the subsystem information is transmitted through the communication network. Because the communication network may be unreliable, the data exchanged between subsystems may be altered, and they may be damaged by malicious network attacks. There are two common types of attacks, denial-of-service attack (DoS) and false data injection (FDI). DoS attack usually blocks the information flow between the sending device and the receiving device, thus increasing the packet loss rate in the communication process. FDI attack will hijack network nodes or physical devices and inject wrong or useless data information into the system, seriously endangering the safe and reliable operation of the system. To deal with the unreliable communication networks, a resilient distributed framework is proposed with information compensation strategy.

Based on the decomposed subsystems, distributed state estimator will be designed to show the improvement of the resiliency compared to the centralized state estimator.

During the distributed state estimator design, the subestimators will work colorably to make a coordination. Denote $\hat{X}_i(t_{k-1})$ as the latest subsystem estimate information of $X_{(i)}(t)$ for time $t \in [t_{k-1}, t_k]$ available to filter i. The communication network can be unreliable, i.e., $\hat{X}_i(t_{k-1})$ is attacked or lost. When the communication network is suffering FDI attack, the exchanged status will change to $\hat{X}_i(t_{k-1}) = \hat{X}_i(t_{k-1}) + X_a(t_{k-1})$. Thus, resilient distributed state estimator is necessary.

In the designed resilient DEKF, the attack is evaluated before each communication between subsystems with following strategy:

$$a = \begin{cases} 1, & \text{if } \|y_i(t_{k-1}) - C_i\hat{X}_i(t_{k-1})\| > \Delta \text{ or } \hat{X}_i(t_{k-1}) \text{ is not received} \\ 0, & \text{if } \|y_i(t_{k-1}) - C_i\hat{X}_i(t_{k-1})\| \leq \Delta. \end{cases} \tag{12.28}$$

$$\hat{X}_i(t_{k-1}) = \begin{cases} \hat{X}_i(t_{k-1}), & \text{if } a = 0 \\ \hat{X}_i(t_{k-2}) + \int_{t_{k-2}}^{t_{k-1}} f_i(\hat{x}_i(t_{k-2}), \hat{X}_i(t), u_i(t), U_i(t)))dt, & \text{if } a = 1. \end{cases} \tag{12.29}$$

In the resilient distributed EKF design, each local filter is designed as a continuous–discrete EKF. The resilient distributed EKF is implemented with the prediction step and update step.

(1) Prediction step:

$$\hat{x}_i(t_k|t_{k-1}) = \hat{x}_i(t_{k-1}) + \int_{t_{k-1}}^{t_k} f_i(\hat{x}_i(t), \hat{X}_i(t_{k-1}), u_i(t), U_i(t))dt, \tag{12.30}$$

$$P_i(t_k|t_{k-1}) = A_i(t_{k-1})P_i(t_{k-1})A_i(t_{k-1})^T + Q_i. \tag{12.31}$$

(2) Update step:

$$K_i(t_k) = P_i(t_k|t_{k-1})C_i^T\left[C_iP_i(t_k|t_{k-1})C_i^T + R_i\right]^{-1}, \tag{12.32}$$

$$\hat{x}_i(t_k) = \hat{x}_i(t_k|t_{k-1}) + K_i(t_k)\left[y_i(t_k) - C_i\hat{x}_i(t_k|t_{k-1})\right], \tag{12.33}$$

$$P_i(t_k) = [I - K_i(t_k)C_i]\,P_i(t_k|t_{k-1}), \tag{12.34}$$

where $\hat{x}_i(t_k|t_{k-1})$ denotes the state prediction at time t_k, and, $P_i(t_{k-1})$ is used to denote the error covariance matrix of $x_{(i)}(t_{k-1})$. $P_i(t_k|t_{k-1})$ refer to the predicted error covariance matrix for time t_k. Q_i and R_i are the covariances of process noise and measurement noise of subsystem i, respectively; $A_i(t_{k-1})$ is the Jacobian of $f_{(i)}$ with respect to $x_{(i)}$ at time t_{k-1}; and $K_i(t_k)$ is the filter gain at t_k.

To show the derivation of the distributed EKF algorithm, the system is discretized at time interval Δ, such that $t_k = k\Delta$:

$$x_i(t_k) = f_i(x_i(t_{k-1}), X_i(t_{k-1}), u_i(t_{k-1}), U_i(t_{k-1})) + w_i(t_{k-1}) \tag{12.35a}$$
$$y_i(t_k) = h_i(x_i(t_k)) + v_i(t_k). \tag{12.35b}$$

Assuming that the estimated error between the estimated value and the real value and the prediction error between the predicted value and the real value are $e_i(t_k) = x_i(t_k) - \hat{x}_i(t_k)$ and $e_i(t_k|t_{k-1}) = x_i(t_k) - \hat{x}_i(t_k|t_{k-1})$, respectively, and the estimation error covariance matrix and the prediction error covariance matrix are $P_i(t_k) = E\{e_i(t_k)e_i^T(t_k)\}$ and $P_i(t_k|t_{k-1}) = E\{e_i(t_k|t_{k-1})e_i^T(t_k|t_{k-1})\}$, respectively, where $E\{\cdot\}$ denotes mathematical expectation, then $\hat{x}_i(t_k|t_{k-1})$ and $\hat{x}_i(t_k)$ are

$$\begin{aligned} \hat{x}_i(t_k|t_{k-1}) &= \hat{x}_i(t_{k-1}) + \int_{t_{k-1}}^{t_k} f_i(\hat{x}_i(t), \hat{X}_i(t_{k-1}), u_i(t), U_i(t)dt \\ \hat{x}_i(t_k) &= \hat{x}_i(t_k|t_{k-1}) + K_i(t_k)[y_i(t_k) - h_i(\hat{x}_i(t_k|t_{k-1}))], \end{aligned} \tag{12.36}$$

where $K_i(t_k)$ is the filter gain at t_k. The Taylor expansion of $y_i(t_k)$ in Eq. (12.35) at $\hat{x}_i(t_k|t_{k-1})$

$$y_i(t_k) = h_i(\hat{x}_i(t_k|t_{k-1})) + C_i(x_i(t_k) - \hat{x}_i(t_k|t_{k-1})) + v_i(t_k). \tag{12.37}$$

So the estimation error and the estimation error covariance matrix are

$$\begin{aligned} e_i(t_k) &= x_i(t_k) - \hat{x}_i(t_k) = x_i(t_k) - \hat{x}_i(t_k|t_{k-1}) - K_i(t_k)[y_i(t_k) - h_i(\hat{x}_i(t_k|t_{k-1}))] \\ &= x_i(t_k) - \hat{x}_i(t_k|t_{k-1}) - K_i(t_k)[C_i(x_i(t_k) - \hat{x}_i(t_k|t_{k-1})) + v_i(t_k)] \\ &= (I - K_i(t_k)C_i)(x_i(t_k) - \hat{x}_i(t_k|t_{k-1})) - K_i(t_k)v_i(t_k), \end{aligned} \tag{12.38}$$

$$\begin{aligned} P_i(t_k) &= E\{e_i(t_k)e_i^T(t_k)\} \\ &= E\{[(I - K_i(t_k)C_i)(x_i(t_k) - \hat{x}_i(t_k|t_{k-1})) - K_i(t_k)v_i(t_k)] \\ &\qquad [(I - K_i(t_k)C_i)(x_i(t_k) - \hat{x}_i(t_k|t_{k-1})) - K_i(t_k)v_i(t_k)]^T\} \\ &= (I - K_i(t_k)C_i)E\{e_i(t_k|t_{k-1})e_i^T(k|t_{k-1})\}(I - K_i(t_k)C_i)^T + K_i(t_k)R_iK_i^T(t_k) \\ &= (I - K_i(t_k)C_i)P_i(t_k|t_{k-1})(I - K_i(t_k)C_i)^T + K_i(t_k)R_iK_i^T(t_k). \end{aligned} \tag{12.39}$$

Because the diagonal element of $P_i(t_k)$ is the square of the estimation error, the trace of the matrix (expressed by $T[\cdot]$) is the mean square deviation, that is,

$$\begin{aligned} T[P_i(t_k)] &= T[P_i(t_k|t_{k-1})] - 2T[K_i(t_k)C_iP_i(t_k|t_{k-1})] \\ &\quad + T[K_i(t_k)(C_iP_i(t_k|t_{k-1})C_i^T + R_i)K_i^T(t_k)]. \end{aligned} \tag{12.40}$$

To make the estimated value closer to the real value, the trace above must be as small as possible. Therefore, it is necessary to obtain an appropriate Kalman gain $K_i(t_k)$ to minimize the trace. The implication is to make the partial derivative of the trace to $K_i(t_k)$ is zero, that is,

$$\frac{dT[P_i(t_k)]}{dK_i(t_k)} = -2(C_i P_i(t_k|t_{k-1}))^T + 2K_i(t_k)(C_i P_i(t_k|t_{k-1})C_i^T + R_i) = 0. \tag{12.41}$$

Furthermore, we have (12.32) and (12.34). The Taylor expansion of $x_i(t_k)$ in Eq. (12.35) at $(\hat{x}_i(t_k), \hat{X}_i(t_{k-1}))$:

$$\begin{aligned} x_i(t_k) = &f_i(\hat{x}_i(t_{k-1}), \hat{X}_i(t_{k-1}), u_i(t_{k-1}), U_i(t_{k-1})) \\ &+ A_i(t_{k-1})(x_i(t_{k-1}) - \hat{x}_i(t_{k-1})) + w_i(t_{k-1}), \end{aligned} \tag{12.42}$$

where $A_i(t_{k-1})$ is the Jacobian of $f_{(i)}$ with respect to $x_{(i)}$ at time t_{k-1}. Discretize $\hat{x}_i(t_k|t_{k-1})$ in Eq. (12.36):

$$\hat{x}_i(t_k|t_{k-1}) = f_i(\hat{x}_i(t_{k-1}), \hat{X}_i(t_{k-1}), u_i(t_{k-1}), U_i(t_{k-1})). \tag{12.43}$$

So the prediction error and its mathematical expectation are

$$\begin{aligned} e_i(t_k|t_{k-1}) &= x_i(t_k) - \hat{x}_i(t_k|t_{k-1}) \\ &= A_i(t_{k-1})(x_i(t_{k-1}) - \hat{x}_i(t_{k-1})) + w_i(t_{k-1}), \end{aligned} \tag{12.44}$$

$$\begin{aligned} &E\{e_i(t_k|t_{k-1})e_i^T(t_k|t_{k-1})\} \\ =&E\{[A_i(t_{k-1})(x_i(t_{k-1}) - \hat{x}_i(t_{k-1})) + w_i(t_{k-1})] \\ &[A_i(t_{k-1})(x_i(t_{k-1}) - \hat{x}_i(t_{k-1})) + w_i(t_{k-1})]^T\} \\ =&A_i(t_{k-1})E\{[x_i(t_{k-1}) - \hat{x}_i(t_{k-1})][x_i(t_{k-1}) - \hat{x}_i(t_{k-1})]^T)\}A_i^T(t_{k-1}) + Q_i. \end{aligned} \tag{12.45}$$

To this end, we get (12.31).

An algorithm is adopted for distributed EKFs to work collaboratively in this work. It is assumed that each local filter shares the state estimates with its interacting subsystem for each sampling periods. The resilient distributed EKF algorithm is implemented:

- Step 1: At $t_0 = 0$, initialize $x_i(0)$, $P_i(t_0)$, $i = 1, \ldots, p$.
- Step 2: For time $t_k > 0$, each local estimator i receives the measured output of the subsystem i, i.e., $y_i(t_k)$.
- Step 3: Each distributed EKF receives the state estimates of the interacting subsystems at the time t_{k-1}.
- Step 4: Check the communication network with (12.28) and set $\hat{X}_i(t_{k-1})$ using (12.29).
- Step 5: Based on the latest $\hat{X}_{(i)}(t_{k-1})$, each EKF i calculates the state estimates $\hat{x}_i(t_k)$, $i = 1, \ldots, p$. The estimate of the entire system state is $\hat{x}(t_k) = \left[\hat{x}_1(t_k)^T \ldots \hat{x}_p(t_k)^T\right]^T$.
- Step 6: At $k = k + 1$, go to Step 2.

12.5 Simulation

In this section, the proposed subsystem decomposition method is validated by decomposing the subsystem model of WWTP system. The resilient distributed EKF under different subsystem methods is tested to show the efficiency of improving the resiliency of the system.

12.5.1 Subsystem Decomposition

In this section, the WWTP system will be divided into two subsystems for distributed state estimation. The input variables are $u = [Q_i, Q_{int}, K_{La3}, K_{La4}, K_{La5}, Q_r, Q_w]^T$. We use the initial conditions shown in Tables 12.1 and 12.2 and $u_s = [18446, 18446, 55338, 240, 240, 84, 18446, 385]^T$ as the working point (x_s, u_s).

Two subsystem models are shown in Table 12.3, in which Decomposition 1 is directly divided according to the physical structure (Zeng et al. 2016) and Decomposition 2 is obtained by the proposed method in this work. As shown in Table 12.3, considering the circulating f low from the secondary clarifier to the first anoxic reactor, Decomposition 1 divides the secondary clarifier and and anoxic section (i.e., reactor 1 and reactor 2) into a subsystem and aerated section (i.e., reactor 3, reactor 4, and reactor 5) into another subsystem. It can be seen that when divided by structure, a reactor is considered as a whole, and the connections between internal states are not considered. While Decomposition 2 considers both the number and strength

Table 12.1 Initial condition of the biological reactor

i	2	3	4	5	1	Units
$S_{I,i}$	30	30	30	30	30	$g\ COD/m^3$
$S_{S,i}$	2.81	1.46	1.15	1.00	0.89	$g\ COD/m^3$
$X_{I,i}$	1149.13	1149.13	1149.13	1149.13	1149.13	$g\ COD/m^3$
$X_{S,i}$	82.13	76.39	64.85	55.69	49.31	$g\ COD/m^3$
$X_{B,H,i}$	2551.77	2553.38	2557.13	2559.18	2559.34	$g\ COD/m^3$
$X_{B,A,i}$	148.39	148.31	148.94	149.53	149.80	$g\ COD/m^3$
$X_{P,i}$	448.85	449.52	450.42	451.31	452.21	$g\ COD/m^3$
$S_{O,i}$	0.004299	0.00006313	1.72	2.43	0.49	$g\ (-COD)/m^3$
$S_{NO,i}$	5.37	3.66	6.54	9.30	10.42	$g\ N/m^3$
$S_{NH,i}$	7.92	8.34	5.55	2.97	1.73	$g\ N/m^3$
$S_{ND,i}$	1.22	0.88	0.83	0.77	0.69	$g\ N/m^3$
$X_{ND,i}$	5.28	5.03	4.39	3.88	3.53	$g\ N/m^3$
$S_{ALK,i}$	4.93	5.08	4.67	4.29	4.13	mol/m^3

Table 12.2 Initial condition of the secondary clarifier

j	X_j	$S_{I,j}$	$S_{S,j}$	$S_{O,j}$	$S_{NO,j}$	$S_{NH,j}$	$S_{ND,j}$	$S_{ALK,j}$
1	6393.98	30	0.89	0.49	10.42	1.73	0.69	4.13
2	356.07	30	0.89	0.49	10.42	1.73	0.69	4.13
3	356.07	30	0.89	0.49	10.42	1.73	0.69	4.13
4	356.07	30	0.89	0.49	10.42	1.73	0.69	4.13
5	356.07	30	0.89	0.49	10.42	1.73	0.69	4.13
6	356.07	30	0.89	0.49	10.42	1.73	0.69	4.13
7	68.98	30	0.89	0.49	10.42	1.73	0.69	4.13
8	29.54	30	0.89	0.49	10.42	1.73	0.69	4.13
9	18.11	30	0.89	0.49	10.42	1.73	0.69	4.13
10	12.50	30	0.89	0.49	10.42	1.73	0.69	4.13
units	gCOD/m^3	gCOD/m^3	gCOD/m^3	g(–COD)/m^3	gN/m^3	g N/m^3	gN/m^3	mol/m^3

Table 12.3 Decomposition of WWTP

Decomposition 1		
Subsystem 1:	States:	All states in the anoxic section (reactor 1 and reactor 2) and the secondary clarifier
	Outputs:	All measured outputs in the anoxic section and the secondary clarifier
	Inputs:	Q_i, Q_{int}, Q_r, Q_w
Subsystem 2:	States:	All states in the aerated section (reactor 3, reactor 4, and reactor 5)
	Outputs:	All measured outputs in the aerated section
	Inputs:	K_{La3}, K_{La4}, K_{La5}
Decomposition 2		
Subsystem 1:	States:	All states in reactor 1, concentration of S_{ALK} in reactors 2–5, concentration of X, S_I, and S_{ALK} in the secondary clarifier
	Outputs:	All measured outputs in reactor 1, S_{ALK} in reactors 2–5, values of X, S_I, and S_{ALK} in the secondary clarifier
	Inputs:	Q_i, K_{La4}, Q_r, Q_w
Subsystem 2:	States:	Concentration of compounds except S_{ALK} in reactors 2–5, concentration of S_S, S_O, S_{NO}, S_{NH}, and S_{ND} in the secondary clarifier
	Outputs:	Measured outputs except S_{ALK} in reactors 2–5, values of S_S, S_O,S_{NO},S_{NH}, and S_{ND} in the secondary clarifier
	Inputs:	Q_{int}, K_{La3}, K_{La5}

of connections between internal variables. In Decomposition 2, Reactor 1, concentration of S_{ALK} in reactors 2–5, concentration of X, S_I, and S_{ALK} in the secondary clarifier are configured as subsystem 1 and subsystem 2 includes the concentration of compounds except S_{ALK} in reactors 2–5, concentration of S_S, S_O, S_{NO}, S_{NH}, and S_{ND} in the secondary clarifier. It can be seen that the concentration of S_{ALK} in all five reactors is taken out alone and put into subsystem 1 because it has little connection with other compounds in the same reactor.

12.5.2 Resilient Distributed State Estimator Design

WWTP is considered critical infrastructure and their resiliency is vital. In this section, the proposed subsystem decomposition and resilient distributed state estimation scheme is tested in the WWTP system to show the improvement of the resiliency compared to a centralized scheme applied to the whole system. We investigate the resiliency analysis under the storm and rain conditions, in which the unreliable communication networks are simultaneously considered.

The random process disturbance of the state equation and the noise in measurement are generated by the normal distribution values with mean value of zero and standard deviation of $w_Q x_0$ and $w_R y_0$, respectively, where w_Q and w_R are parameters and the symbol x_0 represents the initial condition shown in Tables 12.1 and 12.2, and y_0 can be calculated by $y_0 = C x_0$.The initial guess in different estimation schemes is set to be $1.1x_0$. The parameters used in the centralized EKF are $Q = diag((w_Q x_0)^2)$, $R = diag((w_R y_0)^2)$, and $P(0) = Q = diag((w_Q x_0)^2)$, where $diag(V)$ is a diagonal matrix whose diagonal elements are elements of vector v. The parameters used in the distributed EKF are $Q_i = diag((w_Q x_{0,i})^2)$, $R_i = diag((w_R y_{0,i})^2)$, and $P_i(0) = Q_i = diag((w_Q x_{0,i})^2)$, where $x_{0,i}$ and $y_{0,i}$ are the corresponding portion in x_0 and y_0 to subsystem i.

In order to compare the performance of different state estimation schemes, we calculate the error. In order to explain the different magnitude of estimation error in different states, the error of each state is normalized according to the maximum estimation error of all estimation schemes. The Euclidean norm of the normalized estimation error is defined as follows:

$$e(t_k) = \sqrt{\sum_{i=1}^{145} (e_i(t_k))^2}, \tag{12.46}$$

where $e(t_k)$ is the normalized error of 145 states at time instant t_k, and $e_i(t_k)$ is the normalized error of state i, $i = 1, 2, \ldots, 145$, defined as follows:

$$e_i(t_k) = \frac{\hat{x}_i(t_k) - x_i(t_k)}{max(\hat{x}_i - x_i)}, \tag{12.47}$$

where the maximum error of given state i is the maximum error of state i in EKF and distributed EKF methods. This means that the error of each state is normalized based on the maximum estimation error given by two different schemes. The average and maximum value of the normalized estimation error can be defined as follows:

$$Mean|e| = \frac{1}{K}\sum_{k=1}^{K}(e(t_k)) \tag{12.48a}$$

$$Max|e| = max\{e(t_1), e(t_1), \cdots, e(t_k)\}, \tag{12.48b}$$

where K is the total number of samples over the simulation period.

Rain conditions. We adjust w_Q and w_R to test the performance under different interference and noise conditions. The average and maximum values of the estimation error calculated by the three schemes are shown in Table 12.4. Figure 12.3 shows the actual process state trajectories and the estimates given by the centralized EKF and the distributed EKF (see Zhang et al. 2019) under the Decomposition 1 and the Decomposition 2, and Fig. 12.4 shows the trajectories of the Euclidean norms of the normalized estimation errors given by the three different schemes when $w_Q = w_R = 0.1$.

Simulation results in Table 12.4 show that the average estimation error of distributed EKF under Decomposition 2 is always smaller than that of the distributed EKF under Decomposition 1, which shows that the proposed decomposition method makes the internal connection of subsystems closer, which can reduce the state estimation error under the same state estimation scheme. It also shows that when the noise is greater, the distributed EKF may have better performance. It is verified that the proposed subsystem decomposition with weak inter-subsystem interactions can improve the resiliency of the system when applying the distributed state estimation.

Table 12.4 Performance comparison for different schemes under rain conditions

	w_R	w_Q	Centralized EKF	Distributed EKF	
				D1	D2
Mean$\|e\|$	0.2	0.2	4.1954	4.0622	4.0493
	0.2	0.1	3.2232	3.1938	3.1534
	0.1	0.2	4.6826	4.6691	4.6407
	0.1	0.1	3.6538	3.6934	3.6465
	0.2	0.05	2.2295	2.4193	2.3614
	0.05	0.2	4.9480	5.0382	4.9800
	0.05	0.05	2.9801	3.1684	3.0746
Max$\|e\|$	0.2	0.2	6.2810	6.3798	6.2979
	0.2	0.1	8.1007	8.1007	8.1007
	0.1	0.2	7.6567	7.8353	7.7048
	0.1	0.1	8.1256	8.1256	8.1256
	0.2	0.05	9.3052	9.3052	9.3052
	0.05	0.2	8.2770	8.4781	8.3283
	0.05	0.05	9.9826	9.9826	9.9826

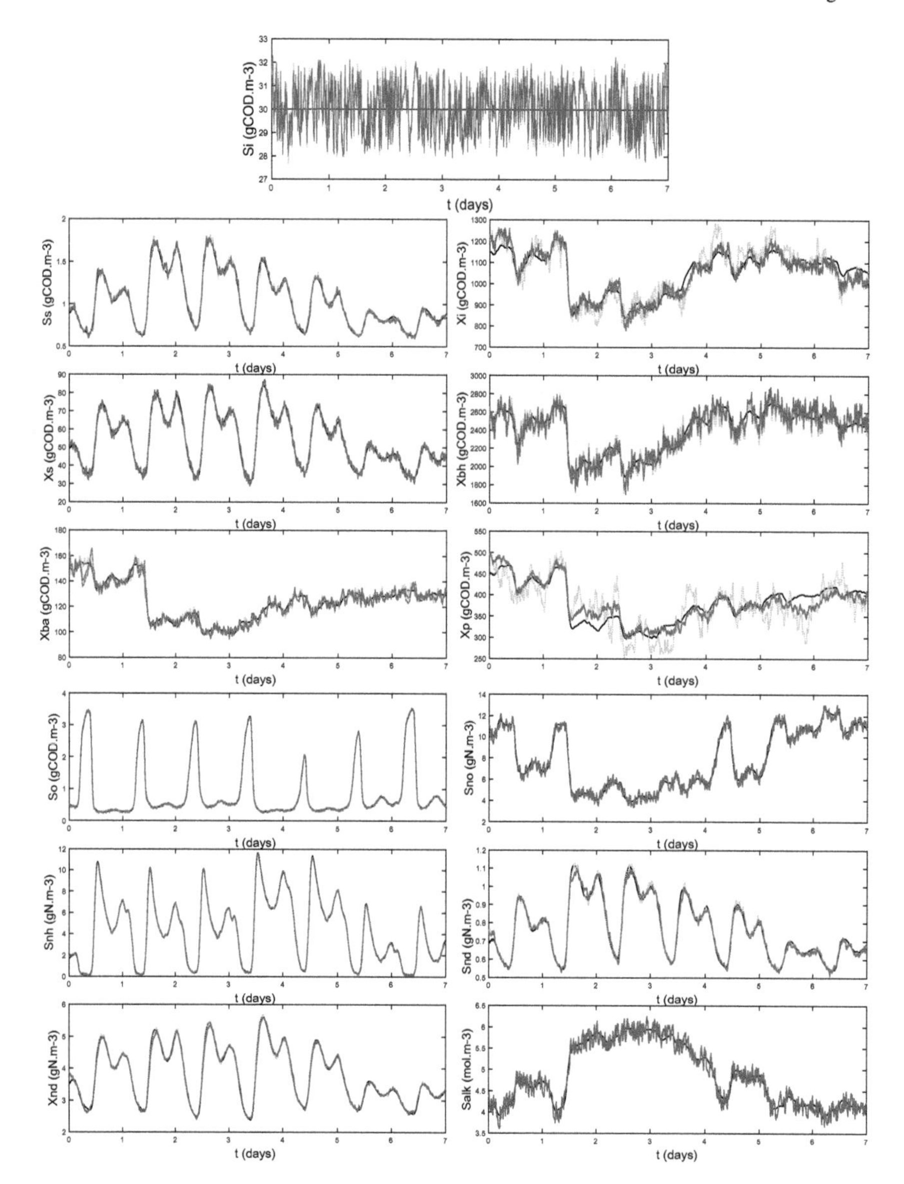

Fig. 12.3 Trajectories of the actual process state (black solid lines) and the estimates given by the centralized EKF (green dashed lines) and the distributed EKF under the Decomposition 1 (red solid lines)and the Decomposition 2 (blue dash-dotted lines) of reactor 5 under rain conditions

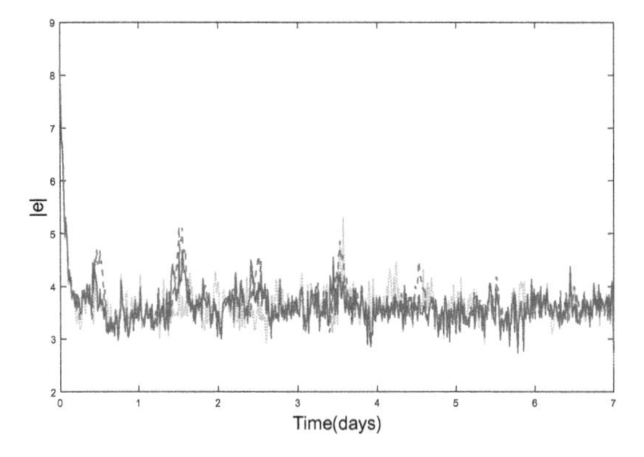

Fig. 12.4 Trajectories of the Euclidean norm of normalized estimation errors given by the EKF (green dashed lines) and the distributed EKF under the Decomposition 1 (red solid lines)and the Decomposition 2 (blue dash-dotted lines) under rain conditions

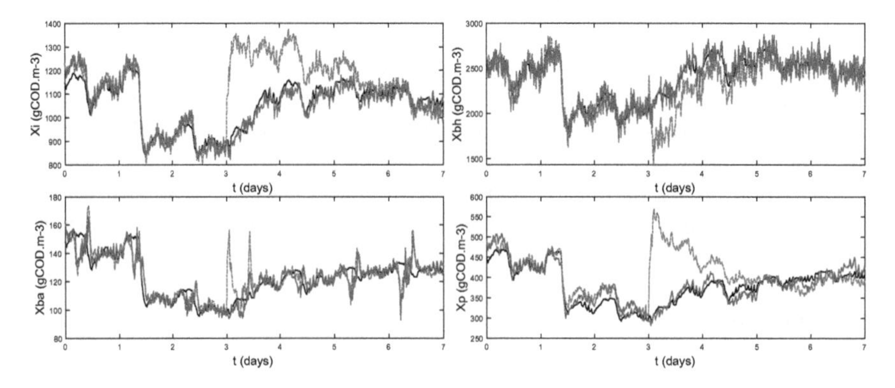

Fig. 12.5 Trajectories of the actual process state (black solid lines) and the DEKF (red dashed lines) and the resilient DEKF (blue solid lines) under the Decomposition 2 of reactor 1 under rain conditions with unreliable communication network $\hat{X}_i(t) = \hat{X}_i(t) + 0.5x_{init}$ during $t = 3$(days) to $t = 3.05$(days)

Furthermore, the condition with unreliable communication network is tested to show the resiliency of the system under the resilient distributed state estimator. The exchanged estimated stated is set as $\hat{X}_i(t) = \hat{X}_i(t) + 0.5x_{init}$ during $t = 3$(days) to $t = 3.05$(days). This means that the exchange information could be attacked or modified. The proposed resilient distributed state estimator is used to construct the states under the reliable communication network. Trajectories of the actual process state and the resilient DEKF under the Decomposition 2 when $w_Q = w_R = 0.1$ are shown in Fig. 12.5. The trajectories of the Euclidean norm of normalized estimation errors given by the Distributed EKF and the resilient distributed EKF under the Decomposition 2 are shown in Fig. 12.6. The results show that the proposed resilient distributed state estimation scheme can improve the resiliency of the system with unreliable communication network.

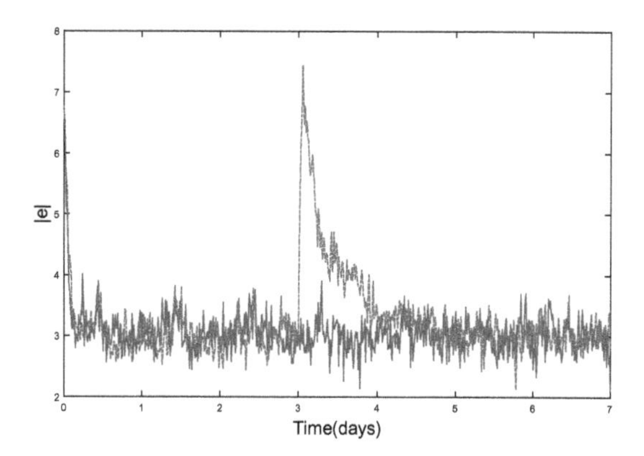

Fig. 12.6 Trajectories of the Euclidean norm of normalized estimation errors given by the DEKF (red dashed lines) and the resilient DEKF (blue solid lines) under the Decomposition 2 under rain conditions with unreliable communication network $\hat{X}_i(t) = \hat{X}_i(t) + 0.5x_{init}$ during $t = 3$(days) to $t = 3.05$(days)

Storm conditions. We further test the resilient distributed EKF under storm conditions. The average and maximum value of the estimation error calculated by the three schemes are shown in Table 12.5. Figure 12.7 shows the actual process state trajectories and the estimates given by the centralized EKF and the distributed EKF under the Decomposition 1 and the Decomposition 2 and Fig. 12.8 shows the trajectories of the Euclidean norms of the normalized estimation errors given by the three different schemes when $w_Q = w_R = 0.1$.

Simulation results in Table 12.5 show that the average estimation error of distributed EKF under Decomposition 2 is always smaller than that of the distributed

Table 12.5 Performance comparison for different schemes under storm conditions

	w_R	w_Q	Centralized EKF	Distributed EKF	
				D1	D2
Mean$\|e\|$	0.2	0.2	4.0668	4.0572	4.0232
	0.2	0.1	3.0962	3.0886	3.0219
	0.1	0.2	4.5115	4.5524	4.5201
	0.1	0.1	3.5968	3.6204	3.5504
	0.2	0.05	2.0707	2.3161	2.2007
	0.05	0.2	4.8841	4.9917	4.9308
	0.05	0.05	2.7948	2.9948	2.9138
Max$\|e\|$	0.2	0.2	6.0169	6.1214	6.0297
	0.2	0.1	7.3005	7.3005	7.3005
	0.1	0.2	6.9390	7.1451	7.0182
	0.1	0.1	7.3580	7.3580	7.3580
	0.2	0.05	8.3908	8.3908	8.3908
	0.05	0.2	7.1152	7.2747	7.1426
	0.05	0.05	8.8528	8.8528	8.8528

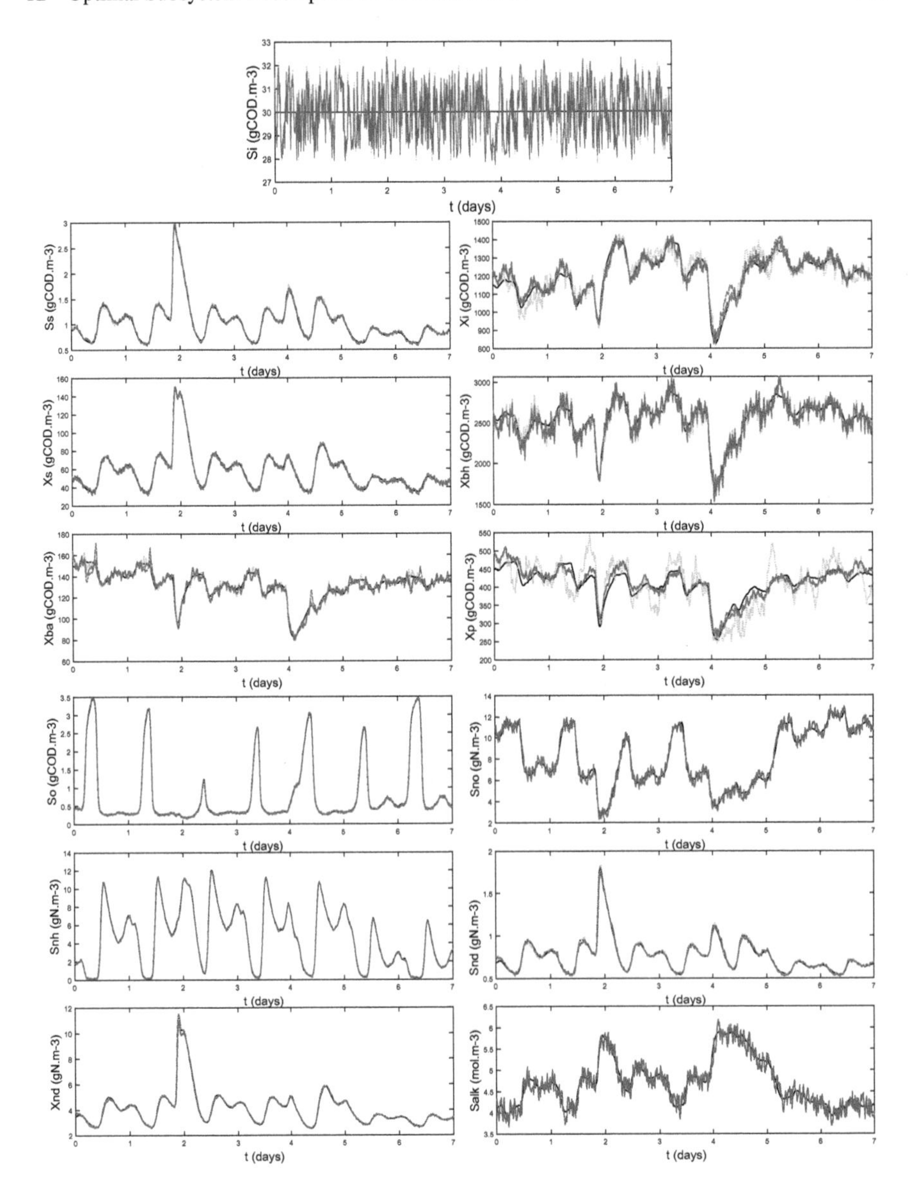

Fig. 12.7 Trajectories of the actual state (black solid lines) and the estimated states given by the centralized EKF (green dashed lines) and the distributed EKF under the Decomposition 1 (red solid lines) and the Decomposition 2 (blue dash-dotted lines) of reactor 5 under storm conditions

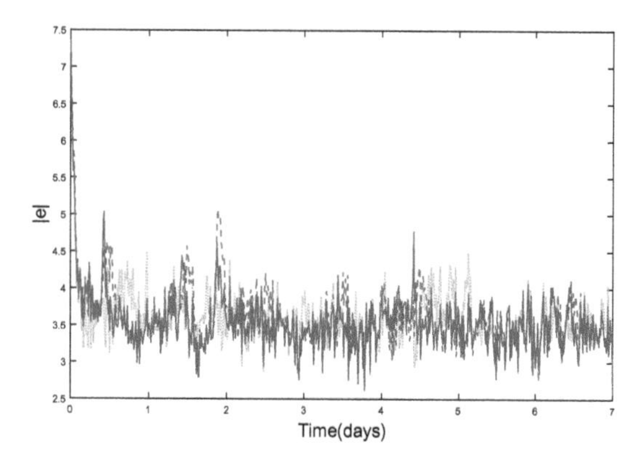

Fig. 12.8 Trajectories of the Euclidean norm of normalized estimation errors given by the EKF (green dashed lines) and the distributed EKF under the Decomposition 1 (red solid lines)and the Decomposition 2 (blue dash-dotted lines) under storm conditions

EKF under Decomposition 1, which shows that the proposed decomposition method makes the internal connection of subsystems closer, which can reduce the state estimation error under the same state estimation scheme. It confirms that the proposed subsystem decomposition with weak inter-subsystem interactions can improve the resiliency of the system when applying the distributed state estimation.

Similarly, the condition with unreliable communication network is tested to show the resiliency of the system under the storm condition. The exchanged estimated stated is set as $\hat{X}_i(t) = \hat{X}_i(t) + 0.5x_{init}$ during $t = 3$(days) to $t = 3.05$(days). Trajectories of the actual process state and the resilient DEKF under the Decomposition 2 when $w_Q = w_R = 0.1$ are shown in Fig. 12.9. The trajectories of the Euclidean norm of normalized estimation errors are shown in Fig. 12.10. The results confirm that the proposed resilient distributed state estimation scheme can improve the resiliency of the system with unreliable communication network.

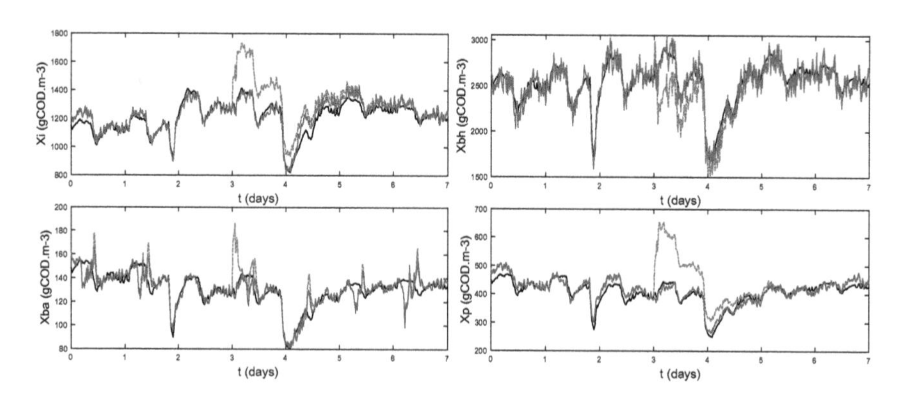

Fig. 12.9 Trajectories of the actual process state (black solid lines) and the DEKF (red dashed lines) and the resilient DEKF (blue solid lines) of reactor 1 under storm conditions with communicate attack which set $\hat{X}_i(t) = \hat{X}_i(t) + 0.5x_{init}$, $t = 3 - 3.05$(days)

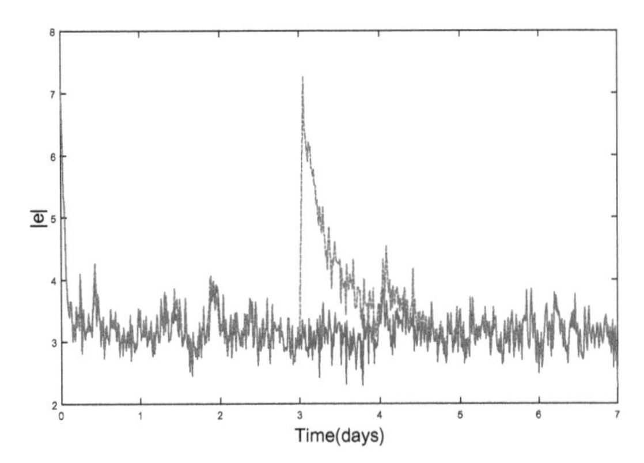

Fig. 12.10 Trajectories of the Euclidean norm of normalized estimation errors given by the DEKF (red dashed lines) and the resilient DEKF (blue solid lines) under the Decomposition 2 under storm conditions with communicate attack which set $\hat{X}_i(t) = \hat{X}_i(t) + 0.5x_{init}$, $t = 3 - 3.05$(days)

12.6 Conclusion

In this work, an optimal subsystem decomposition algorithm is proposed based on the community discovery algorithm with weighted network graph and is applied to a benchmark WWTP system. A resilient distributed state estimator is designed and carried out under the subsystem models which are divided by the physical structure and the subsystem model obtained using the proposed decomposition method. The results show that the subsystem decomposition and distributed state estimation scheme improves the resiliency of the system, compared to a centralized scheme applied to the whole system.

Acknowledgements This work was supported by National Natural Science Foundation of China under Grant no. 61803161, Science and Technology Plan Project of Guangzhou under Grant no. 202102020379, the Project of Department of Education of Guangdong Province under Grant no. 2020KTSCX008.

References

J. Alex, V. Magdeburg, Benchmark Simulation Model no. 1 (BSM1). Germany, IWA: Taskgroup on Benchmarking of Control Stategies for WWTPs (2008)

A.M. Anter, D. Gupta, O. Castillo, A novel parameter estimation in dynamic model via fuzzy swarm intelligence and chaos theory for faults in wastewater treatment plant. Soft. Comput. **24**(1), 111–129 (2020)

I. Baklouti, M. Mansouri, A.B. Hamida, et al., Monitoring of wastewater treatment plants using improved univariate statistical technique. Process Saf. Environ. Prot. (2018). https://doi.org/10.1016/S0957582018300387

M. Barbu, S. Caraman, G. Ifrim et al., State observers for food industry wastewater treatment processes. J. Environ. Prot. Ecol. **12**(2), 678–687 (2011)

J. Busch, D. Elixmann, P. Kühl et al., State estimation for large-scale wastewater treatment plants. Water Res. **47**(13), 4774–4787 (2013)

D. Ding, Q.L. Han, X. Ge, J. Wang, Secure state estimation and control of cyber-physical systems: a survey. IEEE Trans. Syst. Man Cybern. **51**(51), 176–190 (2020)
W.B. Dunbar, Distributed receding horizon control of dynamically coupled nonlinear systems. IEEE T. Automat. Contr. **52**(7), 1249–1263 (2007)
S. Heo, W.A. Marvin, P. Daoutidis, Automated synthesis of controlconfigurations for process networks based on structural coupling. Chem. Eng. Sci. **136**, 76–87 (2015)
A.M.N. Kiss, B. Marx, et al., State estimation of two-time scale multiple models: application to wastewater treatment plant. Control Eng. Pract. **19** (11), 1354–1362 (2011)
X. Qu, P.J.J. Alvarez, Q. Li, Applications of nanotechnology in water and wastewater treatment. Water Res. **47**(12), 3931–3946 (2013)
W. Tang, A. Allman, D.B. Pourkargar, P. Daoutidis, Optimal decomposition for distributed optimization in nonlinear model predictive control through community detection. Ind. Eng. Chem. Res. **111**, 43–54 (2018)
H.F. Wahab, R. Katebi, R. Villanova, Comparisons of nonlinear estimators for wastewater treatment plants, in *20th Mediterranean Conference on Control & Automation (MED)* (2012), pp. 764–769
P.B. Wang, X.M. Ren, D.D. Zheng, Event-triggered resilient control for cyber-physical systems under periodic DoS jamming attacks. Inf. Sci. **577**, 541–556 (2021)
W. Wei, P. Xia, Z. Liu, M. Zuo, A modified active disturbance rejection control for a wastewater treatment process. Chinese J. Chem Eng. **28**(10), 2607–2619 (2020)
X.Y. Yin, J.F. Liu, State estimation of wastewater treatment plants based on model approximation. Comput. Chem. Eng. **111**, 79–91 (2018)
X.Y. Yin, J.F. Liu, Subsystem decomposition of process networks for simultaneous distributed state estimation and control. AIChE J. **65**(3), 904–914 (2019)
X.Y. Yin, K. Arulmaran, J.F. Liu et al., Subsystem decomposition and configuration for distributed state estimation. AIChE J. **62**(6), 1995–2003 (2016)
X.Y. Yin, B. Decardi-Nelson, J.F. Liu, Subsystem decomposition and distributed moving horizon estimation of wastewater treatment plants. Chem. Eng. Res. Des. **134**, 405–419 (2018)
J. Zeng, J. Liu, T. Zou, et al., Distributed extended kalman filtering for wastewater treatment processes. Ind. Eng. Chem. Res. **55**(28) (2016)
L. Zhang, B. Wang, Y. Li, Y. Tang, Distributed stochastic model predictive control for cyber physical systems with multiple state delays and probabilistic saturation constraints. Automatica **129**, 109574 (2021)
L.W. Zhang, X.Y. Yin, J.F. Liu, Complex system decomposition for distributed state estimation based on weighted graph. Chem. Eng. Res. Des. **151**, 10–22 (2019)

Chapter 13
Cyber-Attack Detection for a Crude Oil Distillation Column

H. M. Sabbir Ahmad, Nader Meskin, and Mohammad Noorizadeh

13.1 Introduction

13.1.1 Preliminary

Due to the continuous development of technology, an increasing number of electronic devices are being developed with networking features suitable for connecting to industrial networks. This technological evolution has also made its way to Industrial Control Systems (ICSs) where an increasing number of monitoring and controlling devices have been connected to computer networks facilitating the supervisory level monitoring and control. Evolution in computing and internet technology has encouraged increasing number of ICS to be linked to cyber-world giving rise to a new class of systems called Cyber-Physical System (CPS) which provides several economic and performance-enhancing benefits. However, it also makes ICS more vulnerable to cyber-attacks. The effect of cyber-attacks differs in cyber-physical critical ICS compared to traditional ICT systems as they can cause damage to physical infrastructure posing threats to human health and environment. The complex CPS infrastructure more than ever requires the development of novel security solutions, as these systems are continuously targeted by attacks and intrusions by intelligent adversaries. Some typical examples of attacks in real systems are the Stuxnet worm attack, multiple recent power blackouts in Brazil, and the SQL Slammer worm attack on the Davis–Besse nuclear plant, to name a few (Pasqualetti et al. 2012; Nourian

H. M. S. Ahmad · N. Meskin (✉) · M. Noorizadeh
Qatar University, Doha, Qatar
e-mail: nader.meskin@qu.edu.qa

H. M. S. Ahmad
e-mail: ha1607441@student.qu.edu.qa

M. Noorizadeh
e-mail: m.noorizadeh@qu.edu.qa

M. Abbaszadeh and A. Zemouche (eds.), *Security and Resilience in Cyber-Physical Systems*, https://doi.org/10.1007/978-3-030-97166-3_13

and Madnick 2018; Pasqualetti et al. 2015), further justifying the need to address cyber-security for ICS.

Extensive research has been conducted on security issues from the prospective of network and communication technologies to securely defend network performance against adversaries. These research works have mainly concentrated on designing methodologies to secure communication networks in CPS ignoring interactions between the cyber and physical domain. Traditionally, cyber-security for ICS has been dealt by IT engineers from the prospective of network security. Such approaches primarily aim to secure the communication network to protect the IT infrastructure without considering the physical behavior of the plant and how the ICS is affected by cyber-attack. ICS are characterized by feedback closed-loop control architecture and aim to optimize the system control performance, such as reducing state estimation errors, stabilizing an unstable plant, and enhancing the robustness against uncertainties and noise. Therefore, it is important to guarantee the resiliency of cyber-physical ICS subject to multiple types of malicious attacks. This chapter focuses on the development of cyber-attack detection technique for a Cyber- Physical Distillation Column.

13.1.2 Cyber-Security of Distillation Column

Cyber-security of CPS has become a hot topic of research lately with focus on a wide range of physical plants. In Kundur et al. (2011), Manandhar et al. (2014), He et al. (2017), Kurt et al. (2019), cyber-security for smart grid has been studied and in Abokifa et al. (2019), the effect of cyber-attacks on water distribution systems is investigated. In Li et al. (2019), Kravchik and Shabtai (2018), Lin et al. (2018), Adepu and Mathur (2021), Elnour et al. (2020), different techniques for detecting attacks on a cyber-physical Reverse Osmosis Water Treatment Plant are presented. In Noorizadeh et al. (2021), a hybrid testbed is developed for Tennessee Eastman process and different data-driven detection algorithms are developed and tested. In Elnour et al. (2021), the security of Smart Buildings has been studied. To the best of the author's knowledge, cyber-security for a Crude Oil Distillation Column (DC) is only considered in Sabbir Ahmad and Meskin (2020) where the system dynamics simulated using Aspen Plus Dyanmics was integrated with Simulink and an observer-based attack detection scheme was implemented and validated using computer simulation in Simulink. In this study, first a detailed dynamical model of the DC is presented and a HIL testbed is designed for a cyber-physical DC using hardware from Siemens. Finally, an online real-time distributed detection scheme is proposed based on Unscented Kalman Filter (UKF) scheme implemented directly on PLCs.

In Taqvi et al. (2016), Minh and Pumwa (2012a), George and Francis (2015), Kathel and Jana (2010), Zou et al. (2017), Bendib et al. (2015), Radulescu et al. (2007), Weerachaipichasgul et al. (2010), a set of equations collectively called MESH equations are presented to describe the internal dynamics of a distillation column.

In Taqvi et al. (2017), the column model is described in terms of the relationship between the inputs and outputs which are generated using data from Aspen Plus Dynamics. In Minh and Pumwa (2012a), George and Francis (2015), Kathel and Jana (2010), Zou et al. (2017), a binary continuous distillation column is simulated with the assumption that the molar hold up in each tray including the condenser and reflux drum remains constant and there is negligible vapor holdup in each tray. This assumption neglects the dynamics of liquid and vapor flow rates inside the column due to tray hydraulics which have significant time constants impacting the dynamic performance of the model. In Bendib et al. (2015), the MESH equations are presented without any description for liquid and vapor flow rates dynamics inside the column. The crude feed is considered as a pseudo-binary mixture with a constant relative volatility in Minh and Pumwa (2012a), George and Francis (2015), Kathel and Jana (2010), Zou et al. (2017), Bendib et al. (2015) which is not the case in reality as the volatility varies with temperature and pressure. Finally, the fundamental limitation of using input–output relationship for distillation column simulation is that the internal dynamics which contains information on individual trays inside the column is ignored. Such information can be extremely valuable in several ways, one of which is temperature inferential output product quality measurement. The purity of output product stream can be determined using off-line analyzers which is indeed time-consuming. Time inferential measurement is fast and provides an efficient way of controlling the quality of the products from a distillation column.

As part of this study, the DC plant presented in Minh and Pumwa (2012a, b) is considered and the presented data to design the column in Aspen Plus is used to generate the steady-state data. Then, in order to improve the model accuracy, the DC plant is transported into Aspen Plus Dynamics to observe the effect of various column parameters to include them in the mathematical model. Finally, using the steady-state data, the dynamical model is simulated in real-time using MESH equations given in Minh and Pumwa (2012a, b) inside Simulink environment.

Next, a hybrid Hardware-In-the-Loop (HIL) ICS testbed is developed and implemented for the DC plant using industrial automation hardware from Siemens to make the study resembles a practical ICS. The hybrid HIL testbed contains three layers: (I) Field layer, (II) Control layer, and (III) Supervisory layer, and PROFINET as an industrial communication protocol is used for communication between I/O modules and PLCs. Different types of attack on ICS sensors and actuator such as false data injection attack (Lv et al. 2019; Zhang et al. 2017) (scaling attack, bias injection attack, etc.), Denial of Service (DoS) attack (Meraj et al. 2015), replay attack (AlDairi and Tawalbeh 2017) are emulated inside the testbed using their mathematical representation.

Finally, an online distributed attack detection method for the DC plant is developed and implemented in real-time on the testbed PLCs. The proposed detection algorithm is based on state estimation using UKF. There are various nonlinear state estimators available. As part of this study, three factors are considered while choosing UKF, namely, convergence, implementation simplicity (the estimators are implemented inside the PLCs which have limited mathematical library tool set), and computational complexity (PLCs have limited computational ability). Based on these criteria, UKF is chosen as it is able to provide full system state estimation based on the systems inputs and outputs in the presence of process and measurement noise which provided the main motivation for the choice of this algorithm. Computationally, the algorithm primarily involves basic linear mathematical operations (addition, subtraction, and multiplication) which could be easily implemented inside the PLCs. The fundamental idea is that during normal operation, the estimated measurements will coincide with the actual measurements, while in the presence of any attack, there will be deviation between the estimated and actual measurements. Hence, by computing the residuals corresponding to the difference between the actual and estimated measurements and comparing them with a given threshold, attacks can be successfully detected. Various formulated attack scenarios are emulated inside the testbed and performance of the proposed detection scheme is demonstrated.

This chapter includes seven sections. In Sect. 13.2, the mathematical model and the control system of the DC plant are presented and the details of the developed hybrid testbed are discussed in Sect. 13.3. Next, the mathematical models of various attacks used in this study are provided in Sect. 13.4 and the proposed attack detection algorithm is presented in Sect. 13.5. The results corresponding to different attack scenarios injected in the developed testbed are given in Sect. 13.6. Finally, the summary of the chapter is presented in the conclusion section.

13.2 Distillation Column Design and Modeling

A continuous binary distillation splits a crude feed into two fractions, which are collected from the top and bottom sections of the crude tower. The raw crude is fed to the binary column at the feed section and the column can be divided into two sections, namely, rectifying and stripping section. The rectifying section is located at the top just above the feed and the bottom section is called the stripping section. The original crude feedstock is passed through a preheater which heats the feed to a certain temperature in order to convert it into a two phase fluid before feeding to the distillation column. Inside the column, the temperature gradient causes the relatively volatile lighter components to vaporize and rise to the top of the column, and the less volatile heavier components fall down to the bottom section of the column. The vapor at the top is cooled down by a condenser and collected at the reflux drum where a portion of it is extracted out as distillate and the remaining cooled liquid (known as reflux) is fed back to the column. Similarly, the liquid at the column base is collected in reboiler drum where a portion of it is extracted out as bottoms product and the remaining portion is vaporized by the reboiler and fed back to the column.

13.2.1 Plant Data

The considered DC model is based on a real petroleum project presented in Minh and Pumwa (2012a, b). The plant operates for 24 h and 365 d over a year during which it processes 130,000 tons of raw condensate. Figure 13.1 illustrates the flowsheet of the binary distillation column considered in this work. The plant operating specification is to maintain the product quality within desired range; the purity of the distillate has

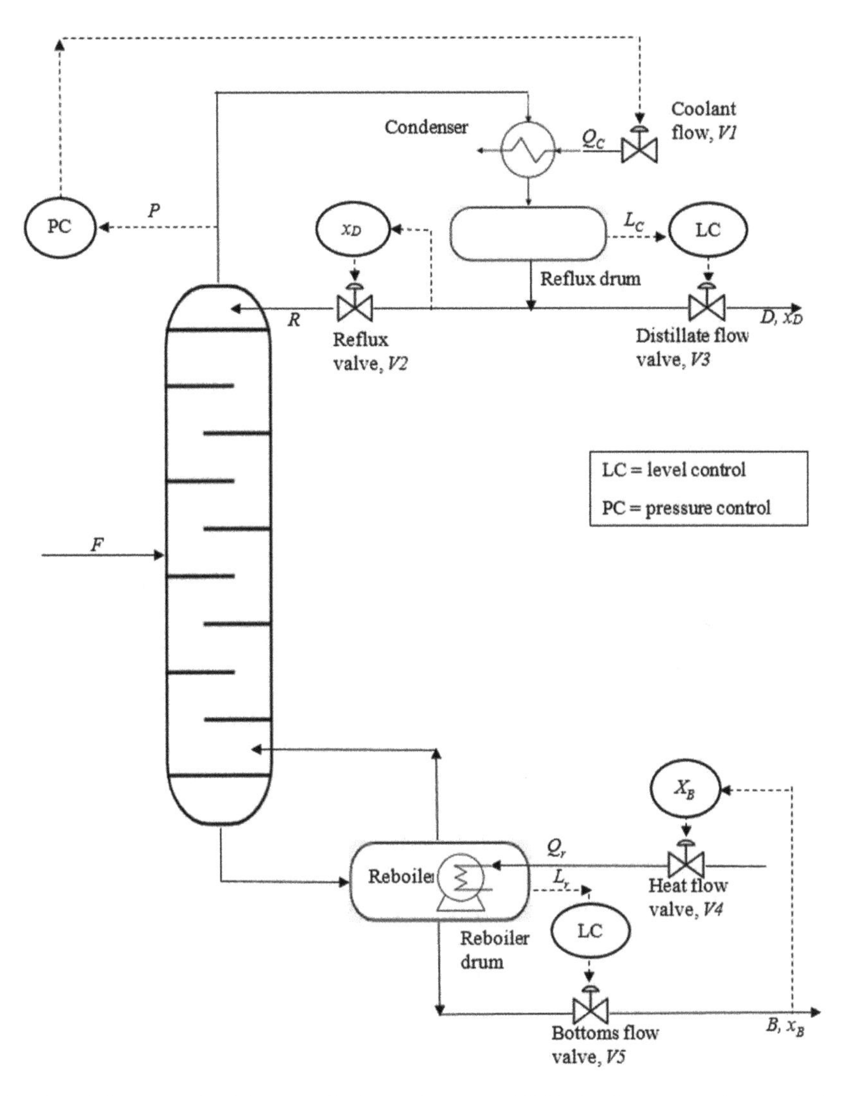

Fig. 13.1 Flowsheet of a binary distillation column, ©2020 IEEE. Reprinted, with permission, from Sabbir Ahmad and Meskin (2020)

Table 13.1 Raw condensate composition

Component	Mole fraction	Component	Mole fraction (%)
Propane	0.00	n-C11H24	1.94
Normal Butane	19.00	n-C12H26	2.02
Iso-Butane	26.65	Cyclopentane	1.61
Iso-Pentane	20.95	Methylclopentane	2.02
Normal Pentane	10.05	Benzene	1.61
Hexane	7.26	Toulen	0.00
Heptane	3.23	O-xylene	0.00
Octane	1.21	E-benzen	0.00
Nonane	0.00	124-Mbenzen	0.00
Normal Decane	0.00		

Table 13.2 Properties of pseudo components

Properties	Ligas	Napthas
Molar weight	54.5–55.6	84.1–86.3
Liquid density (kg/m^3)	570–575	725–735
Feed composition (vol%)	48–52	48–52

to be higher than or equal to 98% and the impurity of the bottoms has to be equal or less than 2%.

Table 13.1 presents the nominal composition of the raw condensate and the actual composition of the raw condensate generally fluctuates around their nominal values. Although the liquid feed consists of multiple components, however, since the aim is to use a binary distillation column, a pseudo-binary mixture is considered consisting of Ligas (iso-butane, n-butane, and propane) and Napthas (iso-pentane, n- pentane, and heavier components). There are 14 trays inside the column with the topmost tray is numbered as the first layer. The properties of the pseudo components are allowed to fluctuate within the range shown in Table 13.2 based on the fluctuation in the condensate composition. Before feeding to the column, the raw crude is passed through a preheater to convert it into two phases which are vapor and liquid phase and fed to trays 7 and 8, respectively.

13.2.2 Distillation Column Design

The column is designed using Redfrac model available in ASPEN Plus where two degrees of freedom are considered for the column design and distillate rate (kmole/hr)

Table 13.3 Column design parameters used in Aspen Plus

Parameter	Value	Parameter	Value
Feed temperature (°C)	118	Feed pressure (atm)	4.6
Condenser pressure (bar)	4	Stage pressure drop (bar)	0.075
Distillate rate (kmole/hr)	93	Reflux rate (kmole/hr)	350
Vapor feed stage	7	Vapor feed rate	185.827
		(kmole/hr)	
Liquid feed stage	8	Liquid feed rate	16.937
		(kmole/hr)	
Ligas concentration	0.513	Ligas concentration	0.127
In vapor feed		In liquid feed	

and reflux rate (kmole/hr) are the two parameters selected for the column design. Table 13.3 presents the data used in ASPEN Plus for the column design.

13.2.3 Dynamic Model of the Distillation Column

The dynamics of the nth tray using mass and balance equations can be written as

$$\frac{dM_n}{dt} = L_{n-1} - L_n + V_{n+1} - V_n \tag{13.1}$$

$$\frac{d(M_n x_{n,i})}{dt} = L_{n-1}x_{n-1,i} - L_n x_{n,i} + V_{n+1}y_{n+1,i} - V_n\, y_{n,i}, \tag{13.2}$$

where M_n is the liquid hold up (kmole) in the nth tray inside the column, L_n and h_n denote the flow rate (kmole/hr) of the liquid flowing down the nth tray and the amount of heat energy that is passed with the liquid from the nth tray, respectively, V_n and H_n denote the vapor flow rate (kmole/hr) at the nth tray and the energy carried by the vapor, respectively, and $x_{n,i}$ and $y_{n,i}$ denote the liquid and vapor mole fraction of the ith component in tray n, respectively, where

$$\sum_{i=1}^{C} x_{n,i} = 1;\ \sum_{i=1}^{C} y_{n,i} = 1, \tag{13.3}$$

with C as the number of components in the feed. Since, it is assumed the feed to be a pseudo-binary mixture, we have $C = 2$ and it is only necessary to consider the molar concentration dynamics of the lighter component based on the summation condition.

By differentiating (13.2) and substituting (13.1), it follows that

$$\frac{\mathrm{d}(x_{n,i})}{\mathrm{d}t} = \frac{L_{n-1}x_{n-1,i} + V_{n+1}y_{n+1,i} - (L_{n-1} + V_{n+1})x_{n,i} + V_n(y_{n,i} - x_{n,i})}{M_n}. \tag{13.4}$$

The column is numbered from top as $n = 1$ for the reflux drum, $n = 2$ for the first tray, $n = f$ for the feed tray, $n = N + 1$ for the bottom tray, and $n = N + 2$ for the reboiler with total of 14 trays inside the column, i.e., $N = 14$.

In order to perform a dynamic simulation to observe the dynamics of the tray hydraulics, the model from Aspen Plus is transported to Aspen Plus Dynamics. It is indeed necessary to include this dynamics since the time constants associated with liquid and vapor flow rates are quite large which will affect the overall response time of the system. Hence, the effect of tray hydraulics is included (as continuous system states) to the liquid and vapor flow rates across every tray in the column by introducing a time constant as follows:

$$dL_n(s) = \frac{1}{\tau_{L_n}s + 1}dL(s) \tag{13.5}$$

$$dV_n(s) = \frac{1}{\tau_{v_n}s + 1}dV(s), \tag{13.6}$$

where $dL = L - L^{\text{nominal}}$, $dL_n = L_n - L_n^{\text{nominal}}$, $dV = V - V^{\text{nominal}}$, and $dV_n = V_n - V_n^{\text{nominal}}$. L_n^{nominal} and V_n^{nominal} are the nominal liquid and vapor flow rates for the nth tray inside the column which have been acquired from Aspen Plus. The time constants are determined from Aspen Plus Dynamics. The initial molar holdup in each tray has been computed using the Francise–Wier formula presented in Wijn (1999). The following assumptions are considered here

- The relative volatility is constant across each tray of the column. This implies that the vapor–liquid equilibrium relationship for the nth tray can be expressed as

$$y_n = \frac{\alpha x_n}{1 + (\alpha - 1)x_n}.$$

- The overhead vapor is totally condensed in a condenser.
- The pressure remains constant at the top of the column and the differential pressure between trays remains constant.
- The holdup of vapor is negligible throughout the system.

The overall model of the DC plant is expressed as follows:

$$\dot{\boldsymbol{x}}(t) = \boldsymbol{f}(\boldsymbol{x}(t), \boldsymbol{u}(t)) + \boldsymbol{w}(t) \tag{13.7}$$

$$\boldsymbol{y}(t) = \begin{bmatrix} x_1(t) \\ x_{16}(t) \end{bmatrix} + \boldsymbol{v}(t), \tag{13.8}$$

where

$$x(t) = [x_1(t), x_2(t), \ldots, x_{16}(t), M_1(t), M_2(t), \ldots, M_{16}(t),$$
$$L_2(t), L_3(t), \ldots, L_{15}(t), V_2(t), V_3(t), \ldots, V_{15}(t)]^T$$

is the state variable of the system, where for brevity, the subscript i is dropped from $x_{n,i}$ due to having only two components in DC, $u(t) = [L_1(t), V_{16}(t)]^T$ is the control input signal, and w and v are the process and measurement noise vector, respectively which have been modeled as Gaussian white noise.

13.2.4 Control of Distillation Column

13.2.4.1 Control Requirement for Distillation Column

In order to control a binary DC plant, at first, it is essential to determine its degree of freedom (DoF). DoF of a process is the number of independent variables that must be specified in order to define the process completely. Consequently, the desired control of a process will be achieved when and only when all degrees of freedom have been specified. Among several available approaches, one of the simple approaches to determine the DoF for a DC plant is to count the number of valves. There are four control valves as shown in Fig. 13.1, one on each of the following streams: distillate, reflux, bottoms, and reboiler vapor, and hence this column has four degrees of freedom. The feed stream is considered being set by the upstream process and consequently it is considered to be a constant. Inventories in any process must be always controlled, and the inventory loops involve liquid levels and pressures. The column has been designed in Aspen Plus to operate under constant pressure at the top of the column with a constant differential pressure between the trays and this implies that the liquid level in the reflux drum and the liquid level in the column base must be controlled. Hence, by considering the two variables that must be allocated for controlling the liquid level in the reflux drum and column base, there exist two remaining degrees of freedom. Thus, there are two and only two additional variables that can (and must) be manipulated to maintain the product quality of distillate and bottoms product.

Generally, a column is designed to operate in the steady state at the values determined from design calculations during normal operation and a column remains at energy and material balance (described by MESH equations) during the steady-state operation. Material balance infers that the sum of products entering the column must be equal (approximately) to the sum of products leaving the column, and energy balance implies that the heat input to the column must be equal (approximately) to heat removed from the system. A column is said to be "stable" when it is under energy and material balance.

Table 13.4 Manipulated and controlled variable pairs for the binary distillation column

Controlled variables	Manipulated variables	Control valve (Fig. 13.1)
Purity of distillate	Reflux flow rate	Reflux flow V_2
Liquid level in reflux drum	Distillate flow rate	Distillate flow V_3
Impurity in bottoms	Reboiler duty	Heat flow V_4
Liquid level in column base	Bottoms flow rate	Bottom flow V_5

The column dynamics arises from the control loops, i.e., if value of a control variable fluctuates from its desired value then the corresponding manipulated variable is adjusted to bring the control variable back to its desired value. Such changes in value of control variables may occur due to various reasons including change in properties of the feed within the range mentioned in Table 13.2.

13.2.4.2 Controller Design for Distillation Column

As mentioned previously, the proposed DC in Fig. 13.1 has four control and four controlled variables. Table 13.4 summarizes the control variables selected to control each of the four controlled variables. The PID controllers for distillation and bottoms product composition control is tuned using model-based PID tuning tools available from MATLAB. A PID controller contains a proportional, integral, and derivative term associated with each is a constant gain, that takes into account tracking error to achieve error convergence. The PID controller is given as

$$u(t) = K_p e(t) + K_i \int e(t)dt + K_d \frac{\mathrm{de(t)}}{dt}, \tag{13.9}$$

where $e(t) = y(t) - y_d(t)$, $y(t)$ is the output and $y_d(t)$ is the set-point.

The levels of the reflux drum and column base are maintained constant by adjusting the distillate and bottoms product flow, respectively, using the feed-forward control as

$$D(t) = V_{14}(t) - L_{\text{refluxflow}}(t), \tag{13.10}$$

$$B(t) = L_1(t) - V_{\text{vaporflow}}(t), \tag{13.11}$$

where $V_{14}(t)$ and $L_1(t)$ represent the vapor (kmole/hr) flowing out of tray 14 into the condenser and liquid (kmole/hr) flowing from tray 1 to the reboiler, respectively, $D(t)$ corresponds to the distillate flow rate (kmole/hr) and $B(t)$ corresponds to the bottoms product flow rate (kmole/hr).

13.3 Testbed Design

The hybrid testbed is designed to implement an ICS for the DC by integrating industrially used hardware in the simulation loop to make the study practically viable. The control objective of the DC plant is to maintain the purity of the distillate from the rectifying section and the bottoms product from the stripping section. Therefore, the DC has two outputs which are controlled using the two inputs which are the reflux flow rate (kmole/hr) and the vapor flow rate (kmole/hr) from the reboiler. The developed hybrid testbed contains two control PLCs: one for the rectifying section regulating the quality of the distillate and the second for the stripping section regulating the bottoms impurity level.

As part of this study, a three-level hybrid HIL Cyber-Physical ICS testbed is designed for the DC as shown in Figs. 13.2 and 13.3. The DC dynamics is simulated in real-time in a PC using Simulink and a data acquisition board (DAQ) is used to generate the measurements as well as receiving the valves commands from the controller. The field layer (Level 0) of the testbed is implemented using ESP-200 Distributed I/O modules from Siemens which are connected to DAQ. The control layer (Level 1) is implemented using Siemens S7-1500 PLCs which are interfaced to the Distributed I/Os using PROFINET which is an industrially used communication network. In addition, the second layer has a supervisory engineering station for supervisory monitoring an control. Finally, a cloud server is included in the testbed in the third layer (Level 2) for remote logging and online monitoring of the testbed. The link between simulator and the ICS is established using Humosoft MF634 DAQ

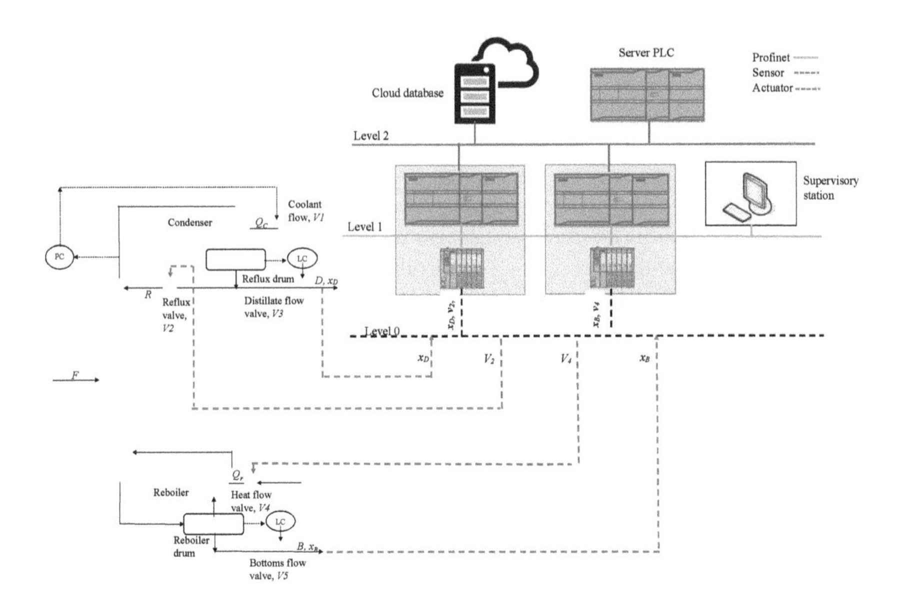

Fig. 13.2 Block level diagram of the DC testbed

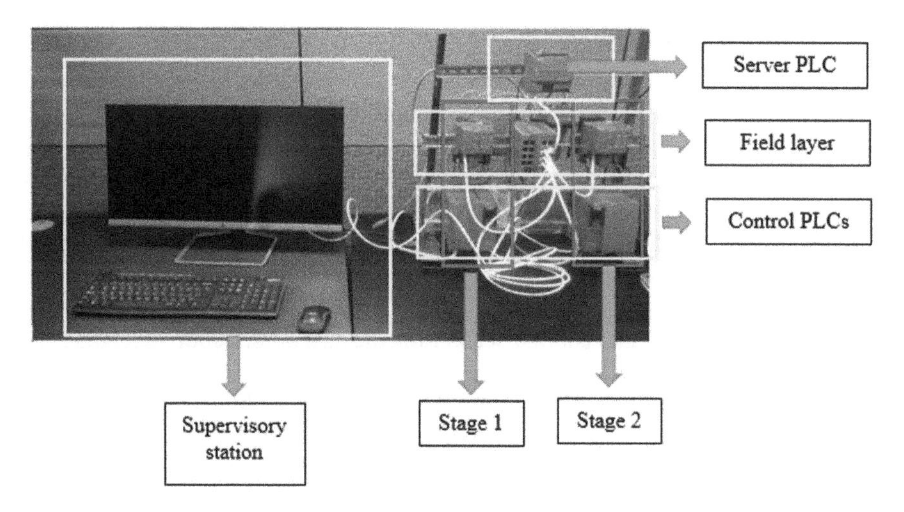

Fig. 13.3 The developed cyber-physical DC testbed

card which is used to extract the sensor measurements and manage the actuator inputs as voltages, and feed them to the distributed I/O modules.

The simulation sampling time for the DC plant simulator in MATLAB/Simulink is set to 3.6 s and the PLC monitors and updates the sensors and actuators every 3.6 s. The control firmware has been implemented in the PLC using an interrupt routine which is set to time-out every 3.6 s to service the feedback control loops in order to fulfill the control objective.

13.4 Attack Modeling

Industrial control systems (ICS) for any physical plant consist of a number of control loops that are responsible for controlling various parameters related to the plant. Each control loop fundamentally contains a controller, sensors, and actuators. Our study assumes that the attacker has managed to sneak through the IT security infrastructure to the control systems operating the plant and is capable of launching attacks on these systems, i.e., sensors and actuators. This is the worst attack scenario possible on the ICS. Figure 13.4 presents a diagram of a networked CPS under attack that has been considered as part of this study.

For any arbitrary attack of time period T_{ai}, let $\psi_i(t)$ and $\hat{\psi}_i(t)$ correspond to the healthy and corrupt data due to attack on the ith sensor/actuator ICS resource. In this case, the attack models can be expressed as follows:

1. **Scaling attack** (Sridhar and Govindarasu 2014): A scaling function is used to generate a false data injection attack whereby the channel data during attack is scaled by a constant factor as expressed below

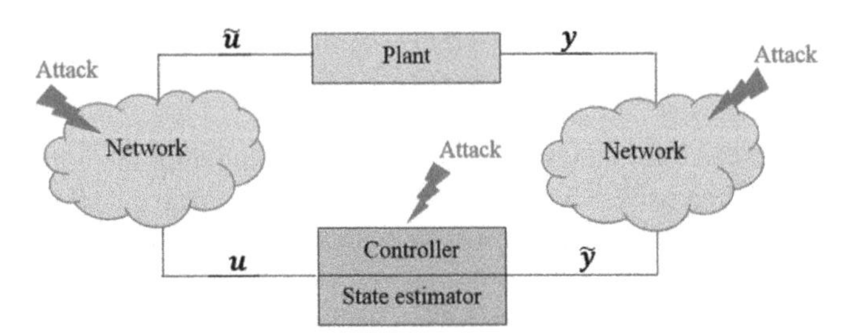

Fig. 13.4 Block level illustration of ICS of Cyber-Physical System under attack, ©2020 IEEE. Reprinted, with permission, from Sabbir Ahmad and Meskin (2020)

$$\hat{\psi}_i(t) = \begin{cases} \psi_i(t), & t \notin T_{ai} \\ \lambda_s \psi_i(t), & t \in T_{ai} \end{cases}, \tag{13.12}$$

where $\lambda_s \in \mathbb{R}$ is a constant.

2. **Bias Injection attack:** In this attack, the true sensor/actuator measurements are modified by adding a constant bias denoted by λ_b, as follows:

$$\hat{\psi}_i(t) = \begin{cases} \psi_i(t), & t \notin T_{ai} \\ \psi_i(t) + \lambda_b, & t \in T_{ai} \end{cases}, \tag{13.13}$$

where $\lambda_b \in \mathbb{R}$ is a constant.

3. **Ramp attack** (Sridhar and Govindarasu 2014): As part of this attack, the true sensor/actuator readings of the targeted resource are modified by adding a ramp function which gradually increases/decreases with time based on the gradient of ramp denoted by λ_r as follows:

$$\hat{\psi}_i(t) = \begin{cases} \psi_i(t), & t \notin T_{ai} \\ \psi_i(t) + \lambda_r t, & t \in T_{ai} \end{cases}. \tag{13.14}$$

4. **Replay attack** (Mo et al. 2015): The replay attack has two stages. At first, the adversary gathers sensor/actuator readings by disclosing the data from the targeted ICS resources. Subsequently, the attacker replays this collected data to the targeted ICS resources.
Stage 1 ($0 \leq t < T_I$): disclosure of resource

$$I_t = I_{t-1} \cup \begin{bmatrix} \gamma_u & 0 \\ 0 & \gamma_y \end{bmatrix} \begin{bmatrix} u(t) \\ y(t) \end{bmatrix}, \tag{13.15}$$

where γ_u and γ_y are the binary incidence matrices mapping the actuator and sensor data channels to the corresponding data gathered by the adversary, T_I is the length of gathering information for the replay attack, and the collected data is stored in I_k.
Stage 2 ($T_I \leq t < 2T_I$): disruption of resource

$$\begin{bmatrix} u(t) \\ y(t) \end{bmatrix} = I_{t-n}. \tag{13.16}$$

5. **DoS attack:** The DoS attack can be launched by jamming the communication channels, flooding packets in the network, and compromising devices to prevent data transfer, etc. As the lack of available sensor/actuator data, the DoS attack can be modeled as follows:

$$\hat{\psi}_i(t) = \begin{cases} \psi_i(t) & t \notin T_{ai} \\ (1 - D_s(t))\psi_i(t) + D_s(t)\psi_i(t - t_n) & t \in T_{ai}, \end{cases} \tag{13.17}$$

where $D_s(t)$ is a binary index and takes a value of 1 to resemble a scenario when a packet is denied and 0 for the normal operation. To encompass energy limitations, it is assumed that, within the attack time horizon T_{ai}, the targeted resource can send at most M data packets, while the attacker can launch DoS attack at most N times where $N < M$. In (13.17), t_n is the number of consecutive packets which are jammed by the attacker and hence can take values from $k_n = \{1, 2, 3, \ldots, N\}$. The attack model sends the last available packet during the DoS attack. DoS attack is able to make the data channels unavailable by jamming the disruption resources.
6. **Bounded random attack** (Manandhar et al. 2014; Sridhar and Govindarasu 2014): This attack involves the addition of randomly generated attack values to the sensor/actuator signal as follows:

$$\hat{\psi}_i(t) = \begin{cases} \psi_i(t) & t \notin T_{ai} \\ \psi_i(t) + N(0, \sigma^2) & t \in T_{ai} \; and \; |\sigma| < \rho, \end{cases} \tag{13.18}$$

where $\rho \in \mathbb{R}$.

It should be noted that the above-presented attack models are applicable for targeting both sensors and actuators.

13.5 Attack Detection Algorithm

13.5.1 UKF Based Attack Detection

The proposed detection method is based on state estimation which is implemented using UKF. A UKF is a state estimation algorithm that estimates the system states based on the system measurements and control inputs in the presence of Gaussian process and measurement noise. The proposed detection scheme is based on the idea of comparing the system measurements against the estimates from UKF and computing the residuals for every measurement upon which a threshold is applied to detect cyber-intrusions.

The proposed column has two control inputs and output measurements which are the reflux flow rate (kmole/hr) and vapor flow rate (kmole/hr), and distillate purity and bottoms impurity concentrations, respectively. Each tray has four associated states which are the molar holdup in the tray, molar concentration of distillate, liquid and vapor flow rate for that particular tray. Additionally, the condenser and reboiler each have two states which are the molar holdup and molar concentration of the distillate and bottoms product. As there are 14 trays besides the condenser and reboiler, hence in total there are 60 states. The system 13.7 is decomposed by separating the rectifying and stripping section dynamics of the column. Hence, the rectifying section dynamics is given as follows:

$$\begin{aligned} \dot{\boldsymbol{x}}_r(t) &= \boldsymbol{f}_1(x_r(t), x_s(t), u(t)) + \boldsymbol{w}_r(t) \\ y_r(t) &= x_1(t) + v_r(t) \end{aligned} \tag{13.19}$$

and the dynamics of the stripping section is given as follows:

$$\begin{aligned} \dot{\boldsymbol{x}}_s(t) &= \boldsymbol{f}_2(x_s(t), x_r(t), u(t)) + \boldsymbol{w}_s(t) \\ y_s(t) &= x_{16}(t) + v_s(t), \end{aligned} \tag{13.20}$$

where

$$\begin{aligned} \boldsymbol{x}_r(t) = [&x_1(t), x_2(t), \ldots, x_8(t), M_1(t), M_2(t), \ldots, M_8(t), \\ &L_2(t), L_3(t), \ldots, L_8(t), V_2(t), V_3(t), \ldots, V_8(t)]^T \\ \boldsymbol{x}_s(t) = [&x_9(t), x_{10}(t), \ldots, x_{16}(t), M_9(t), M_{10}(t), \ldots, M_{16}(t), \\ &L_9(t), L_{10}(t), \ldots, L_{15}(t), V_9(t), V_{10}(t), \ldots, V_{15}(t)]^T, \end{aligned}$$

$x_r(t)$ and $x_s(t)$, and $y_r(t)$ and $y_s(t)$ correspond to the states and outputs, for the rectifying and stripping section, respectively. The continuous states of the rectifying and stripping section include the liquid molar concentration of the lighter components in every tray along with the liquid and vapor flow rate dynamics for every tray inside each section. $f_1(\cdot)$ an $f_2(\cdot)$ represent the vector fields describing the state dynamics for the rectifying and stripping section, respectively, $x_i(t)$, $M_i(t)$, $L_i(t)$ and $V_i(t)$ denote the molar concentration, molar holdup, liquid and vapor flow rate for the ith tray in the column, and $w_r(t)$, $w_s(t)$, $v_r(t)$, and $v_s(t)$ represent the Gaussian white process and measurement noise, for the rectifying and stripping section, respectively.

The distributed scheme is implemented using two UKF, one for the rectifying section and one for the stripping section on their respective control PLC which interact with each other for estimating the overall system states. Based on the estimated state, each PLC computes residuals for its sensor measurements for each of which a threshold is applied for attack detection. Figure 13.5 shows a block diagram of the proposed detection scheme.

As the given model is continuous-time hence Eulers discretization is applied to derive the discrete-time model of the rectifying and stripping section of the column.

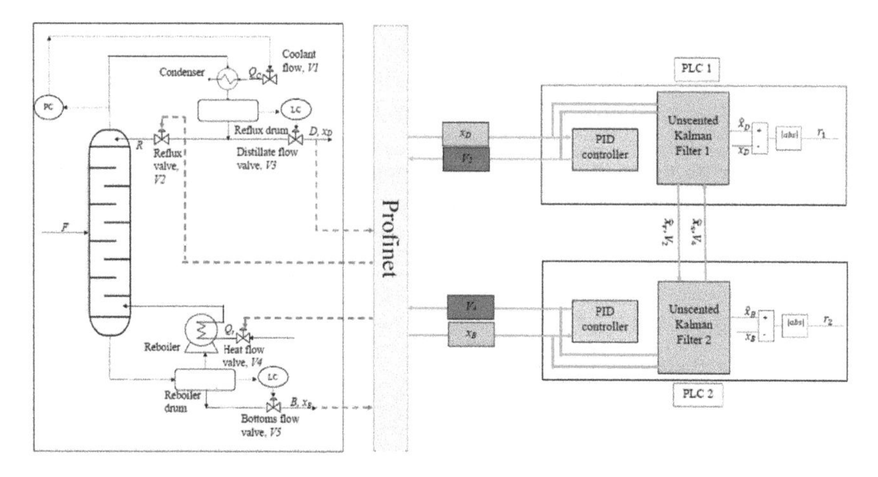

Fig. 13.5 Block level illustration of the detection scheme

The two main steps for implementing UKF for a discrete-time system are given below.

Prediction step:

$$
\begin{aligned}
\boldsymbol{X}^{a}_{k-1} &= \hat{\boldsymbol{x}}^{a}_{k-1} \pm \sqrt{(\Delta+\lambda)P^{a}_{k-1}} \\
\boldsymbol{X}^{x}_{k|k-1} &= f(\boldsymbol{X}^{x}_{k-1}, \boldsymbol{X}^{w}_{k-1}) \\
\hat{\boldsymbol{x}}_{k|k-1} &= \sum_{i=0}^{2\Delta} W_i^{(m)} \boldsymbol{X}^{x}_{i,k|k-1} \\
P_{k|k-1} &= \sum_{i=0}^{2\Delta} W_i^{(c)} [\boldsymbol{X}^{x}_{i,k|k-1} - \hat{\boldsymbol{x}}_{k|k-1}][\boldsymbol{X}^{x}_{i,k|k-1} - \hat{\boldsymbol{x}}_{k|k-1}]^T.
\end{aligned}
$$

Update step:

$$
\begin{aligned}
\boldsymbol{Y}_{k|k-1} &= h(\boldsymbol{X}^{x}_{k|k-1}, \boldsymbol{X}^{v}_{k-1}) \\
\hat{\boldsymbol{y}}_k &= \sum_{i=0}^{2\Delta} W_i^{(m)} \boldsymbol{Y}_{i,k|k-1} \\
P_{\tilde{\boldsymbol{y}}_k, \tilde{\boldsymbol{y}}_{k-1}} &= \sum_{i=0}^{2\Delta} W_i^{(c)} [\boldsymbol{Y}_{i,k|k-1} - \hat{\boldsymbol{y}}_k][\boldsymbol{Y}_{i,k|k-1} - \hat{\boldsymbol{y}}_k]^T \\
P_{\boldsymbol{x}_k, \boldsymbol{y}_k} &= \sum_{i=0}^{2\Delta} W_i^{(c)} [\boldsymbol{X}^{x}_{i,k|k-1} - \hat{\boldsymbol{x}}_{k|k-1}][\boldsymbol{Y}_{i,k|k-1} - \hat{\boldsymbol{y}}_k]^T \\
K &= P_{\boldsymbol{x}_k, \boldsymbol{y}_k} P^{-1}_{\tilde{\boldsymbol{y}}_k, \tilde{\boldsymbol{y}}_k}
\end{aligned}
$$

$$\hat{\boldsymbol{x}}_k = \hat{\boldsymbol{x}}_{k|k-1} + K(\boldsymbol{y}_k - \hat{\boldsymbol{y}}_k)$$
$$P_k = P_{k|k-1} - K P_{\tilde{\boldsymbol{y}}_k,\tilde{\boldsymbol{y}}_k} K^T,$$

where $\boldsymbol{x}^a = [\boldsymbol{x}^T \ \boldsymbol{w}^T \ \boldsymbol{v}^T]^T$, $\boldsymbol{X}^a = [(\boldsymbol{X}^x)^T \ (\boldsymbol{X}^w)^T \ (\boldsymbol{X}^v)^T]^T$, $W_0^{(m)} = \lambda/(\Delta + \lambda)$, $W_i^{(m)} = W_i^{(c)} = 1/\{2(\Delta + \lambda)\}, i = 1, \ldots, 2\Delta, \lambda = \alpha^2(\Delta + K) - \Delta$ is the composite scaling parameter, Δ is the dimension of augmented state, $\hat{\boldsymbol{x}}_k$ is the mean state estimate, $\hat{\boldsymbol{y}}_k$ is the mean output estimate, P_k is the covariance matrix, $\boldsymbol{X}_i, i = 1, \ldots, 2\Delta$, are the sigma points, $P_k^a = \text{diag}(P_k, P_w, P_a)$, and P_w, P_v are the covariance of process and measurement noise, respectively. The parameter α determines the spread of the sigma points around $\hat{\boldsymbol{x}}_k$ and is usually set to a positive value (between 0 and 1) and K is a secondary scaling parameter which is usually set to 0.

The residuals which are used for detection are defined as follows:

$$r_1 = \left| x_D - \hat{x}_D \right| \tag{13.21}$$

$$r_2 = \left| x_B - \hat{x}_B \right|, \tag{13.22}$$

where x_D, x_B, $\hat{x}_D$, and $\hat{x}_B$ correspond to the distillate purity and bottoms impurity measurement (i.e., x_1, x_{16}), and estimated distillate purity and bottoms impurity (i.e., $\hat{x}_1$, $\hat{x}_{16}$), respectively, and r_1 and r_2 denote the residual in distillate purity estimation and bottoms impurity estimation, respectively. The value of the residuals is chosen based on the specification of the measured parameters, i.e., product purity requirement with the aim of detecting the attack as early as possible to limit the potential damage on the product qualities due to an attack without triggering false alarms.

13.5.2 *Detector Design*

Fundamentally, the detection algorithm is implemented using moving window-based monitoring, whereby at each time-instant the window is shifted by one sample. The time-instant is set as the same as the update frequency of the UKF filter (T_s) as 3.6 s. The window length is defined as the number of samples corresponding to a residual that has to be monitored. The window length for this study has been set to ten samples, i.e., 36 s. The length of the window is set as such to reduce the number of false alarms without missing any true positive attack events. A Boolean flag is allocated to each residual at every time-instant indicating the outcome from comparing the residual against a predefined threshold. If the residual exceeds the threshold the flag is set to False and vice versa. In the proposed window-based monitoring, at every time-instant a decision status is assigned to each residual based on the evaluation of the flags in the window. The decision status is binary, and can be either "Healthy (0)" or "Abnormal (1)" which is determined based on the percentage of the flag in each window with given value. In our study, the status is set as Abnormal (1) if 60% of the flags inside the window are set as False. The detection algorithm is implemented inside the PLC as shown in Fig. 13.5.

13.6 Results

This section presents the results of the various attack cases that are used to validate the proposed detection scheme. For all attack cases, the threshold for the residuals defined in (13.21) and (13.22) is set to 0.02 and 0.01, respectively.

13.6.1 Attack on Distillate Purity Measurement

During this attack, the distillate purity is scaled up by 5% with the aim of violating the product quality specification of the distillate. The result for this attack is presented in Fig. 13.6. The attack is detected within 36 s by the residual corresponding to the distillate purity. This is achieved as the UKF is able to estimate the distillate purity correctly in the event of the attack as illustrated in the figure. Besides that the presented results confirm that the scheme successfully detects the attack before the product quality specification is violated.

13.6.2 Attack on Bottoms Impurity Measurement

In this case, the bottoms impurity measurement is targeted using a ramp attack with $\lambda_r = 1.8 \times 10^{-6}$. Figure 13.7 presents the results corresponding to this attack and as can be seen, the attack is successfully detected in 36 s by r_2 before the bottoms impurity requirement could be violated. Principally, in the event of a sensor attack, a discrepancy arises between the estimator output estimate and the actual measurement as illustrated in Fig. 13.7 that facilitates the attack detection. Additionally, the difference between the estimated and actual bottoms impurity during the normal operation is due to the fact that the actual measurements contain noise which is filtered out by the UKF.

13.6.3 Attack on Reflux Flow Rate

The attack is injected by scaling the actual reflux rate down by 20%. In the event of an actuator attack, as both the correct sensor and actuator data is available to the control PLC, hence it is able to detect the attack by monitoring the system measurements which changes abnormally due to the attack as illustrated in Fig. 13.8. From these

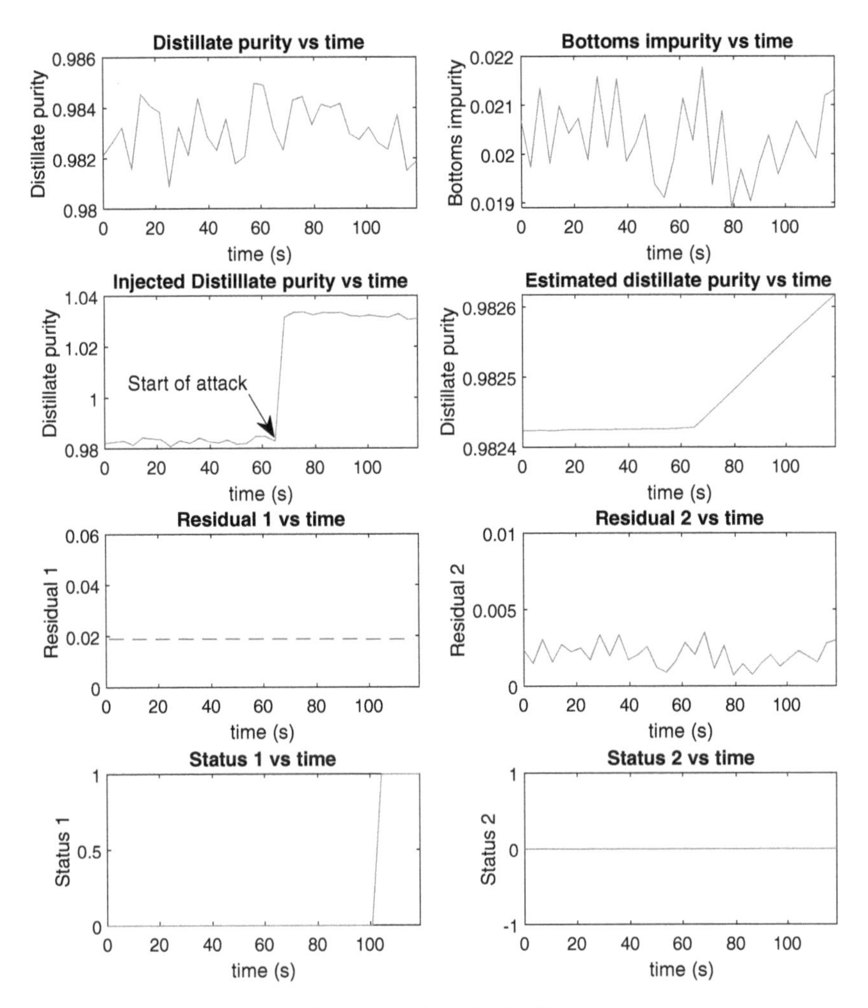

Fig. 13.6 Results illustrating the effect of attack on the distillate purity measurement

results, it can be seen that the attack is successfully detected by both residuals; however, r_1 detected the attack earlier in 1.5 h. As a result of this attack, the distillate product quality requirement is violated.

13.6.4 Attack Case Summary

Besides the presented cases, Table 13.5 summarizes the results for various other attack cases considered as part of this study. The main noticeable observation is the difference between the sensor and actuator attacks detection times where the

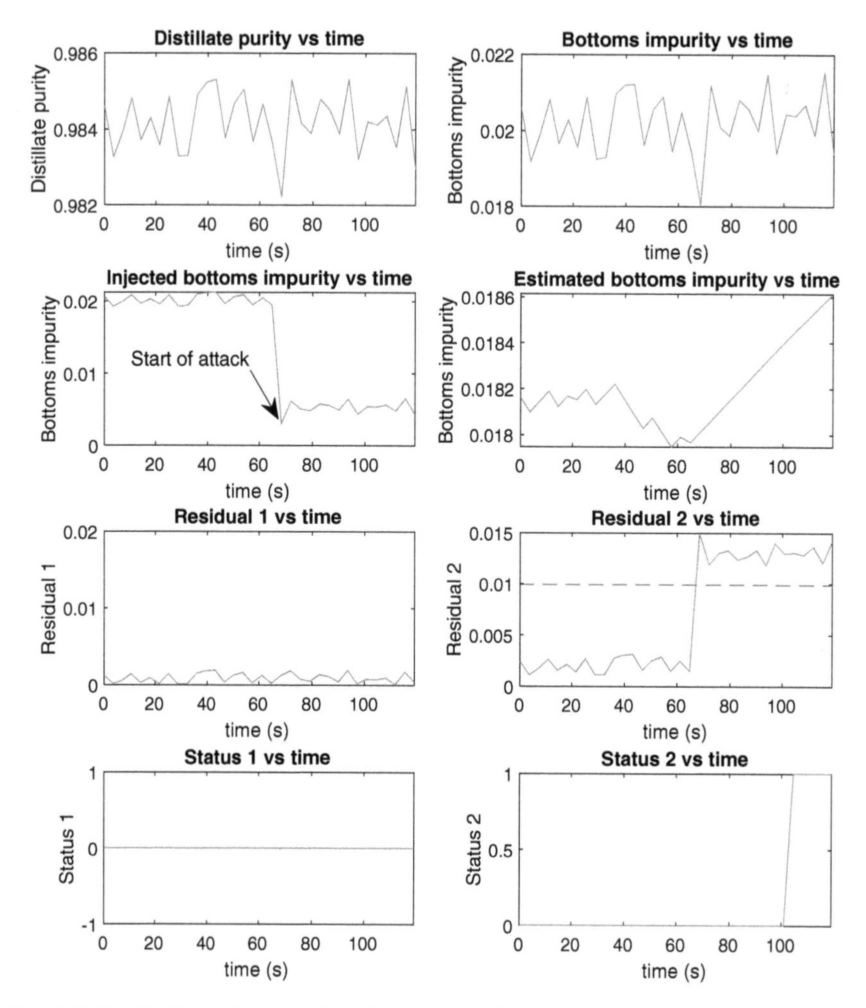

Fig. 13.7 Results illustrating the effect of attack on the bottoms impurity measurement

actuator attacks take longer to be detected. This is due to the fact that the sensor attack directly manipulates a variable of the residual functions, whereas in the case of actuator attacks, the attack is detected using the change in the product quality which takes longer to be appear as the system is relatively slow.

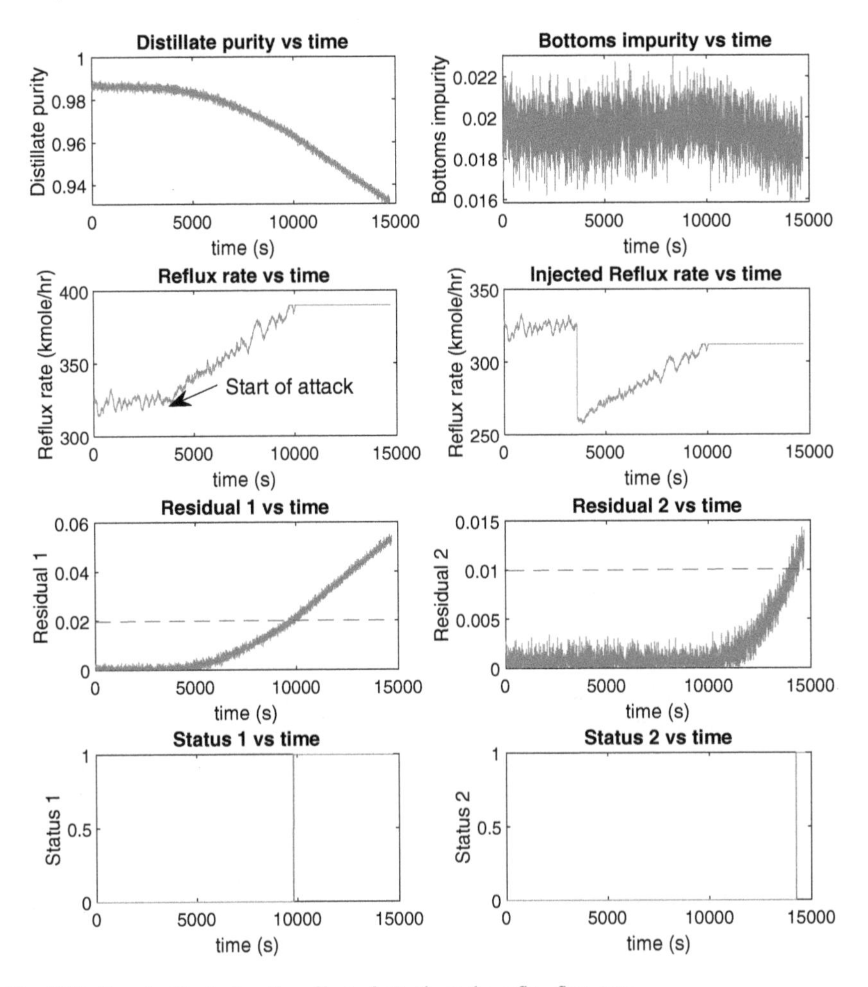

Fig. 13.8 Results illustrating the effect of attack on the reflux flow rate

13.7 Conclusion and Future Work

This chapter addresses the cyber-security of a cyber-physical DC plant by proposing an attack detection technique. A dynamical model of the DC plant is developed which allows for performing simulation study without the necessity of having a physical column. Following that, a hybrid HIL ICS testbed is proposed for the DC plant implemented using industrial hardware from Siemens. A PLC-based online distributed detection scheme is developed based on state estimation using Unscented Kalman Filter and successfully validated for various attack scenarios formulated using the presented attack models. In the proposed model, it is assumed that the column pressure at the top remains constant which is not the case in reality. A feedback control loop is generally used to maintain constant column pressure by

Table 13.5 Summary of detection results for various sensor and actuator attacks

Attack type	Attack name	Targeted resource	Detection time (s)
Sensor attack	Bias injection	Distillate purity	36
		Bottoms impurity	36
	DoS attack	Distillate purity	48
		Bottoms impurity	68
Actuator attack	Replay attack	Reflux rate	4320
		Vapor flow rate	3885
	Constant value attack	Reflux rate	3655
		Vapor flow rate	5139
	Random attack	Reflux rate	4481
		Vapor flow rate	3593

adjusting the condenser duty cycle. Thus, the existing model can be extended by incorporating the column pressure dynamics and an additional feedback control loop can be added to enhance the practicality of the study. As part of this study, a continuous binary DC is considered while there exist other types of columns that are found in industry, e.g., batch distillation column, multi-component distillation column. Hence, further studies can be done to tackle cyber-security for the other distillation column configurations. Furthermore, distillation column is a part of crude processing and there exist various chemical and physical processes both upstream and downstream that is used to convert the raw crude into commercial product. These processes can be included in the future study to make it industrially more feasible.

Acknowledgements This publication was made possible by the Graduate Sponsorship Research Award (GSRA) award (GSRA4-2- 0518-17083) from the Qatar National Research Fund (QNRF), a member of the Qatar Foundation. The authors would also like to acknowledge the financial support received from NATO under the Emerging Security Challenges Division program.

References

A.A. Abokifa, K. Haddad, C. Lo, P. Biswas, Real-time identification of cyber-physical attacks on water distribution systems via machine learning-based anomaly detection techniques. J. Water Resour. Plan. Manag. **145**(1), 04018089 (2019)

S. Adepu, A. Mathur, Distributed attack detection in a water treatment plant: method and case study. IEEE Trans. Dependable Secure Comput. **18**(1), 86–99 (2021)

S.H.M. Ahmad, N. Meskin, Cyber attack detection for a nonlinear binary crude oil distillation column, in *2020 IEEE International Conference on Informatics, IoT, and Enabling Technologies (ICIoT)* (2020), pp. 212–218

A. AlDairi, L. Tawalbeh, Cyber security attacks on smart cities and associated mobile technologies. Procedia Comput. Sci. **109**, 1086–1091 (2017). 8th International Conference on Ambient Systems, Networks and Technologies, ANT-2017 and the 7th International Conference on Sustainable Energy Information Technology, SEIT 2017, 16–19 May 2017, Madeira, Portugal

R. Bendib, H. Bentarzi, Y. Zennir, Investigation of the effect of design aspects on dynamic control of a binary distillation column, in *2015 4th International Conference on Electrical Engineering (ICEE)* (2015), pp. 1–5

M. Elnour, N. Meskin, R. Jain, A dual-isolation-forests-based attack detection framework for industrial control systems. IEEE Access 1 (2020)

M. Elnour, N. Meskin, R. Jain, Application of data-driven attack detection framework for secure operation in smart buildings. Sustain. Cities Soc. **69**, 102816 (2021)

A. George, R.M. Francis, Model reference adaptive control of binary distillation column composition using MIT adaptive mechanism. Int. J. Eng. Res. Technol. **4** (2015)

Y. He, G.J. Mendis, J. Wei, Real-time detection of false data injection attacks in smart grid: a deep learning-based intelligent mechanism. IEEE Trans. Smart Grid **8**(5), 2505–2516 (2017)

P. Kathel, A.K. Jana, Dynamic simulation and nonlinear control of a rigorous batch reactive distillation. ISA Trans. **49**(1), 130–137 (2010)

M. Kravchik, A. Shabtai, Anomaly detection; industrial control systems; convolutional neural networks. CoRR (2018), arXiv:abs/1806.08110

D. Kundur, X. Feng, S. Mashayekh, S. Liu, T. Zourntos, K. Butler-Purry, Towards modelling the impact of cyber attacks on a smart grid. Int. J. Secur. Netw. **6**, 2–13 (2011)

M.N. Kurt, O. Ogundijo, C. Li, X. Wang, Online cyber-attack detection in smart grid: a reinforcement learning approach. IEEE Trans. Smart Grid **10**(5), 5174–5185 (2019)

D. Li, D. Chen, L. Shi, B. Jin, J. Goh, S. Ng, MAD-GAN: multivariate anomaly detection for time series data with generative adversarial networks. CoRR (2019), arXiv:abs/1901.04997

Q. Lin, S. Adepu, S. Verwer, A. Mathur, Tabor: a graphical model-based approach for anomaly detection in industrial control systems, in *Proceedings of the 2018 on Asia Conference on Computer and Communications Security, ASIACCS '18* (Association for Computing Machinery, New York, NY, USA, 2018), pp. 525–536

M. Lv, W. Yu, Y. Lv, J. Cao, W. Huang, An integral sliding mode observer for cps cyber security attack detection. Chaos: Interdiscip. J. Nonlinear Sci. **29**, 043120 (2019)

K. Manandhar, X. Cao, F. Hu, Y. Liu, Detection of faults and attacks including false data injection attack in smart grid using Kalman filter. IEEE Trans. Control Netw. Syst. **1**(4), 370–379 (2014)

T. Meraj, S. Sharmin, A. Mahmud, Studying the impacts of cyber-attack on smart grid, in *2015 2nd International Conference on Electrical Information and Communication Technologies (EICT)* (2015), pp. 461–466

V.T. Minh, J. Pumwa, Modeling and adaptive control simulation for a distillation column, in *2012 UKSim 14th International Conference on Computer Modelling and Simulation* (2012a), pp. 61–65

V. Minh, J. Pumwa, Modeling and control simulation for a condensate distillation column (2012b)

Y. Mo, S. Weerakkody, B. Sinopoli, Physical authentication of control systems: designing watermarked control inputs to detect counterfeit sensor outputs. IEEE Control Syst. Mag. **35**(1), 93–109 (2015)

M. Noorizadeh, M. Shakerpour, N. Meskin, D. Unal, K. Khorasani, A cyber-security methodology for a cyber-physical industrial control system testbed. IEEE Access **9**, 16 239–16 253 (2021)

A. Nourian, S. Madnick, A systems theoretic approach to the security threats in cyber physical systems applied to stuxnet. IEEE Trans. Dependable Secure Comput. **15**(1), 2–13 (2018)

F. Pasqualetti, F. Dorfler, F. Bullo, Attack detection and identification in cyber-physical systems. IEEE Trans. Autom. Control **58**, 2715–2729 (2012)

F. Pasqualetti, F. Dorfler, F. Bullo, Control-theoretic methods for cyberphysical security: geometric principles for optimal cross-layer resilient control systems. IEEE Control Syst. Mag. **35**(1), 110–127 (2015). (Feb)

G. Radulescu, N. Paraschiv, A. Kienle, An original approach for the dynamic simulation of a crude oil distillation plant 2: setting-up and testing the simulator. Revista de Chimie **58** (2007)

S. Sridhar, M. Govindarasu, Model-based attack detection and mitigation for automatic generation control. IEEE Trans. Smart Grid **5**(2), 580–591 (2014)

S.A. Taqvi, L.D. Tufa, S. Muhadizir, Optimization and dynamics of distillation column using aspen plus∘R. Procedia Eng. **148**, 978–984 (2016). Proceeding of 4th International Conference on Process Engineering and Advanced Materials (ICPEAM 2016)

S.A. Taqvi, L.D. Tufa, H. Zabiri, S. Mahadzir, A.S. Maulud, F. Uddin, Rigorous dynamic modelling and identification of distillation column using aspen plus, in *2017 IEEE 8th Control and System Graduate Research Colloquium (ICSGRC)* (2017), pp. 262–267

W. Weerachaipichasgul, P. Kittisupakorn, A. Saengchan, K. Konakom, I.M. Mujtaba, Batch distillation control improvement by novel model predictive control. J. Ind. Eng. Chem. **16**(2), 305–313 (2010)

E. Wijn, Weir flow and liquid height on sieve and valve trays. Chem. Eng. J. **73**(3), 191–204 (1999)

T. Zhang, Y. Wang, X. Liang, Z. Zhuang, W. Xu, Cyber attacks in cyber-physical power systems: a case study with GPRS-based SCADA systems, in *2017 29th Chinese Control And Decision Conference (CCDC)* (2017), pp. 6847–6852

Z. Zou, Z. Wang, L. Meng, M. Yu, D. Zhao, N. Guo, Modelling and advanced control of a binary batch distillation pilot plant. Chin. Autom. Congr. (CAC) **2017**, 2836–2841 (2017)

Chapter 14
A Resilient Nonlinear Observer for Light-Emitting Diode Optical Wireless Communication Under Actuator Fault and Noise Jamming

Ibrahima N'Doye, Ding Zhang, Ania Adil, Ali Zemouche, Rajesh Rajamani, and Taous-Meriem Laleg-Kirati

14.1 Introduction

Special attention has been recently devoted to designing distributed autonomous robotic systems in several mission scenarios in which human operators cannot assess the situation. Distributed autonomous robotic systems (DARS) presents many opportunities beyond supporting human task forces in various applications, such as patrolling in communication-restricted environments, rescue and search and localization of targets, and surveillance of complex environments (GroB et al. 2018). Such distributed autonomous robot systems work well when every node is functional and

The two first authors contributed equally to the chapter.

I. N'Doye (✉) · A. Adil · T.-M. Laleg-Kirati
King Abdullah University of Science and Technology (KAUST), Thuwal 23955-6900, Saudi Arabia
e-mail: ibrahima.ndoye@kaust.edu.sa

A. Adil
e-mail: ania.adil@kaust.edu.sa

T.-M. Laleg-Kirati
e-mail: taousmeriem.laleg@kaust.edu.sa

D. Zhang
Department of Electronic and Computer Engineering, The Hong Kong University of Science and Technology, Clear Water Bay, Kowloon, Hong Kong, China
e-mail: ding.zhang@connect.ust.hk

A. Zemouche
CRAN CNR-UMR 7039, IUT Henri Poincaré de Longwy, Université de Lorraine, Cosnes-et-Romain, France
e-mail: ali.zemouche@univ-lorraine.fr

R. Rajamani
Department of Mechanical Engineering, University of Minnesota, Minneapolis, MN 55455, USA
e-mail: rajamani@me.umn.edu

M. Abbaszadeh and A. Zemouche (eds.), *Security and Resilience in Cyber-Physical Systems*, https://doi.org/10.1007/978-3-030-97166-3_14

trustworthy, and require coordination capabilities at multiple levels, including global allocation tasks and task selection to local spatial coordination to avoid collisions (GroB et al. 2018; Saldana et al. 2018).

Radio-frequency (RF) communication has been the standard method for the autonomous ground robotic network to operate wirelessly for this distributed autonomous platform. However, RF technology presents limitations such as a limited available data rate, and congested spectrum (Borah et al. 2012; Ghassemlooy et al. 2012). Hence, optical wireless communication (OWC) technology is an alternative that can complement RF technology to overcome these limitations (Majumdar and Ricklin 2010; Elgala et al. 2011; Borah et al. 2012; Ghassemlooy et al. 2012). Furthermore, the rapid adaption and decreasing cost of the light-emitting diode (LED) make it a compelling alternative and a promising communication technique to radio-based wireless communication.

Optical wireless communication (OWC) technologies are of great importance in many indoor and outdoor applications. OWC is considered an emerging alternative technology in the communication area as the demand for capacity increases. It carries out flexible networking solutions with cost-effective and high-speed license-free wireless connectivity for several applications (Ghassemlooy et al. 2012; Zhang et al. 2020; N'Doye et al. 2018). In addition, OWC technology provides low latency, low cost and power consumption, and high data rates (Hanson and Radic 2008; Hagem et al. 2011; Lu et al. 2009). The practical applications of the free-space optical communication system have been a great interest of wide field-of-view (FoV) such that NASA technologies for interplanetary FSO communication systems, Facebook's drones Aquila (Facebook 2018), Google's Internet balloons (De-Vaul et al. 2014), and FSO communication in space (Elgala et al. 2017), and military platforms (Calhoun 2003).

Although there is extensive effort to build reliable OWC for mobile networking sensing applications, however, OWC systems' practicality to maintain accurate alignment angle of tracking optical systems in autonomous robot platforms has been, until recently, a significant problem. In addition, the required alignment angle is not directly measured and has to be estimated. On the other hand, the OWC system is often hampered by noise jamming attack that reduces the system capacity of the wireless optical mobile networks. Additionally, one robot can reduce the system capacity and affects all other robots' communication networks when a hardware failure occurs due to malfunctions or high instantaneous torques of the actuator-mechanism flexible on the receiver side. Figure 14.1 illustrates an example of robots' optical communication networks in which jammer intercepts the receiver aperture under actuator fault.

The jamming attack has become an urgent and severe threat in several communications applications (Li et al. 2018). In noise jamming, the jammer intends to limit the legitimate transmission by saturating the receiver with noise through deliberate signals limiting an opponent's communication effectiveness. It can considerably reduce the system capacity. Noisy jamming is less harmful than disguised jamming, which can have a jamming power that is much higher than the signal power. On the other hand, jamming has been widely modeled as Gaussian noise (Li et al. 2018) or

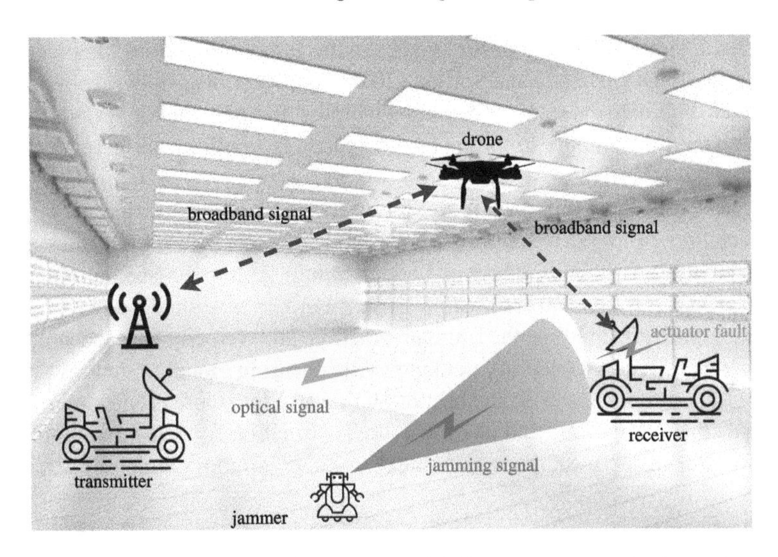

Fig. 14.1 Distributed autonomous robots system with LED optical communication under actuator fault and jamming attack on the optical communication channel (the jamming signal gets trapped inside the photodetector at the receiver side)

non-Gaussian noise (Paul et al. 2019) in free-space optical communications. Different jamming attack strategies that a wireless jammer can generate to interfere with other communications have been proposed in Pelechrinis et al. (2011), Liao et al. (2013), Zou et al. (2016).

Kalman-type filters have been considered industry-standard solutions for motion control problems and navigation systems. However, these filters rely on local linearization assumptions and fail when the initial estimation errors are significant. Furthermore, previous results on the Extended Kalman Filter (EKF)-based algorithm of maintaining active alignment control for LED-based wireless optical communications lack strong theoretical stability guarantees of the convergence of the estimator (Solanki et al. 2016, 2018). Indeed, minor deviations errors in the output measurement can make the EKF system go unstable. In contrast to the stochastic filters frameworks, LMI-based observer design techniques have been widely used for different classes of nonlinear systems (Ha and Trinh 2004; Acikmese and Corless 2011; Wang et al. 2014; Arcak and Kokotovic 2001; Zemouche et al. 2017; Draa et al. 2019). Moreover, these developed LMI-based observer design methods in the literature may fail when applying non-monotonic nonlinear systems (Rajamani et al. 2020). Recently, a novel LMI-based switched-gain observer design method for nonlinear continuous systems has been developed in Rajamani et al. (2020), N'Doye et al. (2020) to tackle the non-monotonicity gap. On the other hand, unknown input observer design techniques have been proposed for estimating states and unknown inputs in the literature. Recently, an estimation algorithm that can detect cyber-attacks on the communication channel with the preceding vehicle and monitor the radar sensor's health was developed in Jeon et al. (2020). The solution decouples the cyber-attack signal from the sensor failures in the estimation error dynamics in Jeon et al. (2020). In Bakhshande and Soffker (2015), a proportional-integral observer was proposed to estimate the states and unknown inputs. In Phanomchoeng and Rajamani (2014), an unknown input estimation method based on nonlinear observer design and a dynamic model inversion was proposed. Another approach is designing an observer for a system represented in descriptor system form (Phanomchocng et al. 2018).

This chapter proposes to track a ground mobile receiver by a vehicle transmitter to establish a point-to-point optical link under actuator fault and noise jamming attack on the optical communication channel. The method derives a constant stabilizing observer gain by providing the angular position and velocity in each monotonic region required for the trajectory tracking while ensuring global asymptotic stability via the Lyapunov function. To the best of our knowledge, there are few works in the literature considering state and unknown input estimation for non-monotonic output functions and in the more general context of noise jamming attack on the optical communication channel.

In this chapter,

- We demonstrate the infeasibility to solutions for the observer design LMIs when the nonlinear functions are all non-monotonic.
- We develop a switched-gain unknown input observer that can detect actuator fault under noise jamming attack on the communication channel.
- We develop conditions on the controller design to guarantee the $\mathscr{H}_\infty$ optimality criterion.
- We project the observer and controller gains design to achieve the asymptotic stability and the $\mathscr{H}_\infty$ performance criterion of the resulting observer-based tracking control, thanks to the certainty-equivalence design.
- We conduct simulation results to analyze the capability of the proposed switched-gain observer-based reference trajectory tracking control to reconstruct the angular position and velocity under actuator fault and noise jamming attack on the optical communication channel.

The chapter is organized as follows. In Sect. 14.2, the LED-based optical communication model is presented, including its state-space and measurement equation. In Sect. 14.4, we formulate our estimation-based reference trajectory tracking problem. In Sect. 14.3, we derive the LED system model representation under actuator fault and noise jamming attack on the communication channel. In Sect. 14.5, simulation results are provided to illustrate the performance of the observer-based tracking under actuator fault and noise jamming attack. Finally, concluding remarks are shown in Sect. 14.6. The proof of the infeasibility of solutions for the observer design LMIs is given in Appendix.

Notation: Matrix A^T represents the transposed matrix of A. The Euclidean norm of a vector $x \in \mathbb{R}^n$ is defined as $\|x\| = \sqrt{x^T x}$. The identity matrix of dimension r is denoted $\mathbb{I}_r$. The blocks induced by symmetry are denoted as $(\star)$. The set $Co(x, y) = \{\lambda x + (1 - \lambda)y, 0 \leqslant \lambda \leqslant 1\}$ is the convex hull of $\{x, y\}$. A vector of the canonical basis of $\mathbb{R}^s$ is denoted as $e_s(i) = \big(\underbrace{0, \ldots, 0, \overbrace{1}^{i\text{ th}}, 0, \ldots, 0}_{s\text{ components}}\big)^T \in \mathbb{R}^s$, $s \geqslant 1$. A positive definite (negative definite) square matrix is denoted as $S > 0$ ($S < 0$).

14.2 LED-Based Optical Channel Modeling

The LED-based optical channel is a two-way communication describing a single LED transmitter and a single photodiode receiver; each end can rotate by an angle in which it establishes and maintains a directed line-of-sight (LOS) optical communication. In this section, we describe the experimental setup for an estimation problem of LED-based optical channel modeling. We discuss the luminous flux model, and finally, we formulate the state-space representation, which takes the form of a dynamical system with a nonlinear output map.

14.2.1 System Setup

The radiation region of the LED source in which the radiation patterns have significant differences can be separated into a near field and far-field by the LED-to-target distance (Ivan and Ching-Cherng 2008). A high-power LED can have 20 mm close to midfield, in which region the radiation pattern is distance-dependent while it will not change in the far-field. As the range of communication is far longer than 20 mm, here we treat radiation pattern as distance-invariant and try to obtain the spatial distribution of LED luminous flux for our specific system setup as shown in Fig. 14.2.

A white high-power LED module commonly used as a mobile phone flash is mounted on a 4-wheels car. The LED module requires a power supply source of 3.5 W and generates light whose wavelength is between 400 and 700 nm with a typical luminous intensity of 245 lux. The power meter device VLP-2000 that measures the strength of the received LED-based optical signal is mounted on another car. The

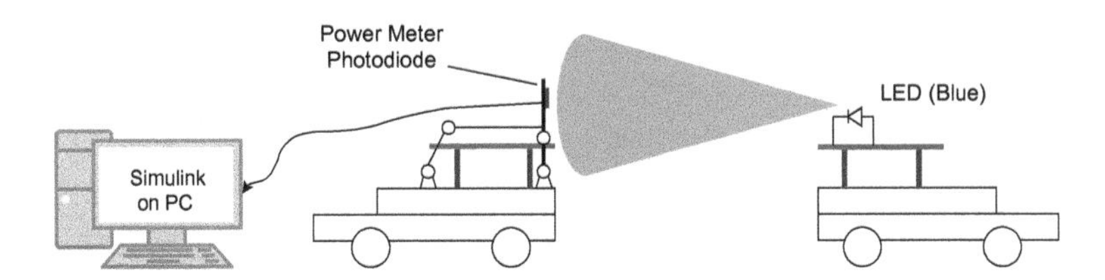

Fig. 14.2 Diagram of the LED-based optical communication

detectable wavelength ranges from 180 to 19 mm with a resolution of 0.001 mW and the uncertainty of this power meter is ±5%.

14.2.2 Luminous Flux Model

The spatial distribution model of the LED mainly describes the effect of relative position and orientation between the transmitter and the receiver on the signal strength (Ghassemlooy et al. 2012; Doniec et al. 2013; Solanki et al. 2018). The relative position between transmitter and receiver is described with three parameters: the distance d between them, the angle θ between the normal direction and the main normal direction of transmitter, and the angle ϕ between the orientation of the receiver and the normal direction. Figure 14.3 illustrates the variables of interest, which include the transmission distance d, the transmission angle θ, and the angle of incidence ϕ.

The power incident on the detector is determined based on the signal irradiance at the relative detector position. The full signal strength model can be formulated as follows (Ghassemlooy et al. 2012; Doniec et al. 2013; Solanki et al. 2016, 2018):

$$P_d = CI(\theta, d) \exp(-cd) g(\phi), \tag{14.1}$$

where P_d is the measurement of power which is proportional to the luminous flux of light that is detected by the receiver, C and c are both constants. The $\exp(-cd)$ portion comes from Beer's law (Miller et al. 2009) which describes the attenuation of power when light travels through medium as an exponential decay; $I(\theta, d)$ is usually in the following form (Ghassemlooy et al. 2012; Doniec et al. 2013; Solanki et al. 2016, 2018)

$$I(\theta, d) = I(0, d) \cos^m(\theta)/d^2, \tag{14.2}$$

where $I(0)$ is the central luminous flux as well as the maximum luminous flux and m is the order of Lambertian emission

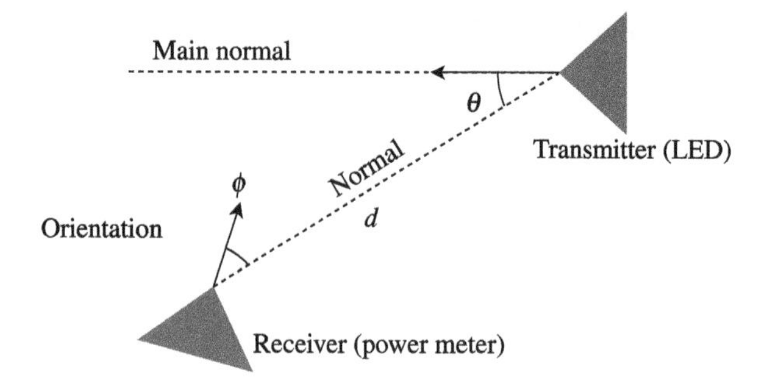

Fig. 14.3 LED optical communication scheme

$$m = \frac{\ln(2)}{\ln\cos(\theta_{1/2})}.$$

In the above formula, $\theta_{1/2}$ is the angle at half the illuminance of an LED. Physically, $I(0, d)\cos^m(\theta)$ represents the radiation pattern of LED source (Ivan and Ching-Cherng 2008) and the reciprocal of d^2 comes from the inverse-square law which describes the geometric dilution of a physical quantity.

14.2.3 Model Calibration

To parameterize the LED-based optical model (14.1), we have conducted experiments to measure the luminous flux of a high-power LED module at different relative positions in clear weather conditions when there is a relative motion between the receiver and the transmitter. We design three experiments to estimate the unknown parameters of the luminous flux model given in (14.1).

14.2.3.1 Measured Signal Strength Versus Transmitter-Receiver Distance

The LED source is fixed at the center of concentric circles as shown in Fig. 14.4, and the main normal direction of the LED is aligned with the symmetric axis. The receiver car is placed at eleven equidistant points to observe the impact of the distance in free-space optical communication. At each endpoint, we took five samples of measured power and computed their means and variances.

As shown in Fig. 14.5, the signal strength declines when the distance between the receiver and the transmitter increases. In addition, the nonlinear model $a\exp(-bx)/x^2$ which combines the effects of absorption, scattering, and geometric dilution fits well with the measured signal strength data

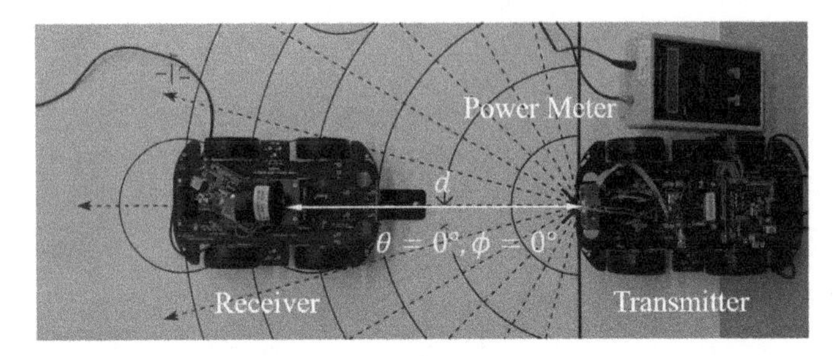

Fig. 14.4 Transmitter–receiver distance setup in free-space. Adapted from figures that were originally published under a CC BY-NC-ND license in N'Doye et al. (2020); 10.1016/j.ifacol.2020.12.1075 by I. N'Doye, D. Zhang, A. Adil, A. Zemouche, R. Rajamani, T.-M. Laleg-Kirati

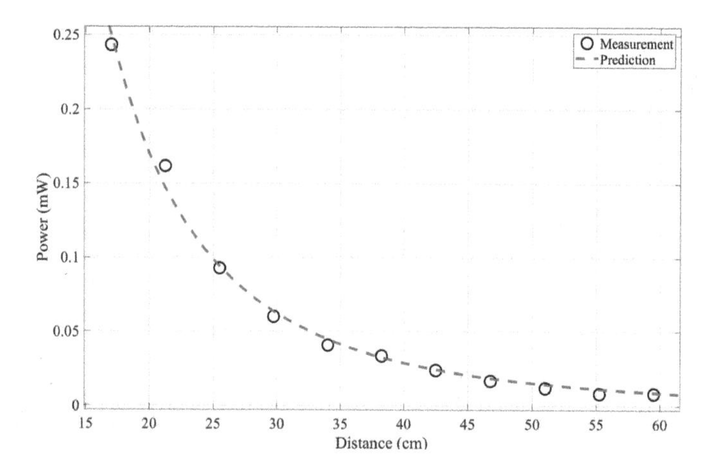

Fig. 14.5 Measured signal strength versus the transmitter–receiver distance in free-space. Adapted from figures that were originally published under a CC BY-NC-ND license in N'Doye et al. (2020); 10.1016/j.ifacol.2020.12.1075 by I. N'Doye, D. Zhang, A. Adil, A. Zemouche, R. Rajamani, T.-M. Laleg-Kirati

$$P_d(d, 0, 0) = a\frac{\exp(-bd)}{d^2}, \tag{14.3}$$

where a and b are the curve fitting parameters defined in Table 14.1.

14.2.3.2 Angular Transmission Intensity Distribution

Assuming that the maximum power at distance d_0 is achieved when $\theta = 0°$, we define the power ratio $\tilde{I}_\theta$ as follows:

$$\tilde{I}_\theta := \frac{I(\theta, d_0)}{I(0, d_0)}. \tag{14.4}$$

Table 14.1 Fitting results

Terms	Model	Parameters						R^2	RMSE
		$a(a_1)$	$b(b_1)$	$c(c_1)$	a_2	b_2	c_2		
Scattering, absorption, dilution	$a\exp(-bx)/x^2$	0.01009	1.972	–	–	–	–	0.9947	0.0058
Receiver orientation	Equation (14.6)	0.9953	0.06298	0.2517	0.2260	−0.1995	0.132	0.9970	0.0205

Using the fact that the angular intensity distribution of the transmitter is rotationally symmetric with the LED's normal ($\theta = 0°$), then we can measure the intensity of all the points at the same radial distance based on spatial power ratio intensity distribution $\tilde{I}_\theta$. Hence, at a unit distance, we assume that $\tilde{I}_\theta$ is known and represents the light intensity for different transmitter angles.

14.2.3.3 Measured Signal Strength Versus Incidence Angle ϕ

High-power LED source is aligned with the center of the detector point and targeted at the main normal direction. To obtain an approximate form of $g(\phi)$, we place the receiver along a circle to maintain the distance d constant and the transmission angle θ constant and known at all times, as illustrated in Fig. 14.6.

In this scenario, θ and the distance d are actually set to $0°$ and 34 cm, respectively. $g(\phi)$ is a unimodal function which represents empirically the power ratio. Assume that at ϕ_0, $g(\phi)$ reaches its maximum $g(\phi_0) = 1$, then we can have

$$\frac{P_d(0.34, 0, \phi)}{P_d(0.34, 0, \phi_0)} = \frac{g(\phi)}{g(\phi_0)} = g(\phi). \tag{14.5}$$

A proper function $g(\phi)$ fitting measured data is composed of two Gaussian terms with six unknowns as shown in Fig. 14.7. The curve fitting could be done using a single Gaussian mode but having one extra Gaussian mode gives significantly better fitting. Using MATLAB curve fitting tool which is based on Least Square method, we can evaluate $g(\phi)$ as follows (N'Doye et al. 2020):

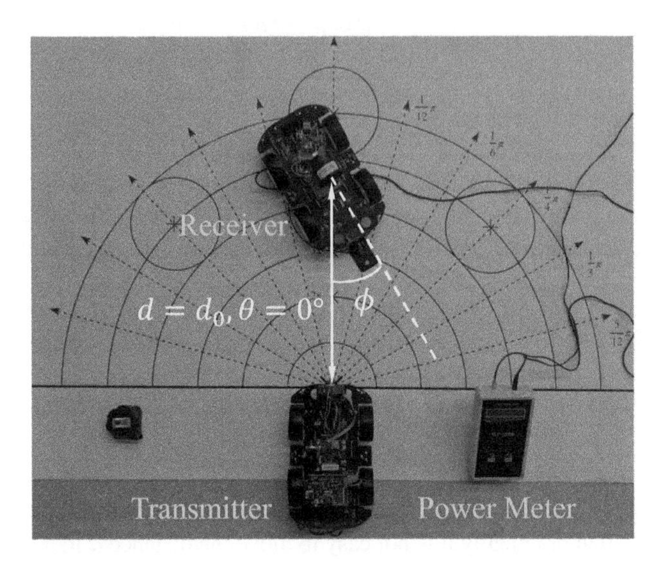

Fig. 14.6 System setup of the received power with respect to the incidence angle in free-space

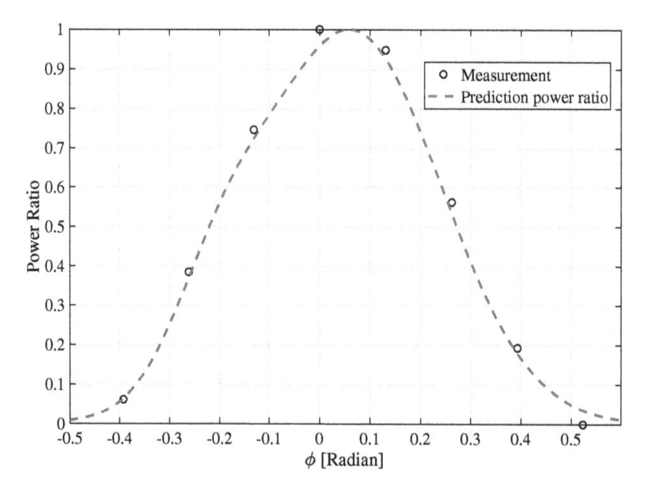

Fig. 14.7 Gaussian curve fitting the incidence angle ϕ. Adapted from figures that were originally published under a CC BY-NC-ND license in N'Doye et al. (2020); 10.1016/j.ifacol.2020.12.1075 by I. N'Doye, D. Zhang, A. Adil, A. Zemouche, R. Rajamani, T.-M. Laleg-Kirati

$$g(\phi) \approx a_1 \exp\left[-\left(\frac{\phi - b_1}{c_1}\right)^2\right] + a_2 \exp\left[-\left(\frac{\phi + b_2}{c_2}\right)^2\right], \tag{14.6}$$

where a_1, a_2, b_1, b_2, c_1, and c_2 are the curve fitting parameters defined in Table 14.1.

Now the resulting luminous flux model is obtained by combining equations (14.3), (14.4) and (14.6) into a compact model.

$$P_d(d, \theta, \phi) = \underbrace{\frac{a \exp(-bd)}{d^2} \tilde{I}_\theta}_{\text{Transmitter}} \underbrace{g(\phi)}_{\text{Receiver}} . \tag{14.7}$$

From (14.7), we can evaluate the luminous flux generated by LED source at given d and θ with ϕ set to 0°, i.e., the receiver's pointing error is set to zero. Then, we transform from polar frame to Cartesian coordinates ($x = d\cos\theta$, $y = d\sin\theta$), and the spacial distribution of LED-based luminous flux in 2-D space is illustrated in Fig. 14.8.

14.2.4 State-Space and Output Measurement Equations

From (14.7), we formulate the state-space representation based on the two variables of interest $\phi \triangleq x_1$, and $\dot{\phi} \triangleq x_2$ that relate to the angles of the receiver. On the other hand, we note that practically it is not easy to move the distance d ideally because it needs to move the whole robot. Besides, controlling the angular velocity of $\dot{\phi} \triangleq x_2$ is

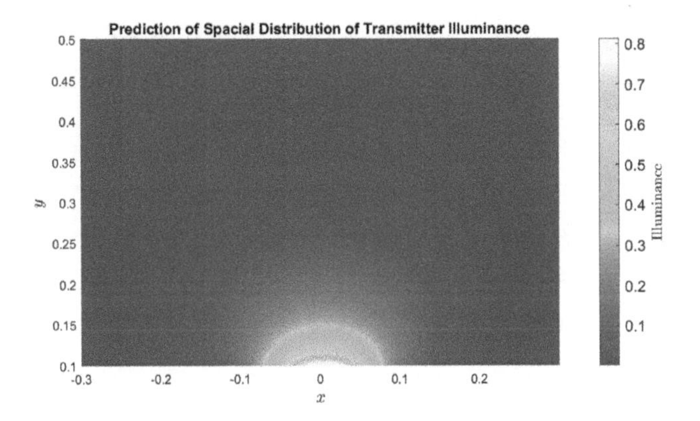

Fig. 14.8 Predicting spacial luminous flux distribution of high-power LED source

more practical. The robot alignment is performed by stabilizing the angular velocity. Since the distance d cannot be adjusted easily and θ fixed, therefore, we define the states as follows:

$$x = \begin{bmatrix} x_1 \\ x_2 \end{bmatrix} = \begin{bmatrix} \phi \\ \dot{\phi} \end{bmatrix}. \tag{14.8}$$

The discrete-time state-space representation can be written as follows:

$$x_{k+1} = \begin{bmatrix} x_{1,k+1} \\ x_{2,k+1} \end{bmatrix} = \begin{bmatrix} x_{1,k} + T_s x_{2,k} + w_{1,k} \\ x_{2,k} + u_k + w_{2,k} \end{bmatrix}, \tag{14.9}$$

where $w_{1,k}$ and $w_{2,k}$ are the process noise inputs which are assumed to be Gaussian, independent and white noise. u_k is the control input which acts on the receiver's angular velocity and T_s is the sampling time.

The measurement $P_{d,k}$ is expressed as

$$y_k \triangleq P_{d,k} = \bar{C}_p g(x_{1,k}) + w_k, \tag{14.10}$$

where $\bar{C}_p = C_p \tilde{I}_\theta \exp{(-cd_0)}/d_0^2$, $g(.)$ is defined in (14.6) and w_k is an additive white Gaussian noise.

14.3 LED System Model Representation Under Actuator Fault and Noise Jamming Attack

Optical wireless communication technologies have significantly advanced in the past decades; however, most optical wireless networks are vulnerable to jamming attacks due to the open nature of the communication channels. On the other hand,

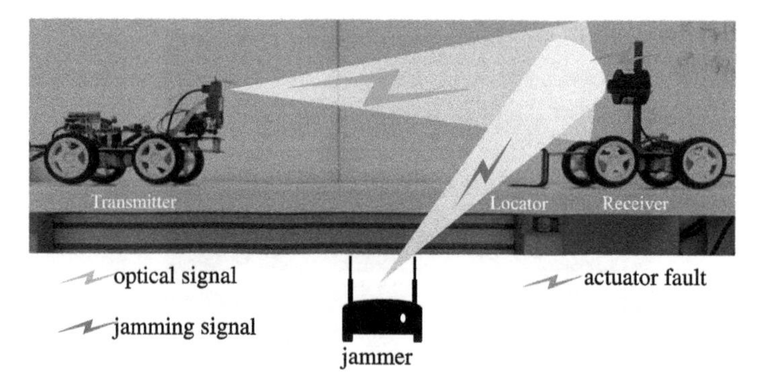

Fig. 14.9 Mobile networking of ground vehicle robots with LED optical communication under actuator fault and jamming attack. The jammer uses a directed line-of-sight signal

the research progress in detecting hostile jamming and designing jamming-resistant wireless networking systems remains limited. In this section, we extend the LED-based optical model described by (14.9) and (14.10) by incorporating a false actuator signal due to any cyber-attack and a jamming sensor attack term. The jamming attack and the actuator fault can easily paralyze the optical wireless communication networks due to the lack of protection mechanisms. Hence, the vulnerability of existing optical wireless communications networks underscores the critical need in developing effective anti-jamming systems in practice. We consider a mobile networking or ground vehicles robot with LED communication in which a situation of an occurrence of actuator fault is considered and a noise jamming attack intend to reduce the system capacity of the LED-based optical wireless communication channels, as illustrated in Fig. 14.9.

The jammer considered as attack intends to reduce the bandwidth or saturate the receiver with false information through deliberate Gaussian noise signals to jam the communication nodes. Besides, a situation of an occurrence of actuator failures is also considered. Since additive bias effects of the actuator-mechanism flexible torque that controls the receiver angular velocity $x_{2,k}$ can occur due to malfunctions. Finally, we have the state-space model

$$\begin{bmatrix} x_{1,k+1} \\ x_{2,k+1} \end{bmatrix} = \begin{bmatrix} x_{1,k} + T_s x_{2,k} + w_{1,k} \\ x_{2,k} + u_k + f_k^a + w_{2,k} \end{bmatrix}, \tag{14.11}$$

where f^a represents the actuator injected false signal.

We introduce an additional receiver on the same robot with a constant shifted angle of $\Delta\phi$ to achieve observability, as illustrated in Fig. 14.10. This shifted angle is added to account for the actual orientation of the receiver. At each movement of the transmitter platform, the states are updated according to the system dynamics. Both ϕ and $\bar{\phi} = \phi \pm \Delta\phi$ can be controlled to 0°, when ϕ is controlled to 0° and reads the wirelessly transmitted data, its orientation is being maintained by using $\bar{\phi}$. The resulting output vector can be written as follows:

Fig. 14.10 Measurements of two receivers ϕ and $\bar{\phi}$

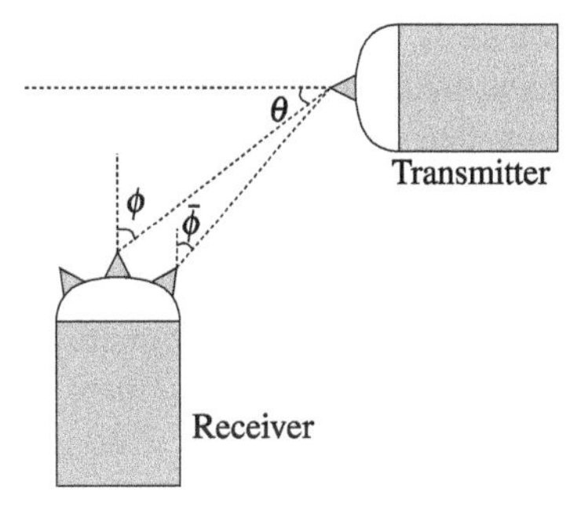

$$y_k = \bar{C}_p \begin{bmatrix} g(x_{1,k}) \\ \underbrace{g(x_{1,k} \pm \Delta\phi)}_{\bar{\phi}} \end{bmatrix} + J_k + w_k, \tag{14.12}$$

where J_k represents the noise jamming signals that intend to reduce the system capacity of the LED-based optical wireless communication channel. Notice that, in practice, J_k can modeled either as non-Gaussian noise (Paul et al. 2019) or Gaussian noise (Li et al. 2018) with known expectation $E\{J_k\} = \bar{\rho}$.

Given the measurement setting, the primary goal is to detect the actuator failure and the noise jamming attack while estimating the angular position $x_{1,k}$ and the angular velocity $x_{2,k}$ based on which the control u_k is designed to drive $x_{2,k}$ towards zero, which corresponds to the maximum light intensity's orientation.

The next section provides the design and stability analysis of the unknown input observer-based reference tracking control design to estimate the states and actuator fault, simultaneously.

14.4 Resilient Observer-Based Tracking Control Design

This section is devoted to a general theory on the unknown input state observer design. This theory is motivated by the LED-based optical communication model described in (14.9) and (14.12).

14.4.1 Problem Formulation

Let us consider the following system

$$\begin{cases} x_{k+1} = Ax_k + Bu_k + Ff_k^a + Ew_k \\ y_k = h(x_k) + GJ_k + Dw_k, \end{cases} \tag{14.13}$$

where $x_k \in \mathbb{R}^n$ is the state vector, $u_k \in \mathbb{R}^m$ is the input vector, $y_k \in \mathbb{R}^p$ is the output measurement, $f_k^a \in \mathbb{R}^r$ is the actuator fault vector, $w_k \in \mathbb{R}^z$ is the disturbance $\mathscr{L}_2$ bounded vector, $J_k \in \mathbb{R}^{\hat{z}}$ is the noise jamming modeled by a Gaussian noise and the matrices $A \in \mathbb{R}^{n\times n}$, $B \in \mathbb{R}^{n\times m}$, $F \in \mathbb{R}^{n\times r}$, $E \in \mathbb{R}^{n\times z}$, $G \in \mathbb{R}^{p\times \hat{z}}$ and $D \in \mathbb{R}^{p\times z}$ are constant. We assume that the actuator bias fault f_k^a and its derivative are bounded. The nonlinear output function $h\colon\ \mathbb{R}^n \longrightarrow \mathbb{R}^p$ is assumed and to be globally Lipschitz.

To simultaneously estimate the unmeasurable state variables and the actuator fault in model (14.13), we augment the state vector and design observers that provide an estimate of $\xi = \begin{bmatrix} x & f^a \end{bmatrix}^\top$. We obtain the following augmented system

$$\begin{cases} \xi_{k+1} = \mathbb{A}\xi_k + \mathbb{B}u_k + \mathbb{E}w_k + \mathbb{F}\Delta f_k^a \\ y_k = h(\xi_k) + \mathbb{G}J_k + \mathbb{D}w_k, \end{cases} \tag{14.14}$$

where $\mathbb{A} = \begin{bmatrix} A & F \\ 0 & I \end{bmatrix}$, $\mathbb{B} = \begin{bmatrix} B \\ 0 \end{bmatrix}$, $\mathbb{E} = \begin{bmatrix} E \\ 0 \end{bmatrix}$, $\mathbb{F} = \begin{bmatrix} 0 \\ I \end{bmatrix}$, $h(\xi_k) = \begin{bmatrix} h(x_k) \end{bmatrix}$, $\mathbb{G} = G$, $\mathbb{D} = D$ and $\Delta f_k^a = f_{k+1}^a - f_k^a$.

Note that designing an extended state observer for the augmented system (14.14) yields estimates of both the original plant state x_k and the actuator fault f_k^a.

Remark 14.1 Note that the augmented transformation (14.14) is used to generate effective methods for unknown input observers for nonlinear systems. For the sake of simplicity, we adopted the extended state observer for unknown input estimation proposed in Chakrabarty and Corless (2019). The choice of this transformation (14.14) is motivated by the fact that we only need to construct an extended state observer to estimate both state and unknown actuator bias fault input as the jamming is considered noise while coping with the switched-gain observer design framework.

Let us consider the following observer structure to estimate the above augmented system

$$\begin{cases} \hat{\xi}_{k+1} = \mathbb{A}\hat{\xi}_k + \mathbb{B}u_k + L(y_k - \hat{y}_k), \\ \hat{y}_k = h(\hat{\xi}_k). \end{cases} \tag{14.15}$$

Matrix L is observer gain parameter to be determined such that the estimation error $e = \xi - \hat{\xi}$ converges towards zero where $\hat{\xi}_k$ is the estimate of ξ_k. Since $h(.)$ is globally Lipschitz, then, there exist $z_i \in Co(\vartheta_i, \hat{\vartheta}_i)$, functions $\phi_{ij}\colon\ \mathbb{R}^{n_i} \longrightarrow \mathbb{R}$, and constants a_{ij}, b_{ij}, such that

$$h(\xi) - h(\hat{\xi}) = \sum_{i,j=1}^{p,n_i} \phi_{ij}(z_i)\mathscr{H}_{ij}\left(\vartheta_i - \hat{\vartheta}_i\right) \tag{14.16}$$

and

$$\vartheta_i = H_i\xi_k,\quad \hat{\vartheta}_i = H_i\hat{\xi}_k,\quad a_{ij} \leqslant \phi_{ij}\left(z_i\right) \leqslant b_{ij},\quad \phi_{ij}(z_i) = \frac{\partial h_i}{\partial \vartheta_i^j}(z_i), \tag{14.17}$$

where

$$\mathscr{H}_{ij} = e_p(i)e_{n_i}^\top(j), \quad \phi_{ij} \triangleq \phi_{ij}\Big(z_i\Big), \quad H_i \in \mathbb{R}^{n_i \times n}.$$

Since $\vartheta_i - \hat{\vartheta}_i = H_i e_k$ and for all $i = 1, \dots, p$ and $j = 1, \dots, n_i$, we can rewrite the nonlinearities as follows:

$$h(\xi_k) - h(\hat{\xi}_k) = \sum_{i,j=1}^{p,n_i} \phi_{ij} \mathscr{H}_{ij} H_i e_k \triangleq \mathbb{C} e_k + \sum_{i,j=1}^{p,n_i} \tilde{\phi}_{ij} \mathscr{H}_{ij} H_i e_k,$$

where

$$\mathbb{C} \triangleq \sum_{(i,j)\in\mathfrak{F}} a_{ij} \mathscr{H}_{ij} H_i, \quad \tilde{\phi}_{ij} \triangleq \phi_{ij} - a_{ij}, \quad \mathfrak{F} \triangleq \Big\{(i, j) : a_{ij} \neq 0\Big\}. \tag{14.18}$$

Then, the dynamic equation of the observation error $e_k = \xi_k - \hat{\xi}_k$ can be written as

$$e_{k+1} = \left(\mathbb{A} - L \sum_{i,j=1}^{p,n_i} \Big[\phi_{ij} \mathscr{H}_{ij} H_i\Big]\right) e_k + \underbrace{[\mathbb{E} - L\mathbb{D}\ \mathbb{F}\ -L\mathbb{G}]]}_{\bar{\mathbb{E}}} \underbrace{\begin{bmatrix} w_k \\ \Delta f_k^a \\ J_k \end{bmatrix}}_{\bar{w}_k}$$

$$e_{k+1} = \underbrace{\left(\mathbb{A} - L\mathbb{C} - \sum_{i,j=1}^{p,n_i} \tilde{\phi}_{ij} L \mathscr{H}_{ij} H_i\right)}_{\bar{\mathbb{A}}} e_k + \bar{\mathbb{E}} \bar{w}_k. \tag{14.19}$$

It follows that

$$0 \leqslant \tilde{\phi}_{ij} \leqslant \tilde{b}_{ij} \triangleq b_{ij} - a_{ij}.$$

The aim is to find the gain matrix L, so that the observation error (14.19) satisfies the following $\mathscr{H}_\infty$ criterion

$$\|e\|_{\ell_2^n} \leq \sqrt{\mu \|\bar{w}\|_{\ell_2^z}^2 + \nu \|e_0\|^2}, \tag{14.20}$$

where $\mu > 0$ is the gain from w to e and $\nu > 0$ is to be determined. To analyze the $\mathscr{H}_\infty$ stability of the error, we use the following quadratic Lyapunov function

$$V_k(e_k) = e_k^\top P e_k, \text{ with } P = P^\top > 0. \tag{14.21}$$

Consequently, the $\mathscr{H}_\infty$ criterion is satisfied if the following inequality holds

$$\mathscr{W}_k \triangleq \Delta V_k + \|e_k\|^2 - \mu \|\bar{w}_k\|^2 \leqslant 0, \tag{14.22}$$

where $\Delta V_k = V(e_{k+1}) - V(e_k)$.

14.4.2 Unknown Input Observer Design Method

This subsection will derive the theoretical results on the unknown input observer design procedure for a class of nonlinear monotonic output equations system. We demonstrate infeasibility to solutions for the unknown input observer design LMIs when the nonlinear LED functions are all non-monotonic. Then, we present a switched-gain observer design methodology that enables stable observers for the non-monotonic output functions of the LED optical communication systems.

The following theorem provides the conditions that guarantee the asymptotic stability of the estimation error system (14.19) in the $\mathscr{H}_\infty$−optimality sense (14.20).

Theorem 14.1 *Assume that there exist symmetric positive definite matrices $P \in \mathbb{R}^{n\times n}$, $\mathbb{S}_{ij} \in \mathbb{R}^{n_i \times n_i}$, $i = 1, \ldots, n$ and matrix $\mathscr{X} \in \mathbb{R}^{p\times n}$, so that the following LMI condition holds*

$$\min(\mu) \textit{ subject to } (14.24) \tag{14.23}$$

$$\begin{bmatrix} \mathbb{M} & \left[\Pi_1^\top \ldots \Pi_p^\top\right] \\ (\star) & -\Lambda\mathbb{N} \end{bmatrix} < 0, \tag{14.24}$$

where

$$\mathbb{M} = \begin{bmatrix} \begin{bmatrix} -P + \mathbb{I} & 0 \\ 0 & -\mu\mathbb{I} \end{bmatrix} & \begin{bmatrix} \mathbb{A}^\top P + \mathbb{C}^\top \mathscr{X} \\ \begin{bmatrix} \mathbb{E}^\top P + \mathbb{D}^\top \mathscr{X} \\ \mathbb{F}^\top P \\ -\mathbb{G}^\top \mathscr{X} \end{bmatrix} \end{bmatrix} \\ (\star) & -P \end{bmatrix}, \tag{14.25}$$

$$\begin{aligned} \Pi_i &= \left[\Pi_{i1}^\top(\mathscr{X}, \mathbb{S}_{i1}) \ldots \Pi_{in_i}^\top(\mathscr{X}, \mathbb{S}_{in_i})\right]^\top, \\ \Pi_{ij}^\top(\mathscr{X}, \mathbb{S}_{ij}) &= \begin{bmatrix} 0 \\ 0 \\ \mathscr{X}^\top \mathscr{H}_{ij} \end{bmatrix} + \begin{bmatrix} H_i^\top \\ 0 \\ 0 \end{bmatrix} \mathbb{S}_{ij}, \end{aligned} \tag{14.26}$$

$$\Lambda = \textit{block-diag}\Big(\Lambda_1, \ldots, \Lambda_p\Big), \tag{14.27}$$

$$\Lambda_i = \textit{block-diag}\left(\frac{2}{\tilde{b}_{i1}}\mathbb{I}_{n_i}, \ldots, \frac{2}{\tilde{b}_{in_i}}\mathbb{I}_{n_i}\right), \tag{14.28}$$

$$\mathbb{N} = block\text{-}diag\Big(\mathbb{N}_1, \ldots, \mathbb{N}_p\Big), \tag{14.29}$$

$$\mathbb{N}_i = block\text{-}diag\Big(\mathbb{N}_{i1}, \ldots, \mathbb{N}_{in_i}\Big), \tag{14.30}$$

then the observation error system in (14.19) *is asymptotically stable and the* $\mathscr{H}_\infty$ *performance criterion* (14.20) *is guaranteed with* $\nu = \lambda_{\max}(P)$. *In addition, the observer gain* L *is computed as*

$$L = -P^{-1}\mathscr{X}^\top.$$

Proof By calculating $\mathscr{W}_k$ along the trajectories of (14.19), we obtain the following equation

$$\begin{aligned}\mathscr{W}_k =& e_k^\top\left[\left(\mathbb{A}-L\mathbb{C}-L\sum_{i,j=1}^{p,n_i}\left[\tilde{\phi}_{ij}\mathscr{H}_{ij}H_i\right]\right)^\top P\left(\mathbb{A}-L\mathbb{C}-L\sum_{i,j=1}^{p,n_i}\left[\tilde{\phi}_{ij}\mathscr{H}_{ij}H_i\right]\right)-P+\mathbb{I}\right]e_k\\ &+\bar{w}_k^\top\left[\bar{\mathbb{E}}^\top P\ \bar{\mathbb{E}}-\mu\mathbb{I}\right]\bar{w}_k\\ &+e_k^\top\left[\left(\mathbb{A}-L\mathbb{C}-L\sum_{i,j=1}^{p,n_i}\left[\tilde{\phi}_{ij}\mathscr{H}_{ij}H_i\right]\right)^\top P\bar{\mathbb{E}}\right]\bar{w}_k\\ &+\bar{w}_k^\top\left[\bar{\mathbb{E}}^\top P\left(\mathbb{A}-L\mathbb{C}-L\sum_{i,j=1}^{p,n_i}\left[\tilde{\phi}_{ij}\mathscr{H}_{ij}H_i\right]\right)\right]e_k.\end{aligned} \tag{14.31}$$

Then, (14.31) can be written as follows:

$$\mathscr{W}_k = \begin{bmatrix} e_k^\top \\ \bar{w}_k^\top \end{bmatrix}\begin{bmatrix} \bar{\mathbb{A}}^\top\bar{\mathbb{A}} - P + \mathbb{I} & \bar{\mathbb{A}}^\top P\bar{\mathbb{E}} \\ (*) & \bar{\mathbb{E}}^\top P\bar{\mathbb{E}} - \mu\mathbb{I}\end{bmatrix}\begin{bmatrix} e_k \\ \bar{w}_k\end{bmatrix}. \tag{14.32}$$

It follows that $\mathscr{W}_k \leqslant 0$ if the following inequality holds

$$\begin{bmatrix} \bar{\mathbb{A}}^\top P\bar{\mathbb{A}} - P + \mathbb{I} & \bar{\mathbb{A}}^\top P\bar{\mathbb{E}} \\ (*) & \bar{\mathbb{E}}^\top P\bar{\mathbb{E}} - \mu\mathbb{I}\end{bmatrix} < 0, \tag{14.33}$$

which is equivalent to

$$\begin{bmatrix} -P + \mathbb{I} & 0 \\ (*) & -\mu\mathbb{I}\end{bmatrix} + \begin{bmatrix} \bar{\mathbb{A}}^\top P \\ \bar{\mathbb{E}}^\top P\end{bmatrix} P^{-1}\left[P\bar{\mathbb{A}}\ P\bar{\mathbb{E}}\right] < 0. \tag{14.34}$$

Using Schur lemma, we deduce that $\mathscr{W}_k < 0$ if the following matrix inequality holds

$$\begin{bmatrix} -P+\mathbb{I} & 0 & \bar{\mathbb{A}}^\top P \\ (*) & -\mu\mathbb{I} & \bar{\mathbb{E}}^\top P \\ (*) & (*) & -P \end{bmatrix} < 0, \tag{14.35}$$

which is equivalent to

$$\begin{bmatrix} \begin{bmatrix} -P+\mathbb{I} & 0 \\ 0 & -\mu\mathbb{I} \end{bmatrix} & \begin{bmatrix} \left(\mathbb{A} - L\mathbb{C} - L\sum_{i,j=1}^{p,n_i}\left[\tilde{\phi}_{ij}\mathscr{H}_{ij}H_i\right]\right)^\top P \\ \left[\mathbb{E} - L\mathbb{D}\ \mathbb{F} - L\mathbb{G}\right]^\top P \end{bmatrix} \\ (\star) & -P \end{bmatrix} < 0. \tag{14.36}$$

Inequality (14.36) can be rewritten as follows:

$$\overbrace{\begin{bmatrix} \begin{bmatrix} -P+\mathbb{I} & 0 \\ 0 & -\mu\mathbb{I} \end{bmatrix} & \begin{bmatrix} \mathbb{A}^\top P - \mathbb{C}^\top L^\top P \\ \begin{bmatrix} \mathbb{E}^\top P - \mathbb{D}^\top L^\top P \\ \mathbb{F}^\top P \\ -\mathbb{G}^\top L^\top P \end{bmatrix} \end{bmatrix} \\ (\star) & -P \end{bmatrix}}^{\mathbb{M}} +$$

$$\sum_{i,j=1}^{p,n_i} \tilde{\phi}_{ij} \left(\overbrace{\begin{bmatrix} H_i^\top \\ 0 \\ 0 \end{bmatrix}}^{\mathbb{Y}_i^\top} \overbrace{\left[0\ 0\ -\mathscr{H}_{ij}^\top L^\top P\right]}^{\mathbb{X}_{ij}} + \mathbb{X}_{ij}^\top \mathbb{Y}_i \right) < 0. \tag{14.37}$$

From Young's inequality, we have

$$\mathbb{Y}_i^\top \mathbb{X}_{ij} + \mathbb{X}_{ij}^\top \mathbb{Y}_i \leqslant \frac{1}{2}\left(\mathbb{X}_{ij} + \mathbb{S}_{ij}\mathbb{Y}_i\right)^\top \mathbb{S}_{ij}^{-1} \overbrace{\left(\mathbb{X}_{ij} + \mathbb{S}_{ij}\mathbb{Y}_i\right)}^{\Pi_{ij}},$$

for any symmetric positive definite matrices $\mathbb{S}_{ij}$. Therefore, from (14.17) and the fact that $a_{ij} = 0$, inequality (14.37) holds if

$$\mathbb{M} + \sum_{i,j=1}^{p,n_i} \left(\Pi_{ij}^T \Big(\frac{2}{\tilde{b}_{ij}} \mathbb{S}_{ij}\Big)^{-1} \Pi_{ij} \right) < 0. \tag{14.38}$$

Hence, by Schur lemma and the change of variable $\mathscr{X} = -L^T P$, inequality (14.38) is equivalent to (14.24). This ends the proof. ■

14.4.3 *Feasibility of* (14.24) *for Non-monotonic Outputs*

The following theorem provides the non-existence of a constant observer gain solution for the non-monotonic LED-based optical communication model.

Theorem 14.2 *Assume that the two following items hold:*

(i) *All the nonlinear output functions* $h_i, i = 1, \ldots, p$, *are non-monotonic.*
(ii) *the system matrix* A *is not Schur stable.*

Then, the LMI (14.24) *is infeasible.*

Proof First, consider the following change of variables

$$\bar{\mathbb{S}}_{ij} \Longleftarrow \frac{2}{\tilde{b}_{ij}} \mathbb{S}_{ij}.$$

Then, the LMI (14.24) in Theorem 14.1 is equivalent to

$$\begin{bmatrix} \mathbb{M} & [\nabla_1^T \ldots \nabla_p^T] \\ (\star) & -\mathbb{N} \end{bmatrix} < 0, \tag{14.39}$$

where

$$\mathbb{M} = \begin{bmatrix} \begin{bmatrix} -P + \mathbb{I} & 0 \\ 0 & -\mu \mathbb{I} \end{bmatrix} & \begin{bmatrix} \mathbb{A}^\top P + \mathbb{C}^\top \mathscr{X} \\ \begin{bmatrix} \mathbb{E}^\top P + \mathbb{D}^\top \mathscr{X} \\ \mathbb{F}^\top P \\ -\mathbb{G}^\top \mathscr{X} \end{bmatrix} \end{bmatrix} \\ (\star) & -P \end{bmatrix}, \tag{14.40}$$

$$\nabla_i = \left[\nabla_{i1}^T(\mathscr{X}, \bar{\mathbb{S}}_{i1}) \ldots \nabla_{in_i}^T(\mathscr{X}, \bar{\mathbb{S}}_{in_i}) \right]^T,$$

$$\nabla_{ij}^T(\mathscr{X}, \bar{\mathbb{S}}_{ij}) = \begin{bmatrix} 0 \\ 0 \\ \mathscr{X}^T \mathscr{H}_{ij} \end{bmatrix} + \frac{\tilde{b}_{ij}}{2} \begin{bmatrix} H_i^T \\ 0 \\ 0 \end{bmatrix} \bar{\mathbb{S}}_{ij}. \tag{14.41}$$

To simplicity the proof, we will use a compact form of (14.39). To this end, we introduce the following notation

$$\mathcal{G} \triangleq \left[\mathcal{H}_{11} \dots \mathcal{H}_{in_1}\ \mathcal{H}_{p1} \dots \mathcal{H}_{pn_p}\right],$$

$$\mathcal{H}^\top \triangleq \Big[\underbrace{H_1^\top \dots H_1^\top}_{n_1\text{times}} \dots \underbrace{H_p^\top \dots H_p^\top}_{n_p\text{times}}\Big].$$

We also define Γ_a and Γ_b under the same form than Λ by replacing $\frac{2}{\tilde{b}_{ij}}$ by a_{ij} and b_{ij}, respectively. Then, LMI (14.39) can be written under the compact form

$$\begin{bmatrix} \mathbb{M} & -\begin{bmatrix} 0 \\ 0 \\ PL\mathcal{G} \end{bmatrix} + \begin{bmatrix} \mathcal{H}^\top \\ 0 \\ 0 \end{bmatrix} \Big(\overbrace{\Gamma_b - \Gamma_a}^{\Lambda^{-1}}\Big)^\top \mathbb{N} \\ (\star) & -\mathbb{N} \end{bmatrix} < 0. \tag{14.42}$$

Since from (i) all the nonlinear functions h_j are non-monotonic, then from the definition of C in (14.18), we deduce that

$$C = \mathcal{G}\Gamma_a\mathcal{H}.$$

It follows that $\mathbb{M}$ in (14.40) can be decomposed as

$$\mathbb{M} = \begin{bmatrix} \begin{bmatrix} -P+\mathbb{I} & 0 \\ 0 & -\mu\mathbb{I} \end{bmatrix} & \begin{bmatrix} \mathbb{A}^\top P + \mathbb{C}^\top \mathcal{X} \\ \begin{bmatrix} \mathbb{E}^\top P + \mathbb{D}^\top \mathcal{X} \\ \mathbb{F}^\top P \\ -\mathbb{G}^\top \mathcal{X} \end{bmatrix} \end{bmatrix} \\ (\star) & -P \end{bmatrix}$$
$$-\begin{bmatrix} 0 \\ 0 \\ PL\mathcal{G} \end{bmatrix} \Gamma_a \left[\mathcal{H}\ 0\ 0\right] - \begin{bmatrix} \mathcal{H}^\top \\ 0 \\ 0 \end{bmatrix} \Gamma_a^\top \begin{bmatrix} 0 \\ 0 \\ PL\mathcal{G} \end{bmatrix}^\top. \tag{14.43}$$

Hence from Schur lemma and the decomposition (14.43), LMI (14.42) is equivalent to (14.44). On the other hand, after some manipulations, the LMI (14.44) is identically written under the form (14.45), which brings out the monotonicity through the term $\Gamma_a^\top \mathbb{N}\Gamma_b + \Gamma_b^\top \mathbb{N}\Gamma_a$.

$$\begin{bmatrix} \begin{bmatrix} -P+\mathbb{I} & 0 \\ 0 & -\mu\mathbb{I} \end{bmatrix} & \begin{bmatrix} \mathbb{A}^\top P + \mathbb{C}^\top \mathscr{X} \\ \begin{bmatrix} \mathbb{E}^\top P + \mathbb{D}^\top \mathscr{X} \\ \mathbb{F}^\top P \\ -\mathbb{G}^\top \mathscr{X} \end{bmatrix} \end{bmatrix} \\ (\star) & -P \end{bmatrix} - \begin{bmatrix} 0 \\ 0 \\ PL\mathscr{G} \end{bmatrix} \Gamma_a [\mathscr{H} \ 0 \ 0] - \begin{bmatrix} \mathscr{H}^\top \\ 0 \\ 0 \end{bmatrix} \Gamma_a^\top \left[0 \ 0 \ \left(PL\mathscr{G}\right)^\top\right]$$

$$+ \left(- \begin{bmatrix} 0 \\ 0 \\ PL\mathscr{G} \end{bmatrix} + \frac{1}{2} \begin{bmatrix} \mathscr{H}^\top \\ 0 \\ 0 \end{bmatrix} \big(\overbrace{\Gamma_b - \Gamma_a}^{\Lambda^{-1}} \big)^\top \mathbb{N} \right) \mathbb{N}^{-1} \left(- \begin{bmatrix} 0 \\ 0 \\ PL\mathscr{G} \end{bmatrix} + \frac{1}{2} \begin{bmatrix} \mathscr{H}^\top \\ 0 \\ 0 \end{bmatrix} \big(\overbrace{\Gamma_b - \Gamma_a}^{\Lambda^{-1}} \big)^\top \mathbb{N} \right)^\top < 0. \tag{14.44}$$

$$\begin{bmatrix} \begin{bmatrix} -P+\mathbb{I} & 0 \\ 0 & -\mu\mathbb{I} \end{bmatrix} & \begin{bmatrix} \mathbb{A}^\top P + \mathbb{C}^\top \mathscr{X} \\ \begin{bmatrix} \mathbb{E}^\top P + \mathbb{D}^\top \mathscr{X} \\ \mathbb{F}^\top P \\ -\mathbb{G}^\top \mathscr{X} \end{bmatrix} \end{bmatrix} \\ (\star) & -P \end{bmatrix} - \frac{1}{2} \begin{bmatrix} \mathscr{H}^\top \\ 0 \\ 0 \end{bmatrix} \overbrace{\left[\Gamma_a^\top \mathbb{N} \Gamma_b + \Gamma_b^\top \mathbb{N} \Gamma_a\right]}^{<0} [\mathscr{H} \ 0 \ 0]$$

$$\underbrace{+ \left(- \begin{bmatrix} 0 \\ 0 \\ PL\mathscr{G} \end{bmatrix} + \frac{1}{2} \begin{bmatrix} \mathscr{H}^\top \\ 0 \\ 0 \end{bmatrix} \left(\Gamma_b + \Gamma_a \right)^\top \mathbb{N} \right) \mathbb{N}^{-1} \left(- \begin{bmatrix} 0 \\ 0 \\ PL\mathscr{G} \end{bmatrix} + \frac{1}{2} \begin{bmatrix} \mathscr{H}^\top \\ 0 \\ 0 \end{bmatrix} \left(\Gamma_b + \Gamma_a \right)^\top \mathbb{N} \right)^\top}_{>0} < 0. \tag{14.45}$$

Hence, if all the nonlinearities are non-monotonic, i.e.,

$$\Gamma_a^\top \mathbb{N} \Gamma_b + \Gamma_b^\top \mathbb{N} \Gamma_a < 0,$$

then the feasibility of (14.45) implies

$$\begin{bmatrix} \begin{bmatrix} -P+\mathbb{I} & 0 \\ 0 & -\mu\mathbb{I} \end{bmatrix} & \begin{bmatrix} \mathbb{A}^\top P + \mathbb{C}^\top \mathscr{X} \\ \begin{bmatrix} \mathbb{E}^\top P + \mathbb{D}^\top \mathscr{X} \\ \mathbb{F}^\top P \\ -\mathbb{G}^\top \mathscr{X} \end{bmatrix} \end{bmatrix} \\ (\star) & -P \end{bmatrix} < 0,$$

and leads necessarily to A Schur stable, which contradicts item (ii) of Theorem 14.2. Then if the matrix A is not Schur stable, the LMI (14.24) is infeasible. This ends the proof. ■

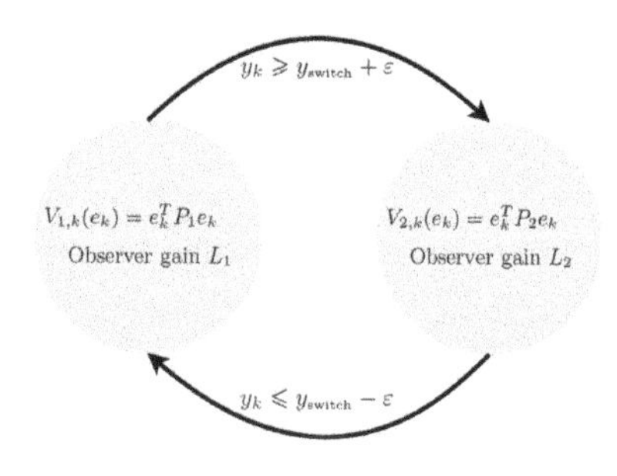

Fig. 14.11 Switched-gain observer with switched gains

14.4.4 A Switched-Gain-Based Observer Solution

It has recently been proven in the continuous-time case (Rajamani et al. 2020) that when all the nonlinear output functions are non-monotonic, a single observer gain that guarantees exponentially stable estimation error over the entire operating range cannot be found. Consequently, the LMI (14.24) is not feasible. However, if we want to keep LMI (14.24) and exploit it for the observer design, then the unique solution is to introduce a switched-gain-based observer (Rajamani et al. 2020) as depicted in Fig. 14.11. To proceed, we consider a switched-gain observer with a constant gain L_i in the region R_i designed using the LMI (14.24) with $\Gamma_{a_{R_i}}$, $\Gamma_{b_{R_i}}$, and the corresponding value of the quadratic Lyapunov positive definite matrix P_i as shown in Fig. 14.11 in the case of switching between two regions. In Fig. 14.11, y_{switch} is the nominal switching point between the two regions, and the parameter ε is the hysteresis added to the switching to ensure a minimum dwell time after each switch (Rajamani et al. 2020). The stability of the switched-gain observer of Fig. 14.11 consisting of different constant observer gain regions needs to be considered. Let the two observers be designed to be asymptotically stable in each of the two regions using the quadratic Lyapunov function analysis of Theorem 14.1. Then, it should be noted that inside each region, a single observer gain is used, and asymptotic stability is guaranteed under the constraint of feasibility of (14.24). Furthermore, the stability of the overall switched system can be guaranteed if the system satisfies a minimum dwell time constraint in each region, according to results from switching system theory (Alessandri et al. 2005; Liberzon 2003; Goebel et al. 2012).

14.4.5 Reference Trajectory Tracking Design

The control objective consists in tracking a given desired trajectory ξ_k^d corresponding to a desired input u_k^d, where (ξ_k^d, u_k^d) is assumed to be an admissible stable solution for the system (14.13) in the absence of dynamics noises. That is the pair (ξ_k^d, u_k^d) satisfies the following dynamic equation

$$\xi_{k+1}^d = \mathbb{A}\xi_k^d + \mathbb{B}u_k^d. \tag{14.46}$$

The observer-based tracking control is given as

$$u_k = -K(\hat{\xi}_k - \xi_k^d) + u_k^d. \tag{14.47}$$

Let us define the trajectory tracking error by

$$\tilde{\xi}_k = \xi_k - \xi_k^d. \tag{14.48}$$

Then, the dynamics of the reference tracking error $\tilde{x}_k$ is written as

$$\tilde{\xi}_{k+1} = \Big(\mathbb{A} - \mathbb{B}K\Big)\tilde{\xi}_k + \mathbb{B}K e_k + \bar{\mathbb{E}}\bar{w}_k. \tag{14.49}$$

In the disturbance-free case, from Barbalat's Lemma (Khalil and Grizzle 2002), the stabilization of (14.49) is ensured by a simple pole assignment of the matrix $\mathbb{A} - \mathbb{B}K$. This is due to the fact that the system is linear and the estimation error e_k converges exponentially towards zero. However, in the disturbance case, i.e., $\bar{w}_k \not\equiv 0$, assuming that the estimation error satisfies the $\mathscr{H}_\infty$ criterion (14.20), the objective is to determine the controller gain K to satisfy the following $\mathscr{H}_\infty$−optimality criterion

$$\|\tilde{\xi}\|_{\ell_2^n} \leq \sqrt{\mu_2\|\bar{w}\|^2_{\ell_2^z} + \nu_2 \left\| \begin{bmatrix} \tilde{\xi}_0 \\ e_0 \end{bmatrix} \right\|^2}, \tag{14.50}$$

where $\mu_2 > 0$ is the gain from $\bar{w}$ to $\tilde{\xi}$ and ν_2 is to determine later. Since the gains L_i are determined by the observer design part in the previous section, then it remains to design the controller gain K. The design procedure we follow in this chapter is borrowed from Draa et al. (2019). Hence for more details we refer the reader to (Draa et al. 2019, Sect. 3). It should be notice that (Draa et al. 2019) concerns continuous-time systems, while in this chapter we deal with discrete-time. However, the extension to discrete-time is straightforward.

Proposition 14.1 *Assume that there exist symmetric positive definite matrices $Y \in \mathbb{R}^{n\times n}$ and matrix $\mathscr{Z}$ of appropriate dimensions, so that the following convex optimization problem holds:*

$$\min(\mu_1) \text{ subject to } (14.52) \tag{14.51}$$

$$\begin{bmatrix} -Y & 0 & Y\begin{bmatrix} A & F \\ 0 & I \end{bmatrix}^\top - \mathscr{Z}\begin{bmatrix} B \\ 0 \end{bmatrix}^\top & Y \\ (\star) & -\mu_1\mathbb{I} & \left[\begin{bmatrix} E \\ 0 \end{bmatrix} - LD\begin{bmatrix} 0 \\ I \end{bmatrix} - LG\right]^\top & 0 \\ (\star) & (\star) & -Y & 0 \\ (\star) & (\star) & (\star) & -\mathbb{I} \end{bmatrix} < 0. \tag{14.52}$$

Then the tracking error $\tilde{x}$ satisfies the $\mathscr{H}_\infty$ performance criterion

$$\|\tilde{\xi}\|_{\ell_2^n} \leq \sqrt{\mu_1 \left\| \begin{bmatrix} \bar{w}_k \\ BKe_k \end{bmatrix} \right\|_{\ell_2^z}^2 + \nu_1 \left\| \tilde{\xi}_0 \right\|^2}, \tag{14.53}$$

with μ_1 given by (14.51)*, $\nu_1 = \lambda_{max}(Y)$, and $K = \mathscr{Z}^T Y^{-1}$.*

Finally, the complete design procedure of the switched-gain observer-based tracking controller can be summarized in the unified proposition below

Proposition 14.2 *Assume that there exist symmetric positive definite matrices P, Y, $\mathscr{S}_i$, $i = 1, \ldots, n$, and matrices $\mathscr{Z}$, $\mathscr{X}$ of appropriate dimensions such that both convex optimization problems* (14.23) *and* (14.51) *hold. Then, the observer-based tracking controller* (14.47) *guarantees the $\mathscr{H}_\infty$ optimality criterion* (14.50) *with μ_2 and ν_2 given by*

$$\mu_2 \triangleq \mu_1 \left[1 + \mu\lambda_{\max}\left(K^T B^T BK\right)\right], \tag{14.54}$$

$$\nu_2 \triangleq \max\left(\mu_1\lambda_{\max}(P)\lambda_{\max}\left(K^T B^T BK\right), \nu_1\right), \tag{14.55}$$

where μ and μ_1 are returned by the convex optimization problems (14.23) *and* (14.51)*, respectively, and $\nu_1 = \lambda_{max}(Y)$.*

Proof See Draa et al. (2019, Proposition 3.1), for the proof in the continuous-time case. ∎

14.5 LED Application Under Actuator Fault and Noise Jamming Attack on the Optical Communication Channel

This section illustrates the theoretical contributions presented in the previous sections. The effectiveness of the discrete-time nonlinear observer-based reference tracking controller is evaluated for the LED-based optical communication system under actuator fault and noise jamming on the optical communication channel. To do so, we augment the state variables and consider the problem of estimating the actuator fault attack f_k^a, the angular position $x_{1,k}$, and the angular velocity $x_{2,k}$ based on which the control u_k is designed to drive the states $x_{2,k}$ towards zero, which gives the orientation with the maximum light intensity. The process dynamics of the LED model (14.9) are linear while the output Eq. (14.12) are nonlinear. It is also clear that $g(.)$ is function of the state $x_{1,k}$. Using the discrete-time nonlinear switched-gain observer (14.14), the nonlinear output functions y_k is monotonic in the operating ranges of $x_{1,k}$ and $\phi \triangleq x_{2,k}$ as illustrated in Fig. 14.12. Hence, a constant observer gain matrix L_i exists

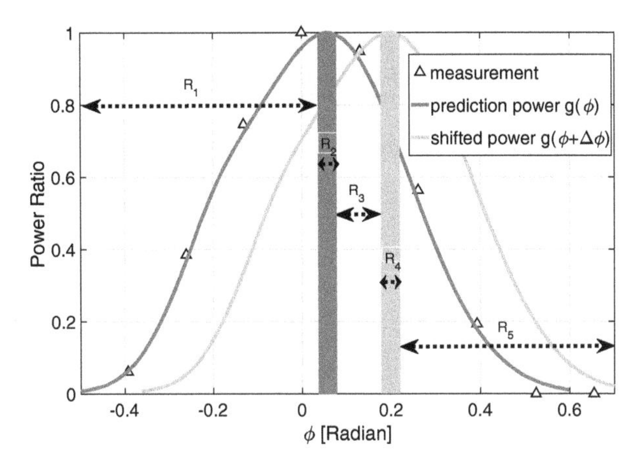

Fig. 14.12 Regions around slope-change points of $g(\phi)$ with $\Delta\phi = -6^\circ$

in the operating ranges of interest. However, it is impossible to find a constant gain matrix of L_i that makes the observer stable for the entire operating range. Therefore, a switched-gain-based observer is needed for the allowable operating regimes.

We divide piece-wise into different regions the nonlinear output functions. In each region, at least one of the output functions is a monotonic function. Figure 14.12 illustrates a piece-wise division of the nonlinear output functions in the operating regions of interest due to the monotonicity concept. We note that the regions' boundaries lie at the slope-change points. For example, R_2 is a narrow region where the nonlinear output function's slope y_1 is close to zero. In this region, only the output y_2 will be used by the observer since y_2 is monotonic. Regions R_1 and R_3 lie on either side of R_2 and both of these regions can utilize both outputs y_1 and y_2. Both y_1 and y_2 are monotonic in these regions. Since each region of interest R_1 through R_5 has monotonic output function properties, as illustrated in Fig. 14.12. Then, a constant stabilizing observer gain exists in each of these regions. Table 14.2 provides the five operating regimes and their corresponding observer gains.

We evaluate the capability of the discrete-time nonlinear switched-gain observer-based controller method to reconstruct the actuator fault and the states variables for tracking the LED optical communication system (14.9)–(14.12). The convex optimization problem in (14.52) is feasible with the controller gain $K = \begin{bmatrix}0.02 \ 0.3\end{bmatrix}$. The actuator fault in the angular velocity of the robot vehicle is considered and is generated as follows:

$$f^a = \begin{cases} 0.002 \times \left[\cos\left(t_k - \frac{2}{5}\right) + 0.25\left(\sin\left(\frac{2\pi}{3}t_k\right) - 1.5\right)\right], & t_k \geqslant 3.5. \\ 0 & \text{otherwise.} \end{cases} \tag{14.56}$$

The actuator fault (14.56) is synchronously sampled at the current sampling time defined as $t_k = k\varepsilon$ where $k \in \mathbb{Z}^+$ is a positive integer and $\varepsilon > 0$ is the sampling

Table 14.2 Operating ranges of ϕ [rad] and corresponding switched-gain observer gains

Region	Left [rad]	Right [rad]	Observer gain
1	−0.5000	0.0083	$\begin{bmatrix} 4.5981 \times 10^{-8} & 2.4859 \\ 1.0061 \times 10^{-6} & 55.1841 \\ 5.0782 \times 10^{-8} & 2.6526 \end{bmatrix}$
2	0.0083	0.1083	$\begin{bmatrix} -0.5733 & 3.5712 \\ -12.9587 & 80.7190 \\ -0.6197 & 3.8602 \end{bmatrix}$
3	0.1083	0.1483	$\begin{bmatrix} -2.2797 & 2.3838 \\ -33.1862 & 34.7009 \\ -1.5286 & 1.5984 \end{bmatrix}$
4	0.1483	0.2483	$\begin{bmatrix} -2.5111 & 0.5538 \\ -65.6161 & 14.4720 \\ -3.1354 & 0.6915 \end{bmatrix}$
5	0.2483	0.5000	$\begin{bmatrix} -2.1796 & -3.3568 \times 10^{-10} \\ -63.8935 & -1.0240 \times 10^{-8} \\ -3.0212 & -4.2906 \times 10^{-10} \end{bmatrix}$

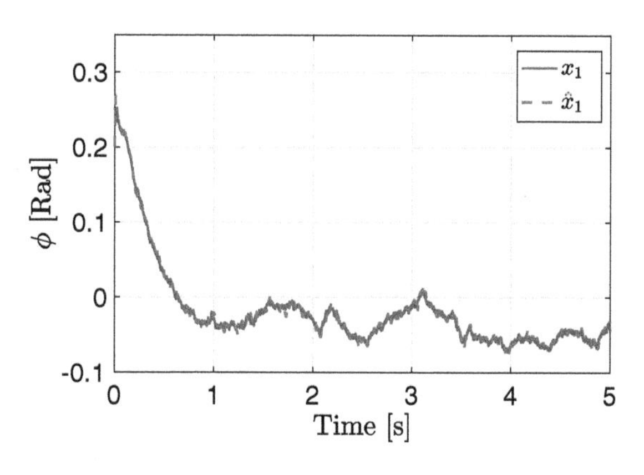

Fig. 14.13 Estimated angular position $\hat{x}_1$ along with the actual angular position x_1

period. We assume that the process noise vector $w_k \sim \mathcal{N}(0, Q)$, measurement noise $w_k \sim \mathcal{N}(0, R)$, and the jamming noise is considered to be non-Gaussian noise.

Figures 14.13, 14.14, 14.16, and 14.17 illustrate the reference trajectory tracking results of the discrete-time switched-gain observer-based controller. Figure 14.15 shows the unknown actuator input fault and its estimate. The proposed nonlinear observer-based control exhibits good estimation performance and maintains a good reference trajectory tracking performance.

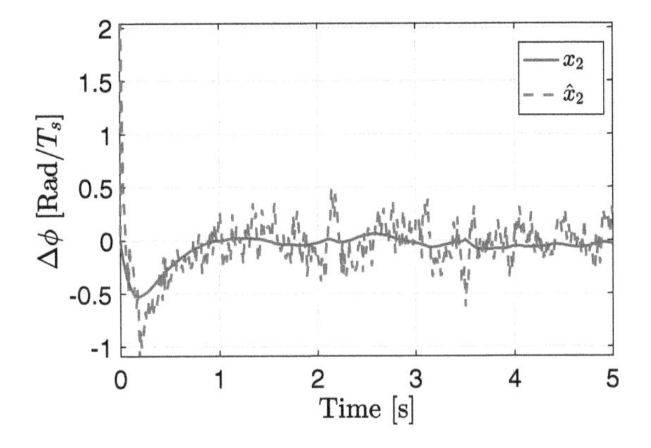

Fig. 14.14 Estimated angular velocity $\hat{x}_2$ along with the actual angular velocity x_2

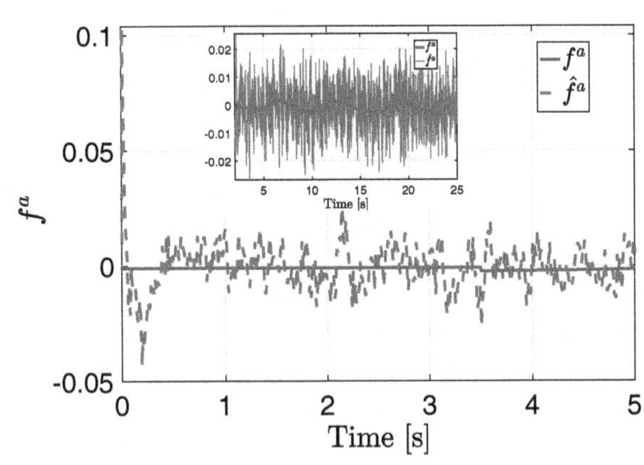

Fig. 14.15 Estimated actuator fault $\hat{x}_3$ along with the actual actuator fault x_3

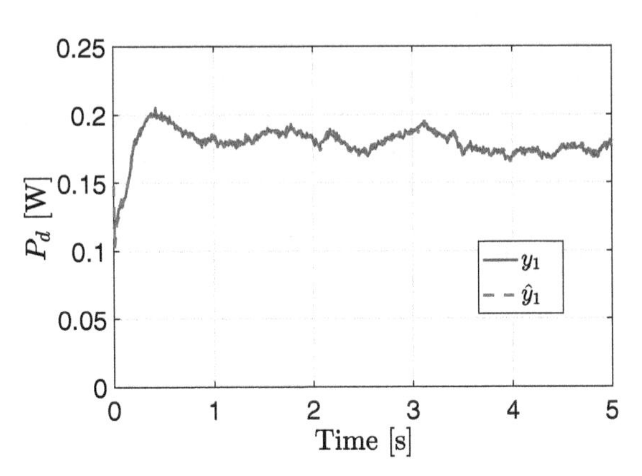

Fig. 14.16 Estimated output power $\hat{y}_1$ along with the actual output power y_1

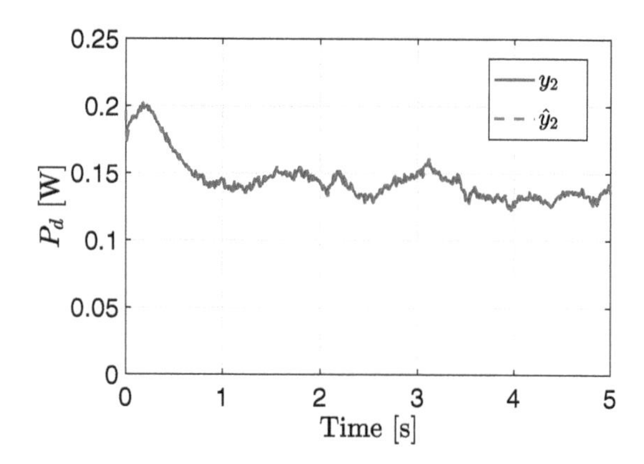

Fig. 14.17 Estimated output power $\hat{y}_2$ along with the actual output power y_2

14.6 Conclusion

In this chapter, we have designed a switched-gain observer-based reference trajectory tracking to estimate the actuator fault and state variables under noise jamming on the optical communication channel where the jammer aims to degrade signal quality. The simultaneous presence of actuator fault and jamming noise poses challenges that arise straight out the estimation and reference trajectory tracking control practice. Specifically, the designed observer-based control is applied to a LED-based optical communication system in which the output nonlinearities are non-monotonic. Based on this framework, sufficient conditions for the asymptotic stability and the $\mathscr{H}_\infty$ performance criterion of the observation error dynamics are guaranteed using Lyapunov-analysis. Our future work will develop a prototype experiment testbed to evaluate the proposed observer design strategy to detect and estimate the unknown inputs and validate the obtained results. Subsequently, the fundamental observer design algorithm developed herein can be extended in the future to enable hostile jamming-attack detection in more complex distributed autonomous robots throughout optical communication architectures.

Acknowledgements This work has been supported by the King Abdullah University of Science and Technology (KAUST) through Base Research Fund (BAS/1/1627-01-01). A. Zemouche would like to thank the ANR agency for the partial support of this work via the project ArtISMo ANR-20-CE48-0015.

References

B. Acikmese, M. Corless, Observers for systems with nonlinearities satisfying incremental quadratic constraints. Automatica **47**(7), 1339–1348 (2011)

A. Alessandri, M. Baglietto, G. Battistelli, Receding-horizon estimation for switching discrete-time linear systems. IEEE Trans. Autom. Control **50**(11), 1736–1748 (2005)

M. Arcak, P. Kokotovic, Observer-based control of systems with slope-restricted nonlinearities. IEEE Trans. Autom. Control **46**(7), 1146–1150 (2001)

F. Bakhshande, D. Soffker, Proportional-integral-observer: a brief survey with special attention to the actual methods using ACC benchmark. IFAC-PapersOnLine **48**(1), 532–537 (2015)

D. Borah, A. Boucouvalas, C. Davis, S. Hranilovic, K. Yiannopoulos, A review of communication-oriented optical wireless systems. EURASIP J. Wirel. Commun. Netw. **91**(3), 226–236 (2012)

Calhoun, Free space optics communication for mobile military platforms (2003), http://calhoun.nps.edu/handle/10945/6160

A. Chakrabarty, M. Corless, Estimating unbounded unknown inputs in nonlinear systems. Automatica **104**(9), 57–66 (2019)

R.W. De-Vaul, E. Teller, C.L. Biffle, J. Weaver, Balloon power sources with a buoyancy trade-off, United States Patent US2014/0 048 646A1 (2014)

N. Doniec, M. Angermann, D. Rus, An end-to-end signal strength model for underwater optical communications. IEEE J. Ocean. Eng. **38**(4), 743–757 (2013)

K.C. Draa, A. Zemouche, M. Alma, H. Voos, M. Darouach, A discrete-time nonlinear state observer for the anaerobic digestion process. Int. J. Robust Nonlinear Control **29**(5), 1279–1301 (2019)

K.C. Draa, A. Zemouche, M. Alma, H. Voos, M. Darouach, Nonlinear observer-based control with application to an anaerobic digestion process. Eur. J. Control **45**, 74–84 (2019)

H. Elgala, R. Mesleh, H. Haas, Indoor optical wireless communication: potential and state of-the-art. IEEE Commun. Mag. **49**(9), 56–62 (2011)

H. Elgala, R. Mesleh, H. Haas, Optical communication in space: challenges and mitigation techniques. IEEE Commun. Surv. Tut. **19**(1), 57–96 (2017)

Facebook, *Harnessing light for wireless communications* (2018), https://code.fb.com/connectivity/harnessing-light-for-wireless-communications

Z. Ghassemlooy, W. Popoola, S. Rajbhandari, *Optical Wireless Communications: System and Channel Modelling with MATLAB*, 1st edn. (CRC Press, Berlin, 2012)

R. Goebel, R.G. Sanfelice, R.A. Teel, *Hybrid Dynamical Systems—Modeling, Stability, and Robustness* (Princeton University Press, New Jersey, 2012)

R. GroB, A. Kolling, S. Berman, E. Frazzoli, A. Martinoli, F. Matsuno, M. Gauci, *Distributed Autonomous Robotic Systems*. Springer Proceedings in Advanced Robotics, vol. 6 (2018)

Q. Ha, H. Trinh, State and input simultaneous estimation for a class of nonlinear systems. Automatica **40**, 1779–1785 (2004)

R. Hagem, D.V. Thiel, S. O'Keefe, A. Wixted, T. Fickenscher, Low cost short-range wireless optical FSK modem for swimmers feedback, in *IEEE Sensors Conference* (Los Angeles, CA, USA, 2011)

F. Hanson, S. Radic, High bandwidth underwater optical communication. Appl. Opt. **47**(2), 277–283 (2008)

M. Ivan, S. Ching-Cherng, Modeling the radiation pattern of LEDs. Opt. Express **16**(3), 1808 (2008)

W. Jeon, Z. Xie, A. Zemouche, R. Rajamani, Simultaneous cyber-attack detection and radar sensor health monitoring in connected ACC vehicles. IEEE Sens. J. **21**(14), 15 741–15 752 (2020)

H.K. Khalil, J.W. Grizzle, *Nonlinear Systems*, vol. 3. (Prentice hall Upper Saddle River, NJ, 2002)

T. Li, T. Song, Y. Liang, *Wireless Communications Under Hostile Jamming: Security and Efficiency* (Springer Nature Singapore, 2018)

H.J. Liao, C.H. Richard-Lun, Y.C. Lin, K.Y. Tung, Intrusion detection system: a comprehensive review. J. Netw. Comput. Appl. **36**(1), 16–24 (2013)

D. Liberzon, *Switching in Systems and Control* (Springer, New York, 2003)

F. Lu, S. Lee, J. Mounzer, C. Schurgers, Low-cost medium-range optical under water modem, in *4th ACMInternationl Workshop Under Water Network* (Los Angeles, CA, USA, 2009), pp. 1–11

A. Majumdar, J. Ricklin, *Free-Space Laser Communications: Principles and Advances* (Springer, Berlin, 2010)

F. Miller, A. Vandome, J. McBrewster, *Beer-Lambert Law* (VDM Publishing, Saarbrucken, Germany, 2009)

I. N'Doye, D. Zhang, M.-S. Alouini, T.-M. Laleg-Kirati, Establishing and maintaining a reliable optical wireless communication in underwater environment. IEEE Access **9**(2), 62 519–62 531 (2018)

I. N'Doye, D. Zhang, A. Zemouche, R. Rajamani, T.-M. Laleg-Kirati, A switched-gain nonlinear observer for LED optical communication, in *21st IFAC World Congress*, Berlin, Germany (2020)

P. Paul, M.R. Bhatnagar, A. Jaiswal, Performance of free space optical communication system under jamming attack and its mitigation over non-Gaussian noise channel, in *2019 IEEE 90th Vehicular Technology Conference* (Honolulu, HI, USA, 2019)

K. Pelechrinis, M. Iliofotou, V.S. Krishnamurthy, Denial of service attacks in wireless networks: the case of jammers. IEEE Commun. Surv. Tutor. **13**(2), 245–257 (2011)

G. Phanomchocng, A. Zemouche, W. Jeon, R. Rajamani, F. Mazenc, Real-time estimation of rollover index for tripped rollovers with a novel unknown input nonlinear observer. Am. Control Conf. (ACC) **19**(2), 5952–5956 (2018)

G. Phanomchoeng, R. Rajamani, Real-time estimation of rollover index for tripped rollovers with a novel unknown input nonlinear observer. IEEE/ASME Trans. Mechatron. **19**(2), 743–754 (2014)

R. Rajamani, W. Jeon, H. Movahedi, A. Zemouche, On the need for switched-gain observers for non-monotonic nonlinear systems. Automatica **114**, 108814 (2020)

D. Saldana, A. Prorok, M.F.M. Campos, V. Kumar, Triangular networks for resilient formations, in *Distributed Autonomous Robotic Systems*, ed. by R. GroB, Springer Proceedings in Advanced Robotics, vol. 6, chap. 7 (2018), pp. 147–158

P.B. Solanki, M. Al-Rubaiai, X. Tan, Extended Kalman filter-aided alignment control for maintaining line of sight in optical communication, in *Proceedings American Control Conference* (2016), pp. 4520–4525

P.B. Solanki, M. Al-Rubaiai, X. Tan, Extended Kalman filter-based active alignment control for LED optical communication. IEEE/ASME Trans. Mechatron. **23**(4), 1501–1511 (2018)

Y. Wang, R. Rajamani, D. Bevly, Observer design for differentiable Lipschitz nonlinear systems with time-varying parameters, in *53th IEEE Conference on Decision and Control*, Los Angeles, CA, USA (2014), p. 2014

A. Zemouche, R. Rajamani, G. Phanomchoeng, B. Boulkroune, H. Rafaralahy, M. Zasadzinski, Circle criterion-based $\mathscr{H}_\infty$ circle observer design for lipschitz and monotonic nonlinear systems—enhanced LMI conditions and constructive discussions. Automatica **85**, 412–425 (2017)

D. Zhang, I. N'Doye, T. Ballal, T.-Y. Al-Naffouri, M.-S. Alouini, T.-M. Laleg-Kirati, Localization and tracking control using hybrid acoustic-optical communication for autonomous underwater vehicles. IEEE Internet Things J. **7**(10), 10 048–10 060 (2020)

Y. Zou, J. Zhu, X. Wang, L. Hanzo, A survey on wireless security: technical challenges, recent advances, and future trends. Proc. IEEE **104**(9), 1727–1765 (2016)

Index

M. Abbaszadeh and A. Zemouche (eds.), *Security and Resilience in Cyber-Physical Systems*, https://doi.org/10.1007/978-3-030-97166-3

Milton Keynes UK
Ingram Content Group UK Ltd.
UKHW020604140823
426834UK00001B/4